DATA COMM

PRAKASH C. GUPTA
General Manager, Telecom District, Rajkot
Department of Telecommunications
Government of India

Prentice-Hall of India Private Limited
NEW DELHI - 110 001
1996

Rs. 225.00

DATA COMMUNICATIONS
by Prakash C. Gupta

PRENTICE-HALL INTERNATIONAL, INC., Englewood Cliffs.
PRENTICE-HALL INTERNATIONAL (UK) LIMITED, London.
PRENTICE-HALL OF AUSTRALIA PTY. LIMITED, Sydney.
PRENTICE-HALL CANADA, INC., Toronto.
PRENTICE-HALL HISPANOAMERICANA, S.A., Mexico.
PRENTICE-HALL OF JAPAN, INC., Tokyo.
SIMON & SCHUSTER ASIA PTE. LTD., Singapore.
EDITORA PRENTICE-HALL DO BRASIL, LTDA., Rio de Janeiro.

ISBN-81-203-1118-3

Published by Prentice-Hall of India Private Limited, M-97, Connaught Circus, New Delhi-110001 and Printed by Bhuvnesh Seth at Rajkamal Electric Press, B-35/9, G.T. Karnal Road, Delhi-110033.

To

My Loving Wife Nutan, and
Daughters Neha and Priyanka

Contents

Foreword

The field of data communications is passing through an exciting stage of development and is becoming increasingly important owing to the enormous growth of computer networking demand. The developments during the last decade have also brought about remarkable changes in the perspective of the subject which now focuses on the integrated approach to data and computer communications.

There are very few books available on data communications which follow the new systems approach and present a coherent treatment of the subject. This book fulfills this demand admirably and presents an in-depth study suitable for undergraduate courses. The book will also be useful to computer professionals who wish to broaden their knowledge of the field of data communications. I have great pleasure in commending this book to the students, teachers and the practising professionals.

P.S. Saran
Member (Services)
Telecom Commission
and
ex-officio Secretary to Government of India

Preface

After the adoption of the OSI reference model as a standard for network architecture, there have been rapid developments in the fields of data and computer communications. These fields, treated so far as separate disciplines, have now a common architecture and a uniform set of terminology. Many protocols have since been standardized and implemented; and many more are in the pipeline for standardization. The subject of data communications is no longer restricted to principles of data transmission. It has a much wider scope and requires a systematic treatment according to the layered architecture of the OSI model. Some of the text books written a decade ago have been updated to accommodate these developments, but much of the information is scattered.

Written as a text book for undergraduate courses in electronics and communications, and computer science, this book presents the subject systematically following the layered architecture approach. It will also be useful for the computer and communications professionals who wish to explore the field of data communications but are not familiar with the terminology and the architecture.

The first four chapters of the book present the review of communication concepts, terminology and basic building blocks of data communications. The reference model for the OSI architecture is introduced in Chapter 5. Each chapter, thereafter, presents a comprehensive analysis of the protocol of a layer of the OSI architecture, with emphasis on the way it performs its functions and provides the services. This approach has been found successful in the courses I have taught. Analytical and mathematical treatment is introduced only to the extent required.

Each chapter includes examples and numerous illustrations, which enhance understanding of the concepts. Some typical exercises have been added at the end of each chapter. In addition, the book includes an extensive glossary, suggestions for further reading and a list of relevant ISO standards for ready reference. With a judicious selection of topics, the text can be covered in a one-semester course. However, I strongly recommend that the sequence of topics as given be retained.

Prakash C. Gupta

CHAPTER 1

Data Communication Concepts and Terminology

Communication, whether between human beings or computer systems, involves the transfer of information from a sender to a receiver. Data communication refers to exchange of digital information between two digital devices. In this chapter, we examine some of the basic concepts and terminology relating to data communication. Data representation, serial/parallel data transmission and asynchronous/synchronous data transmission concepts are discussed first. We then proceed to examine some theoretical concepts of Fourier series, Nyquist's and Shannon's theorems and their application in data transmission. Digital modulation techniques and baud rate are introduced next. We close the chapter with a discussion on the distinction between data transmission and data communication. These terms are used interchangeably in the literature on data communication and are the cause of much confusion and frustration.

1 DATA REPRESENTATION

A binary digit or bit has only two states, "0" and "1" and can represent only two symbols, but even the simplest form of communication between computers requires a much larger set of symbols, e.g.

- 52 capital and small letters,
- 10 numerals from 0 to 9,
- punctuation marks and other special symbols, and
- terminal control characters—Carriage Return (CR), Line Feed (LF).

Therefore, a group of bits is used as a code to represent a symbol. The code is usually 5 to 8 bits long. 5-bit code can have $2^5 = 32$ combinations and can, therefore, represent 32 symbols. Similarly, an 8-bit code can represent $2^8 = 256$ symbols. A code set is the set of these codes representing the symbols. There are several code sets, some are used for specific applications while others are the proprietary code sets of computer manufacturers. The following two code sets are very common:

1. ANSI's 7-bit American Standard Code for Information Interchange (ASCII)
2. IBM's 8-bit Extended Binary-Coded-Decimal Interchange Code (EBCDIC).

EBCDIC is vendor-specific and is used primarily in large IBM computers. ASCII is the most common code set and is used worldwide.

1.1 ASCII—American Standard Code for Information Interchange

ASCII is defined by American National Standards Institute (ANSI) in ANSI X3.4. The corresponding

CCITT recommendation is T.50 (International Alphabet No. 5 or IA5) and ISO specification is ISO 646. It is 7-bit code and all the possible 128 codes have defined meanings (Table 1). The code set consists of the following symbols:

- 96 graphic symbols (columns 2 to 7), comprising 94 printable characters, SPACE and DELete characters
- 32 control symbols (columns 0 and 1).

Table 1 ASCII Code Set

Bit	7	0	0	0	0	1	1	1	1
numbers	6	0	0	1	1	0	0	1	1
	5	0	1	0	1	0	1	0	1
4321		0	1	2	3	4	5	6	7
0000	0	NUL	DLE	SPACE	0	@	P		p
0001	1	SOH	DC1	!	1	A	Q	a	q
0010	2	STX	DC2	"	2	B	R	b	r
0011	3	ETX	DC3	#	3	C	S	c	s
0100	4	EOT	DC4	$	4	D	T	d	t
0101	5	ENQ	NAK	%	5	E	U	e	u
0110	6	ACK	SYN	&	6	F	V	f	v
0111	7	BEL	ETB	'	7	G	W	g	w
1000	8	BS	CAN	(	8	H	X	h	x
1001	9	HT	EM	)	9	I	Y	i	y
1010	A	LF	SUB	*	:	J	Z	j	z
1011	B	VT	ESC	+	;	K	[	k	{
1100	C	FF	FS	,	<	L	\	l	|
1101	D	CR	GS	-	=	M	]	m	}
1110	E	SO	RS	>	N	^	n	~
1111	F	SI	US	/	?	O	—	o	DEL

The binary representation of a particular character can be easily determined from its hexadecimal coordinates. For example, the coordinates of character "K" are (4, B) and, therefore, its binary code is 100 1011.

The control symbols are codes reserved for special functions. Table 2 lists the control symbols. Some important functions and the corresponding control symbols are:

- Functions relating to basic operation of the terminal device, e.g., a printer or a VDU
  - — CR (Carriage Return)
  - — LF (Line Feed)
- Functions relating to error control
  - — ACK (Acknowledgement)
  - — NAK (Negative Acknowledgement)

- Functions relating to blocking (grouping) of data characters
  - — STX (Start of Text)
  - — ETX (End of Text).

DC1, DC2, DC3 and DC4 are user definable. DC1 and DC3 are generally used as X-ON and X-OFF for switching the transmitter.

Table 2 Control Symbols

ACK	Acknowledgement	FF	Form Feed
BEL	Bell	FS	File Separator
BS	Backspace	GS	Group Separator
CAN	Cancel	HT	Horizontal Tabulation
CR	Carriage Return	LF	Line Feed
DC1	Device Control 1	NAK	Negative Acknowledgement
DC2	Device Control 2	NUL	Null
DC3	Device Control 3	RS	Record Separator
DC4	Device Control 4	SI	Shift-In
DEL	Delete	SO	Shift-Out
DLE	Data Line Escape	SOH	Start of Heading
EM	End of Medium	STX	Start of Text
ENQ	Enquiry	SUB	Substitute Character
EOT	End of Transmission	SYN	Synchronous Idle
ESC	Escape	US	Unit Separator
ETB	End of Transmission Block	VT	Vertical Tabulation
ETX	End of Text		

ASCII is often used with an eighth bit called the parity bit. This bit is utilized for detecting errors which occur during transmission. It is added in the most significant bit (MSB) position. We will examine the use of parity bits in detail in the chapter on Error Control.

EXAMPLE 1

Represent the message "3P.bat" in ASCII code. The eighth bit may be kept as "0".

Solution

Bit Positions	8	7	6	5	4	3	2	1
3	0	0	1	1	0	0	1	1
P	0	1	0	1	0	0	0	0
.	0	0	1	0	1	1	1	0
b	0	1	1	0	0	0	1	0
a	0	1	1	0	0	0	0	1
t	0	1	1	1	0	1	0	0

1.2 EBCDIC—Extended Binary Coded Decimal Interchange Code

It is an 8-bit code with 256 possible combinations; however, all combinations are not used and have also not been defined. There is no parity bit for error checking in the basic code set. The graphic symbol subset is approximately the same as ASCII. There are several differences in the control characters. EBCDIC is not the same for all devices. There may be variations even within different models of IBM equipment. Table 3 depicts the version of EBCDIC applicable to IBM 3270. Note the changed bit numbering in EBCDIC. In EBCDIC, the bit numbering starts from the most significant bit (MSB) and in ASCII, it starts from the least significant bit (LSB).

Table 3 EBCDIC—Extended Binary Coded Decimal Interchange Code

	0	0	0	0	0	0	0	0	0	1	1	1	1	1	1	1	1
Bit	1	0	0	0	0	1	1	1	1	0	0	0	0	1	1	1	1
numbers	2	0	0	1	1	0	0	1	1	0	0	1	1	0	0	1	1
	3	0	1	0	1	0	1	0	1	0	1	0	1	0	1	0	1
4567		0	1	2	3	4	5	6	7	8	9	A	B	C	D	E	F
0000	0	NUL	DLE				SP	&	–								0
0001	1	SOH	SBA					/		a	j			A	J		1
0010	2	STX	EUA		SYN					b	k	s		B	K	S	2
0011	3	ETX	IC							c	l	t		C	L	T	3
0100	4									d	m	u		D	M	U	4
0101	5	HT	NL							e	n	v		E	N	V	5
0110	6			ETB						f	o	w		F	O	W	6
0111	7			ESC	EOT					g	p	x		G	P	X	7
1000	8									h	q	y		H	Q	Y	8
1001	9		EM							i	r	z		I	R	Z	9
1010	A					¢	!	!									
1011	B						$		#								
1100	C		DUP		RA	<	*	%	@								
1101	D		SF	ENQ	NAK	(	)	—									
1110	E		FM			+	;	>	=								
1111	F		ITB		SUB	|	¬	?	"								

b_0 b_1 b_2 b_3 b_4 b_5 b_6 b_7

↑ MSB ↑ LSB

1.3 Other Code Sets

The following code sets, though not of much significance to the data processing community today, were used one time or the other:

Baudot Teletype Code. Also called ITA 2 (International Telegraph Alphabet Number 2), it is a 5-bit code and is used in electromechanical teletype machines. 32 codes are possible using 5 bits but in this code there are 58 symbols. The same code is used for two symbols using letter shift/figure shift keys which change the meaning of a code. In telegraphy terminology, binary "1" is called Mark and binary "0" is called Space.

BCDIC—Binary Coded Decimal Interchange Code. It is a six-bit code with 64 symbols.

1.4 Bytes

Byte is a group of bits which is considered as a single unit during processing. It is usually eight bits long though its length may be different. A character code, e.g., 1001011 of ASCII, is a byte having a defined meaning "K", but it should be noted that there may be bytes which are not elements of any standard code set.

2 DATA TRANSMISSION

There is always need to exchange data, commands and other control information between a computer and its terminals or between two computers. This information, as we saw in the previous section, is in the form of bits.

Data transmission refers to movement of the bits over some physical medium connecting two or more digital devices. There are two options of transmitting the bits, namely, Parallel transmission, or Serial transmission.

2.1 Parallel Transmission

In parallel transmission, all the bits of a byte are transmitted simultaneously on separate wires as shown in Fig. 1 and multiple circuits interconnecting the two devices are, therefore, required. It is practical only if the two devices, e.g., a computer and its associated printer are close to each other.

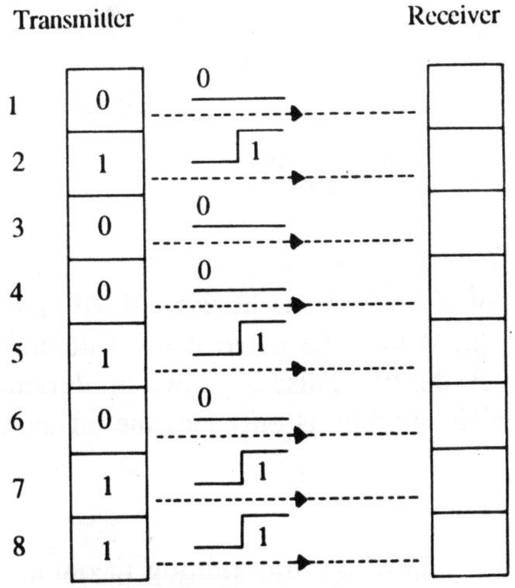

Fig. 1 Parallel transmission.

2.2 Serial Transmission

In serial transmission, bits are transmitted serially one after the other (Fig. 2). The least significant

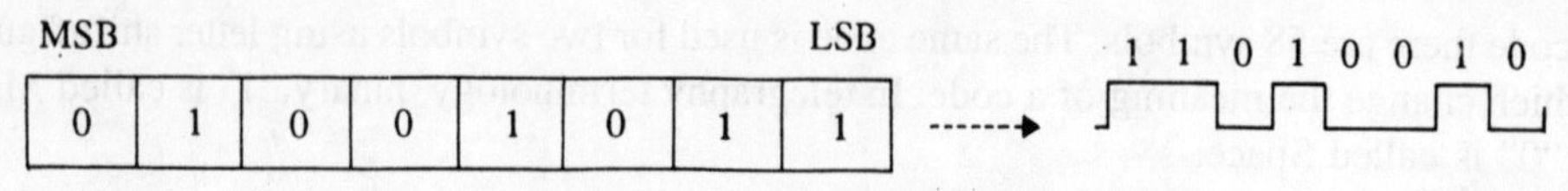

Fig. 2 Serial transmission.

bit (LSB) is usually transmitted first. Note that as compared to parallel transmission, serial transmission requires only one circuit interconnecting the two devices. Therefore, serial transmission is suitable for transmission over long distances.

EXAMPLE 2

Write the bit transmission sequence of the message given in Example 1.

Solution

3	*P*	.	*b*	*a*	*t*
11001100	00001010	01110100	01000110	10000110	00101110

Bits are transmitted as electrical signals over the interconnecting wires. The two binary states "1" and "0" are represented by two voltage levels. If one of these states is assigned 0 volt level, the transmission is termed unipolar and if we choose to represent a binary "1" by , say, a positive voltage + V volts and a binary "0" by a negative voltage $-V$ volts, the transmission is said to be bipolar. Figure 3 shows the bipolar waveform of the character "K". Bipolar transmission is preferred because the signal does not have any DC component. The transmission media usually do not allow the DC signals to pass through.

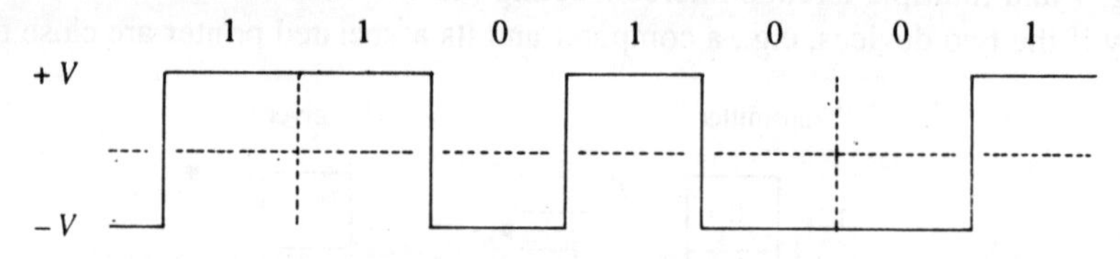

Fig. 3 Bipolar signal.

2.3 Bit Rate

Bit rate is simply the number of bits which can be transmitted in a second. If t_p is the duration of a bit, the bit rate R will be $1/t_p$. It must be noted that bit duration is not necessarily the pulse duration. For example, in Fig. 3, the first pulse is of two-bit duration. Later, we will come across signal formats in which the pulse duration is only half the bit duration.

2.4 Receiving Data Bits

The signal received at the other end of the transmitting medium is never identical to the transmitted signal as the transmission medium distorts the signal to some extent (Fig. 4a). As a result, the receiver has to put in considerable effort to identify the bits. The receiver must know the time instant at which it should look for a bit. Therefore, the receiver must have synchronized clock pulses which mark the location of the bits (Fig. 4b). The received signal is sampled using

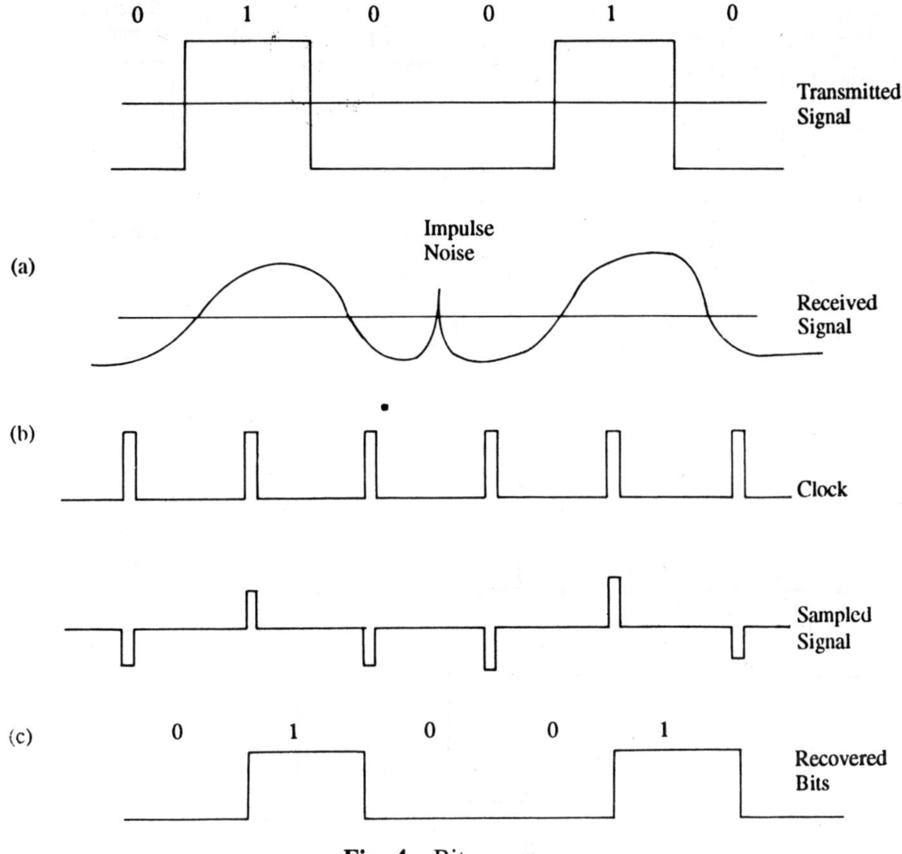

Fig. 4 Bit recovery.

the clock pulses, and depending on the polarity of a sample, the corresponding bit is identified (Fig. 4c).

It is essential that the received signal is sampled at the right instants as otherwise it could be misinterpreted. Therefore, the clock frequency should be exactly the same as the transmission bit rate. Even a small difference will built up as timing error and eventually result in sampling at wrong instants. Figure 5 shows two situations when the clock frequency is slightly faster or slightly slower than the bit rate. When clock frequency is faster, a bit may be sampled twice and may be missed when it is slower.

3 MODES OF DATA TRANSMISSION

There are two methods of timing control for reception of bits. The transmission modes corresponding to these two timing methods are called Asynchronous transmission and Synchronous transmission.

3.1 Asynchronous Transmission

We call an action asynchronous when the agent performing the action does so whenever it wishes. Asynchronous transmission refers to the case when the sending end commences transmission of

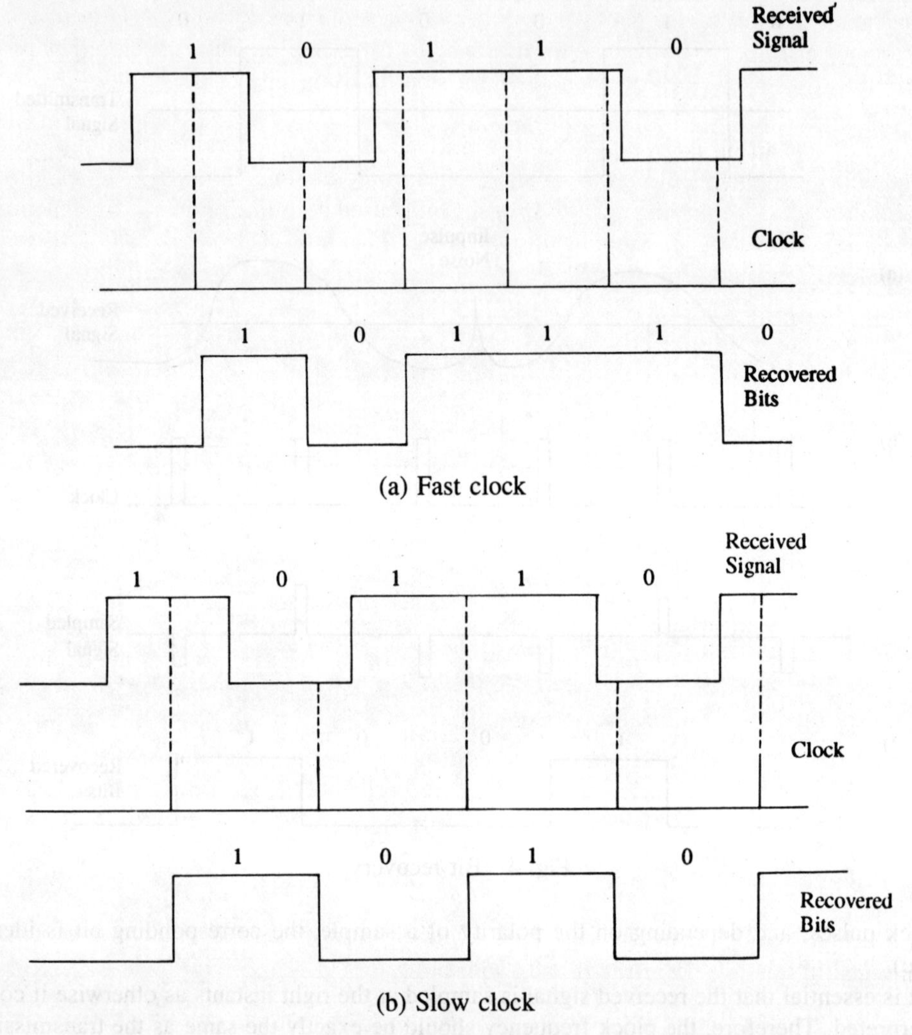

(a) Fast clock

(b) Slow clock

Fig. 5 Timing errors.

bytes at any instant of time. Only one byte is sent at a time and there is no time relation between consecutive bytes, i.e., after sending a byte, the next byte can be sent after arbitrary delay (Fig. 6). In the idle state, when no byte is being transmitted, the polarity of the electrical signal corresponds to "1"

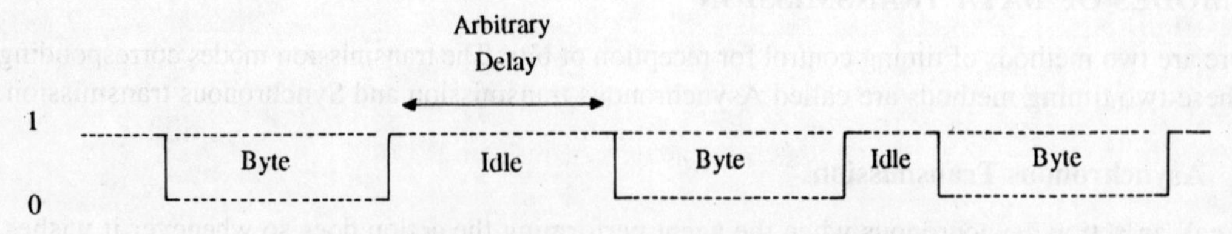

Fig. 6 Asynchronous transmission.

Due to the arbitrary delay between consecutive bytes, the time occurrences of the clock pulses at the receiving end need to be synchronized repeatedly for each byte. This is achieved by providing two extra bits, a start bit at the beginning and a stop bit at the end of a byte.

Start Bit. The start bit is always "0" and is prefixed to each byte. At the onset of transmission of a byte, it ensures that the electrical signal changes from idle state "1" to "0" and remains at "0" for one bit duration. The leading edge of the start bit is used as a time reference for generating the clock pulses at the required sampling instants (Fig. 7). Thus, each onset of a byte results in resynchronization of the receiver clock.

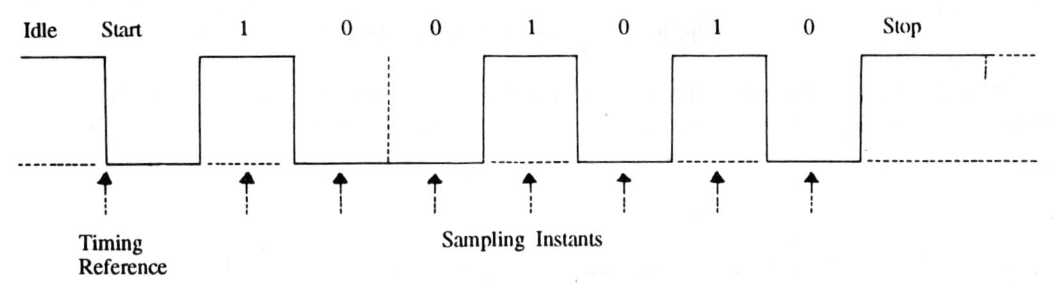

Fig. 7 Start and stop bits.

Stop Bit. To ensure that the transition from "1" to "0" is always present at the beginning of a byte, it is necessary that polarity of the electrical signal should correspond to "1" before occurrence of the start bit. That is why the idle state is kept at "1". But there may be two bytes, one immediately following the other and if the last bit of the first byte is "0", the transition from "1" to "0" will not occur. Therefore, a stop bit is also suffixed to each byte (Fig. 7). It is always "1" and its duration is usually 1, 1.5 or 2 bits.

EXAMPLE 3

Sketch the logic levels for the message "HT" when it is transmitted in asynchronous mode with stop bit equal to one bit. Use ASCII code with parity bit "0"

Solution

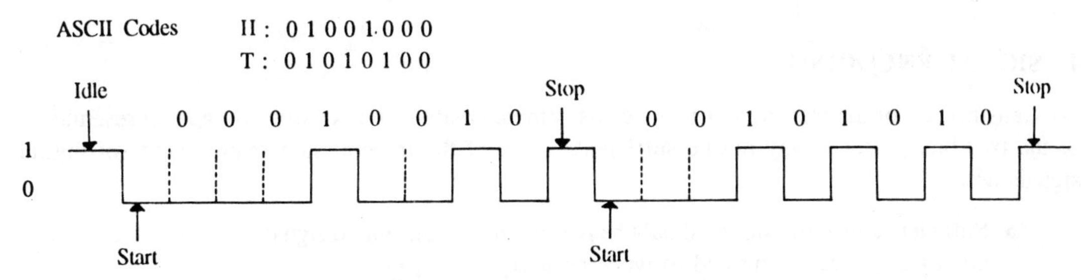

Fig. 8 Asynchronous transmission of "HT"

3.2 Synchronous Transmission

A synchronous action, unlike an asynchronous action, is carried out under the control of a timing source. In synchronous transmission, bits are always synchronized to a reference clock irrespective of the bytes they belong to. There are no start or stop bits. Bytes are transmitted as a block (group of bytes) in a continuous stream of bits (Fig. 9). Even the inter block idle time is filled with idle characters.

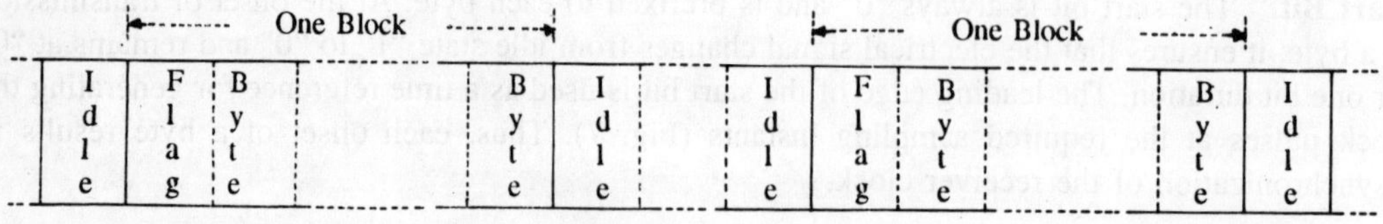

Fig. 9 Synchronous transmission.

Continuous transmission of bits enables the receiver to extract the clock from the incoming electrical signal (Fig. 10). As this clock is inherently synchronized to the bits, the job of the receiver becomes simpler.

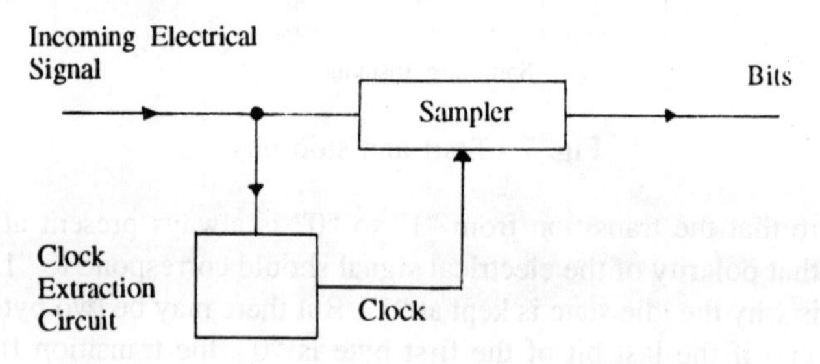

Fig. 10 Bit recovery in synchronous transmission.

There is, however, still one problem. The bytes lose their identity and their boundaries need to be identified. Therefore, a unique sequence of fixed number of bits, called *flag,* is prefixed to each block (Fig. 9). The flag identifies the start of a block. The receiver first detects the flag and then identifies the boundaries of different bytes using a counter. Just after the flag there is first bit of the first byte.

A more common term for data block is *frame*. A frame contains many other fields in addition to the flag. We will discuss frame structures later in the chapter on Data Link Control.

4 SIGNAL ENCODING

For transmission of the bits as electrical signals, simple positive and negative voltage representation of the two binary states may not be sufficient. Some of the transmission requirements of digital signals are:

- Sufficient signal transitions should be present in the transmitted signal for the clock extraction circuit at the receiving end to work properly.
- Bandwidth of the digital signal should match the bandwidth of the transmission medium.
- There should not be any ambiguity in recognizing the binary states of the received signal.

There are several ways of representing bits as digital electrical signals. Two broad classes of signal representation codes are: Non-Return to Zero (NRZ) codes and Return to Zero (RZ) codes.

4.1 Non-Return to Zero (NRZ) Codes

In this class of codes, the signal level remains constant during a bit duration. Figure 11 shows a bit sequence 00101110 along with the clock signal and three examples of NRZ codes.

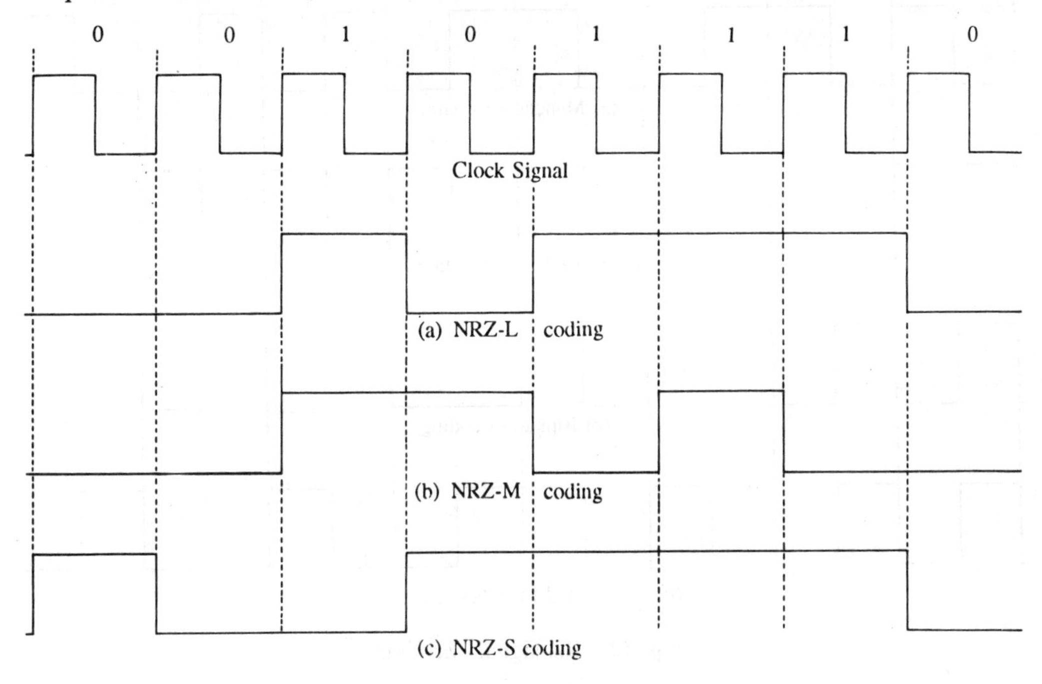

Fig. 11 NRZ signal encoding.

NRZ-L. In NRZ-L (Non Return to Zero-Level), the bit is represented by a voltage level which remains constant during the bit duration.

NRZ-M and NRZ-S. In NRZ-M (Non Return to Zero-Mark), and NRZ-S (Non-Return to Zero-Space), it is a change in signal level which corresponds to one bit value, and absence of a change corresponds to the other bit value. "Mark" or "1" changes the signal level in NRZ-M and "Space" or "0" changes the signal level in NRZ-S. NRZ-M is also called NRZ-I, Non-Return to Zero-Invert on ones.

4.2 Return to Zero (RZ) Codes

We mentioned in the last section that the clock can be extracted from the digital signal if bits are continuously transmitted. However, if there is a continuous string of zeros or ones and if it is coded using one of the NRZ codes, the electrical signal will not have any level transitions. For the receiver clock extraction circuit, it will be as good as no signal.

The RZ codes usually ensure signal transitions for any bit pattern and thus overcome the above limitation of NRZ codes. These codes are essentially a combination of NRZ-L and the clock signal. Figure 12 shows some examples of RZ codes.

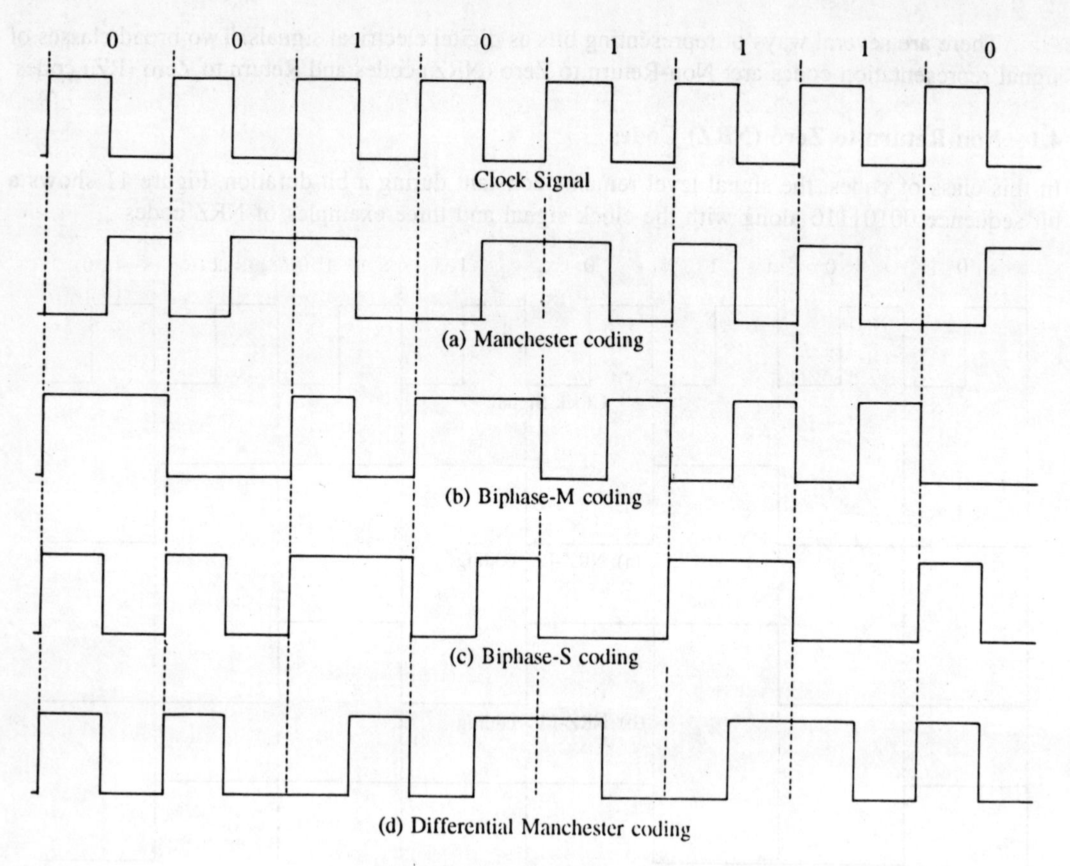

Fig. 12 RZ signal encoding.

Manchester Code. In this code, "1" is represented as logical AND of "1" and the clock. This produces one cycle of clock. For "0", this clock cycle is inverted. Note that whatever be the bit sequence each bit period will have one transition. The receiver clock extraction circuit never faces a dearth of transitions. The Manchester code is widely employed to represent data in local area networks. It is also called Biphase-L code.

Biphase-M Code. In this code there is always a transition at the beginning of a bit interval. Binary "1" has another transition in the middle of the bit interval.

Biphase-S Code. In this code also there is a transition at the beginning of a bit interval. Binary "0" has another transition in the middle of the bit interval.

Differential Manchester Code. In this code there is always a transition in the middle of a bit interval. Binary "0" has additional transition at the beginning of the bit interval.

4.3 Other Signal Codes

Local area networks based on optical fibres use another type of signal code termed 4B/5B. In this code four data bits are taken at a time and coded into 5 bits. There are 32 possible combinations of 5 bits. Of these combinations, 16 codes are selected to represent 16 possible sets of 4 bits. The codes are so selected that there are at least two signal transitions in a group of 5 bits for clock recovery.

5 FREQUENCY SPECTRUM

It was shown by Fourier that any signal can be expressed as the sum of infinite series of sine waves having different frequencies. Figure 13 shows how the sine waves can constitute a pulse train.

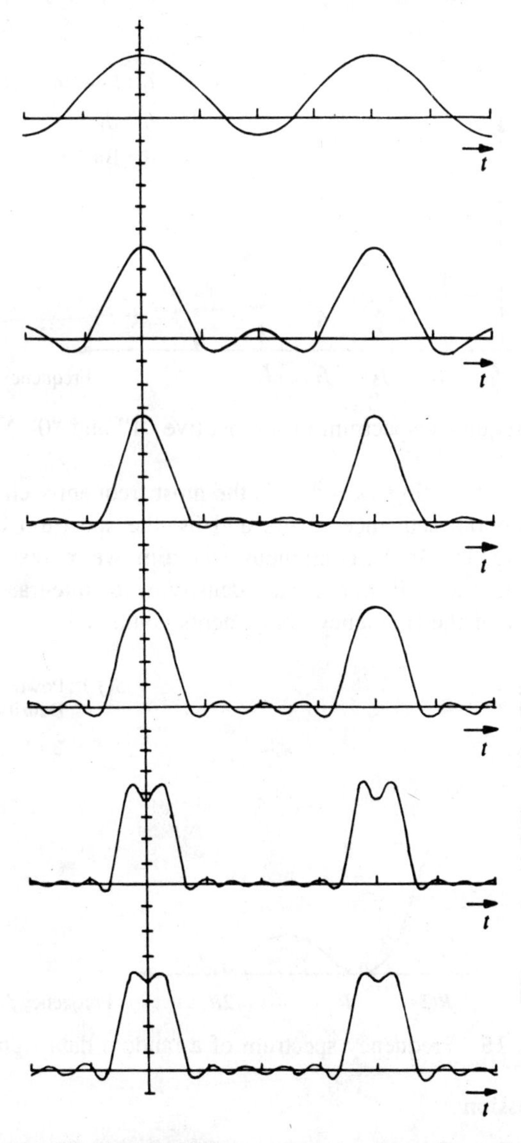

Fig. 13 Sinusoidal components of a pulse train.

Note that as we include more and more frequency components, the resulting signal becomes a better approximation of the original signal. An example of calculating the amplitudes, frequencies and phases of these frequency components using Fourier analysis is given in Appendix A. For periodic signals, the frequency components are harmonic multiples of the fundamental frequency which depends on the bit rate. A repetitive "1" and "0" NRZ pattern has the fundamental frequency which is half the bit rate and odd harmonics of the fundamental frequency. Figure 14 shows the amplitude plot of the frequency components. It appears at first glance that transmitting this signal would require a channel with infinite bandwidth but it is to be noted that the amplitudes of the frequency components decrease as we go up in the spectrum. Therefore, a good likeness of the signal can be obtained by transmitting the significant first few frequency components of the spectrum.

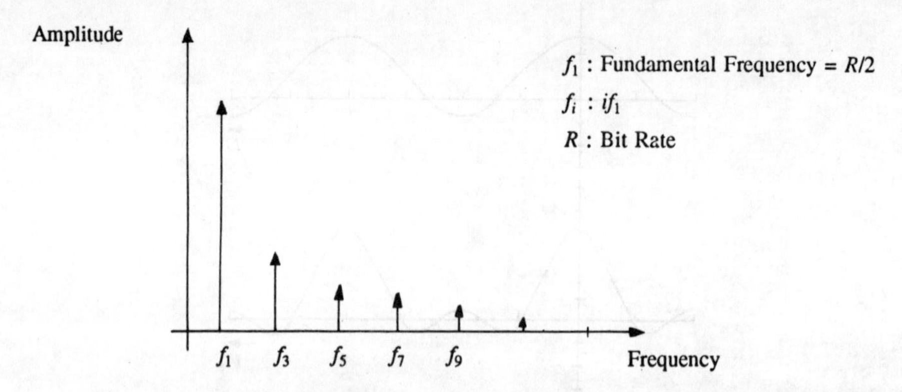

Fig. 14 Frequency spectrum of a repetitive "1" and "0" NRZ signal.

The data signal with alternating 1s and 0s is the most frequently changing data signal. If the data signal consists of random sequences of 1s and 0s, the spectrum becomes continuous and extends upto the origin (Fig. 15). In the continuous spectrum, we represent power spectral density instead of amplitudes on the *Y*-axis. Power spectral density, when integrated over a frequency band, gives the combined power of the frequency components contained in that band.

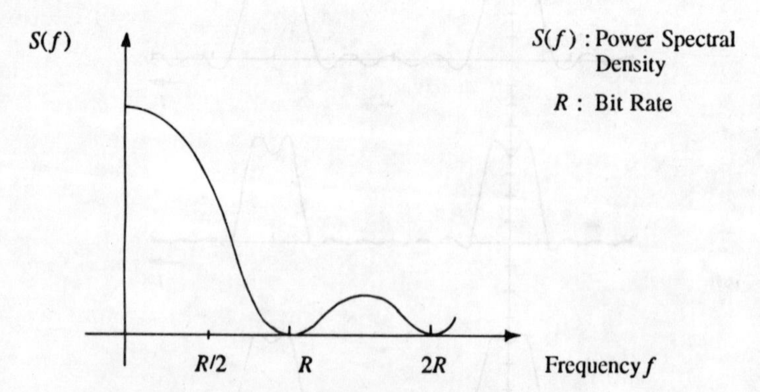

Fig. 15 Frequency spectrum of a random data signal.

5.1 Baseband Transmission

When a digital signal is transmitted on the medium using one of the signal codes discussed earlier, it is called baseband transmission. Baseband transmission is limited to low data rates because at high data rates, significant frequency components are spread over a wide frequency band over which the transmission characteristics of the medium do not remain uniform. For faithful reproduction of a signal, it is necessary that the relative amplitudes and phase relationships of the frequency components are maintained during transmission. For transmitting data at high bit rates we need to use modulation techniques which we will discuss later.

6 TRANSMISSION CHANNEL

A transmission channel transports the electrical signals from the transmitter to the receiver. It is characterized by two basic parameters—bandwidth and signal-to-noise ratio. These parameters determine the ultimate information-carrying capacity of a channel. Nyquist derived the limit of data rate considering a perfectly noiseless channel. Nyquist's theorem states that if B is the bandwidth of a transmission channel which carries a signal having L levels (a binary digital signal has two levels), the maximum data rate R is given by

$$R = 2B \log_2 L$$

The number of levels L can be more than two, as we shall see shortly. Shannon extended Nyquist's work to include the effect of noise. If signal-to-noise ratio (Signal-power divided by noise-power) is S/N, the maximum data rate is given by

$$R = B \log_2 \left(1 + \frac{S}{N}\right)$$

This equation puts a limit on the number of levels L. For example, if bandwidth is 3000 Hz and signal-to-noise ratio is 30 dB (30 dB = 1000), then the maximum data rate can be

$$R = 3000 \log_2 (1 + 1000) \approx 30{,}000 \text{ bits/s.}$$

For the data rate of 30,000 bits per second, the number of levels L can be computed from Nyquist's theorem.

$$30000 = 2 \times 3000 \log_2 L \quad \text{or} \quad L = 32$$

Actual practice is very short of this limit. Very sophisticated equipment is required even for the data rate of 9600 bits per second on a voice grade telephony channel having bandwidth of 3100 Hz.

6.1 Bauds

When bits are transmitted as an electrical signal having two levels, the bit rate and the "modulation" rate of the electrical signal are the same (Fig. 16). Modulation rate is the rate at which the electrical

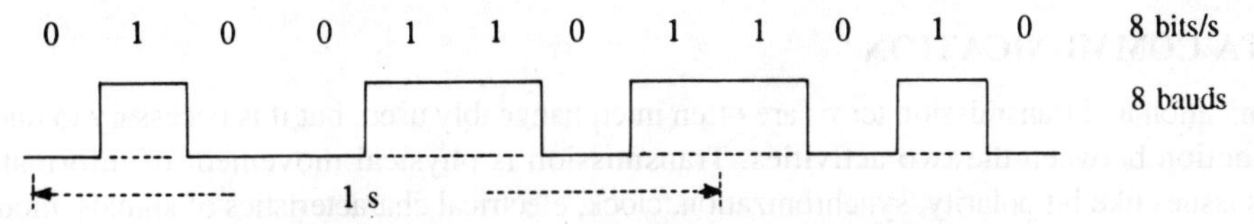

Fig. 16 Baud rate for two-level modulation.

signal changes its levels. It is expressed in bauds ("per second" is implied). Note that there is one to one correspondence between bits and electrical levels.

It is possible to associate more than one bit to one electrical level. For example, if the electrical signal has four distinct levels, two bits can be associated with one electrical level (Fig. 17). In this case, the bit rate is twice the baud rate.

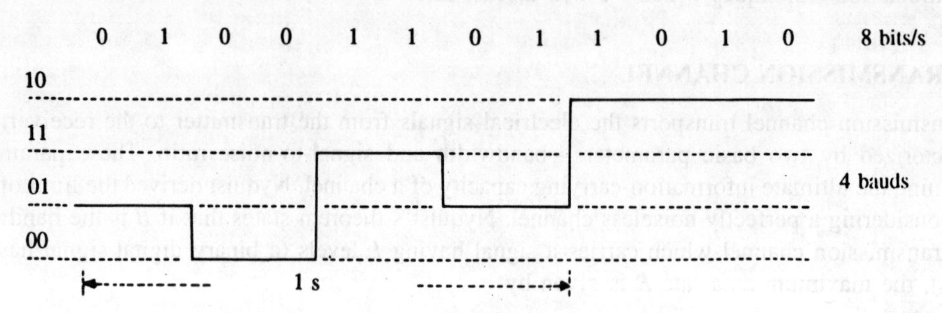

Fig. 17 Baud rate for four-level modulation.

EXAMPLE 4

What is the maximum possible baud rate of a voice channel having bandwidth of 3100 Hz?

Solution

According to Nyquist's theorem,

$$\text{Bit rate } R = 2B \log_2 L$$

$$\text{Baud rate } r = R/\log_2 L = 2B = 6200 \text{ bauds}$$

6.2 Modem

In Fig. 17, the four levels define four states of the electrical signal. The electrical state can also be defined in terms of other attributes of an electrical signal such as amplitude, frequency or phase. The basic electrical signal is a sine wave in this case. The binary signal modulates one of these signal attributes. The sine wave carries the information and is, therefore termed a "carrier". The device which performs modulation is called a modulator and the device which recovers the information signal from the modulated carrier is called a demodulator. In data transmission, we usually come across devices which perform both modulation as well as demodulation functions and these devices are called modems. Modems are required when data is to be transmitted over long distances. In a modem, the input digital signal modulates a carrier which is transmitted to the distant end. At the distant end, another modem demodulates the received carrier to get the digital signal. A pair of modems is, thus, always required.

7 DATA COMMUNICATION

Communication and transmission terms are often interchangeably used, but it is necessary to understand the distinction between the two activities. Transmission is physical movement of information and concerns issues like bit polarity, synchronization, clock, electrical characteristics of signals, modulation, demodulation etc. We have so far been examining these data transmission issues.

Communication has a much wider connotation than transmission. It refers to meaningful exchange of information between the communicating entities. Therefore, in data communications we are concerned with all the issues relating to exchange of information in the form of a dialogue, e.g., dialogue discipline, interpretation of messages, and acknowledgements.

7.1 Synchronous Communication

Communication can be asynchronous and synchronous. In synchronous mode of the communication, the communicating entities exchange messages in a disciplined manner. An entity can send a message when it is permitted to do so.

Entity A	**Entity B**
Hello B!	
	Hello!
Do you want to send data? Go ahead.	
	Yes. Here it is.
Any more data?	
	No.
Bye	
	Bye.

The dialogue between the entities A and B is "synchronized" in the sense that each message of the dialogue is a command or response. Physical transmission of data bytes corresponding to the characters of these messages could be in synchronous or asynchronous mode.

7.2 Asynchronous Communication

Asynchronous communication, on the other hand, is less disciplined. A communicating entity can send message whenever it wishes to.

Entity A	**Entity B**
Hello B!	
	Hello! Here is some data.
Here is some data.	*Here is more data.*
Did you receive what I sent?	
	Yes. Here is more data. Please acknowledge.
Acknowledged, Bye.	
	Bye.

Note the lack of discipline in the dialogue. The communicating entities send messages whenever they please, Here again, physical transmission of bytes of the messages can be in synchronous or asynchronous mode. We will come across many examples of synchronous and asynchronous communication in this book when we discuss protocols. Protocols are the rules and procedures for communication.

8 DIRECTIONAL CAPABILITIES OF DATA EXCHANGE

There are three possibilities of data exchange:

1. Transfer in both directions at the same time.
2. Transfer in either direction, but only in one direction at a time.
3. Transfer in one direction only.

Terminology used for specifying the directional capabilities is different for data transmission and for data communication (Table 4).

Table 4 Terminology for Directional Capabilities

Directional capability	Transmission	Communication
One direction only	Simplex (SX)	One Way (OW)
One direction at a time	Half duplex (HDX)	Two-Way Alternate (TWA)
Both directions at the same time	Full duplex (FDX)	Two-Way Simultaneous (TWS)

9 SUMMARY

Binary codes are used for representing the symbols for computer communications. ASCII is the most common code set used worldwide. The bits of a binary code can be transmitted in parallel or in serial form. Transmission is always serial unless the devices are near each other. Serial transmission mode can be asynchronous or synchronous. Asynchronous transmission is byte by byte transmission and start/stop bits are appended to each byte. In synchronous transmission, data is transmitted in the form of frames having flags to identify the start of a frame. Clock is required in synchronous transmission. Digital signals are coded using RZ codes to enable clock extraction at the receiving end.

A communication channel is limited in its information-carrying capacity by its bandwidth and the noise present in the channel. To make best use of this limited capacity of the channel, very sophisticated carrier-modulation methods are used. Modems are the devices which carry out the modulation and demodulation functions.

Data communication has wider scope as compared to data transmission. Asynchronous and synchronous communication refer to non-disciplined and disciplined exchange of messages respectively.

APPENDIX

Fourier showed that a periodic signal $v(t)$ having time period T can be represented as an infinite trigonometric series.

$$v(t) = a_0 + \sum_{n=1}^{\infty} [a_n \cos(2\pi n f_1 t) + b_n \sin(2\pi n f_1 t)]$$

where

$$f_1 = \frac{1}{T}$$

$$a_0 = \frac{1}{T}\int_{-T/2}^{T/2} v(t)\, dt$$

$$a_n = \frac{1}{T}\int_{-T/2}^{T/2} v(t)\cos(2\pi n f_1 t)\, dt$$

$$b_n = \frac{1}{T}\int_{-T/2}^{T/2} v(t)\sin(2\pi n f_1 t)\, dt$$

Thus, the frequency spectrum consists of a DC component having value a_0 and other frequency components which are harmonic multiples of the fundamental frequency f_1. For the nth harmonic, the amplitude c_n and the phase θ_n are given by

$$c_n = (a_n^2 + b_n^2)^{1/2}$$

$$\theta_n = -\tan^{-1}(b_n/a_n)$$

We can use the above equations to find the spectrum of a data signal. Let us consider an NRZ signal consisting of alternating 1s and 0s. If the bit duration is t_p, the resulting signal is a square wave having time period $T = 2t_p$ (Fig. A1).

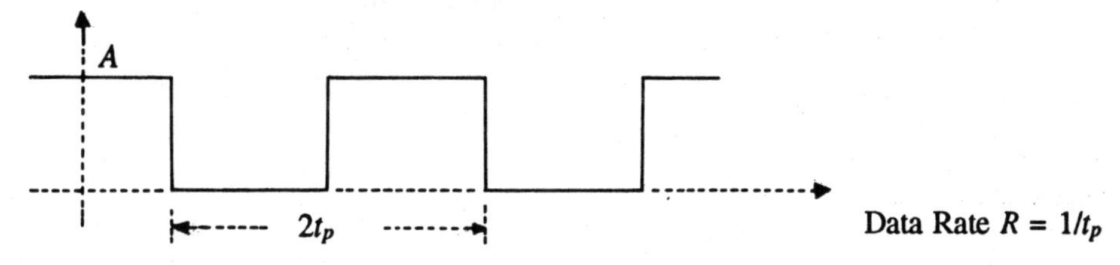

Fig. A.1 Periodic NRZ signal.

The amplitudes of the frequency components can be calculated by using the Fourier series. The spectrum consists of a DC component and odd harmonics of the fundamental frequency $f_1 = 1/2t_p$.

$$a_0 = \frac{A}{2}, \qquad a_n = \frac{A\sin(\pi n/2)}{\pi n/2}, \qquad b_n = 0$$

PROBLEMS

1. (a) Write ASCII code for the word "Data". Assume parity bit is "0".

(b) Write the bit transmission sequence for the above word.

(c) Draw the signal waveform if the word "Data" is transmitted in asynchronous mode using stop bit of one bit duration.

2. Do Problem 1 using IBM's EBCDIC.

3. A system sends a signal that can assume 8 different voltage levels. It sends 400 of these signals per second. What are the baud and bit rates?

4. (a) Using Shannon's theorem, compute the maximum bit rate for a channel having bandwidth 3100 Hz and signal-to-noise ratio 20 dB.

(b) Calculate the number of levels required to transmit the maximum bit rate. What is the baud rate?

5. Sketch the signal waveforms when 00110101 is transmitted in the following signal codes:

(a) NRZ-L
(b) NRZ-M
(c) NRZ-S
(d) Manchester code
(e) Biphase-M
(f) Biphase-S
(g) Differential Manchester code.

6. The following circuit carries out signal encoding. Analyse its operation for the input bit sequence 10110011 and determine the type of signal encoding. Draw the decoder implementation of this encoder using an EXOR gate and one bit delay circuit (Fig. P.6).

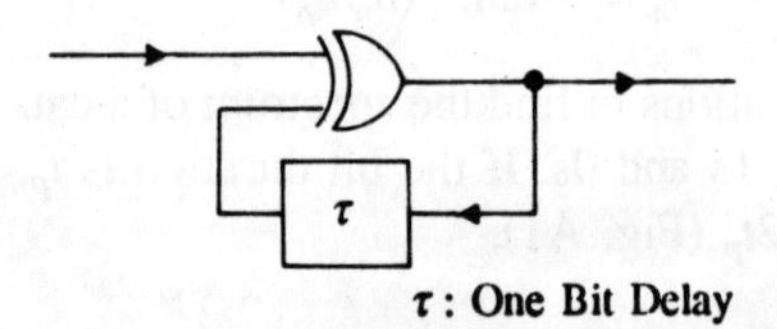

Fig. P.6

CHAPTER 2

Transmission Media

Data transmission is the process of transporting data signals from one location to another. If the terminals are in close proximity, there in really no problem in providing a suitable transmission medium between them but problems arise when the distances are such that it is uneconomic to provide dedicated special-purpose circuits. In this chapter, we examine the problems that must be solved for long-distance data transmission.

First we develop the concepts of transmission line theory and examine two specific examples of transmission lines, namely, balanced pair, and coaxial cables. Having looked at the media characteristics, we move on to principles of data transmission and study the channel characteristics for minimum intersymbol interference. At present, telephone network is the only network which can provide worldwide accessibility. We examine the topology, composition, services, and signal impairments of the telephone network keeping in mind the objective of using this network for data transmission. We close the chapter with relevant CCITT recommendations.

1 TRANSMISSION LINE CHARACTERISTICS

The elementary section of a transmission line of length dx can be modelled as shown in Fig. 1. This model is based on four primary parameters of the transmission line:

1. The series resistance per unit length R which is proportional to the square root of the frequency because of the skin effect.

2. The inductance per unit length L.

3. The capacitance per unit length C, which is related to the permittivity of the dielectric located between the two conductors.

4. The leakage conductance per unit length G, which primarily accounts for the dielectric losses. The insulation losses are generally negligible.

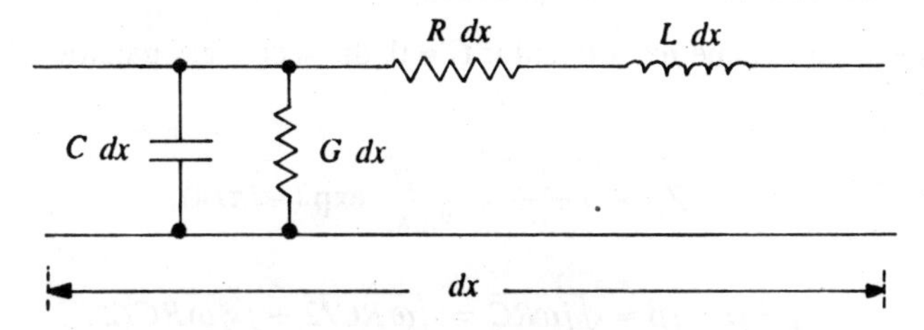

Fig. 1 Primary parameters of a transmission line.

1.1 Secondary Parameters

The primary parameters of a transmission line are of little interest to the user. The secondary parameters which can describe the behaviour of the line and which can be easily measured are, therefore, preferred. These parameters are:

1. The characteristic impedance Z_c which is the input impedance of an infinitely long transmission line. It is given by

$$Z_c = \sqrt{\frac{R + j\omega L}{G + j\omega C}}$$

2. The propagation constant γ whose real part α is the attenuation constant and imaginary part β is the phase constant. It is given by

$$\gamma = \alpha + j\beta = \sqrt{(R + j\omega L)(G + j\omega C)}$$

If a transmission line is terminated in its characteristic impedance, there is no reflected wave and the input impedance of the terminated line is Z_c irrespective of its length. The propagation constant determines the attenuation and the phase change in a sinusoidal wave travelling along the transmission line. α is the attenuation of the signal having frequency ω over a unit distance. β is the phase change in the signal when it travels a unit distance.

1.2 Phase Velocity and Phase Delay

If $\Delta\phi$ is the phase change between points x and $x + \Delta x$, then

$$\Delta\phi = \Delta x \beta$$

The same phase change occurs at point x in time Δt given by

$$\Delta\phi = \Delta t \omega$$

Thus, if a constant phase point is observed along the transmission line, the speed of propagation or the phase velocity is given by

$$v = \Delta x / \Delta t = \omega / \beta$$

The time required to travel unit distance is called phase delay per unit length. It is given by

$$\tau = 1/v = \beta / \omega$$

1.3 Asymptotic Behaviour at Low Frequencies

At low frequencies we can assume $\omega L \ll R$ and if $G \approx 0$, the secondary parameters of the transmission line are given by:

$$Z_c \approx \sqrt{\frac{R}{j\omega C}} = \sqrt{\frac{R}{\omega C}} \exp(-j\pi/4)$$

$$\gamma = \alpha + j\beta \approx \sqrt{j\omega RC} = \sqrt{\omega RC/2} + j\sqrt{\omega RC/2}$$

$$\alpha \approx \sqrt{\omega RC/2}$$

$$\beta \approx \sqrt{\omega RC/2}$$

In conclusion, when $\omega L \ll R$,

- the characteristic impedance is complex and is inversely proportional to the square root of the frequency; and
- the attenuation and phase constants are proportional to the square root of the frequency.

1.4 Asymptotic Behaviour at High Frequencies

At high frequencies we can assume $\omega L \gg R$ and if $G \approx 0$, the secondary parameters of the transmission line are given by

$$Z_c \approx \sqrt{L/C}$$

$$\gamma = \alpha + j\beta \approx \sqrt{-\omega^2 LC + j\omega RC} \approx j\omega\sqrt{LC}\left(1 - j\frac{R}{2\omega L}\right)$$

$$\alpha = \frac{R}{2}\sqrt{L/C}$$

$$\beta = \omega\sqrt{LC}$$

In conclusion, when $\omega L \gg R$,

- the characteristic impedance is real and independent of frequency
- the attenuation constant is proportional to R (and therefore, if the skin effect is taken into account, α will vary as $\sqrt{\omega}$); and
- the phase constant increases linearly with the frequency.

Figure 2 summarizes the results of the preceding sections. Note that α is proportional to the square root of the frequency at low and high frequencies due to entirely different reasons.

2 LINEAR DISTORTIONS

Information is always contained in the shape of a signal and, therefore, the shape should not get distorted when the signal is transmitted. The received signal can at most differ from the transmitted signal in having been multiplied by a constant factor and delayed by a certain time. In the last chapter, we had examined how a signal can be expressed as a series of sinusoidal signals. Thus for distortionless transmission of a signal, it is mandatory that (i) amplitudes of all its frequency components are multiplied by the same factor when they are transmitted, and (ii) all the frequency components are delayed by the same amount.

Conditions (i) and (ii) above imply that the attenuation constant α and the phase delay τ of the transmission line must be independent of frequency. τ being equal to β/ω, the phase constant β should increase linearly with frequency. Figure 3 shows the plots of α and β for distortionless transmission.

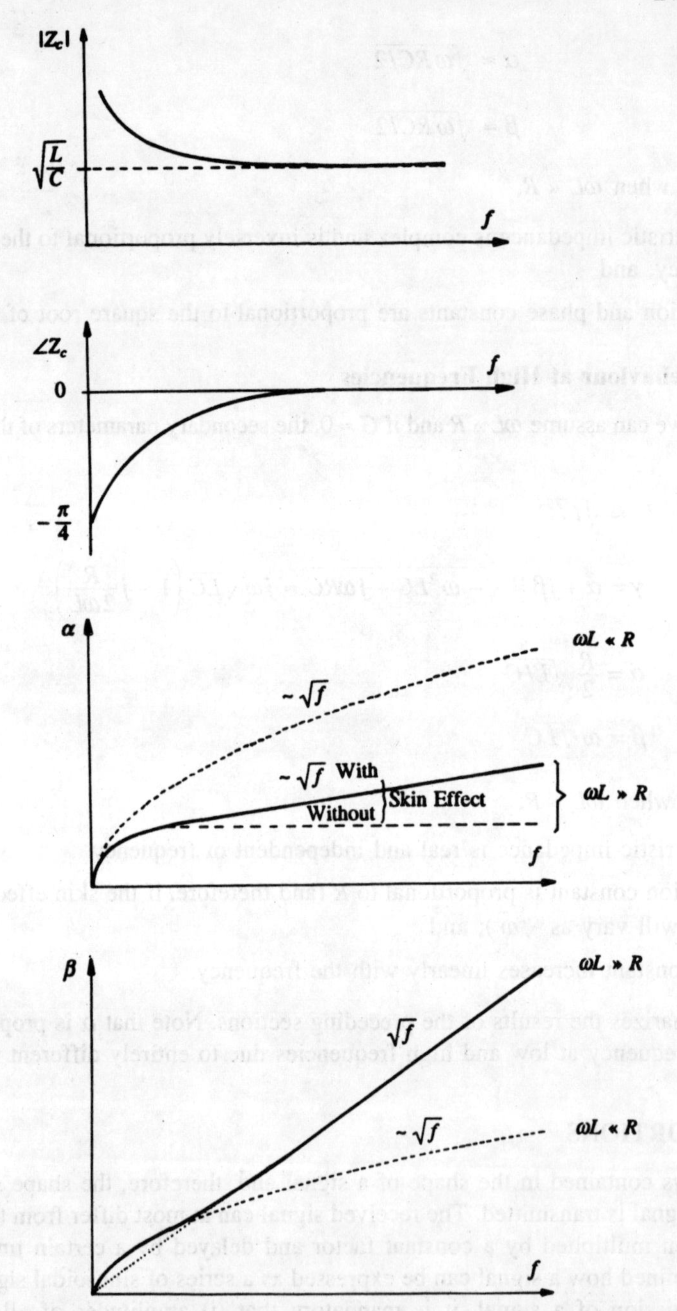

Fig. 2 Asymptotic behaviour of the secondary parameters.

The above conditions for distortionless transmission cannot be perfectly satisfied in practice and, as a result, the transmitted signal gets distorted. A transmission line introduces—attenuation distortion if its attenuation constant α is a function of frequency, and phase distortion if the phase constant β is not a linear function of the frequency.

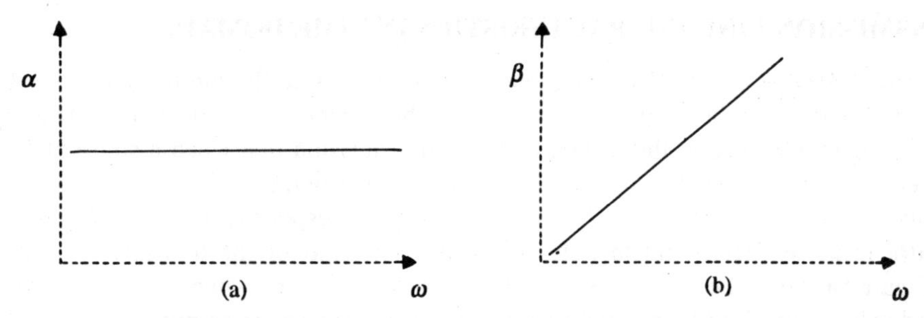

Fig. 3 Attenuation and phase characteristics for distortionless transmission.

These distortions are called linear distortions because the transmission line presents a linear system in which the principle of superposition remains valid. A sinusoidal signal remains a sinusoid and no new spectral components appear at the other end of the transmission line.

2.1 Group Delay

The condition for no phase distortion is in fact very stringent and difficult to implement. Considering that most of the transmitted signals are modulated carriers, we can relax the constant phase delay requirement. It can be shown that phase distortion does not occur in the modulating signal if the slope $d\beta/d\omega$ of phase characteristic is constant. The slope $d\beta/d\omega$ is called group delay and it can be easily measured for any transmission channel.

2.2 Frequency Domain Equalizers

A transmission line always introduces linear distortions because the attenuation constant α is proportional to the square root of the frequency and the phase constant β is not a linear function of frequency. These distortions are corrected by using equalizers. As the name suggests, equalizers make up for the transmission characteristics and minimize the attenuation and phase distortions.

Equalizer is usually provided at the receiving end of the transmission line. There are separate equalizers for correcting attenuation distortion and phase distortion. The attenuation equalizer has a loss characteristic which is inverse of the attenuation characteristic of the transmission line so that the net result is a flat amplitude frequency response (Fig. 4a). As regards the phase distortion, the transmission line is equalized for group delay. The group delay characteristic is made flat using an equalizer with complementary group delay response (Fig. 4b).

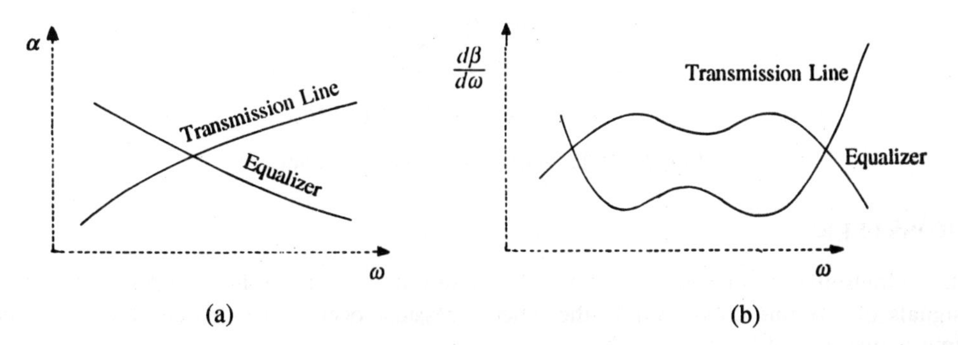

Fig. 4 Equalizer characteristics.

3 TRANSMISSION LINE CHARACTERISTICS IN TIME DOMAIN

The attenuation $A(\omega) = \alpha l$ and the phase change $B(\omega) = \beta l$ describe the behaviour of the transmission line of length l in the frequency domain, i.e., they enable us to calculate the amplitude and phase of the signal obtained at the distant end of a transmission line when a sinusoidal signal of angular frequency ω is applied to it. $A(\omega)$ and $B(\omega)$ together define the transfer function $H(\omega)$ of the line. Inverse Fourier transform of $H(\omega)$ gives the impulse response $h(t)$ of the transmission line. The impulse response $h(t)$ completely describes the behaviour of the transmission line in time domain. Figure 5a shows the received signal at the output of the line when a pulse of time duration T is applied at its input. The shape of the signal depends on the impulse response of the transmission line. Without going into mathematical details, we can quote the results of time domain analysis. A transmission line is said to be short with respect to T when its length l is such that

$$\frac{2T}{RCl^2} \gg 10$$

When the line is "short", the pulse shape is slightly deformed and the pulse duration is retained (Fig. 5b). On the other hand, for a "long" transmission line (when the above inequality is reversed) the pulse is considerably deformed and its duration also gets stretched (Fig. 5c). It may even interfere with the subsequent neighbouring pulses. This type of interference is termed as intersymbol interference (ISI). We will come back to intersymbol interference later and discuss how its effect can be minimized.

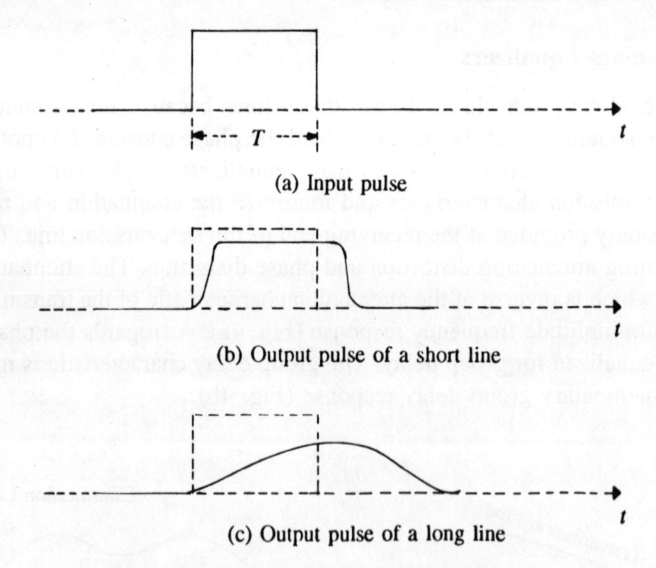

Fig. 5 Pulse distortion in time domain.

4 CROSSTALK

When two transmission lines are very close, they interfere with each other and result in crosstalk, i.e., signals of one line cross over to the other. Crosstalk occurs due to three types of mutual coupling between the lines:

1. Galvanic coupling which is due to a common resistance of the two lines. This phenomenon is noticeable in lines having common return conductor (Fig. 6a).

2. Capacitive coupling which is due to the capacitance between the conductors of the lines (Fig. 6b).

3. Inductive coupling which is due to the mutual inductance of the transmission lines (Fig. 6c).

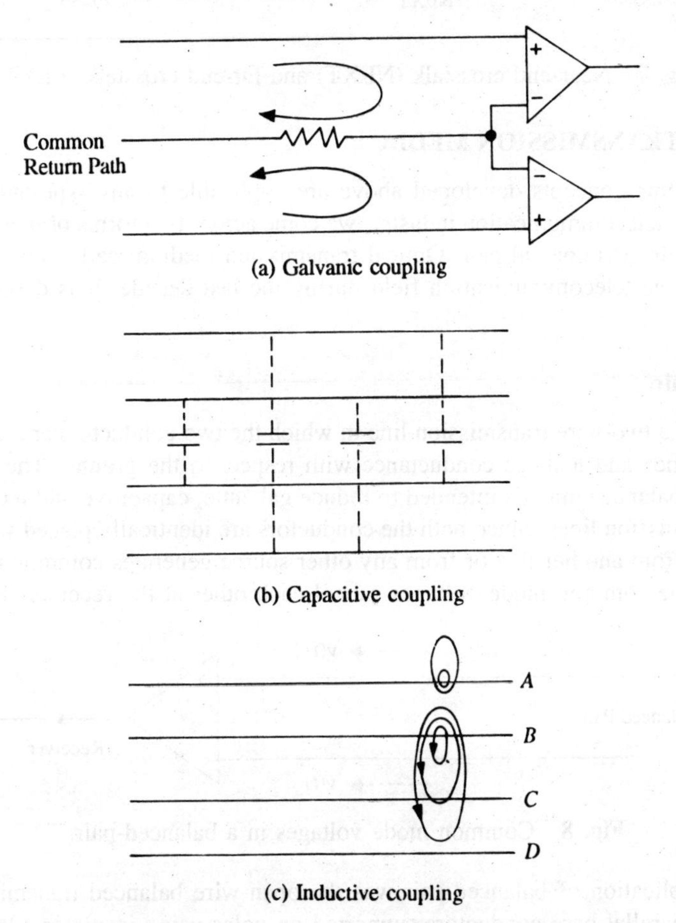

(a) Galvanic coupling

(b) Capacitive coupling

(c) Inductive coupling

Fig. 6 Coupling in transmission lines.

The extent of coupling depends on the geometric configuration of the conductors and the proximity of the transmission lines. The signals coupled to a transmission line due to crosstalk progress towards the far end as well as back to the near end (Fig. 7). The crosstalk which appears at the near end is called near-end crosstalk (NEXT), and the crosstalk which appears at the distant end is called far-end crosstalk (FEXT). NEXT is relatively independent of the length of the transmission line as the first few meters of the transmission line are responsible for the bulk of it. On the other hand, the effect of FEXT increases with the length of the transmission line.

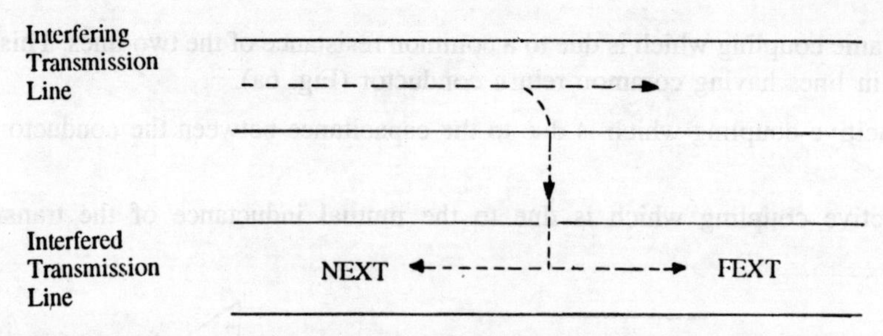

Fig. 7 Near-end crosstalk (NEXT) and far-end crosstalk (FEXT).

5 METALLIC TRANSMISSION MEDIA

The transmission line concepts developed above are applicable to any type and shape of transmission line. In the telecommunication industry, we come across two forms of metallic transmission lines: balanced-pair, and coaxial-pair. Optical transmission medium called optical fibre has also made its entry in the telecommunication field during the last decade. It is described in the next section.

5.1 Balanced-Pair

A balanced-pair is a two-wire transmission line in which the two conductors are identical and have the same capacitance and leakage conductance with respect to the ground. The term "balanced" implies electrical balance which is intended to reduce galvanic, capacitive and inductive couplings between the transmission lines. Since both the conductors are identically placed with respect to the ground, coupling from another line or from any other source generates common mode voltages in the conductors. The common mode voltages cancel each other at the receiver (Fig. 8).

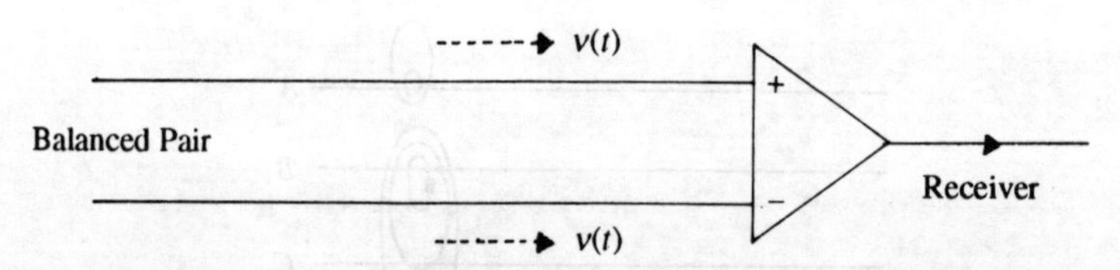

Fig. 8 Common mode voltages in a balanced-pair.

The first application of balanced-pair was the open-wire balanced transmission line which consisted of two parallel bare conductors supported on poles using insulators. It is now obsolete and has been replaced to a large extent by other transmission media.

5.2 Balanced-Pair Cables

The balanced-pair cable consists of a bunch of pairs of insulated copper wires which are stranded, i.e., twisted in such a way as to reduce inductive coupling among the pairs. Balanced-pair cables are used in the telecommunication network as junctions for interconnecting telephone exchanges and as subscriber cables to extend the connection from telephone exchange to subscriber premises. The conductors have diameters ranging from 0.4 to 1.5 mm and the insulating material is paper or polyethylene. The number of pairs in a balanced-pair cable can be from 10 to 1200 or more. The principal characteristics of the balanced-pair cable are:

- The balanced cables are used in the frequency range where $\omega L \ll R$. In this frequency range, α and β are proportional to the square root of the frequency and therefore, attenuation and phase distortions are present.
- The characteristic impedance is complex.

Figure 9 shows variation in the characteristic impedance and attenuation parameter with frequency. Note that the characteristic impedance varies from 600 ohms at 1 kHz to 150 ohms at frequencies higher than 10 kHz. The balanced pairs are terminated accordingly depending on frequency of the signal.

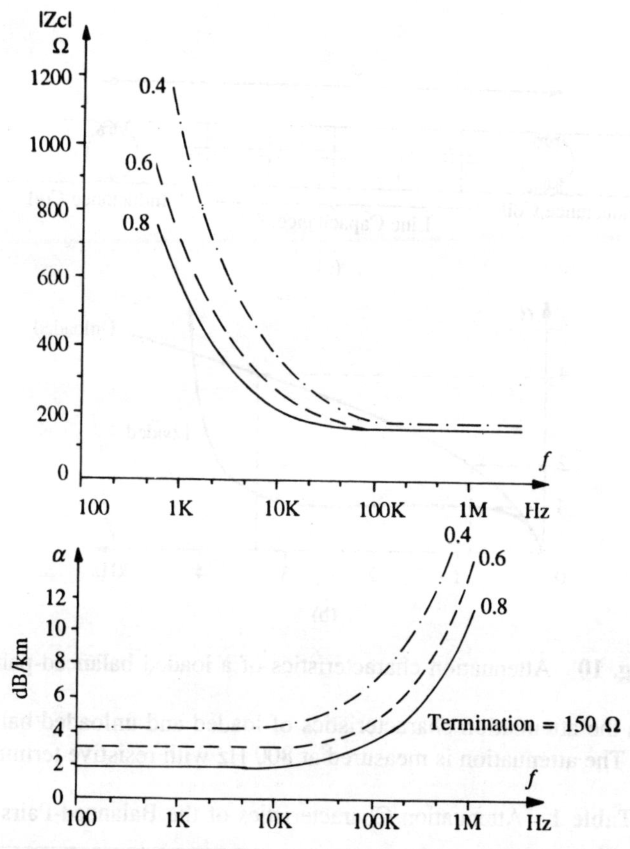

Fig. 9 Transmission characteristics of a balanced-pair.

5.3 Loading of Balanced Pairs

The characteristics of the transmission line are nearly ideal when $\omega L \gg R$, and the skin effect is negligible. We can approach these conditions at low frequencies by artificially increasing the inductance per unit length (L) of the transmission line. When the inductance is increased, the attenuation constant α diminishes and becomes independent of the frequency; and the phase constant β increases and becomes proportional to the frequency.

To increase L, lumped inductances called loading coils are added at regular intervals in the transmission line (Fig. 10a). The transmission line behaves like a low pass filter having reduced constant attenuation in the pass band (Fig. 10b). The attenuation increases rapidly beyond the cutoff frequency f_c given by

$$f_c = \frac{1}{\pi} \frac{1}{\sqrt{L_p Cd}}$$

where L_p is the inductance of the loading coil and d is the separation between adjacent coils. In the telephone network, we generally choose L_p = 88.5 mH and d = 1830 m. Assuming negligible inductance of the line itself, the cutoff frequency calculated from the above relation comes to 4228 Hz.

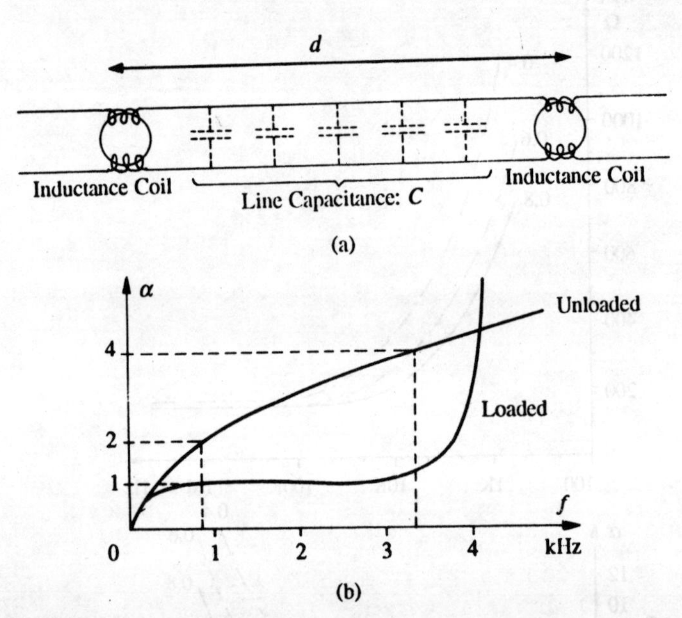

Fig. 10 Attenuation characteristics of a loaded balanced-pair.

Table 1 shows the attenuation characteristics of loaded and unloaded balanced pairs used in telephone networks. The attenuation is measured at 800 Hz with resistive termination of 600 ohms.

Table 1 Attenuation Characteristics of the Balanced-Pairs

Conductor diameter (mm)	Attenuation of unloaded pair (dB/km)	Attenuation of loaded pair (dB/km)
0.6	0.85	0.43
0.8	0.56	0.25
0.9	0.48	0.22
1.0	0.43	0.22
1.2	0.35	0.13
1.4	0.29	0.10
1.5	0.26	0.087

5.4 Coaxial Cable

The coaxial pair is composed of two concentric conductors separated by dielectric discs or continuous material (Fig. 11). The external conductor can be solid or a metallic braid. A coaxial cable may contain one or more coaxial pairs.

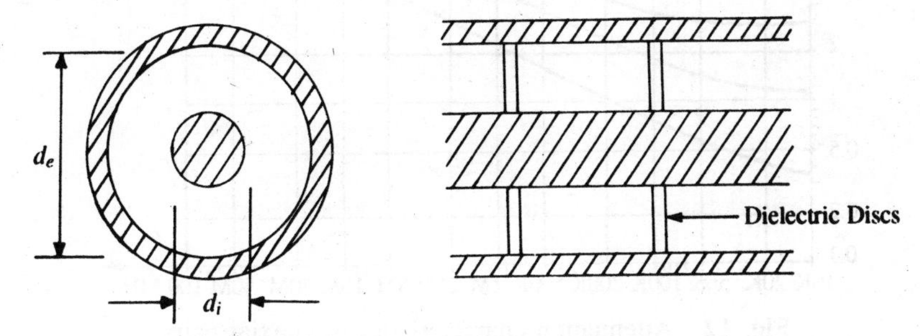

Fig. 11 Construction of a coaxial pair.

The secondary parameters of the coaxial pair can be expressed in terms of d_i, the external diameter of the central conductor; and d_e the internal diameter of the external conductor. The principal characteristics of the coaxial pair are summarized below. It is assumed that $\omega L \gg R$ which is true in the useful frequency range of coaxial pair.

- The characteristic impedance depends on the ratio of diameters d_e/d_i.
- The attenuation constant is inversely proportional to the diameter d_e. For a given d_e, minimum value of α is obtained when $d_e/d_i = 3.6$.
- The attenuation constant is proportional to the square root of the frequency due to skin effect. Thus, there is attenuation distortion.
- The phase constant is a linear function of frequency at high frequencies ($f > 100$ kHz). Thus, there is no phase distortion.

CCITT recommendations for the coaxial pairs are given in Table 2. The attenuation characteristics are shown in Fig. 12. Being a logarithmic plot, the slope of the characteristic is 1/2.

Table 2 CCITT Recommendations for the Coaxial Pairs

	CCITT Recommendations		
	G.623	G.622	G.621
d_i	2.6 mm	1.2 mm	0.7 mm
d_e	9.5 mm	4.4 mm	2.9 mm
d_e/d_i	3.65	3.67	4.14
Z_c	75 ± 1 Ω	75 ± Ω	75 ± 1 Ω

Coaxial cables are used as the long distance transmission medium in telephone networks. Coaxial cables also find application in cable television networks and local area networks for computer communications. They are not used at frequencies lower than 60 kHz due to their degraded phase and crosstalk properties.

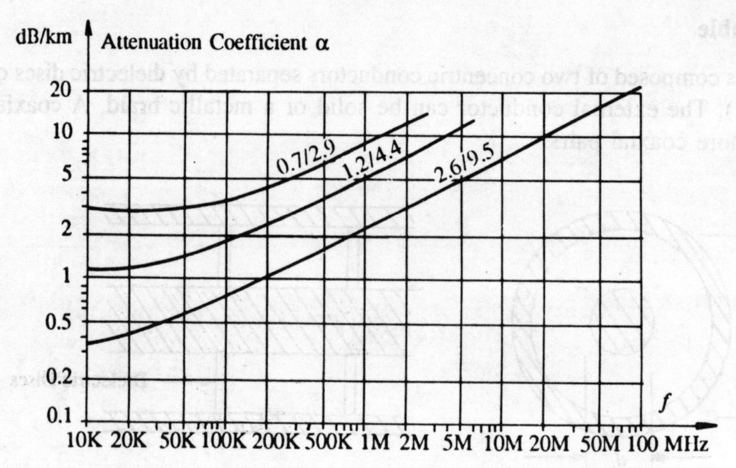

Fig. 12 Attenuation characteristics of coaxial pairs.

6 OPTICAL FIBRE

An optical fibre consists of an inner glass core surrounded by a cladding also of glass but having a lower refractive index. Transmission of digital signals is in the form of intensity-modulated light signal which is trapped in the core. Light is launched into the fibre using a light source (LED or Laser) and is detected at the other end using a photo detector (Fig. 13). Early optical systems used LEDs having wavelength of 870 nm but later 1300 and 1550 nm wavelengths were found more suitable and at present most of the systems use these wavelenghts.

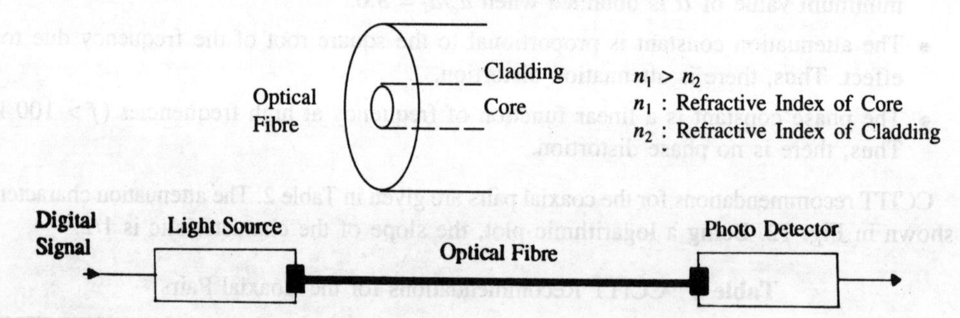

Fig. 13 Optical fibre as a transmission medium.

6.1 Multimode Fibres

Light propagation in the core is based on the phenomenon of total internal reflection which takes place at the core-cladding interface. The refractive index of the cladding being less than that of the core, an oblique light ray in the core is reflected back if it strikes the interface at an angle greater than the critical angle which is determined by the refractive indices of the core and cladding (Fig. 14). It can be shown that in a cylindrical guided medium like optical fibre, a finite number of modes of light propagation can be sustained. Mode refers to the light path which a ray traces depending on its angle of incidence at the interface. The number of the modes depends on the diameter of the core and can be reduced by reducing the core diameter. Typical core and cladding diameters of a multimode fibre are 50 μm and 125 μm respectively.

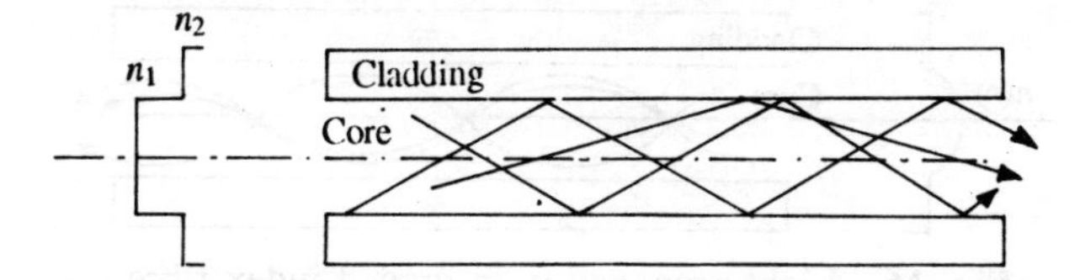

Fig. 14 Light propagation in step index multimode fibre.

6.2 Modal Dispersion

When a pulse of light is incident on the fibre end, several modes of light propagation are generated in the core. These modes propagate through the fibre over paths of different lengths. When they arrive at the other end of the fibre, they are staggered in time and result in stretching of the transmitted pulse. This phenomenon is called modal dispersion and it limits the bit rate a fibre section can support. It can be shown that modal dispersion τ_m is given by

$$\tau_m = \frac{n_1 l}{c n_2}(n_1 - n_2)$$

where n_1 and n_2 are the refractive indices of the core and cladding respectively, c is speed of light in vacuum and l is the length of fibre section. Typical value of modal dispersion for multimode fibre is 50 ns/km.

6.3 Monomode Fibre

If the diameter of the fibre core is so reduced that it can sustain only one mode, modal dispersion can be eliminated. Such a fibre is called monomode or single mode fibre (Fig. 15). Its core diameter is of the order of a few microns. Due to its low dispersion, monomode fibre can support very high bit rates.

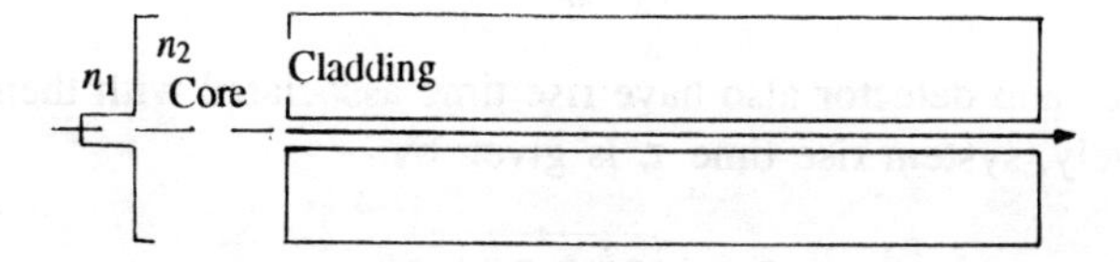

Fig. 15 Light propagation in monomode fibre.

6.4 Graded Index Fibres

Multimode and monomode fibres discussed above are called step index fibres as there is step variation of the refractive index profile. There is another type of fibre called graded index fibre in which the refractive index profile of the core approximates a parabolic shape. It can be shown that the different modes take curved paths in the core of the graded index fibre, periodically converge to common points at the same instant of time and finally emerge from the core of the fibre simultaneously (Fig. 16). Modal dispersion is therefore significantly reduced.

6.5 Chromatic Dispersion

Having reduced the modal dispersion to a very low value in the graded index fibres and having eliminated it in the monomode fibres, further increase in bit rate is limited by another type of

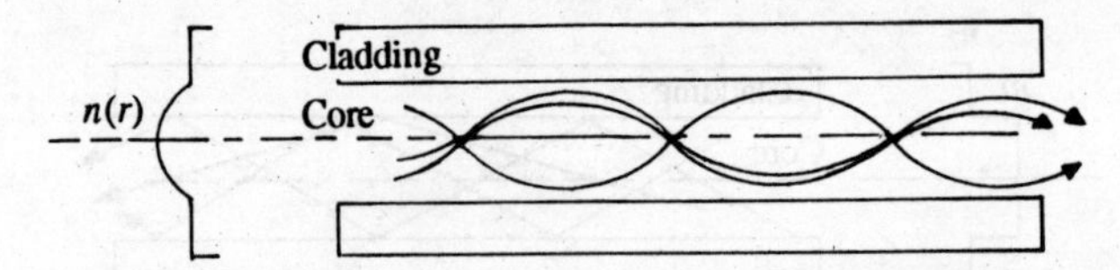

Fig. 16 Light propagation in graded index fibre.

dispersion called chromatic dispersion. It is due to different speeds of propagation of different wavelengths emitted by the source. These wavelengths arrive at the other end of the fibre at different times and cause chromatic dispersion.

LED is not a monochromatic optical source and its spectral width (difference between the highest and the lowest wavelengths emitted) is of the order of 50 nm. Chromatic dispersion can be reduced by using a monochromatic source. Solid state lasers are therefore used along with monomode fibres for very high bit rates. Their spectral width is of the order of 1 nm.

Chromatic dispersion depends on the spectral width of the source, composition of the fibre material and length of the fibre section. It is given by

$$\tau_c = M\Delta\lambda\, l, \qquad M = \frac{d^2 n}{d\lambda^2}$$

where $\Delta\lambda$ is the spectral width of the source, l is the length of fibre section and n is the refractive index. For silica glass and step-index profile, it turns out that at approximately 1300 nm, M becomes zero. Chromatic dispersion is extremely small, ~ 10 ps, at this wavelength.

6.6 Total Dispersion

Total dispersion in a fibre is obtained by adding the squares of modal and chromatic dispersions.

$$\tau = \sqrt{\tau_m^2 + \tau_c^2}$$

The optical source and detector also have rise time associated with them. If their rise times are τ_s and τ_d respectively, system rise time τ_r is given by

$$\tau_r = \sqrt{\tau_s^2 + \tau_d^2 + \tau^2}$$

If the bit rate is R, the maximum design value of τ_r is kept at $0.7/R$ for *NRZ* signal and $0.35/R$ for *RZ* signal.

6.7 Fibre Attenuation

Intrinsic attenuation of the fibre is due to absorption and Rayleigh scattering of light. While absorption losses due to impurities can be controlled and minimised, Rayleigh scattering cannot be overcome. Figure 17 shows the attenuation characteristic of silica glass fibre. In the early seventies, the fibre loss was minimum around λ = 820 nm and therefore, early optical fibre systems operated at this wavelength. The loss was of the order of 5 dB/km. As the fibre manufacturing technology developed, the fibre loss was brought down to about 0.5 dB/km at 1300 nm. Most of the present day systems operate at this wavelength as chromatic dispersion is also minimum at 1300 nm. The wavelength for the near future is 1550 nm at which the loss is less than 0.2 dB/km. Dispersion-

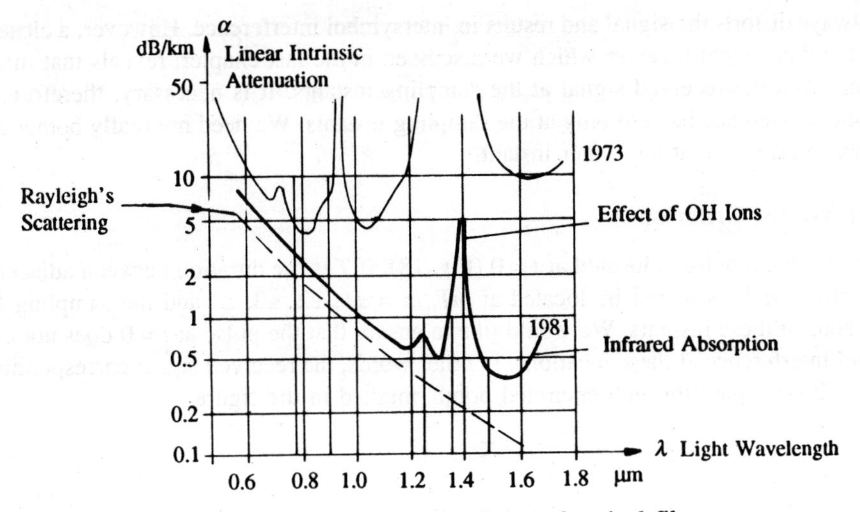

Fig. 17 Attenuation characteristics of optical fibres.

shifted fibres which exhibit minimum chromatic dispersion at this wavelength have already been developed.

6.8 Advantages of Optical Fibre

Some advantages of optical fibres as a transmission medium are:

1. Optical fibres offer very wide bandwidth for transmission of signals. Bit rates in the range of a few giga bits per second are feasible with present technology.

2. Optical signals are not affected by electromagnetic interference.

3. Optical fibres have very low attenuation (0.2 dB/km).

4. Copper cables need repeaters[†] at about every four kilometres while in optical fibres the repeater spacing is 20 to 50 km.

5. Optical transmission is secure as the signals cannot be tapped.

6. Optical fibre cables are very light in weight, very small in size and are easily laid.

7. Glass being an insulator, optical fibres are safe when laid along a high tension power line.

8. Optical fibres are made of silica which is abundantly available as a natural resource. Copper, on the other hand, is rare.

9. Optical fibre has very low sensitivity to temperature and environment.

7 BASEBAND TRANSMISSION OF DATA SIGNALS

Considering that a random binary signal has a frequency spectrum which extends upto infinity, it appears that we require a transmission channel which meets the conditions for distortionless transmission over the infinite frequency spectrum of the baseband signal. A band limited transmission

---

[†]Repeater is an electronic circuit for regenerating a digital signal.

channel always distorts the signal and results in intersymbol interference. However, a closer examination of the data signal receiver which we discussed in the last chapter, reveals that information is extracted from the received signal at the sampling instants. It is necessary, therefore, that the intersymbol interference be zero only at the sampling instants. We need not really bother about the intersymbol interference at any other instant.

7.1 First Nyquist Criterion

Let us assume that a pulse is located at $t = 0$ (Fig. 18). If T is the duration between adjacent binary symbols, other symbols would be located at mT, $m = \pm 1, \pm 2, \pm 3, \ldots$ and the sampling for them would be done at these instants. We would like to ensure that the pulse at $t = 0$ does not cause any intersymbol interference at these locations. In other words, the received signal corresponding to the pulse at $t = 0$ must pass through encircled points marked in the figure.

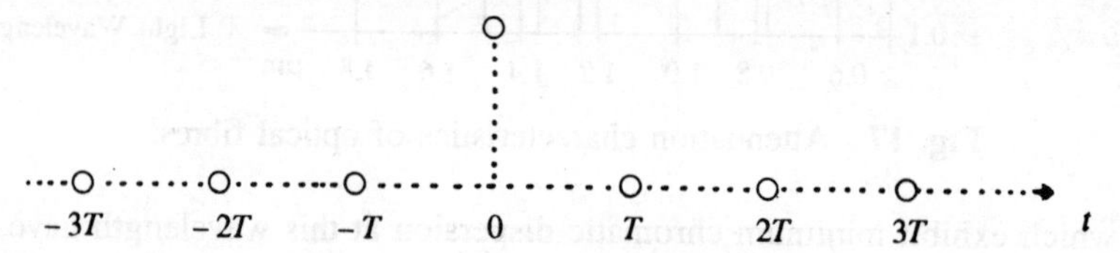

Fig. 18 Instants of zero intersymbol interference for the first Nyquist criterion.

One such signal which satisfies this requirement is sin $(\pi t/T)/(\pi t/T)$ as shown in Fig. 19a. Figure 19b shows the Fourier transform $H(f)$ of the signal. Thus, if the input is an impulse and the transmission channel has the frequency response $H(f)$ as shown in Fig. 19b, the impulse will generate a response shown in Fig. 19a and there will not be any intersymbol interference. Figure 19c shows the channel response for a data signal in which the data symbols are represented as impulse and no impulse. Being a linear system, the principle of superposition is applicable. The overall response is the sum of individual responses. If the input is a stream of finite duration pulses instead of impulses, the channel characteristics need to be slightly modified as we will see later.

Thus, it is sufficient to have a transmission channel having ideal characteristics of a low-pass filter with a cutoff frequency equal to half the bit rate (Fig. 19b). We do not require an ideal channel of infinite bandwidth as inferred earlier. This is called the first Nyquist criterion for zero intersymbol interference.

There are several limitations in realising the transmission channel defined by the first Nyquist criterion:

- Ideal low-pass filter characteristics of the channel are nonrealisable.
- There is significant intersymbol interference around the sampling instants. Even small errors in the sampling instants result in large amount of intersymbol interference.

7.2 Second Nyquist Criterion

We had assumed sin $(\pi t/T)/(\pi t/T)$ response of the transmission channel to arrive at the first Nyquist criterion. The second Nyquist criterion gives realisable solutions for transmission channel characteristics. We arrive at the second criterion by forcing the channel response to pass through some additional points as shown in Fig. 20. The channel response curve which passes through the

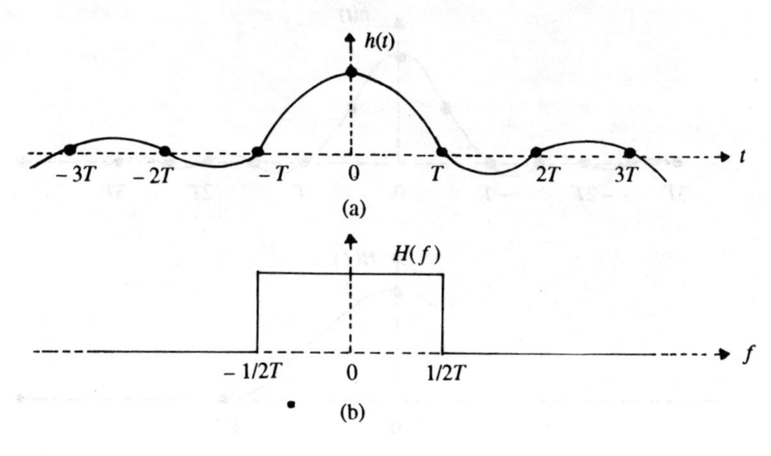

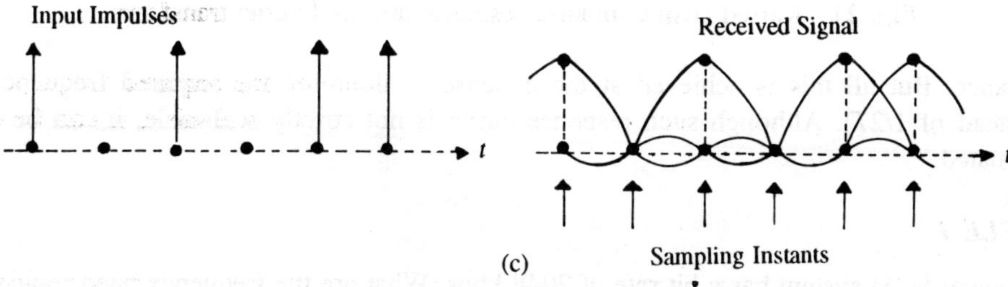

Fig. 19 Sin $(\pi t/T)/(\pi t/T)$ channel.

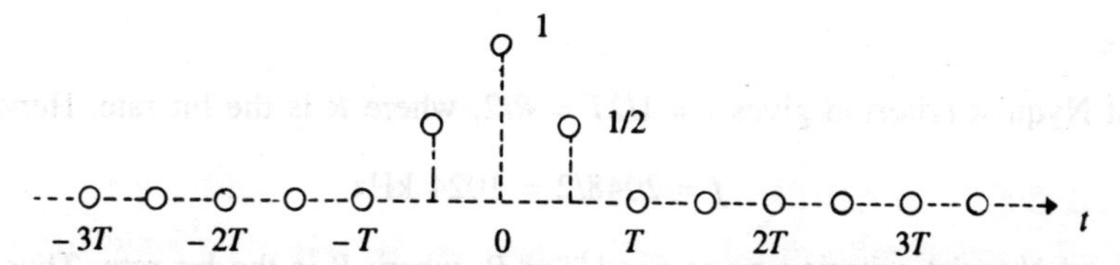

Fig. 20 Instants of zero intersymbol interference for the second Nyquist criterion.

encircled points and its Fourier transform are shown in Fig. 21. This response curve is called raised cosine response as evident from its mathematical representation.

$$h(t) = \frac{\sin(\pi t/T)}{\pi t/T} \; \frac{\cos(\pi t/T)}{1 - (2t/T)^2}$$

$$H(f) = T\cos^2(\pi Tf/2) \quad \text{for } |f| \le 1/T$$

$$= 0 \quad \text{for } |f| > 1/T$$

The above channel characteristics define the second Nyquist criterion for no intersymbol interference at the sampling instants. Due to the second power of time in the denominator, intersymbol interference approaches zero much more rapidly than the sin $(\pi t/T)/(\pi t/T)$ response discussed earlier. Therefore, sensitivity to errors in the sampling instants does not severely degrade the

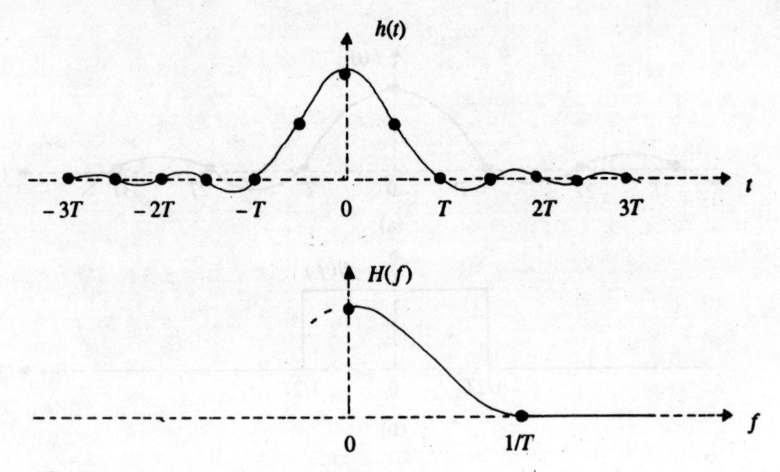

Fig. 21 Raised cosine impulse response and its Fourier transform.

performance. But all this is achieved at the expense of doubling the required frequency band ($1/T$ instead of $1/2T$). Although such response curve is not strictly realisable, it can be closely approximated.

EXAMPLE 1

A 30 channel PCM system has a bit rate of 2048 kbps. What are the frequency band requirements for the PCM signal as per the first and second Nyquist criteria?

Solution

The first Nyquist criterion gives $f = 1/2T = R/2$, where R is the bit rate. Hence,

$$f = 2048/2 = 1024 \text{ kHz}$$

The second Nyquist criterion gives $f = 1/T = R$, where R is the bit rate. Thus,

$$f = 2048 = 2048 \text{ kHz}$$

Raised cosine and sin $(\pi t/T)/(\pi t/T)$ are the two extreme cases of responses for no intersymbol interference at the sampling instants. It is possible to define more channel responses meeting the criterion of no intersymbol interference at the sampling instants and occupying a frequency band greater than $1/2T$ and less than $1/T$. Figure 22 shows several such channel characteristics. The necessary condition for zero intersymbol interference is that the channel frequency response should have an odd symmetry about $f = 1/2T$. In general, we can say that a binary signal having bit rate $R = 1/T$ would occupy a frequency band of $B = (1 + r)R/2$ Hz, where r is the roll-off factor whose value varies from 0 to 1.

EXAMPLE 2

If the bit rate is 9600 bps and raised cosine pulse spectrum with roll-off factor equal to 0.8 is used, determine the frequency band occupied by the signal.

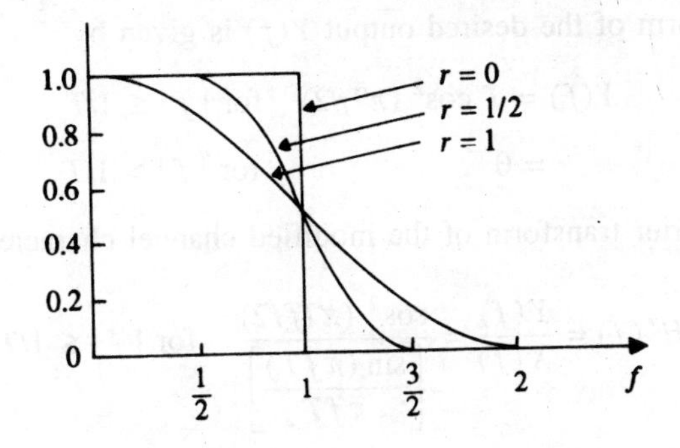

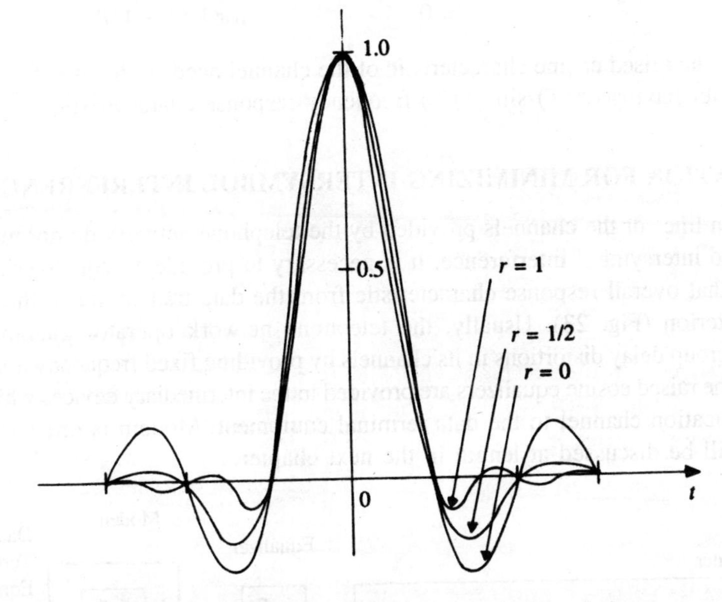

Fig. 22 Generalized raised cosine characteristics.

Solution

$$B = (1 + r)R/2 = (1 + 0.8)9600/2 = 1.8 \times 4800 = 8640 \text{ Hz}$$

7.3 Channel Characteristic for Finite Duration Pulses

The raised cosine response is obtained when an impulse is applied to the channel. In practice, we will not transmit impulses but pulses of finite duration T. Therefore, we need to modify the channel characteristic to ensure that we get the same raised cosine response when a finite duration pulse is applied to it.

The Fourier transform of a pulse of duration T is given by

$$X(f) = T\left[\frac{\sin(\pi f T)}{\pi f T}\right]$$

The Fourier transform of the desired output $Y(f)$ is given by

$$Y(f) = T\cos^2(\pi Tf/2) \quad \text{for } |f| \le 1/T$$
$$= 0 \quad \text{for } |f| > 1/T$$

If $H'(f)$ is the Fourier transform of the modified channel characteristic, then

$$H'(f) = \frac{Y(f)}{X(f)} = \frac{\cos^2(\pi Tf/2)}{\left[\dfrac{\sin(\pi fT)}{\pi fT}\right]} \quad \text{for } |f| \le 1/T$$

$$= 0 \quad \text{for } |f| > 1/T$$

Therefore, the raised cosine characteristic of the channel needs to be modified by introducing an additional filter having $(\pi fT)/\sin(\pi fT)$ frequency response characteristic.

8 EQUALIZATION FOR MINIMIZING INTERSYMBOL INTERFERENCE

The transmission lines or the channels provided by the telephone network do not meet the Nyquist criterion for zero intersymbol interference. It is necessary to provide an equalizer in cascade with the channel so that overall response characteristic from the data transmitter to the receiver meets the Nyquist criterion (Fig. 23). Usually, the telephone network operator guarantees maximum attenuation and group delay distortions in its channels by providing fixed frequency domain equalizers in its network. The raised cosine equalizers are provided in the intermediary devices which interconnect the telecommunication channel to the data terminal equipment. Modem is one such intermediary device and it will be discussed at length in the next chapter.

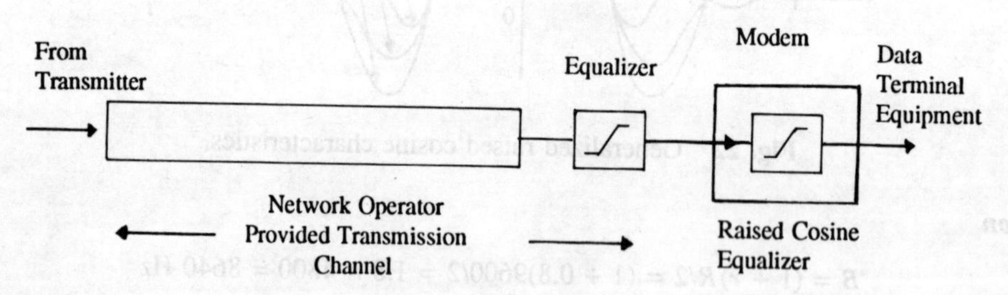

Fig. 23 Equalizers in a transmission system.

The equalizer may have fixed or variable characteristics. Fixed equalizers are used when the transmission channel characteristics are always within the design limits, while variable equalizers are required when the transmission channel characteristics change with time. For example, if the connection between the transmitter and receiver is established through a telephone exchange, characteristics of the transmission channel change every time a new connection is established. Even in the point-to-point circuits, temperature variations cause changes in transmission characteristics of the channels. The variable equalizers are of two types: manual and adaptive.

Manual equalizers are provided with controls to adjust the equalizer for the desired channel response. Most designs use some method of indicating the amount of intersymbol interference which is minimized by adjusting the controls. Adaptive equalizers, on the other hand, have the capability of automatic equalizer adjustment for minimum intersymbol interference. Adaptive equalizers are based on the transversal filter equalizer which is described next.

8.1 Transversal Filter Equalizer

It consists of a tapped delay line having $2N + 1$ taps. The delay between successive taps is equal to the pulse sampling interval T (Fig. 24). Each tap is connected through a variable gain device to a summing amplifier. By adjusting the gain of these devices, it is possible to force intersymbol interference at the sampling instants to zero.

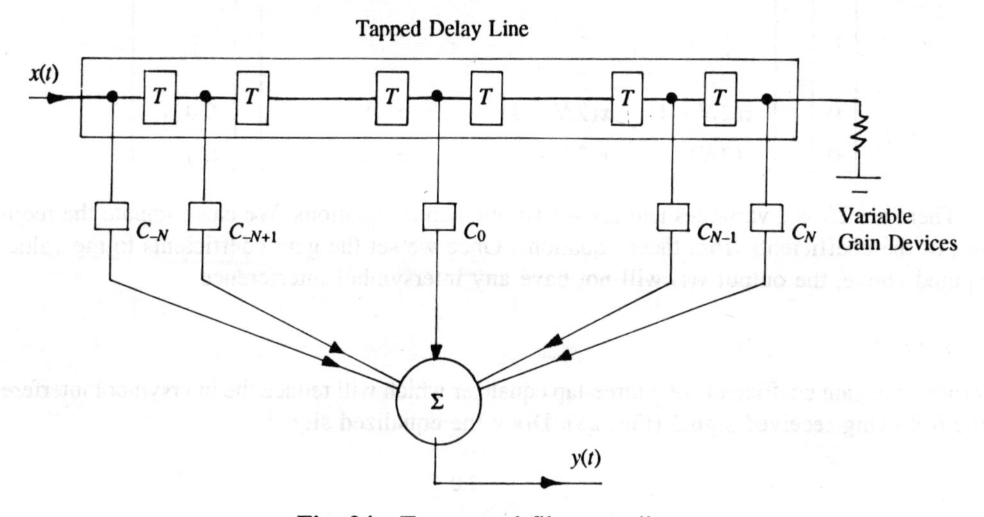

Fig. 24 Transversal filter equalizer.

If the input to the equalizer is $x(t)$, the output $y(t)$ can be written as

$$y(t) = \sum_{i=-N}^{N} C_i \, x[t - (i + N)T]$$

where C_i is the gain associated with the ith tap. Since we are interested in the intersymbol interference at the sampling instants only, let us put $t = (k + N)T$, where $k = 0, \pm 1, \ldots$ in the above equation.

$$y(k) = \sum_{i=-N}^{N} C_i \, x[(k - i)T]$$

where $x(i)$, $-2N \le i \le 2N$ are the values of received signal at the sampling instants. Our objective is to adjust the values of the coefficients C_i, so that the intersymbol interference at the sampling instants become zero in the output. In other words,

$$\begin{aligned} y(k) &= 1 \quad \text{for } k = 0 \\ &= 0 \quad \text{for } k = \pm 1, \pm 2, \pm 3, \ldots, \pm N \end{aligned}$$

The quantity $y(k)$ for $k \neq 0$ represents intersymbol interference. Note that we have specified the output at $2N + 1$ points only assuming that at the other points beyond $k = \pm N$, the residual intersymbol interference will be insignificant. Substituting these values of the output, we get

$$\begin{bmatrix} 0 \\ 0 \\ \vdots \\ 0 \\ 1 \\ 0 \\ \vdots \\ 0 \\ 0 \end{bmatrix} = \begin{bmatrix} x(0) & x(-1) & \dots & x(-2N) \\ x(1) & x(0) & \dots & x(-2N+1) \\ \vdots & \vdots & & \vdots \\ x(N-1) & x(N-2) & \dots & x(-N-1) \\ x(N) & x(N-1) & \dots & x(-N) \\ x(N+1) & x(N) & \dots & x(-N+1) \\ \vdots & \vdots & & \vdots \\ x(2N-1) & x(2N-2) & \dots & x(1) \\ x(2N) & x(2N-1) & \dots & x(0) \end{bmatrix} \begin{bmatrix} C_{-N} \\ C_{-N+1} \\ \vdots \\ C_{-1} \\ C_0 \\ C_1 \\ \vdots \\ C_{N-1} \\ C_N \end{bmatrix}$$

There are $2N + 1$ variables and $2N + 1$ simultaneous equations. We can compute the required values of the coefficients from these equations. Once we set the gain coefficients to the values as computed above, the output $y(t)$ will not have any intersymbol interference.

EXAMPLE 3

Determine the gain coefficients of a three-tap equalizer which will reduce the intersymbol interference of the following received signal (Fig. 25). Draw the equalized signal.

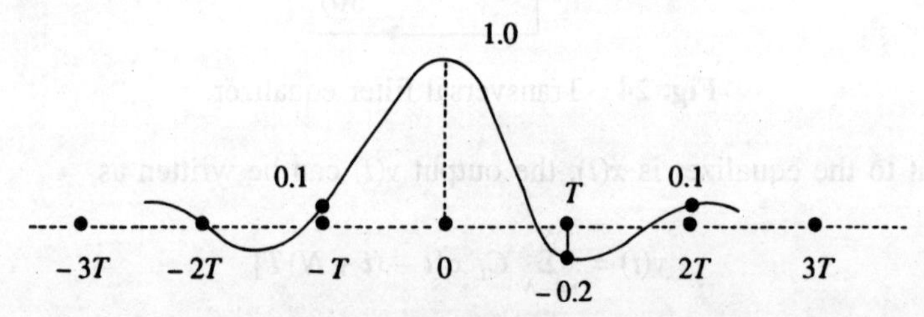

Fig. 25 Received signal of Example 3.

Solution

With the three-tap equalizer we can produce one zero crossing on either side of $t = 0$ in the equalized pulse. The tap gains are given by

$$\begin{pmatrix} 0 \\ 1 \\ 0 \end{pmatrix} = \begin{pmatrix} 1.0 & 0.1 & 0 \\ -0.2 & 1.0 & 0.1 \\ 0.1 & -0.2 & 1.0 \end{pmatrix} \begin{pmatrix} C_{-1} \\ C_0 \\ C_1 \end{pmatrix}$$

Solving for C_{-1}, C_0 and C_1, we get

$$\begin{pmatrix} C_{-1} \\ C_0 \\ C_1 \end{pmatrix} = \begin{pmatrix} -0.09606 \\ 0.9606 \\ 0.2017 \end{pmatrix}$$

The values of the equalized pulse at the sampling instants can be computed using $y(k) = C_{-1}x(k + 1) + C_0x(k) + C_1x(k - 1)$.

$$y(-3) = 0.0$$
$$y(-2) = -0.0096$$
$$y(-1) = 0.0$$
$$y(0) = 1.0$$
$$y(1) = 0.0$$
$$y(2) = 0.056$$
$$y(3) = 0.02$$

The equalized signal is shown in Fig. 26.

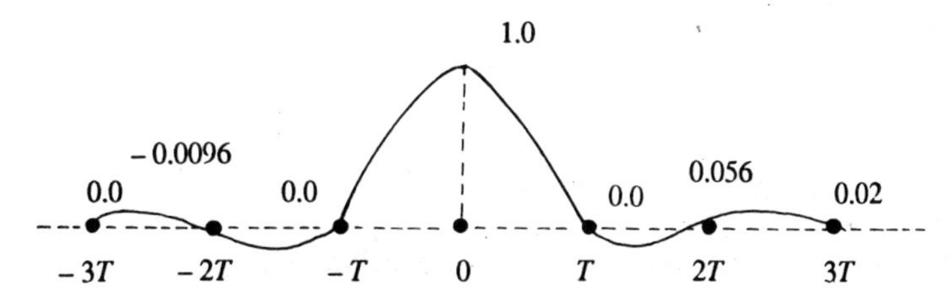

Fig. 26 Equalized signal of Example 3.

8.2 Adaptive Equalization

Transversal filter equalizer can be configured to adapt to the characteristics of the transmission channel automatically. Adaptation involves the following steps:

1. Prior to data transmission a known test sequence called a training sequence is transmitted on the channel.

2. Resulting response sequence $y(k)$ is obtained in the receiver by measuring output of the transversal filter at the sampling instants.

3. An error sequence $e(k)$ is obtained by subtracting the received response sequence from the desired response sequence $d(k)$

$$e(k) = d(k) - y(k)$$

4. The error sequence $e(k)$ is used to determine the gain coefficients. An algorithm is used for optimum setting of the coefficients. It is based on minimizing the sum of the squares of the errors, $\Sigma\, e^2(k)$.

5. The duration of the training sequence is so chosen that the adaptive equalizer converges to the optimum setting.

Adaptive equalizers are provided in high speed data modems. We will examine their operation and the training sequences in Chapter 3.

9 CLOCKED REGENERATIVE RECEIVER

Regeneration is the process of reconstructing the data signal from the received signal which has been attenuated, deformed and possibly contaminated with noise during transmission. The clocked regenerative receiver removes as much disturbances as possible and then interprets the processed signal to extract the digital information. The information is extracted at the sampling instants which are determined by a clock signal. Figure 27 shows functional block schematic of a regenerative receiver. The receiver block schematic which we saw in Chapter 1 was the simplified version of this receiver.

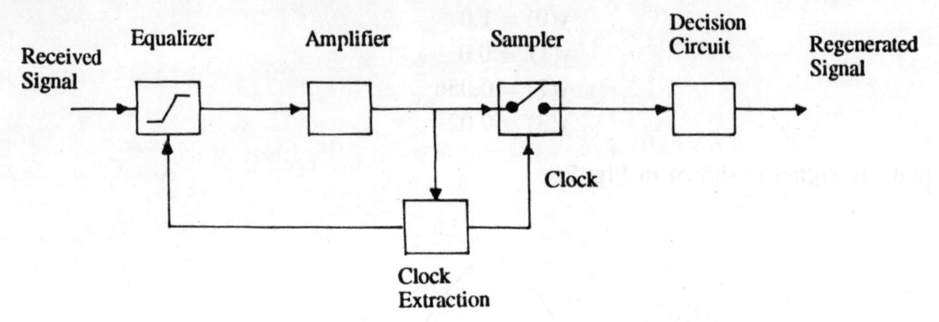

Fig. 27 Clocked regenerative receiver.

The main features of the clocked regenerative receiver are:

- Equalization is carried out to remove intersymbol interference as much as possible.
- The equalized signal is amplified to compensate for the attenuation and restore the signal to the required level.
- Sampling clock is extracted from the signal. This clock is inherently synchronized in frequency and phase to the received signal.
- The equalized and amplified signal is sampled using the clock. The clock determines the precise instants at which the sampling is done.
- The decision circuit maintains a threshold to decide the discrete value of the received signal. It compares the sampled signal with the threshold and generates a reconstructed signal.

Ideally, the sampled value of the signal should depend only on the information present at the sampling instant but it is not so because of the residual intersymbol interference which may persist even after equalization. The overall performance of the data transmission system is determined by ability of the decision circuit to discriminate between the discrete symbols using the defined thresholds. This performance can be monitored using eye patterns.

10 EYE PATTERN

Eye pattern is a convenient method of displaying, on an oscilloscope, the effect of intersymbol interference in a received data signal. Figure 28 shows an example of a response curve which extends over three clock periods in case of a binary unipolar signal. Let us assume that it is generated whenever symbol "1" is transmitted on the channel. The response curve has been divided into three parts, namely, A, B and C. It is obvious from the figure that there is significant intersymbol interference at the adjacent sampling instant following the symbol.

Since the response curve extends over three clock periods, the received signal at any instant is determined by the last three symbols. Table 3 lists all possible combinations of the three bit

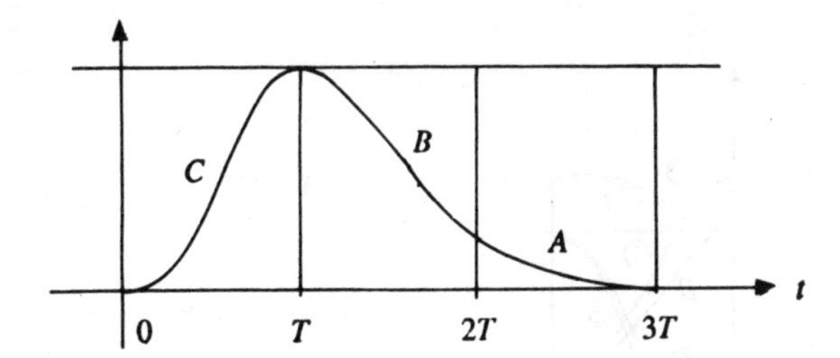

Fig. 28 Response curve with intersymbol interference.

Table 3 Response Curves for Various 3-Bit Input Sequences

Bit sequence	Response parts to be superimposed
0 0 0	0
0 0 1	A
0 1 0	B
0 1 1	$A + B$
1 0 0	C
1 0 1	$A + C$
1 1 0	$B + C$
1 1 1	$A + B + C$

sequences. The resulting signal waveform during a clock period is obtained by superimposing the sections A, B and C of the response curve depending on the bit sequence.

Figure 29a shows the waveforms generated by the various bit sequences. If a random data signal is transmitted, all these sequences will be generated and if the received signal is observed in an oscilloscope, the display on the screen will be overlapping waveforms corresponding to each bit sequence (Fig. 29b). The extracted clock of the data signal is used for horizontal synchronization in the oscilloscope. The display is in the shape of an "eye" and is, therefore, termed an eye pattern. The opening X is meaningful indicator of the quality of equalization. When equalization is poor, the eye opening is small. A perfect equalization, i.e. zero intersymbol interference, appears as maximum opening of the eyes.

The eye pattern is of practical value as it allows (a) fine tuning of the equalizer to get minimum intersymbol interference and (b) adjustment of the phase of the local sampling clock in the regenerative receiver. The phase is so adjusted that the sampling takes place exactly at the maximum eye opening where the discrimination between the discrete levels is maximum.

11 TELEPHONE NETWORK

Public Switched Telephone Network (PSTN), or simply the telephone network, provides a means of voice communication. Human speech signals occupy the frequency band extending from a few tens of Hz to about 8 kHz. Most of the information is contained in the 100–500 Hz frequency range. The higher frequencies serve to give "character" to a voice. As a compromise between high quality of speech which demands full voice bandwidth, and cost which calls for low bandwidth, the telephone network provides voice service over the frequency band 300–3400 Hz.

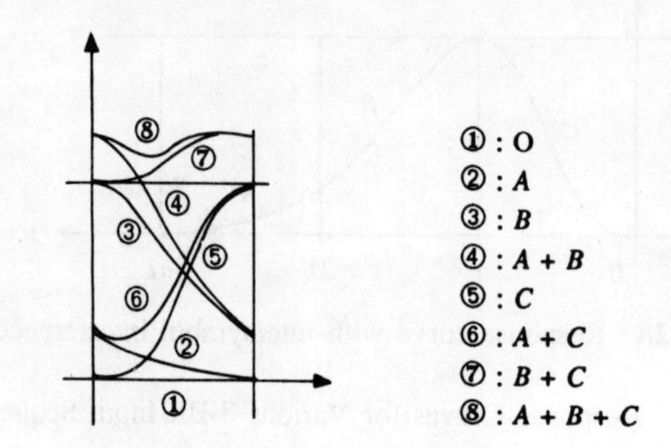

(a) Response curves for various input bit sequences

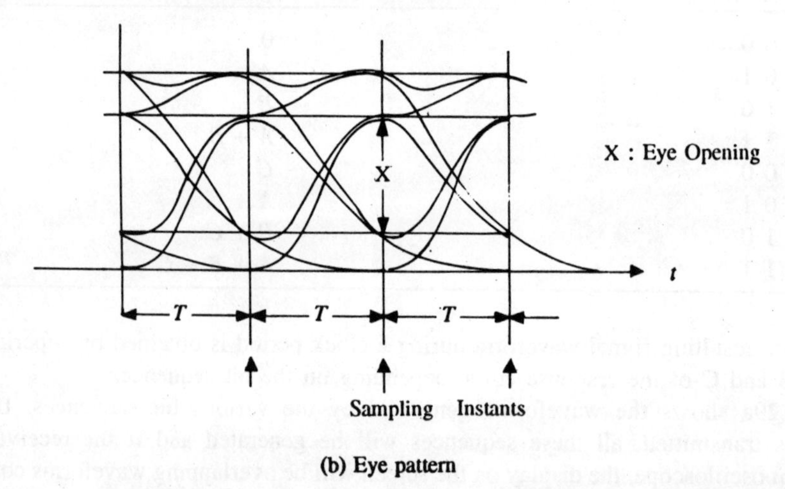

(b) Eye pattern

Fig. 29 Formation of eye pattern.

The telephone network today provides communication services to any corner of the world. The ready availability of the network resources induced the data processing community to make use of this network for transporting digital information as well. To make use of the telephone network for data transmission, we need to study the composition of the network, its signal transmission characteristics and the specifications of the service offered by it. In the following sections, we will come across many abbreviations associated with signal and noise level measurements. These abbreviations are explained in Appendix A at the end of this chapter.

11.1 Network Topology

The concept of a network emerges when several sources and several sinks of information are to be interconnected. The telephone network provides interconnection service to its vast number of subscribers for voice communications. The network comprises transmission channels which are assigned to the users either on a permanent basis or dynamically. In the latter case, switching is necessary. The switching operation interconnects two users when they request for interconnection. Switching is performed at the network exchanges (Fig. 30).

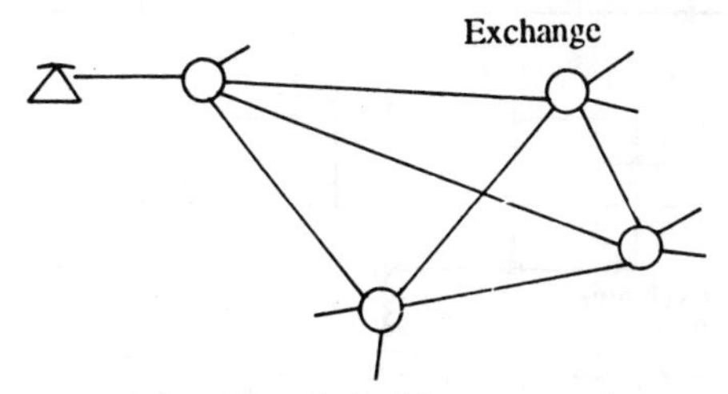

Fig. 30 Switching network.

11.2 Single Exchange Area

Let us examine a simple telephone network consisting of one exchange serving the users or subscribers located within its service area (Fig. 31). The service area is usually less than a hundred square kilometres. The network resources consist of a local network and a telephone exchange.

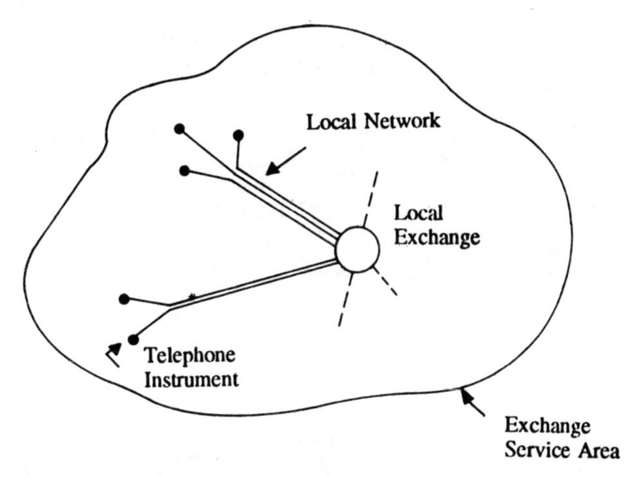

Fig. 31 Single exchange telephone network.

The local network consists of balanced twisted-pair copper cables from the exchange to the subscribers. Each telephone instrument of the network is connected to the exchange through the local network cable and is identified by a number. In the exchange, each pair of the local network is terminated on a distribution frame. This two-wire connection from the exchange to the subscriber provides both-ways voice transmission capability.

The telephone exchange consists of switching equipment which accepts the dialled digits from a telephone instrument and establishes connection with the dialled telephone number. The switching operation in the exchange involves operation of some relay contacts which connect one subscriber line to another. The connection through the exchange provides both-way AC continuity for speech signals. DC signals are blocked by a transmission bridge which couples the subscriber line to the exchange equipment either through capacitors or through transformers (Fig. 32).

Nowadays, electronic exchanges have been introduced in the network. In these exchanges, the speech signals are first converted into digital form and then switched. The digital signal is converted back into the analog form before transmission to the subscriber. For conversion to the digital form, the speech signal is band limited to 300–3400 Hz by the exchange equipment.

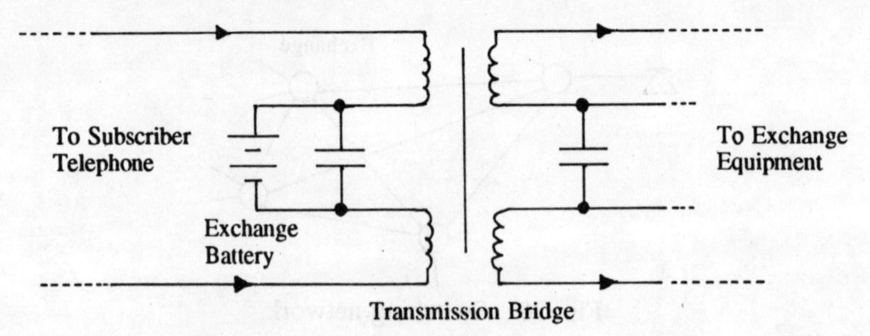

Fig. 32 DC blocking transmission bridge.

11.3 Multiple Exchanges

A single exchange can serve a limited number of subscribers, usually 10,000. It can serve the subscribers located up to a maximum distance of about five kilometres. When the number of subscribers increases and their geographic distribution expands, it becomes more economical to establish additional exchanges. Each exchange has its well defined exchange service area and its local network. It is identified by an exchange code (Fig. 33).

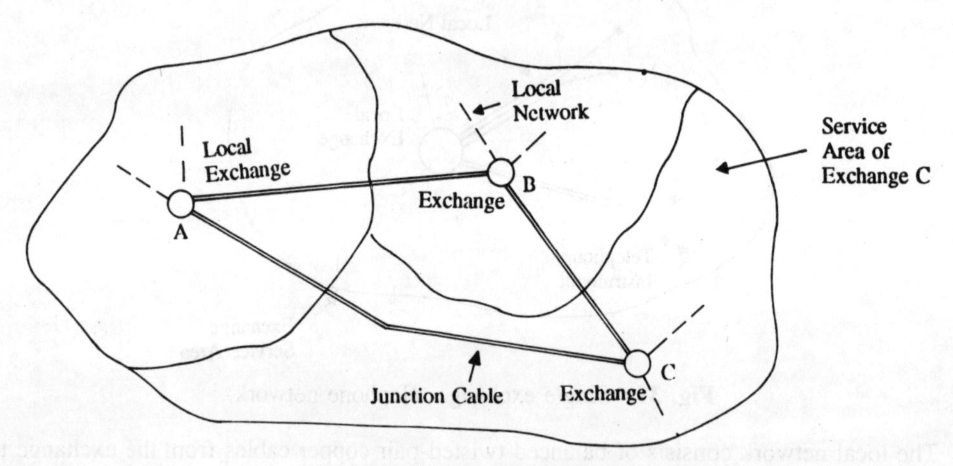

Fig. 33 Multiexchange telephone network.

The exchanges are interconnected through junction cables to carry the signals from subscribers belonging to one exchange to the subscribers belonging to another exchange. The junction cables consist of balanced twisted copper pairs but have lower attenuation than local cables. The junction cables may be loaded, in which case the bandwidth gets restricted to about 4 kHz. Junctions can also be provided using transmission systems which employ analog or digital multiplexing techniques (described later). The transmission systems restrict the voice channel bandwidth to 300–3400 Hz.

11.4 Trunk Automatic Exchanges

An average sized city may have several telephone exchanges. The total number of exchanges in a country may be several hundred or even several thousand. Providing interconnectivity among these exchanges using junction cables poses problems because each exchange has to be connected to every other exchange. The problem is solved by introducing another level of switching. It consists of trunk automatic exchanges (TAX) which are interconnected using trunk circuits (Fig. 34). Just like local exchanges which provide interconnectivity among subscribers, trunk automatic exchanges are connected to the local exchanges and provide city-wise interconnectivity. One trunk automatic exchange is established in a city or a group of adjoining cities. It is identified by a code. When a subscriber wants to establish a connection to a subscriber in another city, he first dials the code of the distant trunk automatic exchange followed by the telephone number of the subscriber. His call is routed through his local exchange, the local trunk exchange, the distant trunk exchange and the distant local exchange.

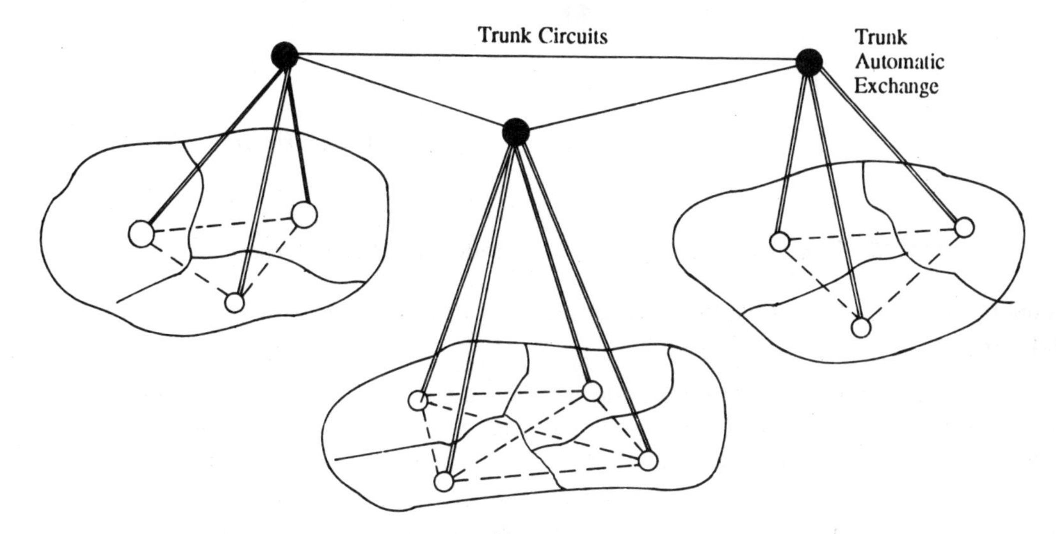

Fig. 34 Nation-wide telephone network with trunk automatic exchanges.

The telephone network may have even more than two levels of hierarchy depending on the number of cities to be interconnected, number of trunk routes and degree of routing flexibility provided in the network.

11.5 International Transit Exchange

The various telephone networks of different countries are interconnected through yet another level in the hierarchy of telephone exchanges. These are called international transit exchanges (Fig. 35). International transit exchanges are interconnected through international trunk circuits. The trunk exchanges of a telephone network have access to one international transit exchange and the subscribers need to dial yet another code to route the call to the international transit exchange of the required country.

12 LONG DISTANCE NETWORK

The circuits which interconnect the trunk automatic exchanges carry the speech signals over long distances ranging from a few hundred to thousands of kilometres. The speech signals need to be

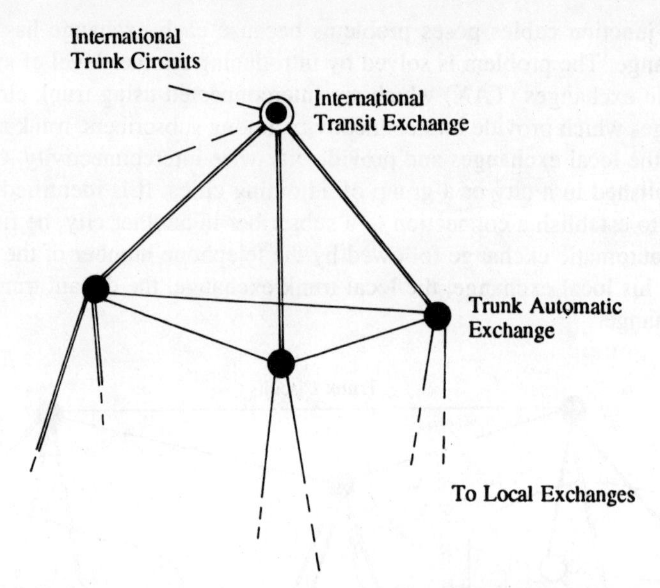

Fig. 35 World-wide telephone network with international transit exchanges.

amplified periodically for transmission over such long distances. There are two basic problems in this regard:

1. An amplifier is a unidirectional device while a two-wire telephone circuit provides both-ways transmission. Therefore, the transmission paths for outgoing and incoming speech signals must be split, i.e., the two-wire circuit must be converted into a four-wire circuit.

2. The trunk circuits are point-to-point and very large in number. Each trunk circuit requires its individual amplifiers to compensate for cable attenuation. Therefore, a very large number of amplifiers are required.

The conversion of a two-wire circuit to a four-wire circuit is carried out by using hybrids. As for the second problem, a cost-effective solution is to multiplex the speech signals so that there is one multiplexed outgoing signal and one multiplexed incoming signal. Of course, the multiplexed signal has a wider bandwidth and, therefore, calls for transmission equipment (amplifiers, equalizers, medium, etc.) having wider bandwidth.

12.1 Hybrids

Hybrids are used for converting a bidirectional two-wire circuit into a four-wire circuit which has separate pairs for 'go' and 'return' directions. A hybrid is a four-port balanced transformer which provides isolation of the opposite ports (Fig. 36). It is designed for 600 ohm terminations at its ports. When properly terminated, a speech signal entering one port is equally divided between the two side ports and no signal flows to the opposite port.

In a two-wire circuit, two hybrids are required, one at each end. Figure 37 shows how two hybrids are connected. The speech signal entering port A of hybrid H_1 is divided equally between ports B and D. Port B is connected to the transmit side of the four wire circuit. The speech signal which appears on this port is transmitted to the other end after amplification. Port D is connected

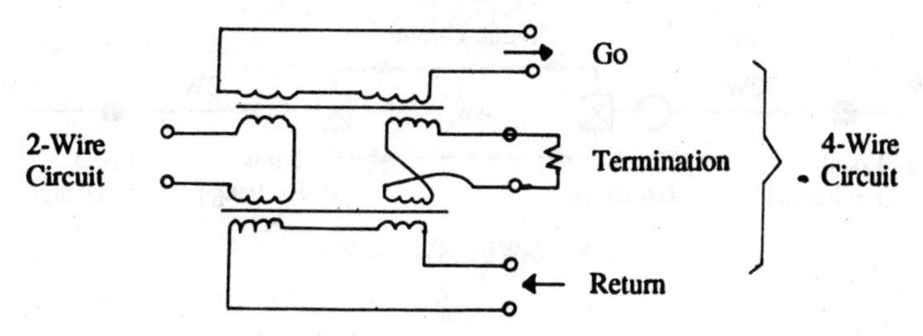

Fig. 36 Hybrid for two-wire/four-wire conversion.

to the output of the receive amplifier and, therefore, the speech signal from A appearing at this port is dissipated.

The speech signal which is received from the distant end appears at port D of the hybrid. It is also equally divided between ports A and C. The signal appearing at port A is transmitted on the two-wire circuit while the signal appearing at port C is dissipated in the resistive termination. If the hybrid is properly terminated, no part of the received signal goes to port B.

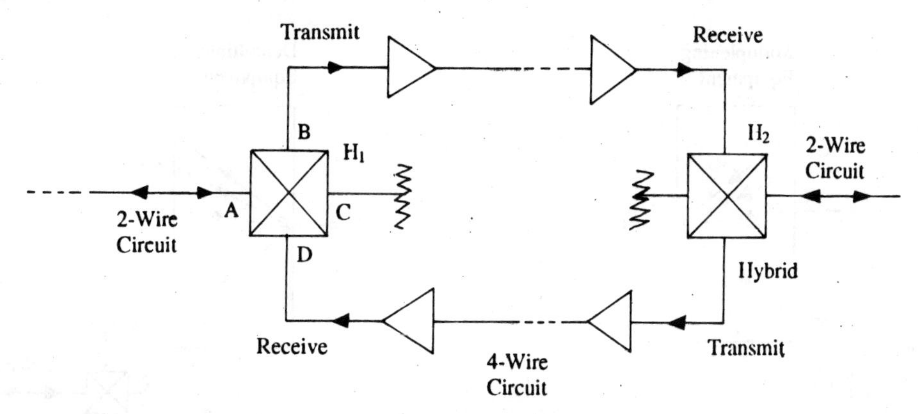

Fig. 37 A two-wire end-to-end circuit using hybrids.

Hybrids are encountered in a telephone network in the junction circuits and the trunk circuits where the speech signals need to be amplified (Fig. 38a). In electronic exchanges, there is need for two-wire to four-wire conversion. Therefore, hybrids are installed in the exchange itself (Fig. 38b).

12.2 Multiplexing

Multiplexing involves grouping of several channels in such a way as to transmit them simultaneously on the same physical transmission medium (e.g., cable or carrier frequency of a radio link) without mixing. At the receiving end, demultiplexing is performed to separate the channels. In the telephone network, each channel provides a bandwidth of 300–3400 Hz for speech signals. Multiplexing and demultiplexing of 'go' and 'return' channels is done separately and, therefore, the multiplexing and demultiplexing equipment comes between the two hybrids of a two-wire circuit (Fig. 39).

There are two basic multiplexing technologies in the telephone network, namely, Frequency Division Multiplexing (FDM), and Time Division Multiplexing (TDM). FDM is the older technology based on analog transmission principles and TDM is the new technology based on digital transmission principles.

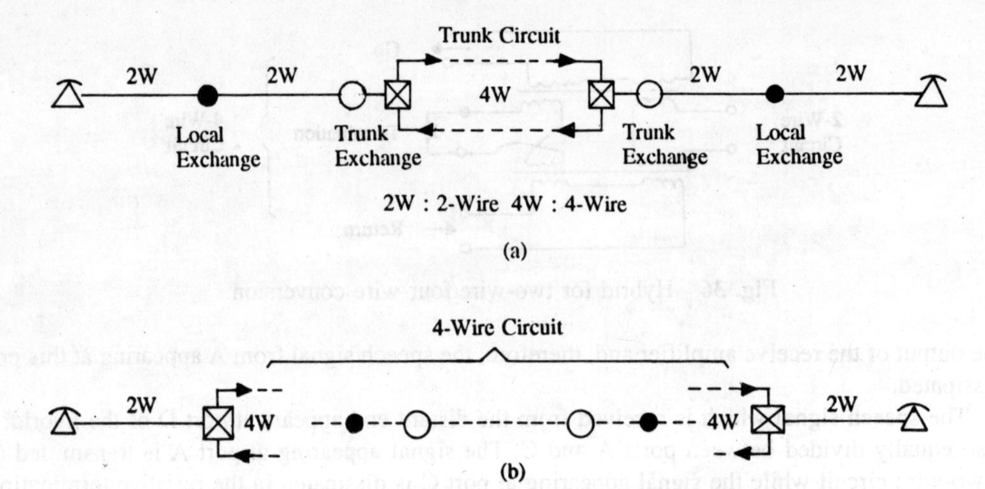

Fig. 38 Use of hybrids in telephone networks.

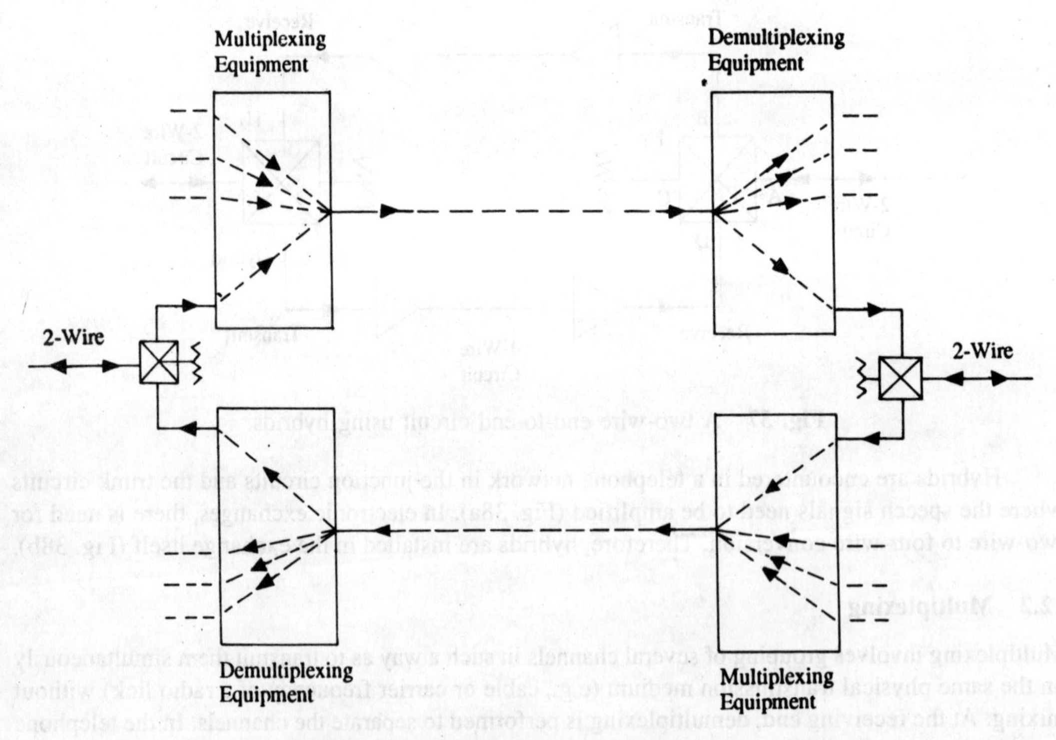

Fig. 39 Multiplexing and demultiplexing of telephone channels.

12.3 Frequency Division Multiplexing (FDM)

Frequency division multiplexing involves translation of the speech signal from the frequency band 300–3400 Hz to a higher frequency band. Each channel is translated to a different band and then all the channels are combined to form a frequency division multiplexed signal. In FDM, the speech channels are stacked at intervals of 4 kHz to provide a guard band between adjacent channels.

Frequency translation is done using a carrier which is modulated by the speech signal using suppressed carrier amplitude modulation. Of the two sidebands generated in this process, the upper sideband is separated using a bandpass filter. This process translates the speech channel to frequency band from f_c to $f_c + 4$ kHz, f_c being the carrier frequency. By using different carrier frequencies at intervals of 4 kHz, we can stack a number of speech channels one after the other. Figure 40 shows how three speech channels can be multiplexed. The frequency band of a speech channel is usually represented as a small right-angled triangle. The lower end corresponds to 0 Hz and the upper end corresponds to 4000 Hz. The speech signal lies within this band from 300 to 3400 Hz.

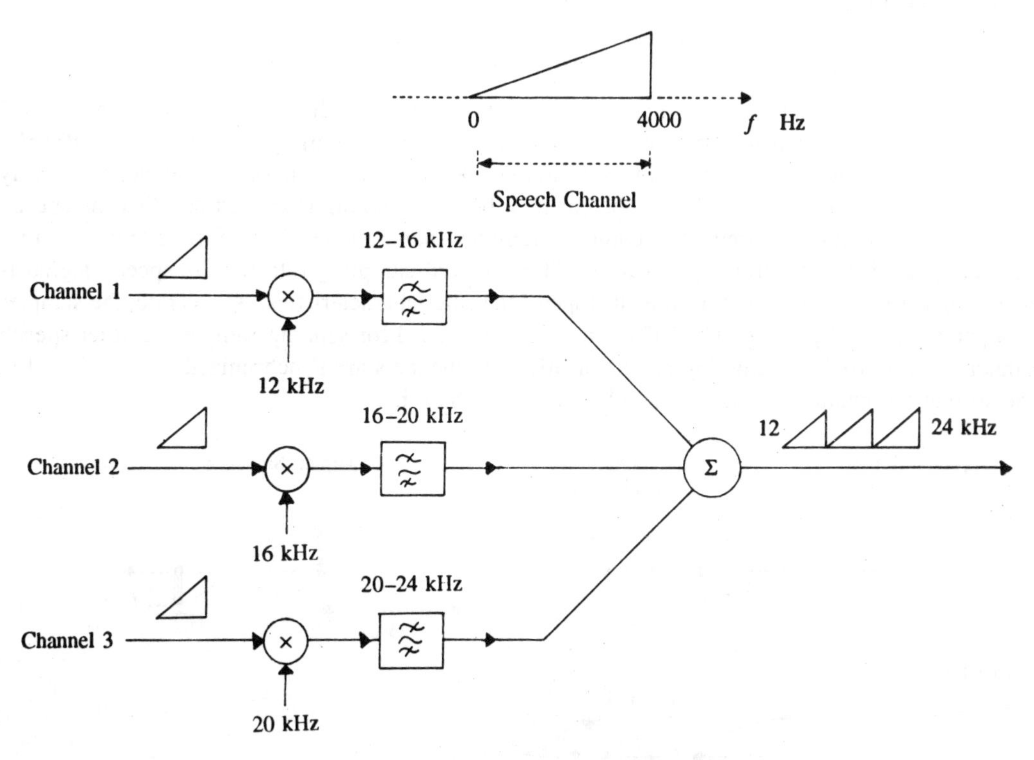

Fig. 40 Frequency division multiplexing.

To facilitate interconnection among the different telecommunication systems in use worldwide, CCITT has recommended a standard frequency translation plan. The smallest multiplexed unit consists of 12 channels and is termed a group. The groups are multiplexed to form supergroups and so on. In other words, each bigger multiplexed unit is composed of several immediately lower multiplexed units. Table 4 shows this hierarchy.

At the demultiplexer, the reverse process is carried out. Each higher level is translated to several signals of the next lower level. Finally each group is demultiplexed to 12 speech channels.

12.4 Time Division Multiplexing (TDM)

The basis of time division multiplexing is the sampling theorem which states that a signal containing

Table 4 FDM Hierarchy

Name	Number of channels	Bandwidth	Composition
Group	12	48	12 Speech channels
Supergroup	60	240	5 Groups
Master group	300	1232	5 Supergroups
Supermaster group	900	3872	3 Master groups

frequency components less than f_{max} can be entirely determined by its equidistant samples taken at the rate f_s, such that

$$f_s \geq 2f_{max}$$

The theorem emphasises that an analog signal has high time redundancy which can be utilised for transmitting additional information. The voice channel has the highest frequency of 3400 Hz. If it is sampled at the rate of more than 6800 samples per second, the speech signal can be entirely reconstructed from the samples. It is to be ensured, before sampling is carried out, that the speech signal does not contain any frequency component higher than 3400 Hz. Therefore, a low pass filter is always provided just before the sampler. The standard sampling rate for the speech signal is 8000 samples/second to allow for gradual slope of the filter characteristics. At this rate, the samples are separated by 125 μs (Fig. 41a). This time can be utilised for sending samples of other speech channels (Fig. 41b). The sampling clocks of different channels are synchronized and staggered in time so that the channels generate samples one after the other.

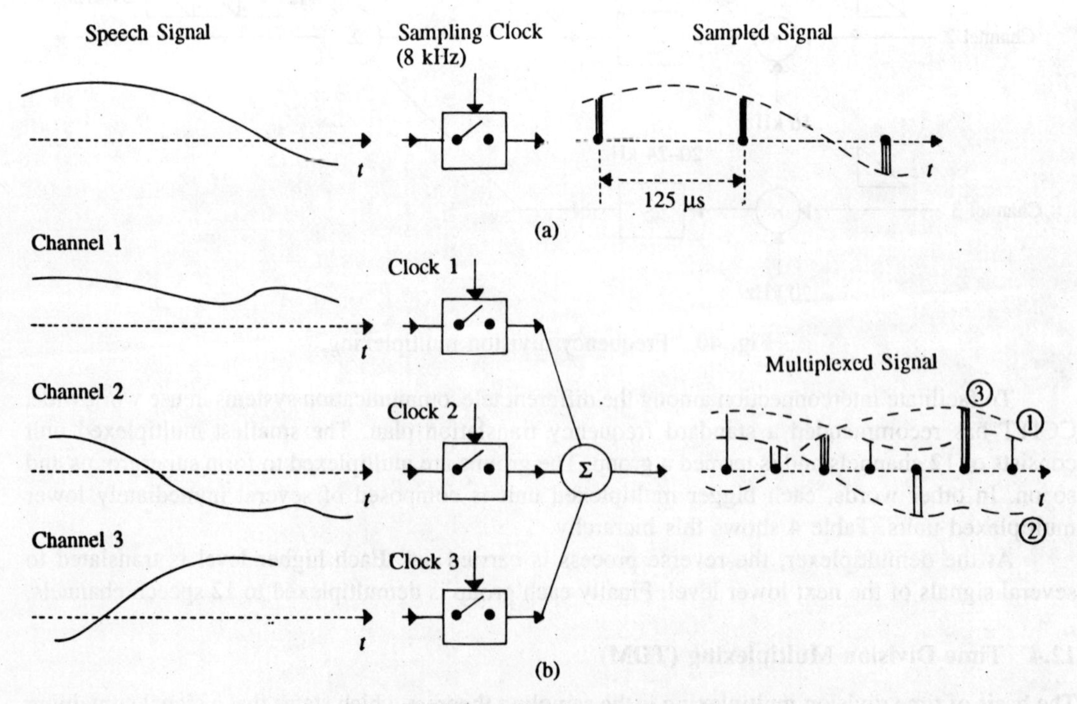

Fig. 41 Time division multiplexing.

Demultiplexer. Demultiplexing involves separating the samples of different channels. Since multiplexing has been done sequentially, the demultiplexer can separate the samples of different channels but it faces the problem of identifying the channels. Therefore, a flag is also multiplexed along with the samples. The flag has a unique attribute which can be identified by the demultiplexer. The first sample just after the flag belongs to channel one (Fig. 42). The flag followed by one complete cycle of sampling is termed a 'frame'.

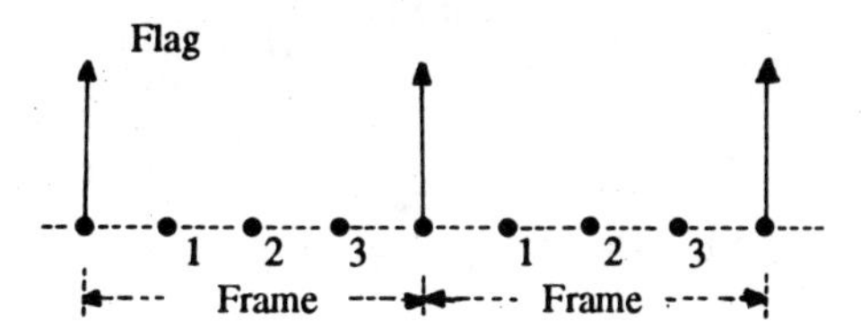

Fig. 42 Identification of samples of a TDM signal using flags.

Once the samples have been separated, the speech signal can be reconstructed by passing the samples through a low pass filter with the cut-off at 3400 Hz.

12.5 Pulse Code Modulation (PCM)

Although in pulse form, the time division multiplexed signal is still an analog signal as the sample levels can have infinite possible values. Pulse code modulation is used for converting these analog samples to digital form. This process involves two stages namely, Quantization, and Coding which we now discuss.

Quantization. Quantization is approximation of the level of the sample by the nearest value drawn from a finite assortment of discrete levels. For example, if the set consists of discrete levels 0, 1, 2, ... 7 volts and the sample level is 3.2 volts, it is approximated by a discrete level of 3 volts.

Note that by approximating the level of 3.2 volts by 3 volts, we have introduced some error because the receiver will later generate the sample of 3 volts when it receives the coded sample. This error is called quantization error. Quantization error is serious at low levels. An error of 0.1 volt in 5 volts is ≈ 2 per cent while the same error in 1 volt is 10 per cent. Therefore, a non-uniform quantization law is adopted in the telephone channels. It ensures equal quantization error percentages at all the levels.

Coding. Coding involves converting the discrete level of the sample after quantization to the binary code of fixed length, e.g., 3 volts may be coded as 011. The number of bits in the code is determined by the total number of discrete levels. In telephony, 256 discrete levels are used and they are coded using eight bits. As the sampler generates 8000 samples per second and each sample is coded into eight bits, the bit rate of a digitalized speech channel is $8 \times 8000 = 64$ kbps.

12.6 30 Channel PCM Signal

CCITT has recommended standards for PCM channels. The basic PCM signal provides for multiplexing of 30 speech channels. Eight bit codes of the samples of these channels are time-division multiplexed. Figure 43 shows the format of the 30-channel PCM frame.

The 30-channel PCM frame consists of 32 time slots. Each time slot (TS) contains an eight

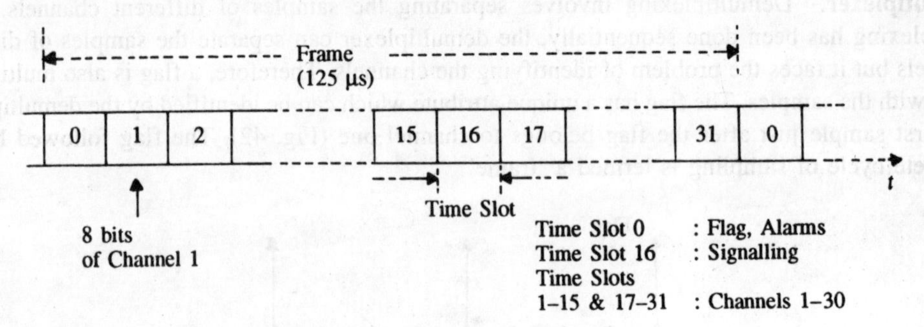

Fig. 43 30 channel PCM frame.

bit code. The frame starts with time slot TS-0 which contains the flag and alarm signals in alternative frames. TS-1 to TS-15 and TS-17 to TS-31 contain sample codes of the 30 speech channels. TS-16 contains the signalling information derived from the dialled digits. Each TS-16 contains signalling information of two channels.

Since the sampling rate is 8000 per second, the frames are generated at the same rate. The number of bits in a frame being 32 × 8 = 256, the bit rate of 30 channel PCM signal is 256 × 8000 = 2.048 Mbps.

12.7 Digital Hierarchy

Just like the analog hierarchy, CCITT has recommended standards for the digital multiplexing hierarchy. Each higher order multiplexed signal is derived from four immediately lower order multiplexed signals. Table 5 shows the digital hierarchy. The gross bit rate of each higher order multiplexed signal is not exact multiple of the lower order signal because additional flags and alarm signals need to be introduced at each stage of multiplexing.

Table 5 Digital Hierarchy

Order	Number of channels	Bit rate	Composition
1	30	2.048	30 × 64 kbps
2	120	8.448	4 × 2.048 Mbps
3	480	34.368	4 × 8.448 Mbps
4	1920	139.264	4 × 34.368 Mbps
5	7680	564.992	4 × 139.264 Mbps

13 TRANSMISSION MEDIA FOR LONG DISTANCE NETWORK

Multiplexed speech channels are transmitted either by using cable transmission media or radio transmission systems. The coaxial cable has been the most commonly used transmission medium for the multiplexed trunk circuits for many decades. It provides reliable long distance transmission of both analog and digital signals. Balanced pairs have also been used in the past but only for very low capacity systems. The radio systems provide wireless communication links for the trunk circuits. Depending on the frequency of radio signals, radio systems are categorized as follows:

- HF (High Frequency) systems
- VHF (Very High Frequency) systems
- UHF (Ultra High Frequency) systems
- Microwave systems which are subdivided into
  - — terrestrial microwave systems, and
  - — satellite communication systems.

In all the radio systems, the multiplexed signal is made to modulate a radio carrier which is transmitted as an electromagnetic wave. At the receiving end, an antenna picks up the radio carrier and the received radio signal is demodulated to get the multiplexed signal. HF, VHF and UHF radio systems are low channel capacity systems. They provide one to thirty speech channels per radio carrier.

Terrestrial microwave systems are high capacity systems and provide the channel capacities of the order of 300 to 2700 channels per radio carrier. The satellite communication systems also operate in the microwave frequency range but provide extreme flexibility in terms of channel capacity and geographic location. Satellite earth stations having capacities as low as a single channel per carrier and as high as 1800 channels per carrier are used in the telephone network.

However, during the past few years, optical fibre technology has become cost effective and optical fibre cables are fast replacing all other types of transmission media. Technological developments are taking place at a very fast pace in this field and in the not too distant future, the information transport network including the switching exchanges will employ optical technology.

14 ECHO IN TRANSMISSION SYSTEMS

Introduction of hybrids in the transmission systems gives rise to a very serious problem, particularly when the link is through a satellite. Theoretically speaking, a hybrid should not cause any leakage of the signal to the opposite port if it has been properly terminated at all the four ports. In actual practice, however, the impedance at the port towards the subscriber (port A in Fig. 37) is highly uncertain. It depends on the distance between location of the subscriber's telephone instrument with respect to the hybrid. The distance is not fixed and varies from subscriber to subscriber. As a result, the hybrid causes leakage of the received signal to the opposite port (Fig. 44). This signal travels back to the subscriber who originated the signal and is heard as an echo.

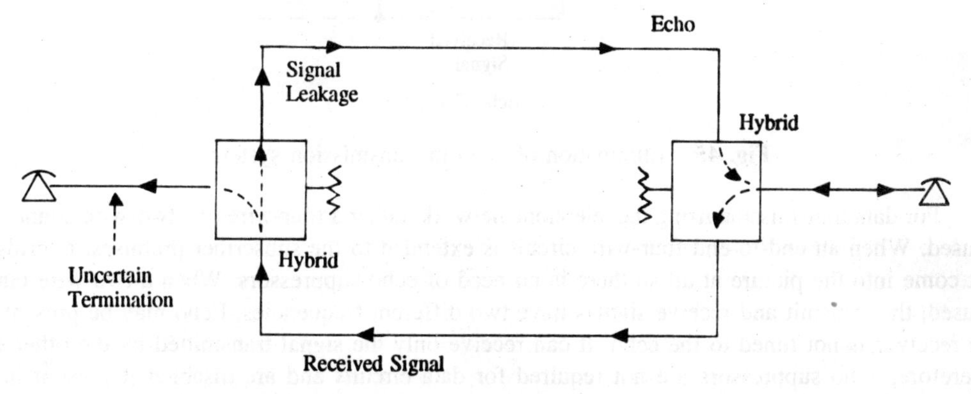

Fig. 44 Echo generation in transmission systems.

The echo is present in all types of transmission systems but as the delay of the echo is very small, ~20 ms in the terrestrial systems, it is not heard distinctly. In case of a satellite link, the delay is of the order of 500 ms and there is a distinct echo in the channel.

Echo is a nuisance needs to be eliminated from the telephone circuits. Echo suppressors/cancellors are installed in the network for this purpose. In echo suppressors, the transmit port of the hybrid is isolated when a signal is being received so that the signal through the hybrid does not travel back (Fig. 45a). In echo cancellors, this is achieved by subtracting the simulated echo from the transmit signals (Fig. 45b).

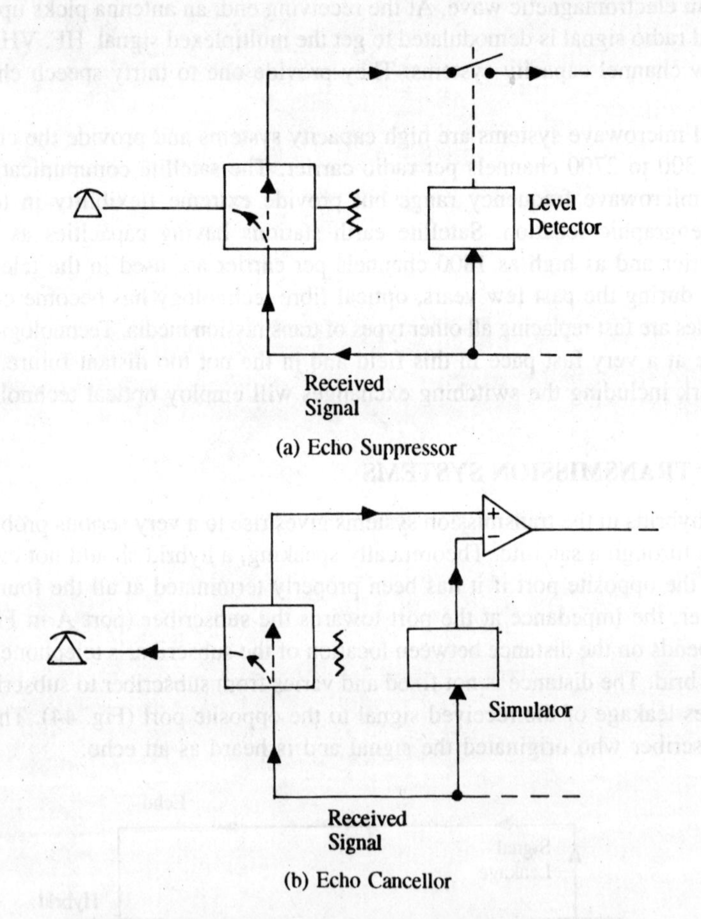

(a) Echo Suppressor

(b) Echo Cancellor

Fig. 45 Elimination of echo in transmission systems.

For data transmission using the telephone network, either a four-wire or a two-wire connection is used. When an end-to-end four-wire circuit is extended to the subscriber premises, hybrids do not come into the picture at all so there is no need of echo suppressors. When a two-wire circuit is used, the transmit and receive signals have two different frequencies. Echo may be present but the receiver is not tuned to the echo. It can receive only the signal transmitted by the other end. Therefore, echo suppressors are not required for data circuits and are disabled if present in the transmission link.

15 NOISE IN TRANSMISSION SYSTEMS

Noise can be generally described as any received signal, which, when interpreted by the receiver, delivers incoherent information of no interest to the receiver. Noise gets added to the signal during its transmission and degrades the quality of the signal. Background noise is always present even in the absence of a useful signal. Sources of background noise are:

- Thermal noise
- Intrinsic noise of the electronic devices, e.g., shot noise
- Atmospheric disturbances
- Electromagnetic interference
- Cosmic sources.

Lightning discharge and rain attenuation are the two sources of atmospheric noise. Electromagnetic interference is caused by other radio systems, discharges in commutator motors and spark plugs of vehicles. All celestial bodies generate electromagnetic interference which is picked up by the radio system antennae. Sources of thermal noise and shot noise are present within the telecommunication equipment.

Signal distortion due to nonlinearities in a telecommunication system results in supplementary noise which appears only when the signal is present. Two important causes of such noise are intermodulation and quantization. While discussing PCM, we mentioned quantization error. It appears as noise superimposed on the useful signal.

15.1 Intermodulation Noise

Non-linear distortion is caused by the non-linear input/output characteristics. For example, if the input level is very high, an amplifier may be driven into saturation and the output level no longer remains proportional to the input level. It is possible to express the output as a series of powers of the input for a non-linear characteristic:

$$y(t) = a_1x(t) + a_2x^2(t) + a_3x^3(t)...$$

If the input is a sinusoidal signal of frequency f, the output of a non-linear stage contains, in addition to the frequency f, its multiples $2f$, $3f$, etc. If the input consists of several sinusoidal components, which is normally the case, the non-linearity generates intermodulation products in addition to the harmonics. Intermodulation products are the sum and differences of the input frequencies and of their harmonics. For example, if the input contains two frequencies, f_1 and f_2, intermodulation products will comprise $f_1 \pm f_2$, $2f_1 \pm f_2$, $2f_2 \pm f_1$, etc. Intermodulation products are categorized as second order, third order and so on. $f_1 \pm f_2$ are the second order intermodulation products, $2f_1 \pm f_2$ and $2f_2 \pm f_1$ are third order intermodulation products.

Figure 46 gives a qualitative picture of the output spectrum when the input signal has a frequency band from f_1 to f_2. Note that the second order harmonics and second order intermodulation products can be removed by filtering the output signal, but some of the third order intermodulation products occupy the same frequency band as the original signal. They cannot be removed by filtering the output and manifest themselves as noise in the original signal. That is why third order nonlinearity (the term $a_3x^3(t)$ above) is the most critical.

Non-linear distortions occur at almost all stages of transmission system except where only passive linear components are involved. Note that intermodulation noise is generated only when

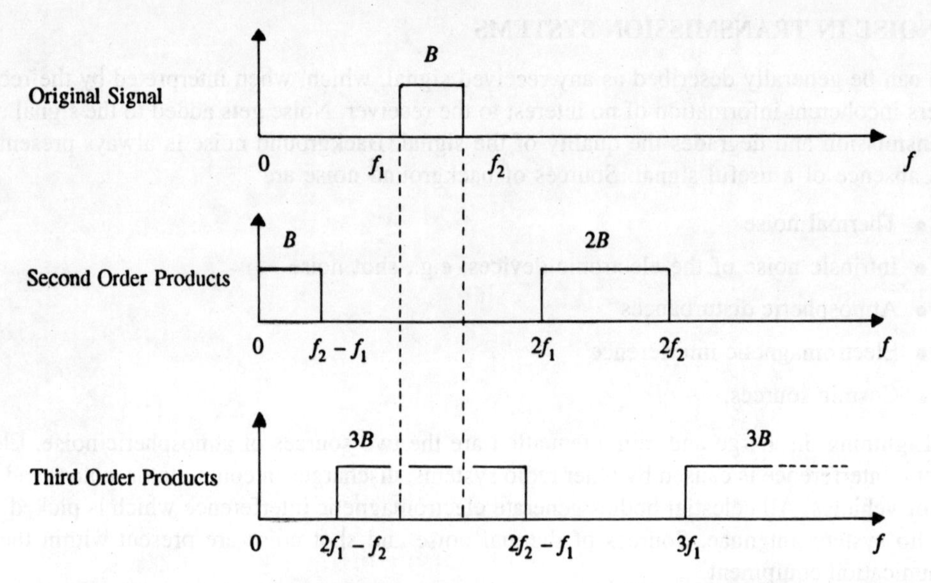

Fig. 46 Intermodulation products due to non-linear distortion.

signals are actually present and it may cause crosstalk if its spectral components lie in the frequency band of another channel. This is particularly true for FDM channels. Intermodulation noise can be kept within specified limits by ensuring that signal levels do not exceed the specified level.

EXAMPLE 4

An amplifier has the following input/output characteristic:

$$y(t) = a_1x(t) + a_2x^2(t) + a_3x^3(t)$$

Write all the spectral components of the output when two tones $\cos(2\pi f_1 t)$ and $\cos(2\pi f_2 t)$ are applied at the input. $f_1 = 1000$ Hz, $f_2 = 1100$ Hz.

Solution

1. Output for the term $a_1x(t)$: $a_1 \cos(2\pi f_1 t) + a_1 \cos(2\pi f_2 t)$
   Spectral frequency components: f_1, f_2

   : 1000, 1100 Hz

2. Output for the term $a_2x^2(t)$: $a_2\{\cos(2\pi f_1 t) + \cos(2\pi f_2 t)\}^2$

   $$= a_2\{\cos^2(2\pi f_1 t) + \cos^2(2\pi f_2 t) + 2\cos(2\pi f_1 t)\cos(2\pi f_2 t)\}$$

   Spectral frequency components: $2f_1, 2f_2, f_1 + f_2, f_2 - f_1$

   : 2000, 2200, 2100, 100 Hz

3. Output for the term $a_3x^3(t)$: $a_3\{\cos(2\pi f_1 t) + \cos(2\pi f_2 t)\}^3$

   $$= a_3\{\cos^3(2\pi f_1 t) + \cos^3(2\pi f_2 t) + 3\cos^2(2\pi f_1 t)\cos(2\pi f_2 t) + 3\cos(2\pi f_1 t)\cos^2(2\pi f_2 t)\}$$

However,

$$\cos^3 \alpha = \frac{3}{4} \cos \alpha + \frac{1}{4} \cos 3\alpha$$

$$\cos^2 \alpha \cos \beta = \frac{1}{2} \cos \beta + \frac{1}{2} \cos 2\alpha \cos \beta$$

$$= \frac{1}{2} \cos \beta + \frac{1}{4} \cos (2\alpha + \beta) + \frac{1}{4} \cos (2\alpha - \beta)$$

Spectral frequency components : f_1, $3f_1$, f_2, $3f_2$, $2f_1 + f_2$, $2f_1 - f_2$, $2f_2 + f_1$, $2f_2 - f_1$
: 1000, 3000, 1100, 3300, 3100, 900, 3200, 1200 Hz

15.2 Thermal and Shot Noise

Thermal noise constitutes the most important source of noise in telecommunication systems. It is generated by the random motion of electrons in a conductor. Its power spectral density function is flat in the range of frequencies of interest in telecommunications. For this reason, thermal noise is also called "white noise". Without going into its mathematical representation and analysis, we will quote the result of the analysis:

"The maximum thermal noise which a resistance R can deliver in the frequency band B Hz at absolute temperature T is given by $P_n = kTB$, where k is the Boltzmann's constant (1.38×10^{-23} J/K)."

EXAMPLE 5

What is the maximum available thermal noise from a resistive termination at ambient temperature of 290 K in the bandwidth 3.1 kHz?

Solution

Maximum available thermal noise from the resistive termination is kTB watts. Substituting the values, we get

$$P_n = 1.38 \times 10^{-23} \times 290 \times 3100$$

$$= 1.24 \times 10^{-17} \text{ watt}$$

$$= -140 + 0.93 = -139.07 \text{ dBm}$$

The shot noise is generated in the active electronic components due to discrete and random emission of electrons which constitute current in these devices. For example, shot noise current is generated in a forward biased diode when free electrons from the n-side accelerate towards the p-side. Mean square value of the shot noise current in a forward biased diode is given by

$$\overline{I_n^2} = 2q\,(I + I_s)B$$

where q is the electronic charge (1.6×10^{-19} Coulomb), I is the forward current in amperes, I_s is the saturation current in amperes, and B is the bandwidth in Hz.

We find that all lossy components and active devices of an electronic circuit generate noise. Thus each stage of the transmission system will add up to the noise of the signal.

15.3 Psophometric Weighting

The annoying effect of noise on the ear is different at different frequencies. To account for this subjective effect, the background noise is measured using a special voltmeter called a psophometer. It incorporates a filter which simulates the sensitivity of the ear to noise at different frequencies. The attenuation characteristic of this filter, called the psophometric curve, has been standardized by CCITT Recommendation P.53. The psophometric curve for voice is shown in Fig. 47. Noise measured using a psophometric filter is termed weighted noise and is expressed in dBmp. When a noise with flat power spectral density is measured in the frequency band 300–3400 Hz, the weighted noise level is 2.5 dB less than the unweighted noise level.

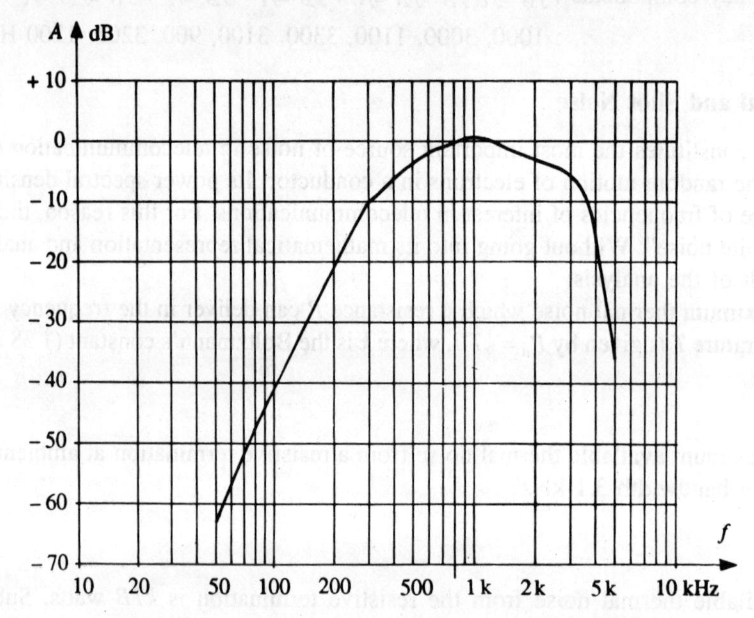

Fig. 47 Psophometric weighting curve for voice.

15.4 Signal to Noise Ratio

The quality of a signal is determined by its level with respect to the level of the noise which contaminates it. It is expressed as the ratio of signal power (P_s) to noise power (P_n) and is termed the Signal-to-Noise Ratio (SNR). If psophometrically weighted noise level is used, the signal-to-noise ratio is called weighted SNR. CCITT has recommended weighted SNR of ≥ 50 dB for speech channels. It includes contribution of all types of noise present in the telecommunication link.

$$\frac{S}{N} = \frac{P_s}{P_n}$$

$$\frac{S}{N} = (P_s)_{\text{dBm}} - (P_n)_{\text{dBm}}$$

EXAMPLE 6

Calculate the weighted SNR of a voice channel if the signal level is 0 dBm and the noise level is – 30 dBm.

Solution

$$\text{Weighted noise level} = -30 - 2.5 = -32.5 \text{ dBmp}$$

$$\text{(S/N) weighted} = 0 - (-32.5) = 32.5 \text{ dB}$$

15.5 Companders

Companders are used in the telephone network whenever there is need to improve SNR without actually increasing the signal level. Companders reduce the dynamic range of the telephone signals by using a compressor at the transmitting end and restore it to its original value by using an expander at the receiving end. The word compander is derived from the terms compressor and expander. We will shortly see how the SNR improvement is achieved by reducing the dynamic range.

The usual dynamic range of a telephone signal is from – 45 dBm to 5 dBm with the average signal level of – 15 dBm. The compressor used in telephone network reduces it to half. It employs a variable gain amplifier. The gain is unity for the nominal input signal level of 0 dBm. For the input level is less than 0 dBm, say $-x$ dBm, the gain is set automatically at $x/2$ so that the output level becomes $-x/2$ dBm. Similarly, if the input level is more than 0 dBm, say x dBm, the gain is set at $-x/2$ so that the output level is $x/2$ dBm. This is illustrated in Fig. 48a.

The operation of the expander used at the receiving end is the reverse of the compressor operation. The received signal at 0 dBm is unaffected and passed as it is. The levels above 0 dBm are amplified and below 0 dBm are attenuated to restore them to their original values. For example, if the input to the expander is x dBm, its output will be $2x$ dBm. If the input is $-x$ dBm, the output will be $-2x$ dBm.

Noise is added to the signal during its transmission through the channel (Fig. 48b) and it is to be noted that it has not passed through the compressor. It is a known fact that noise level is usually much below 0 dBm. Let us assume that it is $-P_n$ dBm. The negative sign has been used to emphasize the fact that the level is below 0 dBm. When the noise passes through the expander circuit, its level is reduced to $-2P_n$ in the absence of any speech signal. Thus, the noise level is significantly reduced at the receiving end.

When the speech signal is also present, the noise advantage is not so significant. Let us assume that the signal level is – 3 dBm and the noise level is – 20 dBm at the expander input. When added together, we get

$$-3 \text{ dBm} = 0.5 \text{ mW}$$

$$-20 \text{ dBm} = 0.01 \text{ mW}$$

$$0.5 + 0.01 \approx 0.5 \text{ mW} = -3 \text{ dBm}$$

The expander which would have given the loss of 20 dB had only the noise been present, gives a loss of only 3 dB to restore the signal level. Therefore, the noise level after the expander circuit is – 23 dBm, and there is only marginal improvement. Therefore, the compander is more effective in reducing noise levels when the channel is idle.

It has been observed by subjective tests that noise causes more annoyance to the listener when it is present during intersyllabic pauses and when the channel is idle. In telephone circuits, the speech signal is present on average for 25 per cent of the time in any one direction. The channel is idle for the rest of the time either because the other party is speaking or due to intersyllabic pauses. Therefore companders can be effective in the telephony channels. Tests have indicated that for speech signals, a compander gives a subjective improvement of 16 dB in the signal quality.

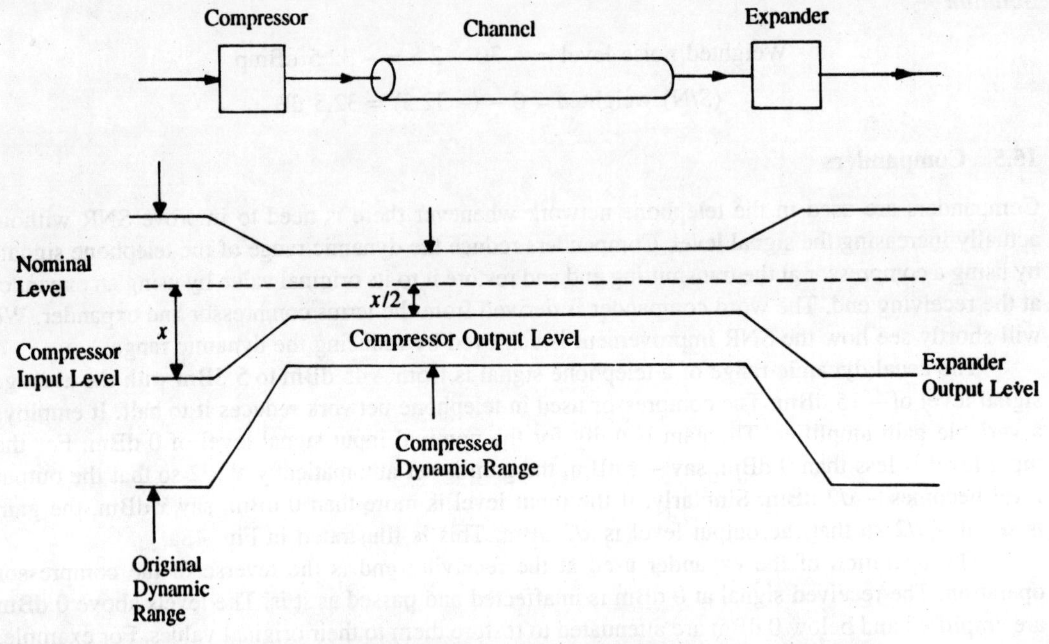

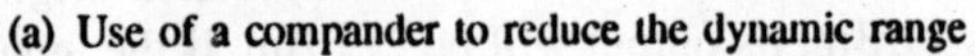
(a) Use of a compander to reduce the dynamic range

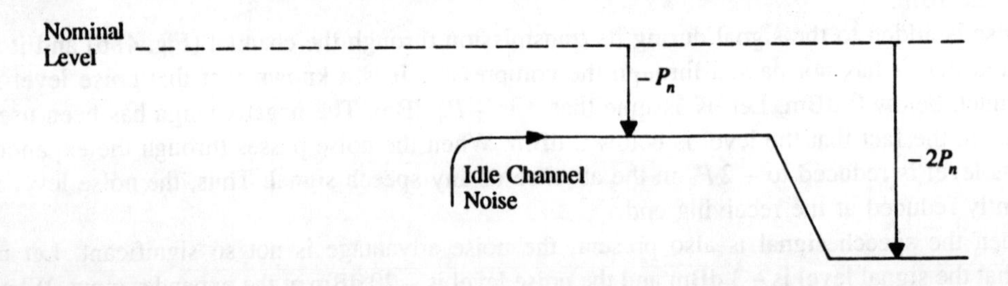

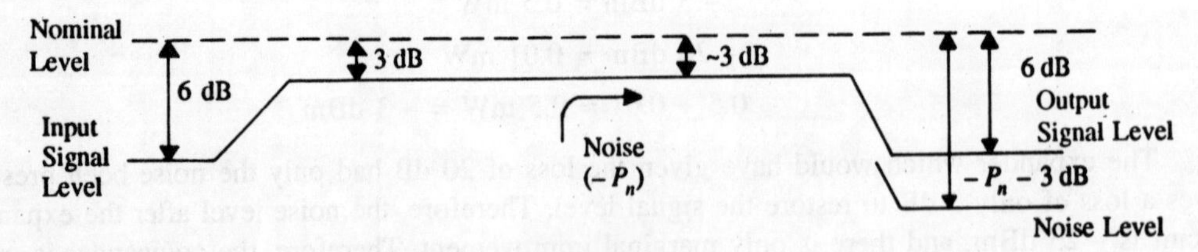

(b) Noise reduction in the received signal

Fig. 48

When a speech channel is used for data transmission, the modem transmits the carrier continuously. Unlike speech signals, there are no intersyllabic pauses and, therefore, the compander is ineffective. It may, on the other hand, introduce errors because it changes the amplitude of the data carrier. The compander needs to be disabled whenever a speech channel is used for data transmission.

16 SIGNAL IMPAIRMENTS

There are several signal processing stages in the telephone network, e.g., switching, filtering, amplification, frequency translation or quantization etc. At every stage there is some impairment of the signal quality. We have already examined some of the major impairments, e.g., linear distortions, noise, etc. Other impairments of the signal, described below, are not very critical for transmission of the speech signal but may be the cause of poor bit error rate for a data signal.

Impulse Noise. Impulse noise is characterized by high amplitude peaks (hits) of short duration. It is caused by bad electrical contacts, dial pulses, crosstalk, relay contacts, etc. It can also originate from external sources such as power lines and lightning.

Impulse noise is generally not objectionable in voice communication as an impulse hit is never more than 10 ms in duration. Typical duration of the impulse hit is 4 ms (Fig. 49). It has been empirically determined that an impulse hit will not produce transmission errors in a data signal unless it comes within 6 dB of the signal level.

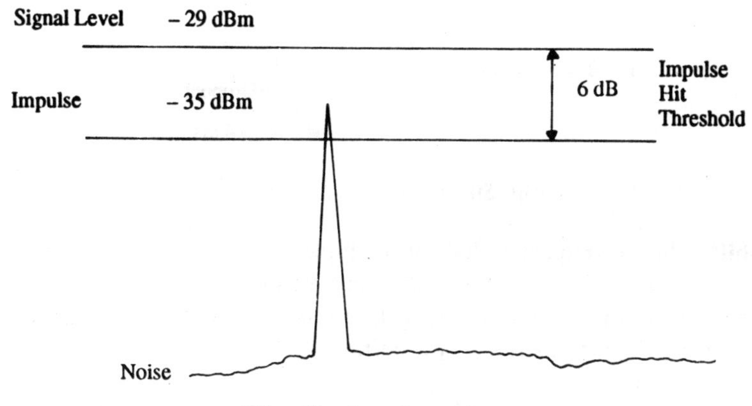

Fig. 49 Impulse noise.

Gain Hits and Dropouts. A gain hit is a sudden random change in the signal level of more than ±3 dB and lasting more than 4 ms (Fig. 50). The signal returns to the original level within 200 ms. Gain hits are due to change in end-to-end gain of the channel which is caused by the transients produced during switch-over of radio channels.

A dropout is a decrease in channel gain of more than 12 dB that lasts longer than 4 ms (Fig. 50). Dropouts are caused by deep fades in the radio channels caused by atmospheric conditions.

Phase Hits. Phase hits are sudden random changes in the phase of a transmitted signal. The hits lasting more than 4 ms and greater than ± 20° peak are recorded for measurement.

Phase Jitter. Phase jitter is a form of incidental phase modulation produced in a tone when transmitted on a speech channel. It is caused by the low frequency ripple in the power supplies of the telecommunication equipment. It is also caused by the jitter present in the carriers used for translating the baseband. Its frequency is generally less than 300 Hz.

Single Frequency Interference. Single frequency interference is the presence of one or more unwanted tones in the speech channel. They are caused by crosstalk and intermodulation.

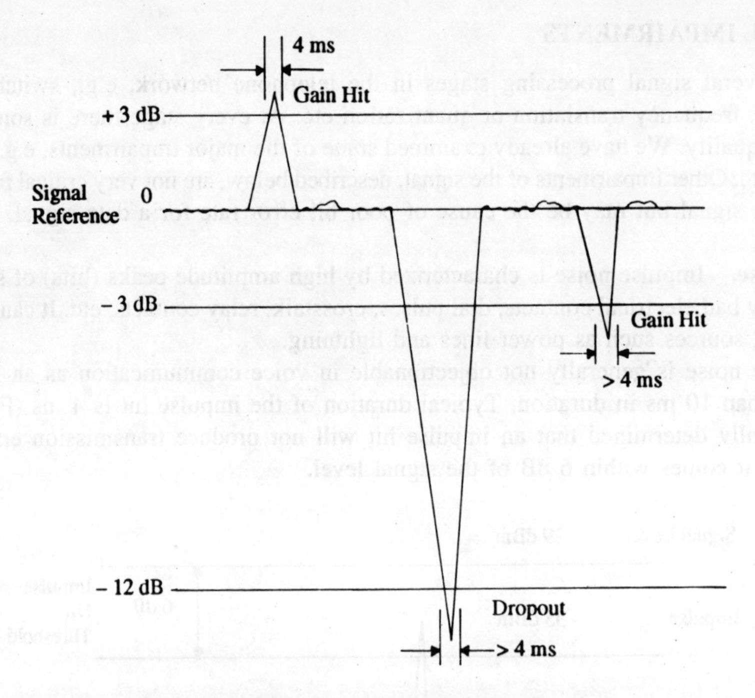

Fig. 50 Gain hits and dropouts.

Frequency Shift. In the frequency division multiplexing equipment, if the carrier frequencies used for translating the frequency bands in the transmit side are not exactly the same as those in the receive side, a tone transmitted on a speech channel will suffer a change in frequency. This change in frequency is termed as frequency shift.

17 NETWORKING OPTIONS

Having considered the characteristics of various transmission media, required channel characteristics for data transmission and configuration of the telephone network, let us now examine the networking options and their features for data transmission. Switched data networks based on message and packet switching technologies are not covered here. We will discuss these options later when these technologies have been thoroughly understood.

There are basically two networking options:

1. Dedicated transmission medium which can be either installed by the user, or implemented on a telephone network
2. Switched connections through telephone exchanges.

A dedicated transmission medium is usually costlier than a switched connection. Some of its features not found in switched connections are as follows:

- It is available all the time and there is no call establishment overhead.
- Its transmission characteristics (attenuation and group delay distortions) can be specially conditioned for the purpose of data transmission.

- Options for four-wire or two-wire circuits are available. Switched connections are always two-wire connections.
- It can support much higher bit rates than switched connections.

17.1 Dedicated Transmission Medium

Dedicated transmission medium can provide point-to-point and point-to-multipoint connectivity. The transmission medium is user owned within a building premises or an office complex. When the dedicated connections extend across public roads and utilities, the local telephone operator provides dedicated connections by leasing the telephone cable network facilities. We discuss below the characteristic features of the dedicated connections which depend on the transmission media and the manner in which the connection is built up.

1. Point-to-point or point-to-multipoint connections within a premises.
   - The transmission medium is self installed by the user.
   - Depending on the bit rate, distances, and the number of terminals to be interconnected, twisted-pair, shielded twisted pair, coaxial cable or optical fibre may be used.
   - Available bandwidth is not limited to 300 to 3400 Hz as offered by the telephone network.
   - DC continuity and wire polarity of the transmission medium, if required, can be established.
   - Bit rates as high as 100 Mbps are possible.
2. Point-to-point connnections within an exchange area.
   - The transmission medium is twisted-pair cable.
   - Connection from subscriber A to subscriber B is established by providing a jumper between the respective pairs of the subscribers at the distribution frame in the exchange.
   - Wire polarity is difficult to establish and maintain but is not impossible.
   - Exchange switching equipment and multiplexing equipment are not involved. Full transmission medium bandwidth is available. Usually there are no loading coils in the local network.
3. Point-to-point connections extending from one exchange area to another nearby exchange area.
   - Connection is extended using the junction between the two exchanges and their local networks.
   - Junctions may be simple twisted-pair cables and may have loading coils. The bandwidth is reduced due to the loading coils.
   - Junctions may include speech channel multiplexing equipment to increase the number of channels. Digital TDM or analog FDM techniques may be used. In either case, the bandwidth available is 300 to 3400 Hz.
   - Balanced-pair cable, coaxial cable, optical fibre or terrestrial microwave systems are used along with the multiplexing equipment.
   - DC continuity and wire polarity cannot be established.
4. Point-to-point connections extending from one city to another.

- The cities are interconnected through trunk circuits using coaxial cable, optical fibre cable, terrestrial or satellite radio systems.
- In this case, speech channel multiplexing equipment is invariably involved. Digital TDM or analog FDM techniques may be used. In either case, the bandwidth available is 300 to 3400 Hz.
- DC continuity and wire polarity cannot be established.
- The satellite-based trunk circuits introduce about 250 ms one way delay. Terrestrial connections may involve delays of the order of 20 ms.
- Satellite-based trunk circuits introduce significant and noticeable echo. Echo suppressors and companders are installed in the satellite circuits.

17.2 Switched Connections

Switched connections may extend from one telephone exchange to another in the same city, in different cities or in different countries. Depending on the locations, switched connections may involve only local exchanges, local and trunk exchanges or local, trunk and international transit exchanges. In all these cases, the following features of the service can be assumed:

- Available channel bandwidth is 300–3400 Hz
- Connection is always a two-wire circuit
- DC continuity, wire polarity cannot be established
- If trunk circuits are involved and are provided through satellite, one-way propagation delay will be of the order of 250 ms
- Special methods have to be deployed to disable echo suppressors and companders.

From the above options for data transmission, it is evident that the telephone network effectively offers 300–3400 Hz bandwidth on a telephone channel, be it dedicated leased circuit or switched connection. Only in the case of leased circuits within the same exchange area, full bandwidth of the balanced pair can be made available. 300–3400 Hz frequency band is too small to support the baseband data transmission at the rate required by the data processing community. Therefore, instead of two-level transmission, multilevel transmission is used. It allows transmission of higher bit rates within the same bandwidth. Digital modulation methods are employed to convert the data signal into an analog signal which can be transmitted over the bandwidth offered by the telephone network. Modulation and demodulation functions are carried out by intermediary devices called modems. We will study modems in detail in the next chapter.

18 CCITT RECOMMENDATIONS FOR LEASED CIRCUITS

CCITT recommendations M.1020 and M.1025 are for the point-to-point leased circuits to be used for data transmission. These recommendations specify the transmission characteristics of the circuits.

18.1 CCITT M.1020 Recommendation

This recommendation is for leased circuits intended to be used for data transmission using modems not equipped with equalizers.

Nominal overall loss	:	≤ 28 dB at 800 Hz
Variation in overall loss		
— short term, a few seconds	:	≤ 3 dB
— daily and seasonal variations	:	≤ 4 dB
Attenuation distortion	:	Limits as shown in Fig. 51
Group delay distortion	:	Limits as shown in Fig. 52
Random circuit noise	:	≤ – 38 dBm0 for leased circuit longer than 10,000 km
Impulse noise > – 21 dBm0	:	≤ 18 impulses in 15 minutes
Gain hits > ± 2 dB	:	10 in 15 minutes
Phase jitter	:	≤ 15° peak to peak
Signal/Quantization noise	:	≥ 22 dB
Single tone interference	:	Should not exceed the level 3 dB below the circuit noise
Frequency shift	:	≤ ± 5 Hz
Non-linear distortion measured with 700 Hz tone at – 15 dBm0	:	Levels of the received harmonics should be 25 dB below the received level of 700 Hz

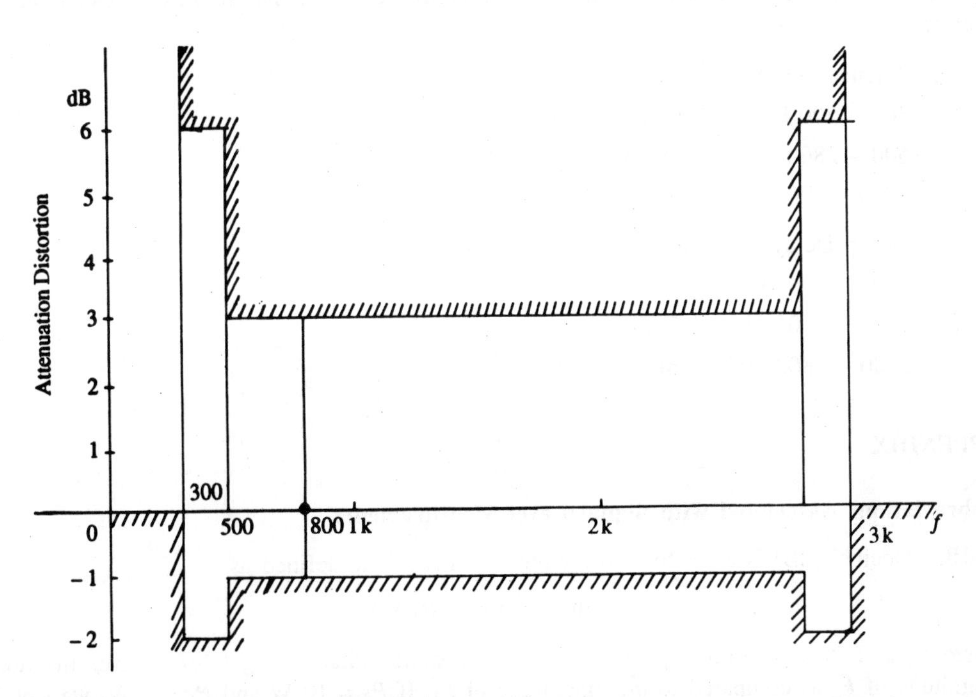

Fig. 51 Attenuation distortion limits as per CCITT M.1020 recommendation.

18.2 CCITT M.1025 Recommendation

This recommendation is for leased circuits intended to be used for data transmission using modems equipped with equalizers. The limits for all the parameters are the same as M.1020 except those

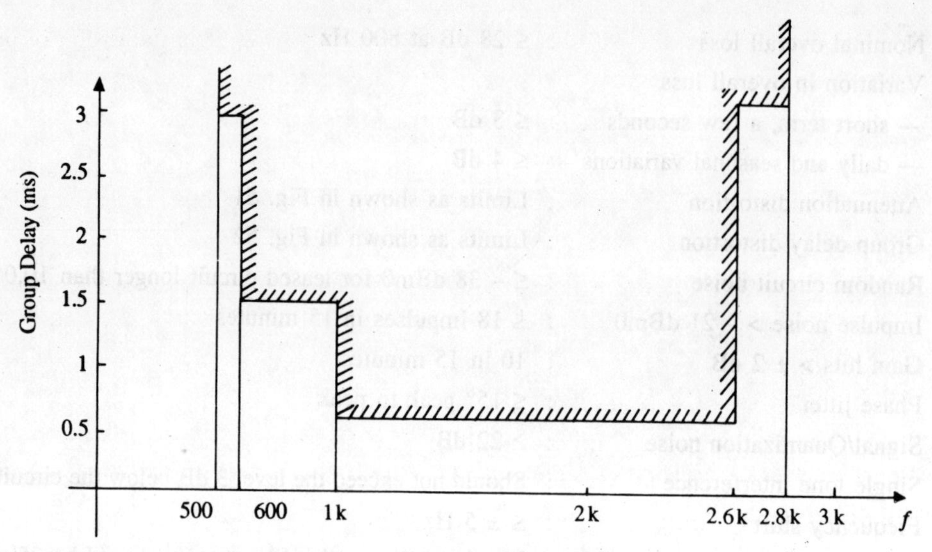

Fig. 52 Group delay distortion limits as per CCITT M.1020 recommendation.

for attenuation and group delay distortions which are more liberal. The figures indicated below are relative to those at 800 Hz.

1. Attenuation Distortion

   300 – 500 Hz 12 dB to – 2 dB

   500 – 2800 Hz 8 dB to – 2 dB

   2800 – 3000 Hz 12 dB to – 2 dB

2. Group Delay Distortion

   500 – 1000 Hz ≤ 3 ms

   1000 – 2600 Hz ≤ 1.5 ms

   2600 – 2800 Hz ≤ 3 ms

APPENDIX

Abbreviations Associated with Signal Level Measurements

1. dB. Decibel (dB) is a relative power measurement. It is defined as

$$\text{dB} = 10 \log (P_2/P_1)$$

where P_2 and P_1 are absolute power levels in the same units. 10 log (P_2/P_1) gives the relative magnitude of P_2 as compared to the magnitude of P_1. If P_2 = 10 W and P_1 = 5 W, we say P_2 is 3 dB higher than P_1 because

$$10 \log (10/5) = 10 \times 0.3 = 3 \text{ dB}$$

When P_2 is less than P_1, 10 log (P_2/P_1) will become negative. $-x$ dB indicates that P_2 is less than P_1 by x dB.

2. dBm. dBm is the power level in dB with reference to 1 mW. It is an absolute unit of power. For example, a power level of 100 mW can be expressed in dBm as

$$100 \text{ mW} = 10 \log (100 \text{ mW}/1 \text{ mW}) = 10 \times 2 = 20 \text{ dBm}$$

3. dBm0. dBm0 is the difference of actual power level at any point in the telecommunication system and the nominal power level at that point. If the measured power level is 4 dBm at a point where the nominal level should be – 5 dBm, we say the power level is 9 dBm0.

4. dBmp. This is the absolute unit of power used for psophometrically weighted noise levels with respect to 1 mW. Weighted noise level is 2.5 dB lower than the unweighted noise level in the speech channel of 300–3400 Hz when the noise is white.

PROBLEMS

1. Impulse response of a channel is given by

$$h(t) = \sin (t/T)/(t/T)$$

Sketch the output of the channel when impulses corresponding to input data 1011 are applied to the channel. Assume "1" is represented by a unit positive impulse and "0" by a unit negative impulse.

2. A single impulse input to a transmission channel results in received samples of

0.2 0.5 1.0 0.5 0.2

where the middle sample is in the sampling interval corresponding to the transmitted impulse and the other entries are in the adjacent intervals. Determine the tap gains of a three tap transversal filter equalizer which will eliminate the inter-symbol interference.

3. Impulse response of a channel is shown below. Draw the eye pattern if input is a random bit pattern. Assume "1" is represented by a impulse and "0" by absence of impulse.

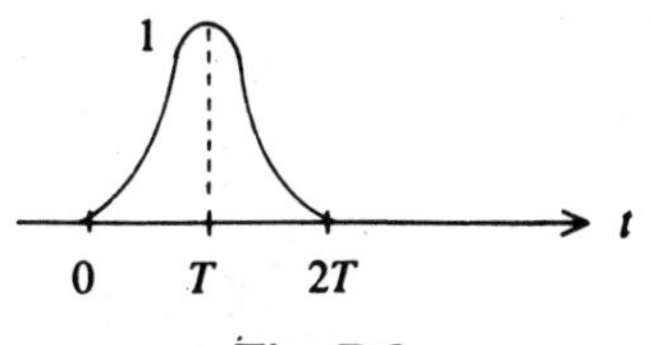

Fig. P.3

4. Do Problem 3 if "1" is represented by a unit positive impulse and "0" by a unit negative impulse.

5. A signal having three frequency components f_1, f_2 and f_3 of equal amplitudes is applied to an amplifier having input/output characteristic defined by

$$y(t) = a_1 x(t) + a_2 x^2(t) + a_3 x^3(t)$$

Write down all the frequency components of the output. If f_1 = 1000 Hz, f_2 = 2000 Hz, and f_3 = 3000 Hz, are there any intermodulation products in the frequency band 1000–2000 Hz?

6. (a) In a voice channel, the unweighted white noise level is 0.001 mW/kHz. What is the psophometrically weighted noise level in dBm?

(b) Express the following levels in dBm:

(i) 100 W
(ii) 0.1 W
(iii) 1.0 mW
(iv) 1.0 nW
(v) 1.0 pW

7. The refractive indices of the core and the cladding are 1.5 and 1.48 respectively in a multimode fibre. Calculate the modal dispersion if its length is 40 km.

8. Calculate the chromatic dispersion in a fibre of length 10 km when the spectral width and the wavelength of the source are 50 nm and 820 nm respectively. M = 110 ps/nm km at 820 nm.

CHAPTER 3

Modems and Data Multiplexers

In Chapter 2 we examined the transmission requirements for data signals. We also saw that the only alternative to the data transmission available today is the telephone network which provides wordwide accessibility. But the network is meant primarily for voice communication and supports only analog voice band service having bandwidth of 300 Hz to 3400 Hz. Baseband transmission of digital signals using analog voice band service has several limitations. To overcome these limitations, some intermediary devices are used to utilize the service in the best possible way. These devices are modems and data multiplexers.

We begin this chapter by examining of various digital modulation methods which are used in the modems. We then proceed to describe operation of the modem. Besides modulation and demodulation, there are many additional functions which are performed by the modems. We examine all these functions and familiarize ourselves with the modem terminology. There are number of CCITT recommendations on the modems. We take a brief look at the features of the CCITT modems and some non-standard modems. We next examine the various data multiplexing techniques. Frequency division and time division multiplexers are discussed in brief while the statistical time division multiplexer is discussed in considerable detail.

1 DIGITAL MODULATION METHODS

There are three basic types of modulation methods for transmission of digital signals. These methods are based on the three attributes of a sinusoidal signal, amplitude, frequency and phase. The corresponding modulation methods are called: Amplitude Shift Keying (ASK), Frequency Shift Keying (FSK) and Phase Shift Keying (PSK).

In addition, a combination of ASK and PSK is employed at high bit rates. This method is called Quadrature Amplitude Modulation (QAM). We will discuss these modulation methods and later examine their application in the standard modems.

1.1 Amplitude Shift Keying (ASK)

Amplitude Shift Keying (ASK) is the simplest form of digital modulation. In ASK, the carrier amplitude is multiplied by the binary "1" or "0" (Fig. 1). The digital input is a unipolar NRZ signal.

The amplitude modulated carrier signal can be written as

$$v(t) = d \sin (2\pi f_c t)$$

where f_c is the carrier frequency and d is the data bit variable which can take values "1" or "0", depending on the state of the digital signal.

The frequency spectrum of the ASK signal consists of the carrier frequency with upper and

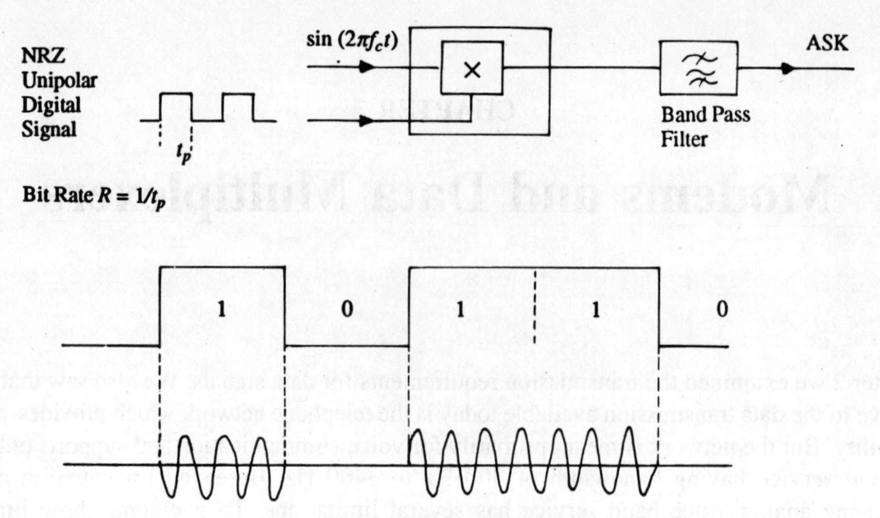

Fig. 1 Amplitude shift keying.

lower side bands (Fig. 2). For random unipolar NRZ digital signal having bit rate R, the first zero of the spectrum occurs at R Hz away from the carrier frequency.

The transmission bandwidth B of the ASK signal is restricted by using a filter to

$$B = (1 + r)R$$

where r is a factor related to the filter characteristics and its value lies in the range 0–1.

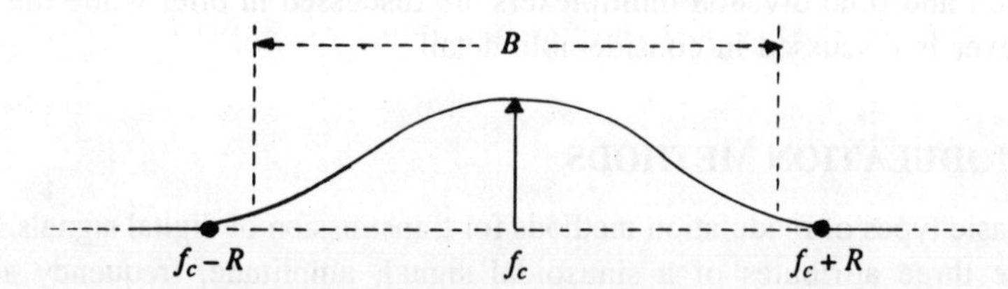

Fig. 2 Frequency spectrum of an ASK signal.

ASK is very sensitive to noise and finds limited application in data transmission. It is used at very low bit rates, of less than 100 bps.

1.2 Frequency Shift Keying (FSK)

In Frequency Shift Keying (FSK), the frequency of the carrier is shifted between two discrete values, one representing binary "1" and the other representing binary "0" (Fig. 3). The carrier amplitude does not change. FSK is relatively simple to implement. It is used extensively in low speed modems having bit rates below 1200 bps.

The instantaneous value of the FSK signal is given by

$$v(t) = d \sin (2\pi f_1 t) + \bar{d} \sin (2\pi f_0 t)$$

where f_1 and f_0 are the frequencies corresponding to binary "1" and "0" respectively and d is the data signal variable as before.

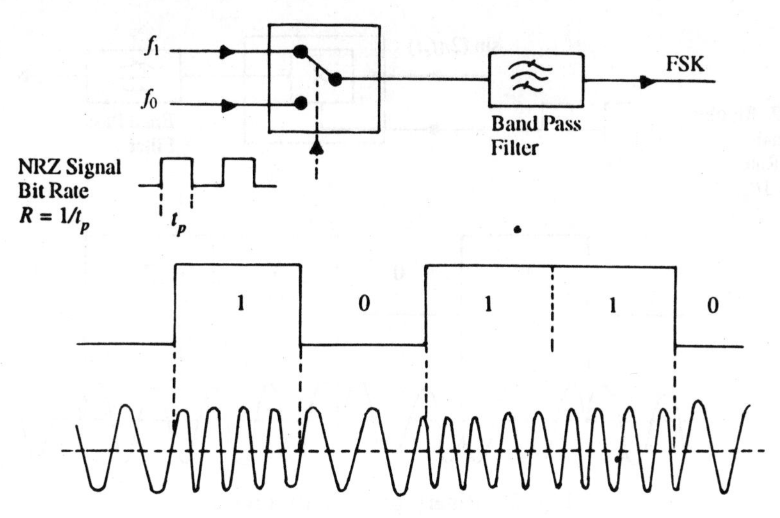

Fig. 3 Frequency shift keying.

From the above equation, it is obvious that the FSK signal can be considered to be comprising two ASK signals with carrier frequencies f_1 and f_0. Therefore, the frequency spectrum of the FSK signal is as shown in Fig. 4.

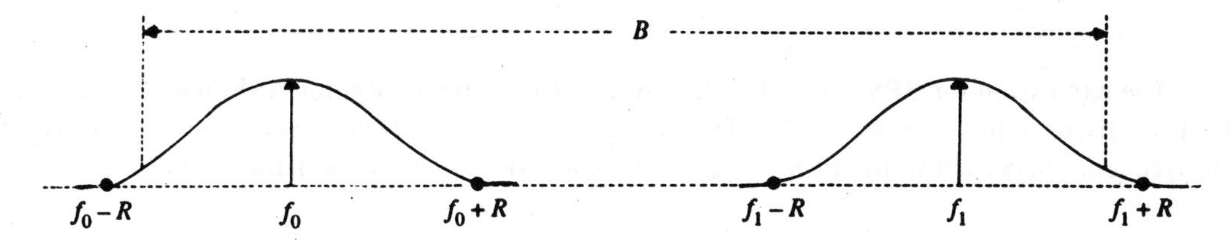

Fig. 4 Frequency spectrum of a FSK signal.

To get an estimate of the bandwidth B for the FSK signal, we need to include the separation between f_1 and f_0 and significant portions of the upper side band of carrier f_1 and of the lower side band of carrier f_0.

$$B = |f_1 - f_0| + (1 + r)R$$

The separation between f_1 and f_0 is kept at least $2R/3$. CCITT Recommendation V.23 specifies $f_1 = 2100$ Hz and $f_0 = 1300$ Hz for bit rate of 1200 bps. FSK is not very efficient in its use of the available transmission channel bandwidth.

1.3 Phase Shift Keying (PSK)

Phase Shift Keying (PSK) is the most efficient of the three modulation methods and is used for high bit rates. In PSK, phase of the carrier is modulated to represent the binary values. Figure 5 shows the simplest form of PSK called Binary PSK (BPSK). The carrier phase is changed between 0 and π by the bipolar digital signal. Binary states "1" and "0" are represented by the negative and positive polarities of the digital signal.

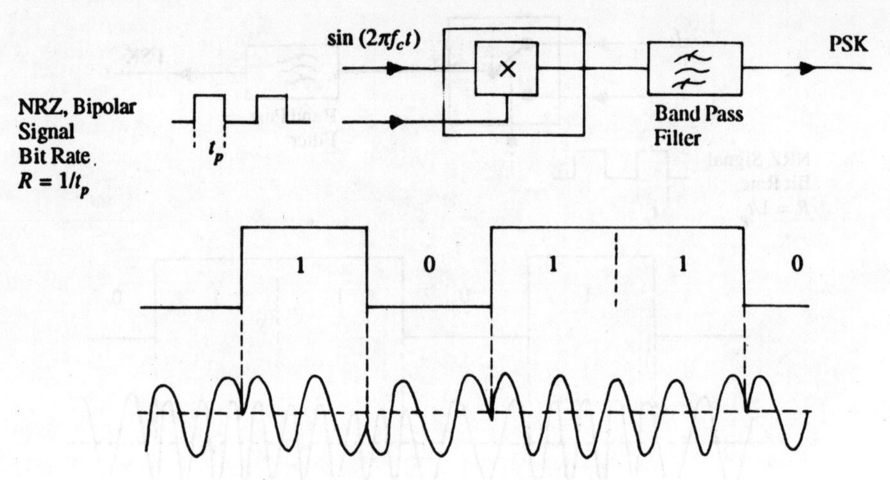

Fig. 5 Binary phase shift keying.

The instantaneous value of the BPSK signal can be written as

$$v(t) = \sin (2\pi f_c t) \qquad \text{when } d = 1 \text{ for binary state "0"}$$

$$v(t) = -\sin (2\pi f_c t) = \sin (2\pi f_c t + \pi) \qquad \text{when } d = -1 \text{ for binary state "1"}$$

In other words,

$$v(t) = d \sin (2\pi f_c t) \qquad d = \pm 1$$

The expression for BPSK signal is very similar to the expression for ASK signal except that the data variable d takes the values $\pm$ 1. The carrier gets supressed due to bipolar modulation signal. The frequency spectrum of the PSK signal for random NRZ digital modulating signal is shown in Fig. 6.

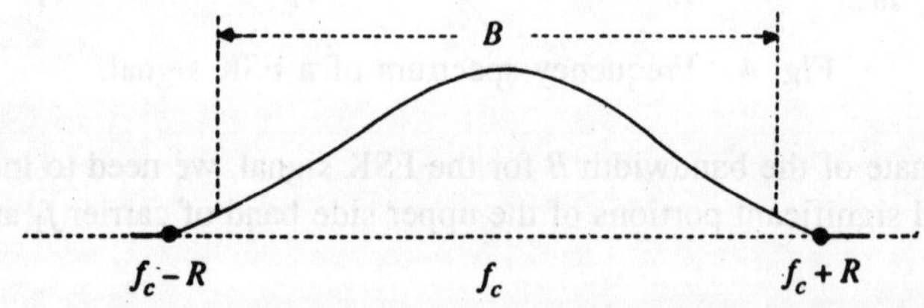

Fig. 6 Frequency spectrum of a BPSK signal.

The estimate of bandwidth B of the BPSK signal can be obtained as before.

$$B = (1 + r)R \qquad 0 < r < 1$$

where the parameter r depends on transmission filter characteristics. The BPSK signal requires less bandwidth as compared to the FSK signal.

2 MULTILEVEL MODULATION

We have so far considered the schemes for two-level modulation, i.e., the bit rate and the baud rate are the same. Due to the limited bandwidth of the telephone voice channel, the maximum bit rate which can be achieved using any of the above two-level modulation methods does not meet the requirements of the data processing community. Keeping the baud rate the same, the bit rate can be increased using multilevel modulation as we saw in Chapter 1. The data bits are divided into groups of two or more bits and each group is assigned a specific state of the sinusoidal signal. Any of the three attributes of the signal—amplitude, frequency or the phase, can be used to represent the groups of the data bits. ASK being very sensitive to noise and FSK being very expensive from the bandwidth point of view, multilevel modulation is used only with PSK. Instead of two phase states, 0 and π as in BPSK, the carrier is allowed to have four or more phase states (Fig. 7).

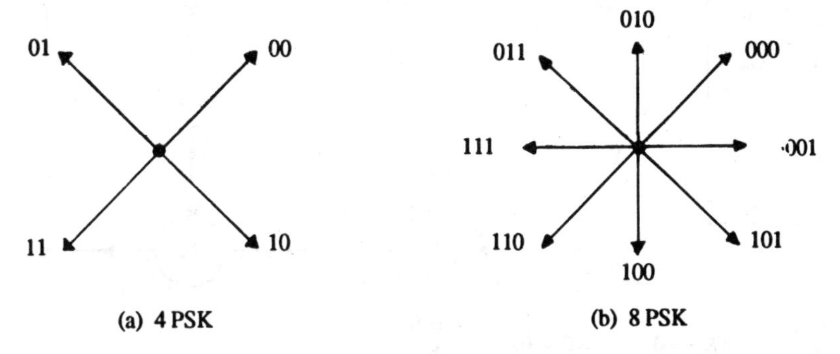

Fig. 7 Multilevel PSK.

In the four-state PSK (or simply 4 PSK), two bits are associated with each phase state. Therefore, the bit rate is twice the baud rate. Similarly, three bits are associated with each phase state in 8 PSK and, therefore, its bit rate is three times the baud rate.

2.1 Gray Code

A very important point to be noted in Fig. 7 is the sequence of assignment of codes to the phase states. The codes of adjacent states differ in only one bit position and the resulting sequence is not in usual binary count. This sequence is called Gray code. When a PSK signal is transmitted, the received signal does not have the precisely defined phase states shown in the Fig. 7 due to noise and phase distortion. For example, 000 state may be received at 15° instead of 45°. The receiver has to take a decision whether the received code is 000 or 001. As the received phase is nearer to the adjacent code 001, it decodes the phase as 001. It makes a mistake but introduces only one bit error because the adjacent codes differ in only one bit position. Had we used the binary count sequence, the receiver would have introduced three errors by decoding the received phase as 111. Thus phase errors result in reduced bit error rate when the Gray code is used.

2.2 4 PSK Modulator

Figure 8 shows the schematic of a 4 PSK modulator. It consists of two BPSK modulators. The carrier frequency of one of the modulators is phase shifted by $\pi/2$ radians. The data bits are taken in groups of two bits called dibits and two bipolar digital signals are generated, one from the first bit of the dibits and the other from the second bit of the dibits. Outputs of the modulators are added so that the phase of the resultant carrier is the vectorial addition of the respective phasors of the two modulated carriers.

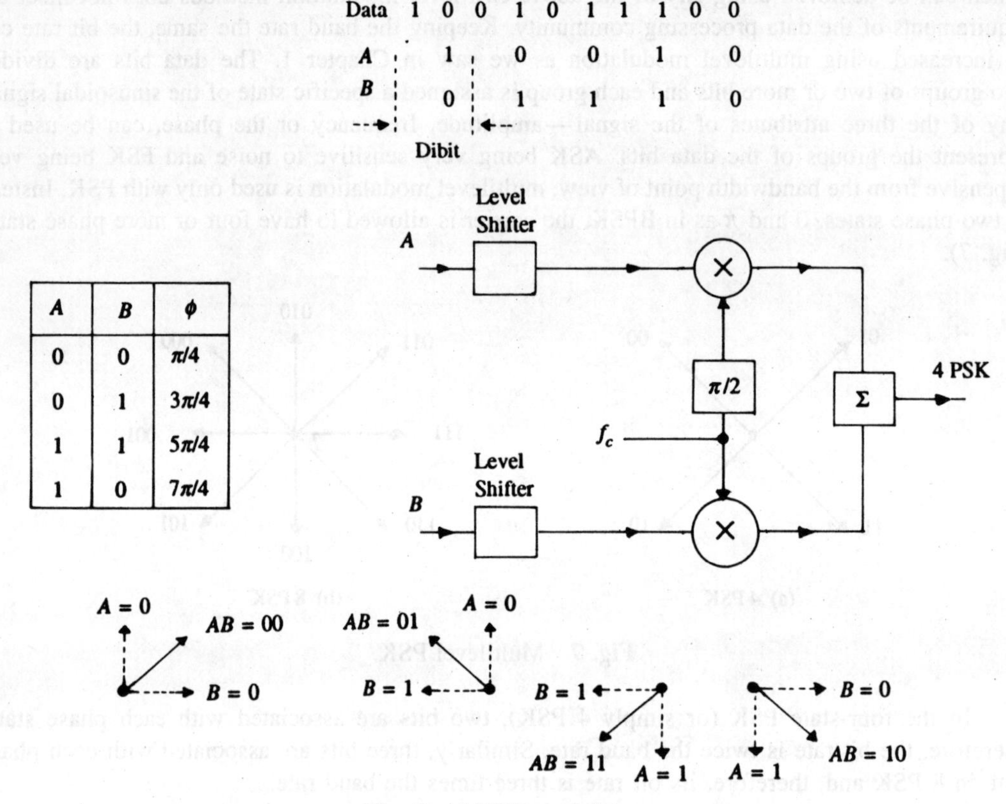

A	B	φ
0	0	π/4
0	1	3π/4
1	1	5π/4
1	0	7π/4

Fig. 8 4 PSK modulator.

2.3 4 PSK Demodulator

Figure 9 shows a 4 PSK demodulator. The reference carrier is recovered from the received modulated carrier. As in the modulator, a $\pi/2$ phase shifted carrier is also generated. When these carriers are multiplied with the received signal, we get

$$\sin(2\pi f_c t + \phi)\sin(2\pi f_c t) = \frac{1}{2}\cos(\phi) - \frac{1}{2}\cos(4\pi f_c t + \phi)$$

and

$$\sin(2\pi f_c t + \phi)\sin(2\pi f_c t + \pi/2) = \frac{1}{2}\cos(\phi - \pi/2) - \frac{1}{2}\cos(4\pi f_c t + \phi + \pi/2)$$

where ϕ is the phase of the received carrier.

The multiplier outputs are passed through low pass filters to remove the $2f_c$ frequency component and are applied to the comparators which generate the dibits. Table 1 gives the outputs of low pass filters for various values of input phase ϕ.

In the above demodulation method, we have assumed availability of the phase coherent carrier at the receiving end, i.e., the recovered carrier at the receiving end being in phase with the carrier at the transmitting end. But it is quite possible that the phase of the recovered carrier is out by $\pi/2$ or π. And if this happens, the demodulator operation will be upset.

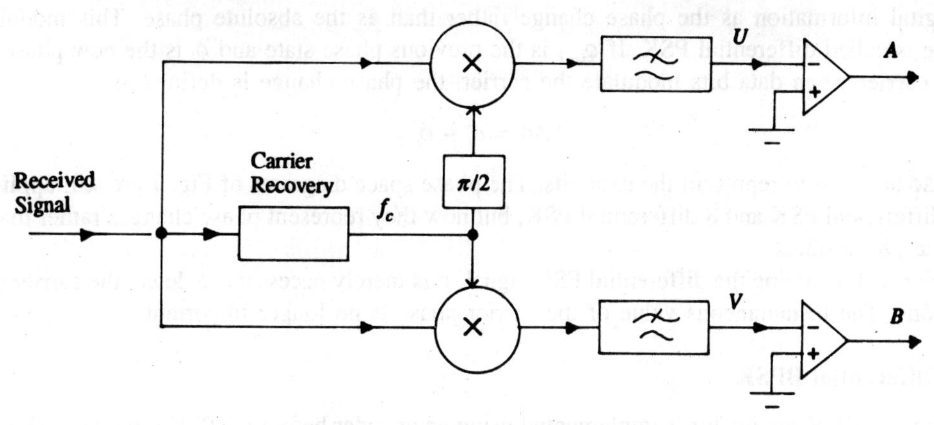

Fig. 9 4 PSK demodulator.

Table 1 4 PSK Demodulator Outputs

ϕ	U	V	A	B
$\pi/4$	0.35	0.35	0	0
$3\pi/4$	0.35	− 0.35	0	1
$5\pi/4$	− 0.35	− 0.35	1	1
$7\pi/4$	− 0.35	0.35	1	0

EXAMPLE 1

1. What are the phase states of the carrier when the bit stream

1 0 1 1 1 0 0 1 0 0

is applied to 4 PSK modulator shown in Fig. 8.

2. If the recovered carrier at the demodulator is out of phase by π radians, what will be the output when the above 4 PSK carrier is applied to the demodulator shown in Fig. 9.

Solution

1. Modulator input	1	0	1	1	1	0	0	1	0	0
Phase states of the transmitted carrier	$7\pi/4$		$5\pi/4$		$7\pi/4$		$3\pi/4$		$\pi/4$	
2. Relative phase with respect to the recovered carrier	$3\pi/4$		$\pi/4$		$3\pi/4$		$7\pi/4$		$5\pi/4$	
Output of the demodulator (Table 1)	0	1	0	0	0	1	1	0	1	1

3. DIFFERENTIAL PSK

The problem of generating the carrier with a fixed absolute phase can be circumvented by encoding the digital information as the phase change rather than as the absolute phase. This modulation scheme is called differential PSK. If ϕ_{t-1} is the previous phase state and ϕ_t is the new phase state of the carrier when data bits modulate the carrier, the phase change is defined as

$$\Delta\phi = \phi_t - \phi_{t-1}$$

$\Delta\phi$ is coded to represent the data bits. The phase space diagrams of Fig. 7 are still applicable for 4 differential PSK and 8 differential PSK, but now they represent phase changes rather than the absolute phase states.

For demodulating the differential PSK signal, it is merely necessary to detect the carrier phase variations. The instantaneous value of the carrier phase is no longer important.

3.1 Differential BPSK

Differential BPSK modulator is implemented using an encoder before a BPSK modulator (Fig. 10). The encoder logic is so designed that the desired phase changes are obtained at the modulator output.

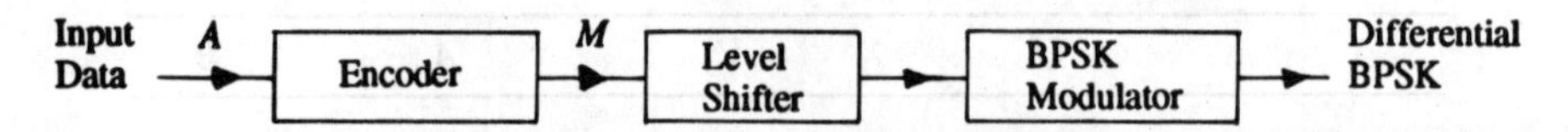

Fig. 10 Differential BPSK modulator.

Table 2 shows the relation between the input data bits and the phase states of the carrier at the modulator output. Knowing that the carrier phase is 0 for binary "0" at the modulator input and π for binary "1" at the modulator input, we can write the logic table for the encoder. It is easily implemented using a JK flip flop in the toggle mode.

Table 2 Encoder Logic of Differential BPSK Modulator

A	ϕ_{t-1}	ϕ_t	$\Delta\phi$	M_{t-1}	M_t
0	0	0	0	0	0
0	π	π	0	1	1
1	0	π	π	0	1
1	π	0	π	1	0

EXAMPLE 2

Write the phase states of the differential BPSK carrier for input data stream 100110101. The starting phase of the carrier can be taken as 0.

Solution

A		1	0	0	1	1	0	1	0	1
$\Delta\phi$		π	0	0	π	π	0	π	0	π
ϕ	0	π	π	π	0	π	π	0	0	π

Figure 11 shows the demodulation scheme for differential BPSK signal. The received signal is delayed by one bit and multiplied by the received signal. In other words, the carrier phase states

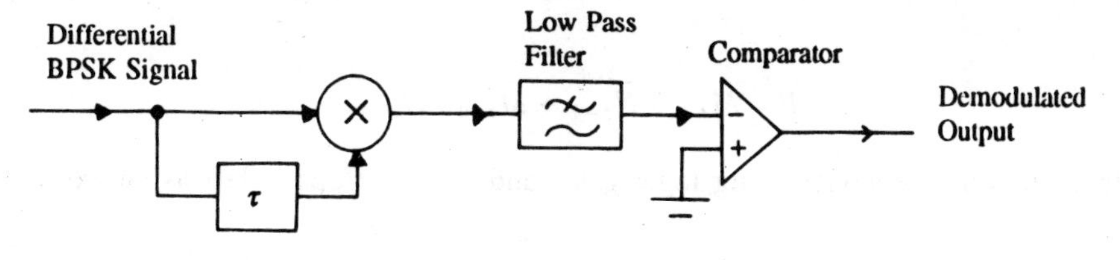

Fig. 11 Differential BPSK demodulator.

of the adjacent bits are multiplied. Adjacent phase states may be in phase or π out of phase. If they are in phase, the multiplier output is positive and if they are out of phase, the multiplier output is negative.

$$\sin^2 (2\pi f_c t) = \sin^2 (2\pi f_c t + \pi) = \frac{1}{2} - \frac{1}{2} \cos (4\pi f_c t)$$

$$\sin (2\pi f_c t) \sin (2\pi f_c + \pi) = -\frac{1}{2} + \frac{1}{2} \cos (4\pi f_c t)$$

The low pass filter allows only the DC component to pass through. Thus polarity of the signal at the filter output reflects the phase change. The comparator generates the demodulated data signal.

The differential demodulator does not require phase coherent carrier for demodulation. Also, note that there is no decoder corresponding to the encoder in the modulator. If a phase-coherent demodulator is used in place of the differential demodulator, a decoder will be required at the output of the demodulator.

3.2 Differential 4 PSK

Just like differential BPSK modulator, differential 4 PSK modulator can also be implemented using an encoder before a 4 PSK modulator as shown in Fig. 12.

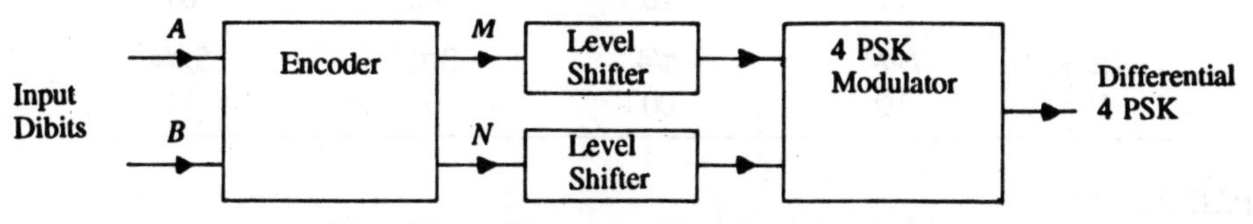

Fig. 12 Differential 4 PSK modulator.

The encoder logic is so designed that its outputs M and N modulate the carrier to produce the required phase changes in the carrier. Table 3a shows the relation between the input dibit AB and the phase changes of the modulated carrier. This modulation scheme has been standardized in CCITT recommendation V.26. Table 3b shows the relation between MN bits and the corresponding phase of the modulated carrier. Table 3c gives the encoder logic derived from Tables 3a and 3b. From Table 3c, it can be shown that

$$M_t = A \cdot B + \bar{A} \cdot B \cdot P + A \cdot \bar{B} \cdot \bar{P}$$

$$N_t = A \cdot B + \bar{A} \cdot B \cdot \bar{P} + A \cdot \bar{B} \cdot P$$

where

$$P = M_{t-1} \cdot \overline{N}_{t-1} + \overline{M}_{t-1} \cdot N_{t-1}$$

Implementation of encoder using logic gates and JK flip flops is left as an exercise to the reader.

Table 3a Modulator Logic

A	B	$\Delta\phi$
0	0	0
0	1	$\pi/2$
1	1	π
1	0	$3\pi/2$

Table 3c Absolute Phase Changes

M	N	ϕ
0	0	$\pi/4$
0	1	$3\pi/4$
1	1	$5\pi/4$
1	0	$7\pi/4$

Table 3b Encoder Logic of Differential 4 PSK Modulator

	$AB = 00$ $\Delta\phi = 0$	$AB = 01$ $\Delta\phi = \pi/2$	$AB = 11$ $\Delta\phi = \pi$	$AB = 10$ $\Delta\phi = 3\pi/2$
ϕ_{t-1} $(MN)_{t-1}$	ϕ_t $(MN)_t$	ϕ_t $(MN)_t$	ϕ_t $(MN)_t$	ϕ_t $(MN)_t$
$\pi/4$ 00	$\pi/4$ 00	$3\pi/4$ 01	$5\pi/4$ 11	$7\pi/4$ 10
$3\pi/4$ 01	$3\pi/4$ 01	$5\pi/4$ 11	$7\pi/4$ 10	$\pi/4$ 00
$5\pi/4$ 11	$5\pi/4$ 11	$7\pi/4$ 10	$\pi/4$ 00	$3\pi/4$ 01
$7\pi/4$ 10	$7\pi/4$ 10	$\pi/4$ 00	$3\pi/4$ 01	$5\pi/4$ 11

EXAMPLE 3

The following bit stream is applied to the differential 4 PSK modulator described in Table 3. Write the carrier phase states taking the initial carrier phase as reference.

1 0 1 1 1 1 0 0 0 1

Solution

Bit stream		1 0	1 1	1 1	0 0	0 1
$\Delta\phi$		$\frac{3\pi}{2}$	π	π	0	$\frac{\pi}{2}$
ϕ	0	$\frac{3\pi}{2}$	$\frac{\pi}{2}$	$\frac{3\pi}{2}$	$\frac{3\pi}{2}$	0

3.3 16 Quadrature Amplitude Modulation (QAM)

We can generalize the concept of differential phase shift keying to M equally spaced phase states. The bit rate will become $n \times$ (baud rate), where n is such that $2^n = M$. This is called M-ary PSK or simply MPSK. The phase states of the MPSK signal are equidistant from the origin and are separated by $2\pi/M$ radians (Fig. 13). As M is increased, the phase states come closer and result in degraded error rate performance because of reduced phase detection margin. In practice, differential PSK is used up to $M = 8$.

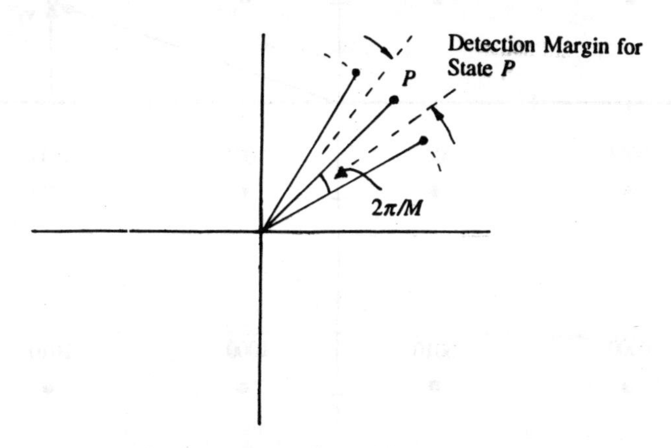

Fig. 13 Phase states of M-ary PSK.

Quadrature Amplitude Modulation (QAM) is one approach in which separation of the phase states is increased by utilizing combination of amplitude and phase modulations. Figure 14 shows the states of 16 QAM. There are sixteen states and each state corresponds to a group of four bits. Unlike PSK, the states are not equidistant from the origin, indicating the presence of amplitude modulation.

Note that each state can be represented as the sum of two carriers in quadrature. These carriers can have four possible amplitudes $\pm v_1$ and $\pm v_2$. Figure 15 shows block schematic of the modulator for 16 QAM. The odd numbered bits at the input are combined in pairs to generate one of the four levels at the D/A output which modulates the carrier. The even numbered bits are combined in a similar manner to modulate the other $\pi/2$ phase shifted carrier. The modulated carriers are combined to get the 16 QAM output.

It can be shown that 16 QAM gives better performance than does 16 PSK. Out of the basic modulation methods PSK comes closest to Shannon's limit for bit rate which we studied in Chapter 1. QAM displays further improvement over PSK.

4 MODEM

The term 'Modem' is derived from the words, MOdulator and DEModulator. A modem contains a modulator as well as a demodulator. The digital modulation/demodulation schemes discussed above are implemented in the modems. Most of the modems are designed for utilizing the analog voice band service offered by the telecommunication network. Therefore, the modulated carrier generated by a modem "fits" into the 300–3400 Hz bandwidth of the speech channel.

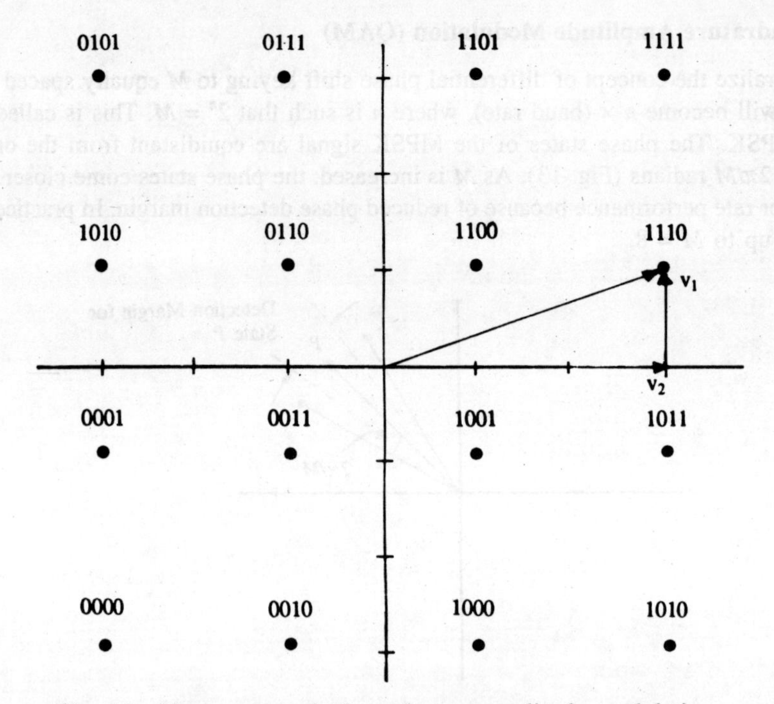

Fig. 14 Phase states of 16 quadrature amplitude modulation.

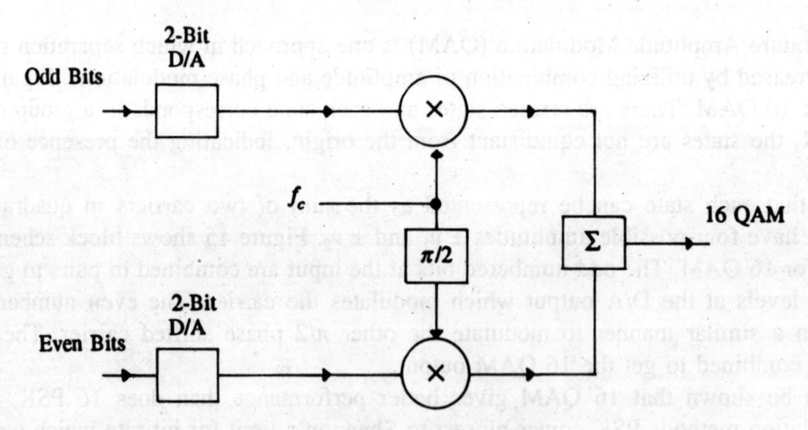

Fig. 15 16 QAM modulator.

A typical data connection set up using modems is shown in Fig. 16. The digital terminal devices which exchange digital signals are called Data Terminal Equipment (DTE). Two modems are always required, one at each end. The modem at the transmitting end converts the digital signal from the DTE into an analog signal by modulating a carrier. The modem at the receiving end demodulates the carrier and hands over the demodulated digital signal to the DTE.

The transmission medium between the two modems can be a dedicated leased circuit or a switched telephone circuit. In the latter case, modems are connected to the local telephone exhanges. Whenever data transmission is required, connection between the modems is established through the

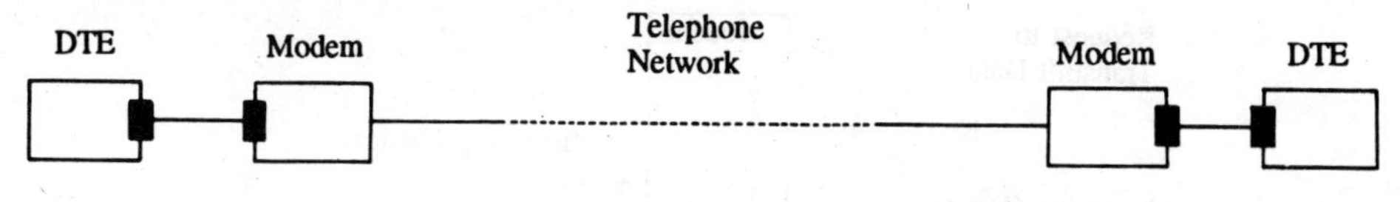

Fig. 16 A data circuit implemented using modems.

telephone exchanges. Modems are also required within a building to connect terminals which are located at distances usually more than 15 metres from the host.

Broadly, a modem comprises a transmitter, a receiver and two interfaces (Fig. 17). The digital signal to be transmitted is applied to the transmitter. The modulated carrier which is received from the distant end is applied to the receiver. The digital interface connects the modem to the DTE which generates and receives the digital signals. The line interface connects the modem to the transmission channel for transmitting and receiving the modulated signals. Modems connected to telephone exchanges have additional provision for connecting a telephone instrument. The telephone instrument enables establishment of the telephone connection.

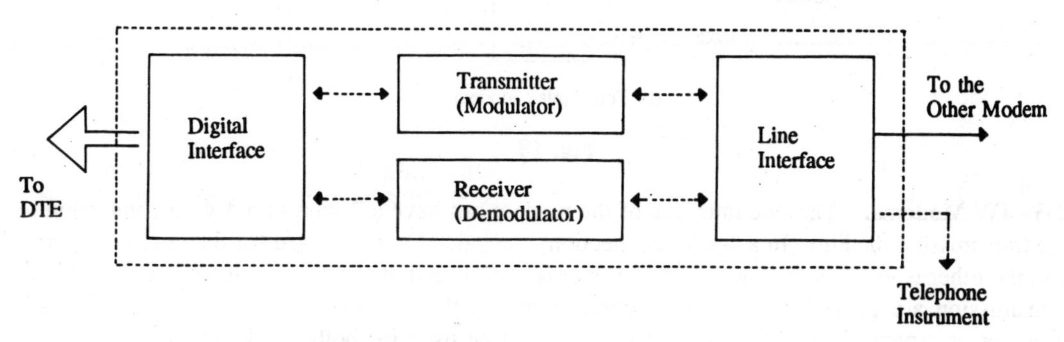

Fig. 17 Building blocks of a modem.

The transmitter and receiver in a modem comprise several signal processing circuits which include a modulator in the transmitter and a demodulator in the receiver.

4.1 Types of Modems

Modems can be of several types and they can be categorized in a number of ways. Categorization is usually based on the following basic modem features:

- Directional capability—Half duplex modem and full duplex modem.
- Connection to the line—2-wire modem and 4-wire modem.
- Transmission mode—Asynchronous modem and synchronous modem.

Half Duplex and Full Duplex Modems. A half duplex modem permits transmission in one direction at a time. If a carrier is detected on the line by the modem, it gives an indication of the incoming carrier to the DTE through a control signal of its digital interface (Fig. 18a). So long as the carrier is being received, the modem does not give clearance to the DTE to transmit.

A full duplex modem allows simultaneous transmission in both directions. Thus, there are two carriers on the line, one outgoing and the other incoming (Fig. 18b).

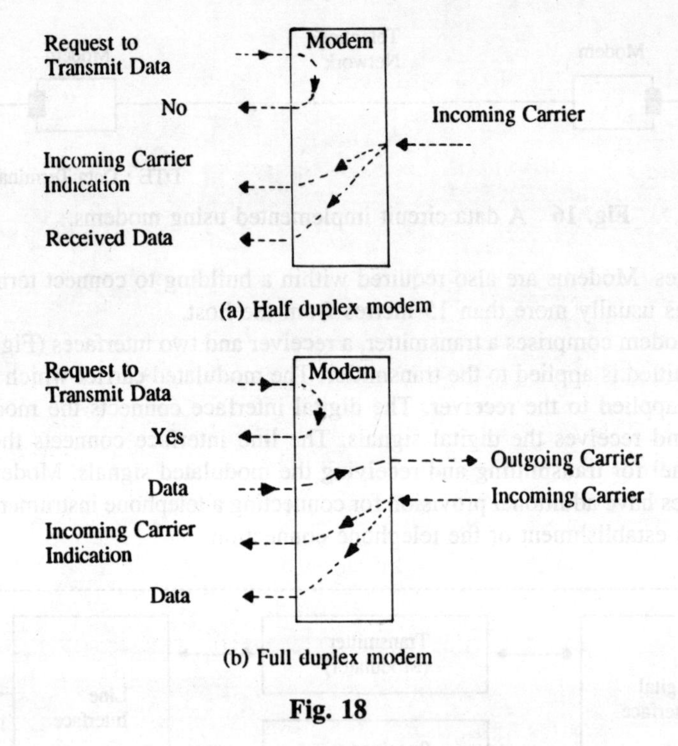

Fig. 18

2W–4W Modems. The line interface of the modem can have a 2-wire or a 4-wire connection to the transmission medium. In a 4-wire connection, one pair of wires is used for the outgoing carrier and the other is used for the incoming carrier (Fig. 19). Full duplex and half duplex modes of data transmission are possible on a 4-wire connection. As the physical transmission path for each direction is separate, the same carrier frequency can be used for both the directions.

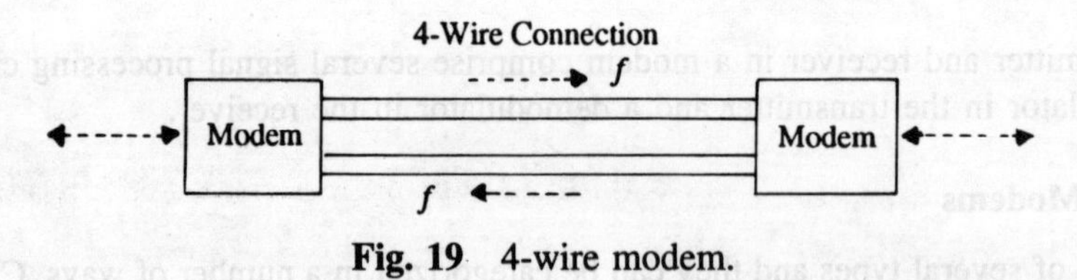

Fig. 19 4-wire modem.

A leased 2-wire connection is cheaper than a 4-wire connection because only one pair of wires is extended to the subscriber's premises. The data connection established through telephone exchanges is also a 2-wire connection. For the 2-wire connections, modems with a 2-wire line interface are required. Such modems use the same pair of wires for outgoing and incoming carriers. Half duplex mode of transmission using the same frequency for the incoming and outgoing carriers can be easily implemented (Fig. 20a). The transmit and receive carrier frequencies can be the same because only one of them is present on the line at a time.

For full duplex mode of operation on a 2-wire connection, it is necessary to have two transmission channels, one for the transmit direction and the other for the receive direction (Fig. 20b). This is achieved by frequency division multiplexing of two different carrier frequencies. These carriers are placed within the bandwidth of the speech channel (Fig. 20c). A modem transmits

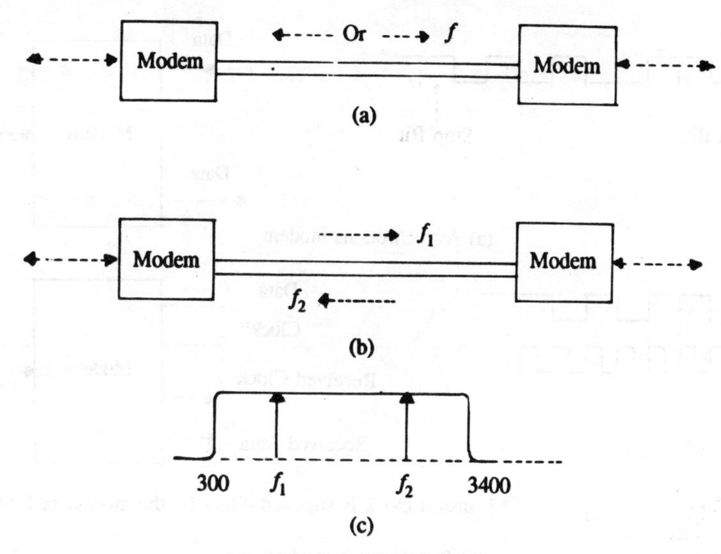

Fig. 20 2-wire modems.

data on one carrier and receives data from the other end on the other carrier. A hybrid is provided in the 2-wire modem to couple the line to its modulator and demodulator (Fig. 21).

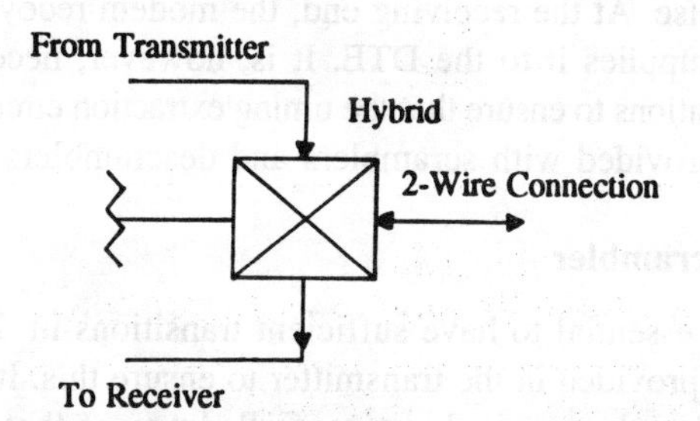

Fig. 21 Line interconnection in a 2-wire full duplex modem.

Note that available bandwidth for each carrier is reduced to half. Therefore, the baud rate is also reduced to half. There is a special technique which allows simultaneous transmission of incoming and outgoing carriers having the same frequency on the 2-wire transmission medium. Full bandwidth of the speech channel is available to both the carriers simultaneously. This technique is called echo cancellation technique and is implemented in high speed 2-wire full duplex modems.

Asynchronous and Synchronous Modems. Modems for asynchronous and synchronous transmission are of different types. An asynchronous modem can only handle data bytes with start and stop bits. There is no separate timing signal or clock between the modem and the DTE (Fig. 22a). The internal timing pulses are synchronized repeatedly to the leading edge of the start pulse.

A synchronous modem can handle a continuous stream of data bits but requires a clock signal (Fig. 22b). The data bits are always synchronized to the clock signal. There are separate clocks for the data bits being transmitted and received.

For synchronous transmission of data bits, the DTE can use its internal clock and supply the

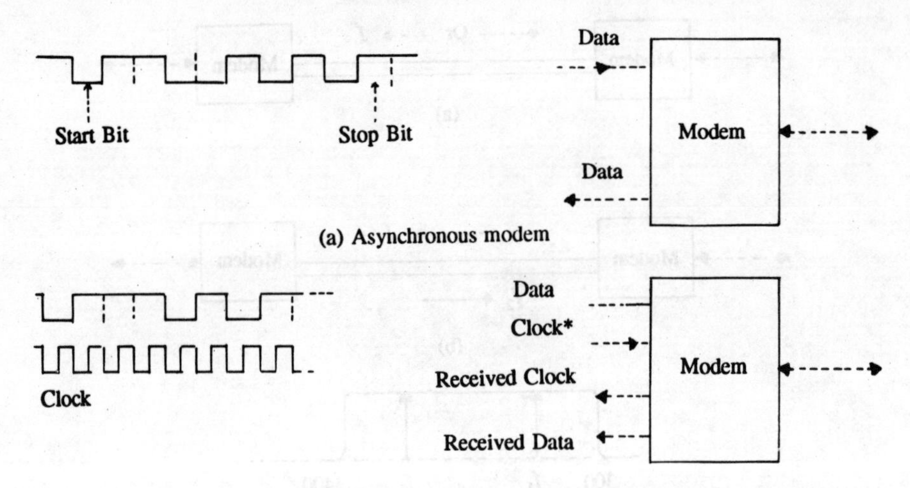

(a) Asynchronous modem

(b) Synchronous modem

Fig. 22

same to the modem. Else, it can take the clock from the modem and send data bits on each occurrence of the clock pulse. At the receiving end, the modem recovers the clock signal from the received data signal and supplies it to the DTE. It is, however, necessary that the received data signal contains enough transitions to ensure that the timing extraction circuit remains in synchronization. High speed modems are provided with scramblers and descramblers for this purpose.

4.2 Scrambler and Descrambler

As mentioned above, it is essential to have sufficient transitions in the transmitted data for clock extraction. A scrambler is provided in the transmitter to ensure this. It uses an algorithm to change the data stream received from the terminal in a controlled way so that a continuous stream of zeros or ones is avoided. The scrambled data is descrambled at the receiving end using a complementary algorithm.

There is another reason for using scramblers. It is often seen in data communications that computers transmit "idle" characters for relatively long periods of time and then there is a sudden burst of data. The effect is seen as repeating errors at the beginning of the data. The reason for these errors is sensitivity of the receiver clock phase to certain data patterns. If the transmission line has poor group delay characteristic in some part of the spectrum and the repeated data pattern concentrates the spectral energy in that part of the spectrum, the recovered clock phase can be offset from its mean position. Drifted clock phase results in errors when the data bits are regenerated. This problem can be overcome by properly equalizing the transmission line but the long term solution is to always randomize the data before it is transmitted so that pattern sensitivity of the clock phase is avoided. The scramblers randomize the data and thus avoid the errors due to pattern sensitivity of the clock phase.

The scrambler at the transmitter consists of a shift register with some feedback loops and exclusive OR gates. Figure 23 shows a scrambler used in the CCITT V.27 4800 bps modem.

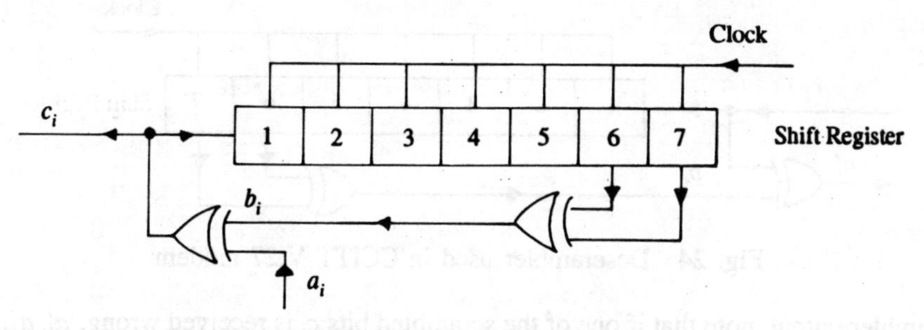

Fig. 23 Scrambler used in CCITT V.27 modem.

For the *i*th pulse, the output c_i can be obtained as

$$b_i = c_{i-6} + c_{i-7}$$

$$c_i = a_i + b_i = a_i + c_{i-6} + c_{i-7}$$

If we represent one-bit delay using a delay operator x^{-1}, the above equation can be rewritten as follows:

$$c_i = a_i + c_i(x^{-6} + x^{-7})$$

$$c_i\,(1 + x^{-6} + x^{-7}) = a_i$$

$$c_i = a_i/(1 + x^{-6} + x^{-7})$$

Note that in modulo-2 arithmetic, addition and subtraction operations are the same. Thus, a scrambler effectively divides the input data stream by polynomial $1 + x^{-6} + x^{-7}$. This polynomial is called the generating polynomial. By proper choice of the polynomial, it can be assured that undesirable bit sequences are avoided at the output. The generating polynomials recommended by CCITT for scramblers are given in Table 4.

Table 4 CCITT Generating Polynomials

CCITT recommendations	Generating polynomial
V.22, V.22*bis*	$1 + x^{-14} + x^{-17}$
V.27	$1 + x^{-6} + x^{-7}$
V.29, V.32	$1 + x^{-18} + x^{-23}$
V.26*ter*	
V.32	$1 + x^{-5} + x^{-23}$

To get back the data sequence at the receiving end, the scrambled data stream is multiplied by the same generating polynomial. The descrambler is shown in Fig. 24.

$$b_i = c_{i-6} + c_{i-7}$$

$$a_i' = c_i + b_i = c_i + c_{i-6} + c_{i-7} = c_i\,(1 + x^{-6} + x^{-7}) = a_i$$

In the above analysis, we have assumed that there was no transmission error. If an error occurs in the scrambled data, it is reflected in three data bits after descrambling. In the expression

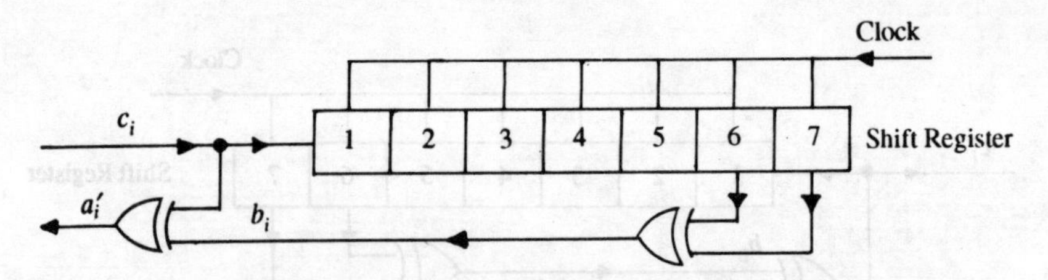

Fig. 24 Descrambler used in CCITT V.27 modem.

for descrambler output, note that if one of the scrambled bits c_i is received wrong, a'_i, a'_{i+6} and a'_{i+7} will be affected as c_i moves along the shift register. Therefore, scramblers result in increased error rate but their usefulness outweighs this limitation.

4.3 Block Schematic of a Modem

With this background, we can now describe the detailed block schematic of a modem. The modem design and complexity vary depending on the bit rate, type of modulation and other basic features as discussed above. Low speed modems upto 1200 bps are asynchronous and use FSK. Medium speed modems from 2400 to 4800 bps use differential PSK. High speed modems which operate at 9600 bps and above employ QAM and are the most complex. Medium and high speed modems operate in synchronous mode of transmission.

Figure 25 shows important components of a typical synchronous differential PSK modem. It must, however, be borne in mind that this design gives the general functional picture of the modem. Actual implementation will vary from vendor to vendor.

Digital Interface. The digital interface connects the internal circuits of the modem to the DTE. On the DTE side, it consists of several wires carrying different signals. These signals are either from the DTE or from the modem. The digital interface contains drivers and receivers for these signals. A brief description of some of the important signals is given below.

- Transmitted Data (TD) signal from the DTE to the modem carries data to be transmitted.
- Received Data (RD) signal from the modem carries the data received from the other end.
- DTE Ready (DTR) signal from the DTE and indicates readiness of the DTE to transmit and receive data.
- Data Set Ready (DSR) signal from the modem indicates its readiness to transmit and receive data signals.
- Request to Send (RTS) signal from the DTE seeks permission of the modem to transmit data.
- Clear to Send (CTS) signal from the modem gives clearance to the DTE to transmit its data. CTS is given as response to the RTS.
- Received line signal detector signal from the modem indicates that the incoming carrier has been detected on the line interface.
- Timing signals are the clock signals from the DTE to the modem and from the modem to the DTE for synchronous transmission.

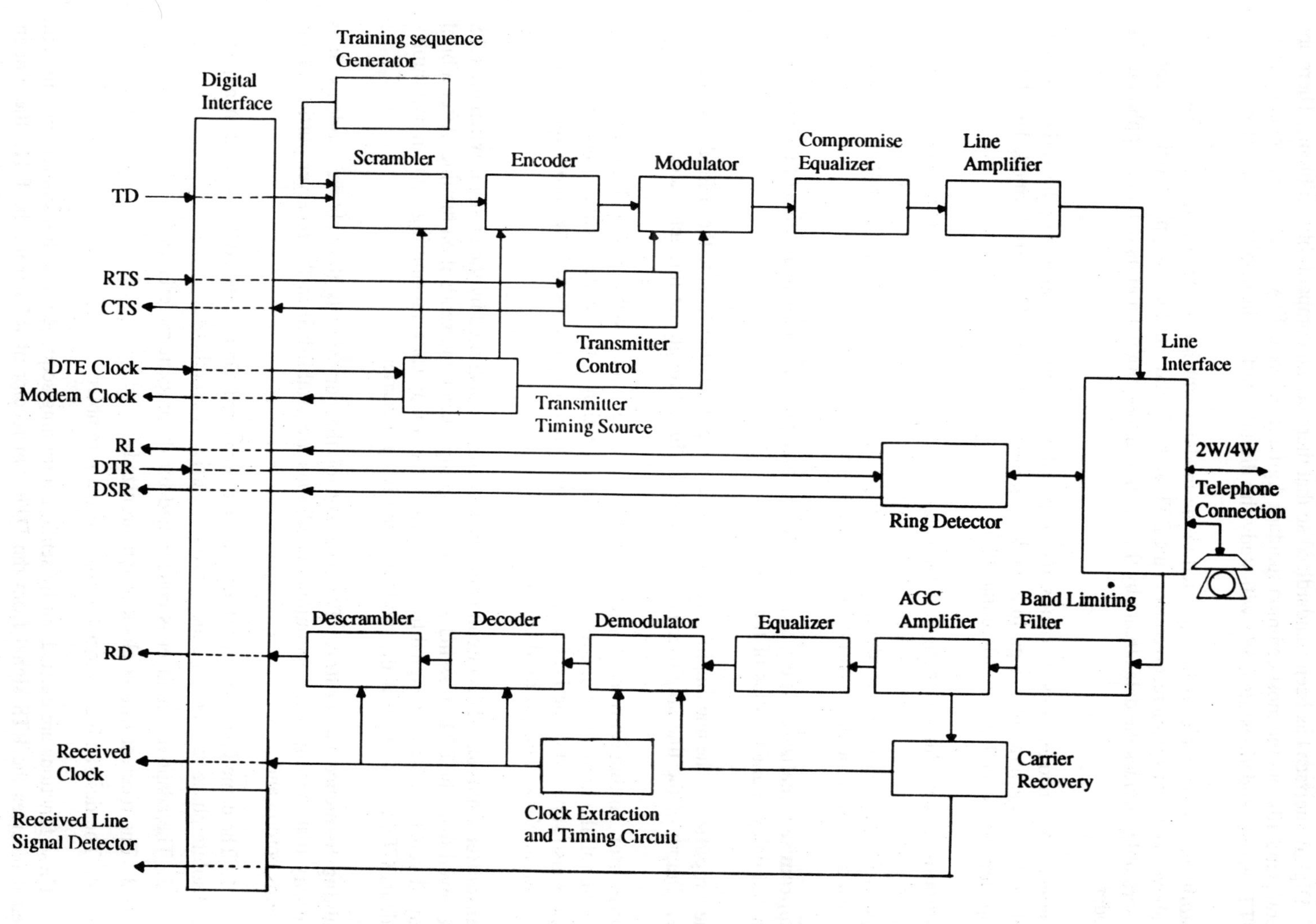

Fig. 25 Components of a differential PSK modem.

Digital interface has been standardized so that there are no compatibility problems. There are several standards, but the most common standard digital interface is EIA 232D. There are equivalent CCITT recommendations also. We will study the digital interface in detail in Chapter 6.

Scrambler. A scrambler is incorporated in the modems which operate at data rates of 4800 bps and above. The data stream received from the DTE at the digital interface is applied to the scrambler. The scrambler divides the data stream by the generating polynomial and its output is applied to the encoder.

Encoder. An encoder consists of a serial to parallel converter for grouping the serial data bits received from the scrambler, e.g., in a modem employing 4 PSK, dibits are formed. The data bit groups are then encoded for differential PSK.

Modulator. A modulator changes the carrier phase as per the output of the encoder. A pulse shaping filter precedes the modulator to reduce the intersymbol interference. Raised cosine pulse shape is usually used. The modulator output is passed through a band pass filter to restrict the bandwidth of the modulated carrier within the specified frequency band.

Compromise Equalizer. It is a fixed equalizer which provides pre-equalization of the anticipated gain and delay characteristics of the line.

Line Amplifier. The line amplifier is provided to bring the carrier level to the desired transmission level. Output of the line amplifier is coupled to the line through the line interface.

Transmitter Timing Source. Synchronous modems have an in-built crystal clock source which generates all the timing references required for the operation of the encoder and the modulator. The clock is also supplied to the DTE through the digital interface. The modem has provision to accept the external clock supplied by the DTE.

Transmitter Control. This circuit controls the carrier transmitted by the modem. When the RTS is received from the DTE, it switches on the outgoing carrier and sends it on the line. After a brief delay, it generates the CTS signal for the DTE so that it may start transmitting data. In half duplex modems CTS is not given if the modem is receiving a carrier.

Training Sequence Generator. For reception of the data signals through the modems, it is necessary that the following operational conditions are established in the receiver portion of the modems beforehand:

1. The demodulator carrier is detected and recovered. Gain of the AGC amplifier is adjusted and absolute phase reference of the recovered carrier is established.
2. The adaptive equalizer is conditioned for the line characteristics.
3. The receiver timing clock is synchronized.
4. The descrambler is synchronized to the scrambler.

These functions are carried out by sending a training sequence. It is transmitted by a modem when it receives the RTS signal from the DTE. On receipt of RTS from the DTE, the modem transmits a carrier modulated with the training sequence of fixed length and then gives the CTS signal to the DTE so that it may commence transmission of its data. From the training sequence, the modem at the receiving end recovers the carrier, establishes its absolute phase reference, conditions its adaptive equalizer and synchronizes its clock and descrambler. The composition of the training sequence depends on the type of the modem. We will examine some of the training sequences while discussing the modem standards later.

Line Interface. The line interface provides connection to the transmission facilities through coupling transformers. The coupling transformers isolate the line for DC signals. The transmission facilities provide a two-wire or four-wire connection between the two modems. For a four-wire connection, there are separate transformers for the transmit and receive directions. For a 2-wire connection, the line interface is equipped with a hybrid.

Receive Band Limiting Filter. In the receive direction, the band limiting filter selects the received carrier from the signals present on the line. It also removes the out-of-band noise.

AGC Amplifier. Automatic Gain Control (AGC) amplifier provides variable gain to compensate for carrier-level loss during transmission. The gain depends on the received carrier level.

Equalizer. The equalizer section of the receiver corrects the attenuation and group delay distortion introduced by the transmission medium and the band limiting filters. Fixed, manually adjustable or adaptive equalizers are provided depending on speed, line condition and the application. In high speed dial up modems, an adaptive equalizer is provided because characteristics of the transmission medium change on each instance of call establishment.

Carrier Recovery Circuit. The carrier is recovered from the AGC amplifier output by this circuit. The recovered carrier is supplied to the demodulator. An indication of the incoming carrier is given at the digital interface.

Demodulator. The demodulator recovers the digital signal from the received modulated carrier. The carrier required for demodulation is supplied by the carrier recovery circuit.

Clock Extraction Circuit. The clock extraction circuit recovers the clock from the received digital signal. The clock is used for regenerating the digital signal and to provide the timing information to the decoder. The receiver clock is also made available to the DTE through the digital interface.

Decoder. The decoder performs a function complementary to the encoder. The demodulated data bits are converted into groups of data bits which are serialized by using a parallel to serial converter.

Descrambler. The decoder output is applied to the descrambler which multiplies the decoder output by the generating polynomial. The unscrambled data is given to the DTE through the digital interface.

4.4 Additional Modem Features

As mentioned above, modems vary in design and complexity depending on speed, mode of transmission, modulation methods and their application. The driving force for the developments in modems has been the high cost of the transmission medium. By more efficient utilization of the available bandwidth and increasing the effective throughput, the high cost of transmission can be neutralized. Echo cancellers and secondary channel are the two additional features of modems in this direction. For ease of operation, modems are also equipped with test loops. We will take a brief look at these features of modems also.

Echo Canceller. Full duplex transmission of data on 2-wire leased or dial up connection is implemented by dividing the available frequency band for the two carriers. This effectively reduces the available bandwidth for each carrier to half and limits the data speed to about 2400 to 4800 bps. Echo cancellation makes it possible to use the same carrier frequency and the entire frequency band for both the carriers simultaneously.

Transmit and receive carrier frequencies being the same, it becomes essential for the transmitted carrier not to appear at the local receiver input. The line-coupling hybrid gives about 15 dB loss across the opposite ports. Thus the transmitted carrier with 15 dB loss appears at the receiver input of the modem. This signal is referred to as near-end echo (Fig. 26). It has high amplitude and very short delay.

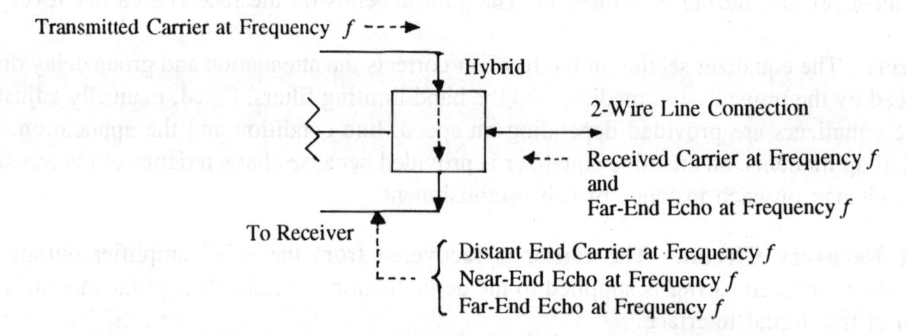

Fig. 26 Echoes present in a 2-wire full duplex modem.

There is another type of echo which is called the far-end echo. Far-end echo is caused by the hybrids present in the interconnecting telecommunication link. It is characterized by low amplitude but long delay. For terrestrial connections, the delay can be of the order 40 ms and for the satellite based connections, it is of the order of half a second.

The echo being at the same carrier frequency as the received carrier, interferes with the demodulation process and needs to be removed. For this purpose, an echo canceller is built into the high-speed modems. It generates a copy of the echo from the transmitted carrier and subtracts it from the received signals (Fig. 27).

The echo canceller circuit consists of a tapped-delay line with a set of coefficients which are adjusted to get the minimum echo at the receiver input. This adjustment is carried out when the training sequence is being transmitted.

Secondary Channel. We have seen that a DTE needs to exchange RTS/CTS signals with the modem before it transmits data. On receipt of the RTS signal, the modem gives the CTS after a certain delay. During this period, it transmits the training sequence so that the modem at the other end may detect the carrier, extract the clock, synchronize the descrambler and condition the equalizers.

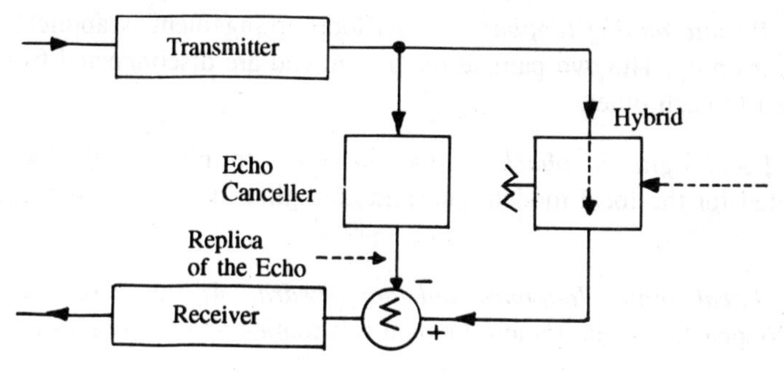

Fig. 27 Echo canceller.

If the mode of operation is half duplex, each reversal of the direction of transmission involves RTS-CTS delay and thus, reduces the effective throughput. In most of the data communication situations, the receiver sends short acknowledgements for every received data frame and for transmitting these acknowledgements the direction of transmission must be reversed. To avoid frequent reversal of direction of transmission, a low speed secondary channel is provided in the modems (Fig. 28).

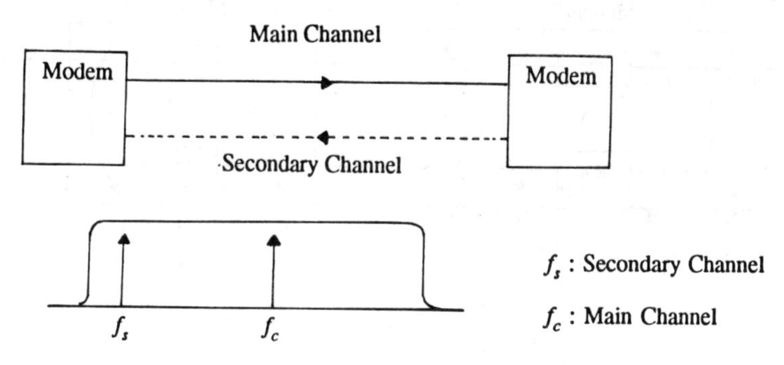

Fig. 28 Secondary channel.

The secondary channel operates at 75 bps and uses FSK. The secondary channel has its own RTS, CTS and other control signals which are available at the digital interface of the modem. It should be noted that the main channel is used in half duplex mode for data transmission and the DTEs are configured to send the acknowledgements on the secondary channel.

Test Loops. Modems are provided with the capability for locating faults in the digital connection from DTE to DTE. The testing procedure involves sending a test data and looping it back at various stages of the connection. The test pattern can be generated by the modem internally or it can be applied externally using modem tester. The common test configurations are shown in Fig. 29.

- *Loop 1: Digital loopback.* This loop is set up as close as possible to the digital interface.
- *Loop 2: Remote digital loopback.* This loop checks the line and the remote modem. It can be used only in full duplex modems.
- *Loop 3: Local analog loopback.* The modulated carrier at the transmitter output of the local modem is looped back to the receiver input. The loopback may require some attenuators to adjust the level.

- *Loop 4: Remote analog loopback.* This loop arrangement is applicable for 4-wire line connections only. The two pairs at the distant end are disconnected from the modem and connected to each other.

- *Loop 5: Local digital loopback and loopforward.* In this case, the local digital loopback is provided for the local modem and remote digital loopback is provided for the remote modem.

- *Loop 6: Local analog loopback and loopforward.* In this case, the local modem has analog loopback and the remote modem has remote analog loopback.

The test configurations can be set up by pressing the appropriate switches provided on the modems. The digital interface also provides some control signals for activating the loop tests. When in the test mode, the modem indicates its test status to the local DTE through a control signal in the digital interface.

All modems do not have provision for all these tests. Test features are specific to the modem type. Test loops 1 to 4 have been standardized by CCITT in their Recommendation V.54.

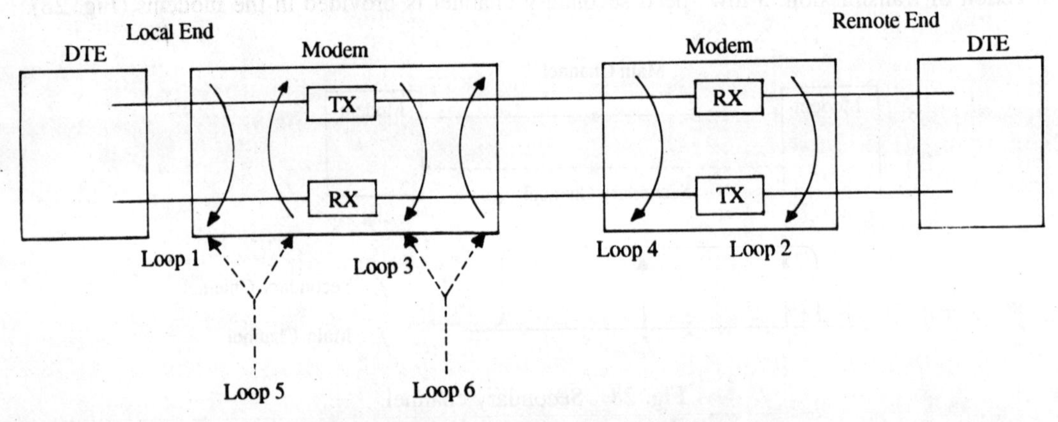

Fig. 29 Test loops in modems.

5 STANDARD MODEMS

It is essential that modems conform to international standards because similar modems supplied by different vendors must work with each other. CCITT has drawn up modem standards which are internationally accepted. We will discuss the main features of the CCITT modems. The reader is urged to refer to the CCITT recommendations for detailed description of these modems.

5.1 CCITT V.21 Modem

This modem is designed to provide full duplex asynchronous transmission over the 2-wire leased line or switched telephone network. It operates at 300 bps.

Modulation. It utilises FSK over the following two channels:

1. Transmit channel frequencies (originating modem)

   Space 1180 Hz, Mark 980 Hz.

2. Receive channel frequencies (originating modem)

Space 1850 Hz, Mark 1650 Hz.

The channel selection for the transmit and receive directions can be done through the digital interface by switching on the appropriate control circuit.

5.2 CCITT V.22 Modem

This modem provides full duplex synchronous transmission† over 2-wire leased line or switched telephone network. It transmits data at 1200 bps. As an option, it can also operate at 600 bps.

Scrambler. A scrambler and a descrambler having the generating polynomial $1 + x^{-14} + x^{-17}$ are provided in the modem.

Modulation. Differential 4 PSK over two channels is utilised in this modem. The dibits are encoded as phase changes as given in Table 5. The carrier frequencies are

Low channel 1200 Hz
High channel 2400 Hz

Table 5 Modulation Scheme of CCITT V.22 Modem

A	B	$\Delta\phi$
0	0	$\pi/2$
0	1	0
1	1	$3\pi/2$
1	0	π

At 600 bps, the carrier phase changes are $3\pi/2$ and $\pi/2$ for binary "1" and "0" respectively.

Equalizer. Fixed compromise equalizers shared equally between the transmitter and receiver are provided in the modem.

Test Loops. Test loops 2 and 3 as defined in Recommendation V.54 are provided in the modem. For self-test, an internally generated binary pattern of alternating "0"s and "1"s is applied to the scrambler. At the output of the descrambler, an error detector identifies the errors and gives visual indication.

5.3 CCITT V.22*bis* Modem

This modem provides full duplex synchronous transmission† on a 2-wire leased line or switched telephone network. The bit rates supported are 2400 or 1200 bps at the modulation rate of 600 bauds.

†As an alternative, the recommendation provides for accepting from the terminal asynchronous data which is suitably converted for transmission synchronously. The internal clock of the modem is used for this purpose.

Scrambler. The modem incorporates a scrambler and a descrambler having the generating polynomial $1 + x^{-14} + x^{-17}$.

Modulation. At 2400 bps, the modem uses 16 QAM having a constellation as shown in Fig. 30. From the scrambled data stream quadbits are formed. The first two bits of the quadbits are coded as quadrant change as given in Table 6. The last two bits of the quadbits determine the phase within a quadrant as shown in Fig. 30.

Table 6 Quadrant Changes Determined by the First Two Bits of Quadbits (CCITT V.22*bis* Modem)

		First two bits of quadbits			
		00	01	11	10	
Last quadrant	1	2	1	4	3	Next quadrant
	2	3	2	1	4	
	3	4	3	2	1	
	4	1	4	3	2	

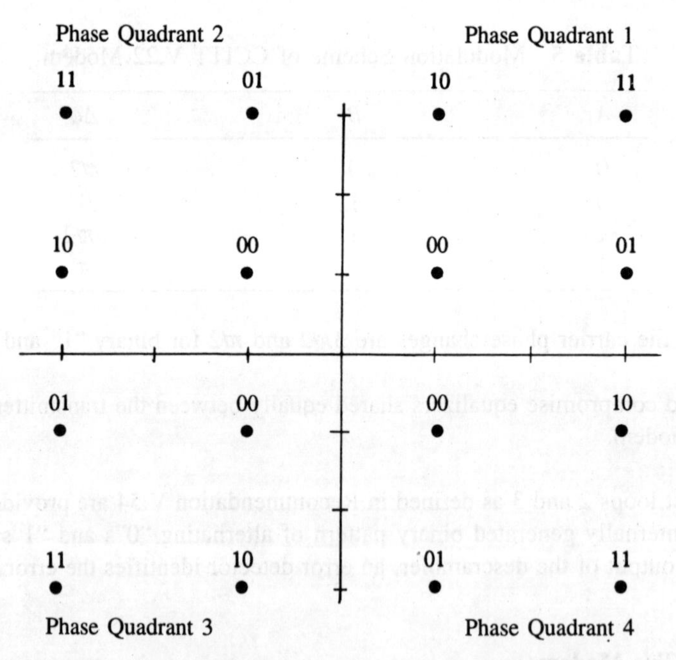

Fig. 30 Phase states of CCITT V.22*bis* 16 QAM modem.

At 1200 bps, the dibits are formed from the scrambled data stream and coded as quadrant change as shown above. In each quadrant, the phase state corresponding to "01" is transmitted.

The following two carriers are used for transmit and receive directions. The calling modem uses the low channel to transmit data.

Low channel carrier	1200 Hz
High channel carrier	1800 Hz

Equalizer. A fixed compromise equalizer is provided in the modem transmitter. The modem receiver is equipped with an adaptive equalizer.

Test Loops. Test loops 2 and 3 as defined in Recommendation V.54 are provided in the modem. For self-test, an internally generated binary pattern of alternating "0"s and "1"s is applied to the scrambler. At the output of the descrambler, an error detector identifies the errors and gives visual indication.

5.4 CCITT V.23 Modem

The modem is designed to operate in full duplex asynchronous transmission mode over a 4-wire leased line. It can also operate in half duplex over a 2-wire leased line and switched telephone network.

The modem can operate at two speeds—600 bps and 1200 bps. It is equipped with the secondary channel which operates at 75 bps.

Modulation. The modem employs FSK over two channels. The frequencies are:

Transmit frequencies (originating modem)

Space 1180 Hz, Mark 980 Hz

Receive frequencies (originating modem)

Space 1850 Hz, Mark 1650 Hz

Secondary channel frequencies

Space 450 Hz, Mark 390 Hz

5.5 CCITT V.26 Modem

This modem operates in full duplex synchronous mode of transmission on a 4-wire leased connection. It operates at 2400 bps. It also includes a secondary channel having a bit rate of 75 bps.

Modulation. Differential 4 PSK is employed to transmit data at 2400 bps. The carrier frequency is 1800 Hz. The modulation scheme has two alternatives *A* and *B* (Table 7). The secondary channel frequencies are the same as in V.23.

Table 7 Modulation Scheme of CCITT V.26 Modem

	A	*B*
Dibit	$\Delta\phi$	$\Delta\phi$
00	0	$\pi/4$
01	$\pi/2$	$3\pi/4$
11	π	$5\pi/4$
10	$3\pi/2$	$7\pi/4$

5.6 CCITT V.26*bis* Modem

It is a half duplex synchronous modem for use in the switched telephone network. It operates at a nominal speed of 2400 bps or at a reduced speed of 1200 bps. It includes a secondary channel which operates at the speed of 75 bps.

Modulation. The modem uses the differential 4 PSK for transmission at 2400 bps. The modulation scheme is the same as for V.26, alternative *B*. At 1200 bps, the modem uses differential BPSK with phase changes $\pi/2$ and $3\pi/2$ for binary "0" and "1" respectively. The frequencies of the secondary channel are the same as in V.23.

Equalizer. A fixed compromise equalizer is provided in the receiver.

5.7 CCITT V.26*ter* Modem

It is a full duplex synchronous modem for use in 2-wire leased line or switched telephone network. It uses an echo cancellation technique for channel separation. As an option, the modem can accept asynchronous data from the DTE. If asynchronous option is used, the modem converts the asynchronous data suitably for synchronous transmission. The modem operates at a nominal speed of 2400 bps with fall-back at 1200 bps.

Modulation. The modem uses differential 4 PSK for transmission at 2400 bps. The carrier frequency is 1800 Hz in both directions. The modulation scheme is the same as for V.26, alternative *A*. At 1200 bps, differential BPSK is used. The phase changes corresponding to binary "0" and "1" are respectively 0 and π radians respectively.

Equalizer. A fixed compromise equalizer or an adaptive equalizer is provided in the receiver. No training sequence is provided for convergence of the adaptive equalizer.

Scrambler. The modem incorporates a scrambler and a descrambler. The generating polynomial for the call-originating modem is $1 + x^{-18} + x^{-23}$. The generating polynomial of the answering modem for transmission of its data is $1 + x^{-5} + x^{-23}$.

Test Loops. Test loops 2 and 3 as defined in Recommendation V.54 are provided in the modem.

5.8 CCITT V.27 MODEM

This modem is designed for full duplex/half duplex synchronous transmission over a 4-wire or 2-wire leased connection which is specially conditioned as per M.1020. It operates at the bit rate of 4800 bps with modulation rate of 1600 baud. It includes a secondary channel which operates at 75 bps.

Scrambler. The modem incorporates a scrambler and a descrambler having the generating polynomial $1 + x^{-6} + x^{-7}$.

Modulation. The modem uses differential 8 PSK for transmission at 4800 bps. The modulation scheme is given in Table 8. The carrier frequency is 1800 Hz. The secondary channel is the same as in V.23.

Table 8 Modulation Scheme of CCITT V.27 Modem

Tribit values	Phase change
001	0
000	$\pi/4$
010	$\pi/2$
011	$3\pi/4$
111	π
110	$5\pi/4$
100	$3\pi/2$
101	$7\pi/4$

Equalizer. A manually adjustable equalizer is provided in the receiver. The transmitter has provision to send scrambled continuous binary "1"s for the equalizer adjustment. The modem has means for indicating correct adjustment of the equalizer.

5.9 CCITT V.27*bis* Modem

This modem is designed for full duplex/half duplex synchronous transmission over 4-wire/2-wire leased connection not necessarily conditioned as per M.1020. Its speed, modulation scheme and other features are the same as in V.27. The principal differences are given below:

1. It can operate at a reduced rate of 2400 bps. At 2400 bps, the modem uses differential 4 PSK. The modulation scheme is the same as in V.26, alternative *A*.
2. An automatic adaptive equalizer is provided in the receiver.
3. A training sequence generator is incorporated in the transmitter.

The training sequence used in V.27*bis* modem is shown in Table 9. It comprises three segments whose durations have been expressed in terms of Symbol Intervals (SI). One SI is equal to 1/baud rate. The figures shown within brackets are for the 2-wire connection and for the 4-wire connection worse than M.1020 conditioning.

Table 9 Training Sequence of CCITT V.27*bis* Modem

	Segment 1	Segment 2	Segment 3
Duration (SI)	14(58)	58(1074)	8
Type of line signal	Continuous 180° phase reversals	Differential BPSK carrier	Differential 8/4 PSK carrier

The first segment consists of continuous phase reversals of the carrier. It enables AGC convergence and carrier recovery. During the second segment, the adaptive equalizer is conditioned. Differential BPSK carrier is transmitted during this interval. The modulating sequence is generated from every third bit of a PRBS having the generating polynomial $1 + x^{-6} + x^{-7}$. The phase changes in the carrier are 0 and π radians for binary "0" and "1" respectively. The third segment of the training sequence synchronizes the descrambler. It consists of scrambled binary "1"s.

5.10 CCITT V.27*ter* Modem

This modem is designed for use in the switched telephone network. It is similar to V.27*bis* modem in most respects. It incorporates additional circuits for auto answering, ring indicator etc.

5.11 CCITT V.29 Modem

This modem is designed for point-to-point full duplex/half duplex synchronous operation on 4-wire leased circuits conditioned as per M.1020 or M.1025. It operates at a nominal speed of 9600 bps. The fall-back speeds are 7200 and 4800 bps.

Scrambler. The modem incorporates a scrambler and a descrambler having the generating polynomial $1 + x^{-18} + x^{-23}$.

Modulation. The modem employs 16 state QAM with modulation rate of 2400 baud. The carrier frequency is 1700 Hz. The scrambled data at 9600 bps is divided into quadbits. The last three bits are coded to generate differential eight-phase modulation identical to Recommendation V.27. The first bit along with the absolute phase of the carrier determines its amplitude (Fig. 31). The absolute phase is established during transmission of the training sequence.

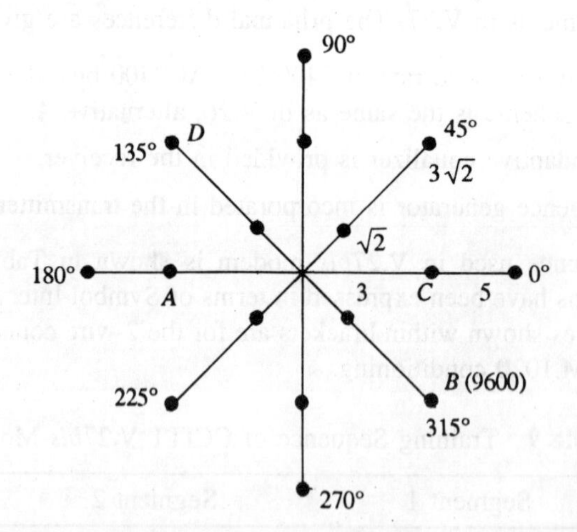

Fig. 31 Phase states of CCITT V.29 16 QAM modem at 9600 bps.

At the fallback rate of 7200 bps, tribits are formed from the scrambled 7200 bps bit stream. Each tribit is prefixed with a zero to make the quadbit. At the fallback rate of 4800 bps, dibits are formed from the scrambled 4800 bps bit stream. These dibits constitute the second and third bits of the quadbits. The first bit of the quadbits is zero as before and the fourth bit is modulo 2 sum of the second and third bits. The phase state diagrams for the modem operation at 7200 and 4800 bps are shown in Fig. 32a and Fig. 32b respectively.

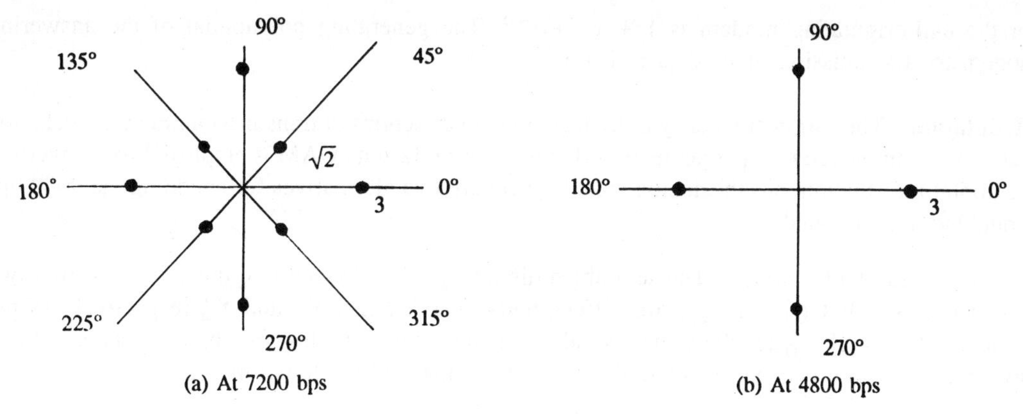

Fig. 32 Phase states of CCITT V.29 modem.

Equalizer. An adaptive equalizer is provided in the receiver.

Training Sequence. The training sequence is shown in Table 10. It consists of four segments which provide for clock synchronization, establishment of absolute phase reference for the carrier, equalizer conditioning and descrambler synchronization.

Table 10 Training Sequence of CCITT V.29 Modem

Segment	Signal type	Duration (Symbol intervals)
1	No transmitted energy	48
2	Alternations	128
3	Equalizer conditioning pattern	384
4	Scrambled binary 1s	48

The second segment consists of two alternating signal elements *A* and *B* (Fig. 31). This sequence establishes absolute phase of the carrier.

The third segment consists of the equalizer conditioning signal which consists of elements *C* and *D* (Fig. 31). Whether *C* or *D* is to be transmitted is decided by a pseudo-random binary sequence at 2400 bps generated using the generating polynomial $1 + x^{-6} + x^{-7}$. The element *C* is transmitted when a "0" occurs in the sequence. The element *D* is transmitted when a "1" occurs in the sequence.

The fourth segment consists of a continuous stream of binary "1"s which is scrambled and transmitted. During this period descrambler synchronization is achieved.

5.12 CCITT V.32 Modem

This modem is designed for full duplex synchronous transmission on 2-wire leased line or switched telephone network. It can operate at 9600 and 4800 bps. The modulation rate is 2400 bauds.

Scrambler. The modem incorporates a scrambler and a descrambler. The generating polynomial for the call-originating modem is $1 + x^{-18} + x^{-23}$. The generating polynomial of the answering modem for transmission of its data is $1 + x^{-5} + x^{-23}$.

Modulation. The carrier frequency is 1800 Hz in both directions of transmission. Echo cancellation technique is employed to separate the two channels. 16 or 32 state QAM is employed for converting the digital information into the analog signal. There are two alternatives for encoding the 9600 bps scrambled digital signal.

Nonredundant Coding. The scrambled digital signal is divided into quadbits. The first two bits of each quadbit Q_{1n} and Q_{2n} are differentially encoded into Y_{1n} and Y_{2n} respectively as per Table 11. $Y_{1(n-1)}$, $Y_{2(n-1)}$ are the previous values of the Y bits. The last two bits are taken without any change and the encoded quadbit $Y_{1n}Y_{2n}Q_{3n}Q_{4n}$ is mapped as shown in Fig. 33.

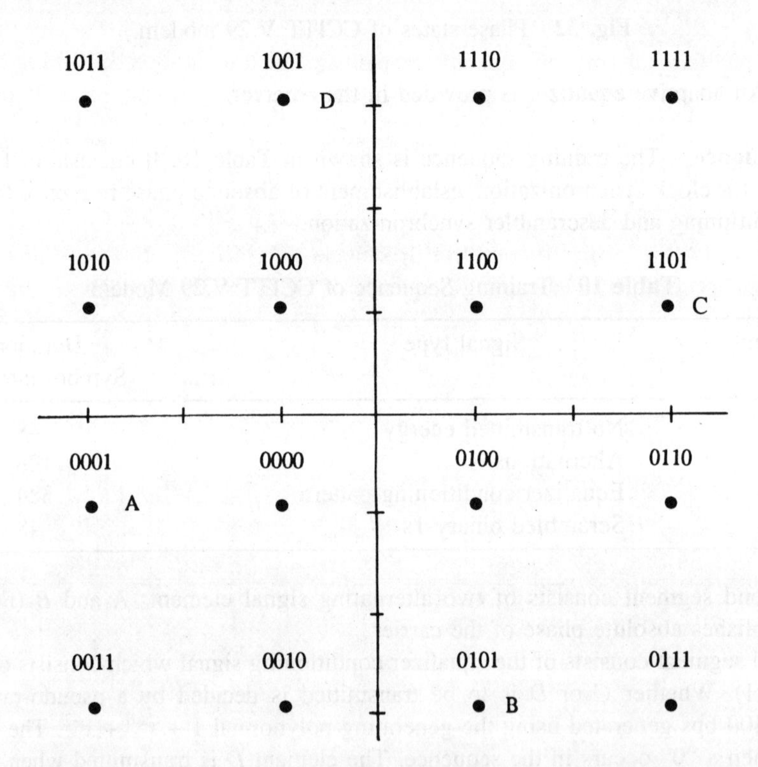

Fig. 33 Phase states of CCITT V.32 modem at 9600 bps when non-redundant coding is used.

At 4800 bps, the scrambled data stream is grouped into dibits which are differentially encoded as per Table 11 and mapped on a subset ABCD of the phasor states (Fig. 33).

Trellis Coding. Trellis coding enables detection and correction of errors which are introduced in the transmission medium. We will study the principles of error control using trellis coding in the next chapter. Here, suffice it to say that some additional bits are added to a group of data bits for detecting and correcting the errors. There are several coding algorithms for error control and trellis coding is one of them. It is implemented using convolution encoders.

Table 11 Differential Encoding Scheme of the First Two Bits of Quadbits (CCITT V.32 Modem)

	$Q_{1n}Q_{2n}$ 00	$Q_{1n}Q_{2n}$ 01	$Q_{1n}Q_{2n}$ 10	$Q_{1n}Q_{2n}$ 11
Y_1Y_2 $(n-1)$	$Y_{1n}Y_{2n}$	$Y_{1n}Y_{2n}$	$Y_{1n}Y_{2n}$	$Y_{1n}Y_{2n}$
00	01	00	11	10
01	11	01	10	00
10	00	10	01	11
11	10	11	00	01

In trellis coded V.32 modem, quadbits formed from the scrambled data stream are converted into groups of five bits using a convolution encoder. The coding scheme is as under:

- The first two bits Q_{1n} and Q_{2n} of the quadbit are differentially encoded into Y_{1n} and Y_{2n} as given in Table 12.
- From Y_{1n} and Y_{2n}, Y_{0n} is generated using the convolution encoder.
- Y_{0n}, Y_{1n} and Y_{2n} form the first three bits of the five bit code. The last bits of the code are Q_{3n} and Q_{4n} bits of the quadbit.

Table 12 Differential Encoding Scheme of the First Two Bits of Quadbits (CCITT V.32 Trellis Coded Modem)

	$Q_{1n}Q_{2n}$ 00	$Q_{1n}Q_{2n}$ 01	$Q_{1n}Q_{2n}$ 10	$Q_{1n}Q_{2n}$ 11
Y_1Y_2 $(n-1)$	$Y_{1n}Y_{2n}$	$Y_{1n}Y_{2n}$	$Y_{1n}Y_{2n}$	$Y_{1n}Y_{2n}$
00	00	01	10	11
01	01	00	11	10
10	10	11	01	00
11	11	10	00	01

The phase state diagram of the V.32 trellis coded modem is shown in Fig. 34.

Equalizer. An adaptive equalizer is provided in the receiver.

Training Sequence. A training sequence is provided in the modem for adaptive equalization, echo cancellation, data rate selection, and for the other functions described earlier. It consists of the following five segments:

1. Alterations between states A and B (Fig. 34) for 256 symbol intervals.
2. Alterations between states C and D (Fig. 34) for 16 symbol intervals.
3. Equalizer and echo canceller conditioning signal of 1280 symbol intervals.

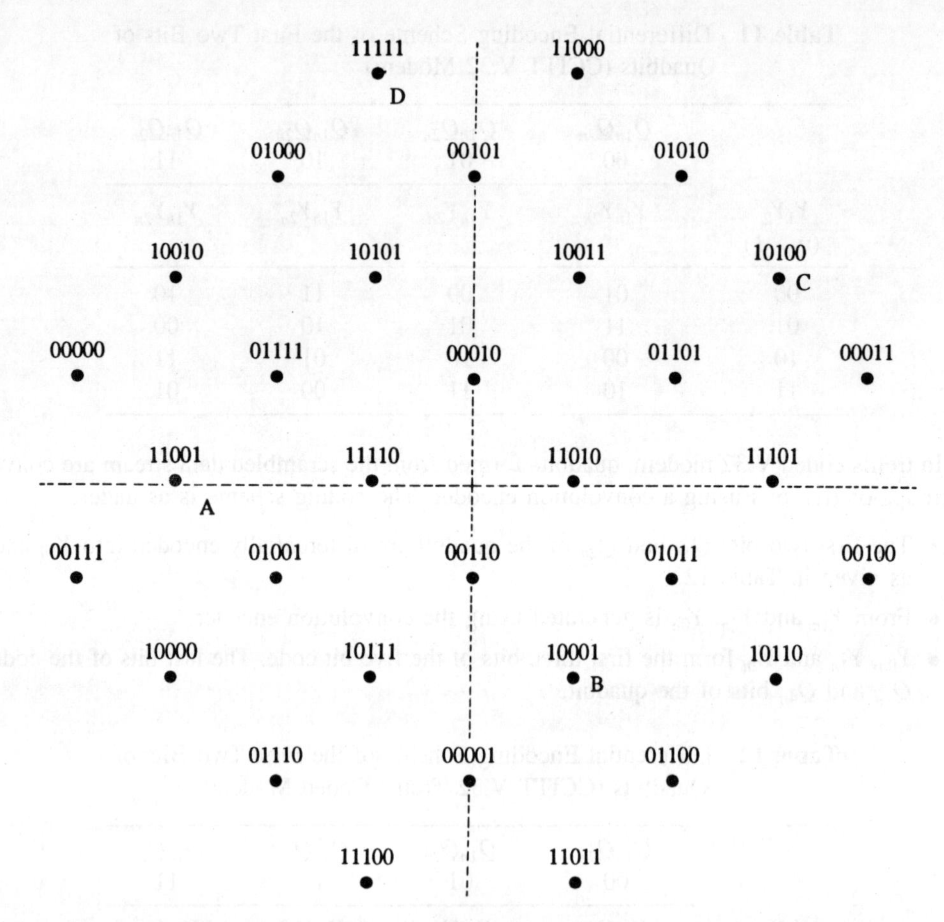

Fig. 34 Phase states of CCITT V.32 modem at 9600 bps when trellis coding is used.

4. Data rate indicating sequence which is delimited by a rate signal ending sequence of eight symbol intervals.

5. Sequence of scrambled binary "1"s of 128 symbol intervals.

Test Loops. Test loops 2 and 3 as defined in Recommendation V.54 are provided in the modem.

5.13 CCITT V.33 Modem

This modem is designed for full duplex synchronous transmission on 4-wire leased connections conditioned as per M.1020 or M.1025. It operates at 14,400 bps with modulation rate of 2400 bauds. The fallback speed is 12,000 bps.

Scrambler. The modem incorporates a scrambler and a descrambler. The generating polynomial for the call-originating modem is $1 + x^{-18} + x^{-23}$.

Modulation. The carrier frequency is 1800 Hz in both directions of transmission. 128 state QAM using trellis coding is employed for converting the digital information into an analog signal. The scrambled data bits are divided into groups of six bits. The first two bits of each six-bit group are encoded into three bits using the differential encoder followed by a convolution encoder as described in V.32. Seven bit code words are thus formed and these codes are mapped on the 128 state phase diagram as shown in Fig. 35.

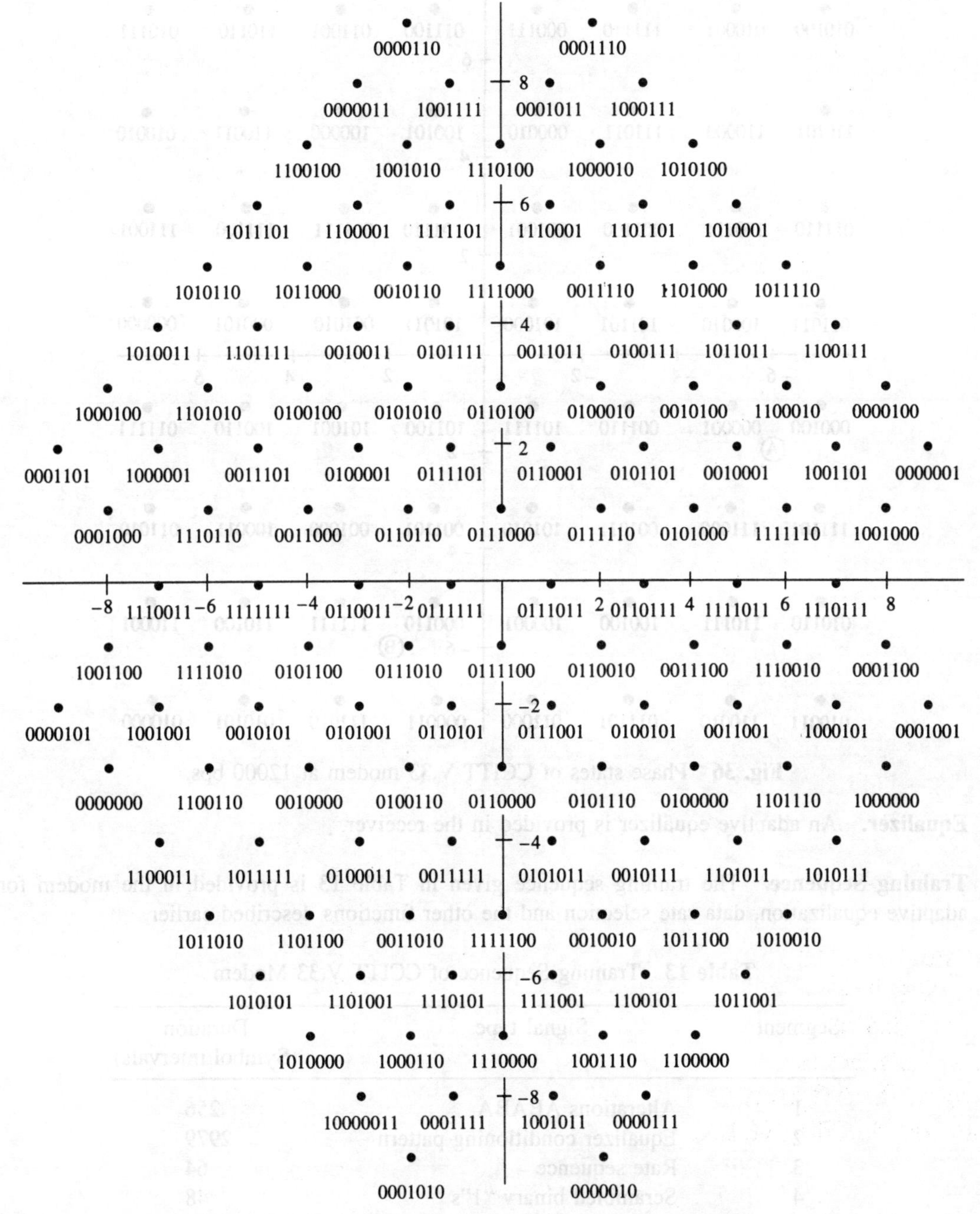

Fig. 35 Phase states of CCITT V.33 modem at 14400 bps.

At the fallback speed of 12,000 bps, five-bit groups are formed and the first two bits of each group are coded into three bits using the same scheme as above. The six-bit codes so generated are mapped as shown in Fig. 36.

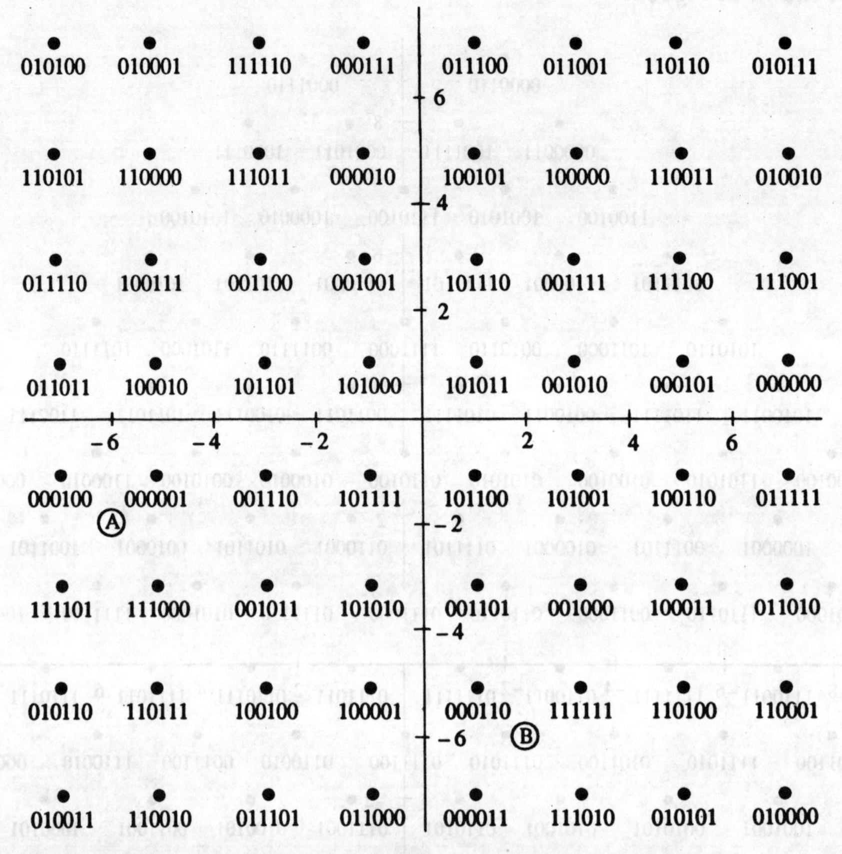

Fig. 36 Phase states of CCITT V.33 modem at 12000 bps.

Equalizer. An adaptive equalizer is provided in the receiver. .

Training Sequence. The training sequence given in Table 13 is provided in the modem for adaptive equalization, data rate selection and the other functions described earlier.

Table 13 Training Sequence of CCITT V.33 Modem

Segment	Signal type	Duration (Symbol intervals)
1	Alterations ABABA	256
2	Equalizer conditioning pattern	2979
3.	Rate sequence	64
4	Scrambled binary "1"s	48

States A and B are shown in the phase state diagrams. For details of the training sequence, the reader is advised to refer to the CCITT recommendation.

6 LIMITED DISTANCE MODEMS AND LINE DRIVERS

The CCITT modems discussed above are designed to operate on the speech channel of 300 to 3400 Hz provided by the telecommunication network. Filters are provided in the network to restrict the bandwidth to this value primarily to pack more channels on the transmission media. The copper pair as such provides much wider frequency pass band as we saw in the last chapter. Limited-distance Modems (LDM) are designed for the entire frequency band of the non-loaded copper transmission line. Their application is limited to short distances as the media distortions and attenuation increase with the distance. The distance limitation is, of course, a function of bit rate and cable characteristics. The longer the distance, the slower must the transmission speed be because sophisticated equalization techniques required for long distance operation are not provided in the LDMs. Some typical figures are 20 kilometres at 1200 bps and 8 kilometres at 19,200 bps on 26-gauge cable. LDMs usually require 4-wire unloaded connection between modems.

Another class of modems which fall under the category of LDMs are the baseband modems. A baseband modem does not have a modulator and demodulator and utilizes digital baseband transmission. It has the usual interfaces and other circuits including equalizers to compensate for the transmission distortions of the line.

Line drivers as modem substitutes provide transmission capabilities usually limited to within buildings where the terminals are separated from the host at distances which cannot be supported by the digital interface. A line driver converts the digital signal to low-impedance balanced signal which can be transmitted over a twisted pair. For the incoming signals, a line driver also incorporates a balanced line receiver. Line drivers usually require DC continuity of the transmission medium.

7 GROUP BAND MODEMS

We have so far concentrated on data modems designed to operate in the frequency band, 300 to 3400 Hz. Use of such modems is restricted to 19,200 bps primarily due to the bandwidth limitations. The telecommunication network also provides group band service which extends from 60 kHz to 108 kHz. The modems designed to operate over this frequency band are called group band modems. Basic features of the CCITT V.36 group band modem are as follows:

- This modem provides synchronous transmission at bit rates 48, 56, 64 and 72 kbps.
- Single sideband amplitude modulation of carrier at 100 kHz is used. The carrier at 100 kHz is also transmitted along with the modulated signal.
- The modem has provision for injecting external group reference pilot at 104.08 kHz.
- An optional speech channel occupying the frequency band 104 to 108 kHz is integrated into the modem.
- The modem incorporates a scrambler and a descrambler.

For bit rates higher than 72 kHz, CCITT has specified the V.37 group band modem. It supports 96 kbps, 112 kbps, 128 kbps and 144 kbps bit rates.

8 DATA MULTIPLEXERS

A modem is an intermediary device which is used for interconnecting terminals and computers when the distances involved are large. Another data transmission intermediary device is the data multiplexer which allows sharing of the transmission media. Multiplexing is adopted to reduce the cost of transmission media and modems. Figure 37 shows a simple application of data multiplexers. In the first option, 16 modems and eight leased lines are required for connecting eight terminals to the host. In the second option, the terminals and the host are connected using two data multiplexers. The modem requirement is reduced to two and the leased line requirement is reduced to one.

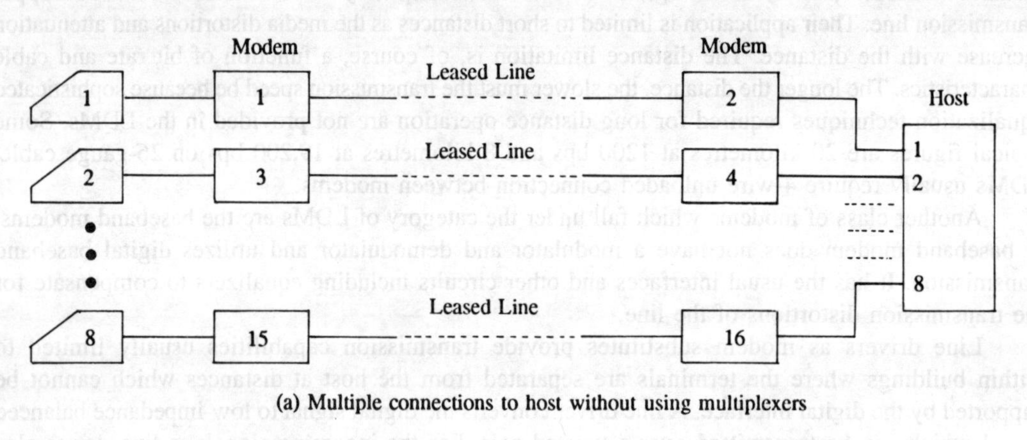

(a) Multiple connections to host without using multiplexers

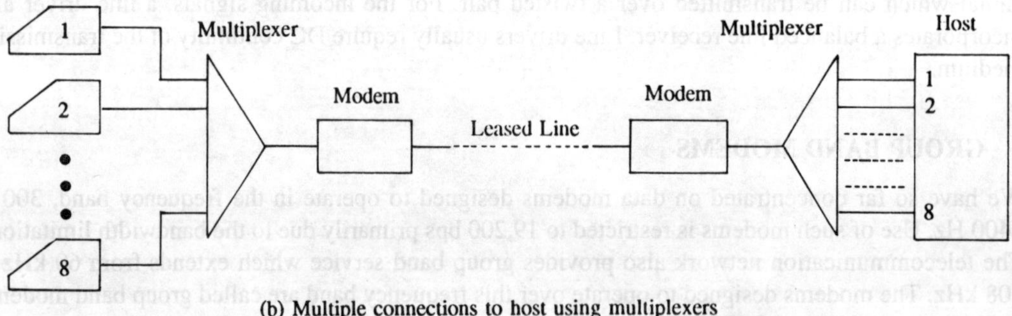

(b) Multiple connections to host using multiplexers

Fig. 37 Use of multiplexers for sharing media and modems.

The multiplexer ports which are connected to the terminals are called terminal ports and the port connected to the leased line is called the line port. A multiplexer has a built-in demultiplexer also for the signals coming from the other end. The terminal port for incoming and outgoing signals is the same. One of the several wires of the terminal port carries the outgoing signal and another carries the incoming signal.

Besides consideration of economy, the other benefit of multiplexing is centralized monitoring of all the channels. Data multiplexers can be equipped with diagnostic hardware/software for monitoring the performance of individual data channels. However, there is possibility of catastrophic failure. If any of the multiplexers or the leased line fails, all the terminals will be cut off from the host.

8.1 Types of Data Multiplexers

Like speech channel multiplexing, data multiplexers use either frequency division multiplexing (FDM) or time division multiplexing (TDM). In FDM, the line frequency band is divided into sub-channels. Each terminal port is assigned one sub-channel for transmission of its data. In TDM, the sub-channels are obtained by assigning time intervals (time slots) to the terminals for use of the line. Time slot allotment to the sub-channels may be fixed or dynamic. A time division multiplexer with dynamic time slot allotment is called Statistical Time Division Multiplexer (STDM or Stat Mux).

In the following sections we will briefly introduce the frequency division and time division multiplexers. Stat Mux is more powerful and common than these two types of multiplexers. It is described in considerable detail. The reader will find many new concepts and terminology to which he has not been introduced so far. In order to appreciate the operation of Stat Mux, it is first necessary to understand data link protocols. The reader is strongly advised to read the section on Stat Mux only after reading the chapter on Data Link Layer.

8.2 Frequency Division Multiplexers (FDM)

The leased line usually provides speech channel bandwidth of 300–3400 Hz. Therefore, most of the multiplexers are designed for this band. For frequency division multiplexing, the frequency band is divided into several sub-channels separated by guard bands. The sub-channels utilize frequency shift keying for modulating the carrier. Aggregate of all sub-channels is within the speech channel bandwidth and is an analog signal. Therefore, the multiplexer does not require any modem to connect it to the line. A four-wire circuit is always required for outgoing and incoming channels.

Bandwidths of the sub-channels depend on the baud rates. Frequency division data multiplexers provide baud rates from 50 to 600 bauds. The number of sub-channels varies from thirty-six to four depending on baud rate (Table 14).

Table 14 Frequency Division Multiplexers

Data rate (bps)	Number of sub-channels	Total capacity (bps)
50	36	1,800
75	24	1,800
110	18	1,980
150	12	1,800
600	4	2,400

Multidrop operation of the frequency division multiplexer is shown in Fig. 38. Each remote transmits and receives a different frequency as determined by the remote single channel units. The multiple line unit which is connected to the host separates the signals received on the line. It also carries out frequency division multiplexing of the outgoing signals.

Frequency division multiplexers are not much in use. Their major limitations are:

1. Production costs are high because of analog components.
2. Total capacity is limited to 2400 bps due to large wasted bandwidth in the guard bands.

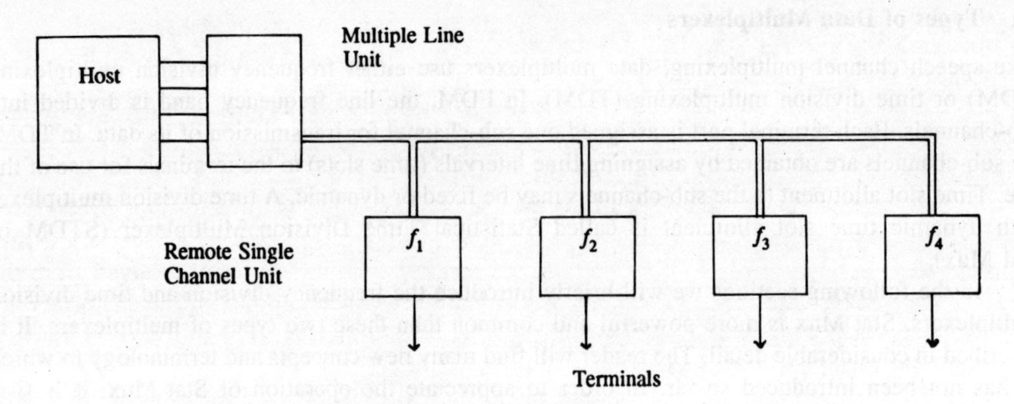

Fig. 38 Multidrop application of frequency division multiplexers.

3. They usually require a conditioned line.

4. Most multiplexers do not allow mixing of bit rates of the sub-channels, i.e., all the sub-channels have the same bit rate.

5. They are inflexible. If the sub-channel capacity has to be changed, hardware modifications are required. Complete replacement of sub-channel cards is usually necessary.

One advantage of frequency division multiplexers is that they are robust. Failure of one channel does not affect other sub-channels.

8.3 Time Division Multiplexers (TDM)

A time division multiplexer uses a fixed assignment of time slots to the sub-channels. One complete cycle of time slots is called a frame and the beginning of a frame is marked by a synchronization word (Fig. 39). The synchronization word enables the demultiplexer to identify the time slots and their boundaries. The first bit of the first time slot follows immediately after the synchronization word.

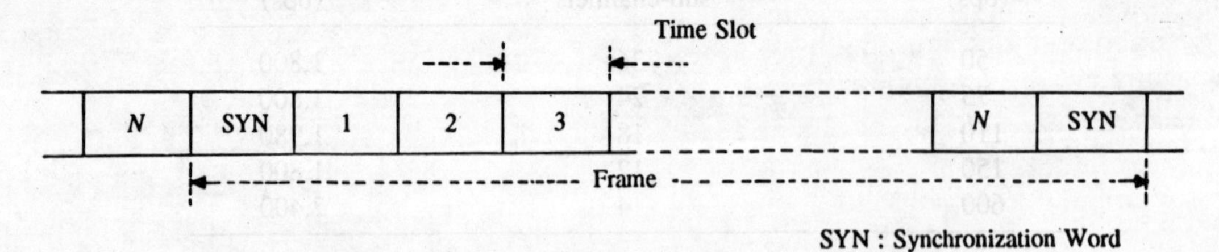

Fig. 39 Frame format of a time division multiplexer.

If all the sub-channels have the same bit rates, all the time slots have the same lengths. If the multiplexer permits speed flexibility, the higher speed sub-channels have longer time slots. The frame format and time slot lengths are, however, fixed for any given configuration or number of sub-channels and their rates. Since the frame format is fixed, time slots of all the sub-channels are always transmitted irrespective of the fact that some of the sub-channels may not have any data to send.

Bit and Byte Interleaved TDM. Time division multiplexers are of two types:

1. Bit interleaved multiplexer
2. Byte interleaved multiplexer.

In the bit interleaved multiplexer, each time slot is one bit long. Thus, the user data streams are interleaved taking one bit from each stream. Bit interleaved multiplexers are totally transparent to the terminals.

In the byte interleaved multiplexer, each time slot is one byte long. Therefore, the multiplexed output consists of a series of interleaved characters of successive sub-channels. Usually, a buffer is provided at the input of each of its ports to temporarily store the character received from the terminal. The multiplexer reads the buffers sequentially. The start-stop bits of the characters are stripped during multiplexing and again reinserted after demultiplexing. It is necessary to transmit a special "idle" character when a terminal is not transmitting.

The bit rate at the output of the multiplexer is slightly greater than the aggregate bit rate of the sub-channels due to the overhead of the synchronization word. Another feature of TDMs is that even though the multiplexed output is formatted, there is no provision for detecting or correcting the errors.

Time division multiplexers permit the mixing of bit rates of the sub-channels. Their line capacity utilization is also better than frequency division multiplexers. A line bit rate of 9600 bps is possible.

9 STATISTICAL TIME DIVISION MULTIPLEXERS

Statistical time division multiplexer, Stat Mux in short, uses dynamic assignment of time slots for transmitting data. If a sub-channel has data waiting to be transmitted, the Stat Mux allots it a time slot in the frame (Fig. 40). Duration of the time slot may be fixed or variable. There is need to identify the time slots and their boundaries. Therefore, some additional control fields are required. When we examine the Stat Mux protocols later we will see how the time slots are identified.

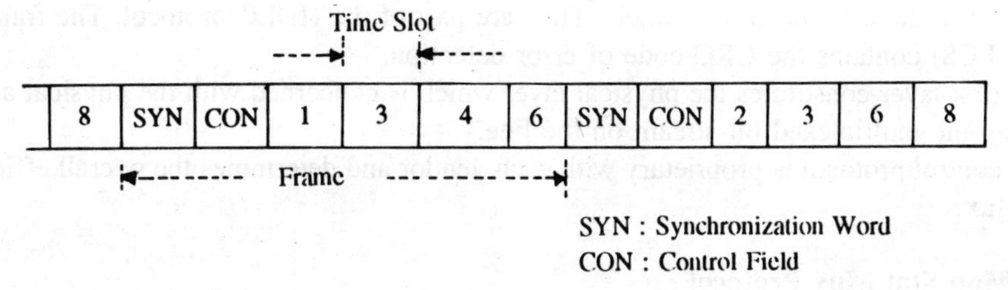

Fig. 40 Frame format of a statistical time division multiplexer.

Dynamic assignment allows the aggregate bit rates of the sub-channels to be more than the line speed of the Stat Mux considering that all the terminals will not generate traffic all the time. If sufficient aggregate traffic is assured at the input, the Stat Mux permits full utilization of the line capacity. It is not so in TDMs, where the line time is wasted if a time slot is not utilized by a sub-channel though another sub-channel may have data to send.

9.1 Stat Mux Buffer

A Stat Mux is configured to handle an aggregate sub-channel bit rate which is more than the line rate. It must have a buffer so that it may absorb the input traffic fluctuations maintaining a constant flow of multiplexed data on the line. The Stat Mux maintains a queue in the buffer to maintain sequence of the data bytes. Buffer size may vary from vendor to vendor but 64 kbyte is typical. This buffer is usually shared by both the directions of transmission, i.e., by the multiplexer and the demultiplexer portions of a Stat Mux. To guard against the overflow, the sub-channel traffic is flow-controlled.

9.2 Stat Mux Protocol

Some of the important issues which need to be addressed to have dynamic time slot allotment are:

1. In simple time division multiplexer, the location of time slot with respect to the synchronization word identifies the time slot because fixed frame format is used. But in Stat Mux, the frame has variable format. Therefore, some mechanism to identify the time slots is required.
2. Lengths of the time slots are variable. There is need to define time slot delimiters.

Therefore, a Stat Mux protocol which defines the format of the Stat Mux frame is required. There are several proprietary protocols but none of them is standard. We will discuss two common Stat Mux protocols, Bit Map and Multiple-character.

The Stat Mux has a well-defined frame structure and has built-in buffer to temporarily store data. Therefore, it is possible to enhance its capability by implementing a data link protocol for error control. A commonly implemented data link protocol is HDLC.

9.3 Layered Architecture

Figure 41a shows the three-layer architecture of a Stat Mux. The control sublayer generates a multiplexed data frame with a control field to identify the data fields. It is handed over to the data link sublayer which adds a header and a trailer to it. The resulting frame structure in case of HDLC protocol is shown in Fig. 41b. The information field of the HDLC frame contains the frame received from the control sublayer. Note that the address and control fields of the HDLC frame have nothing to do with the sub-channels. They are part of the HDLC protocol. The frame check sequence (FCS) contains the CRC code of error detection.

The first layer constitutes the physical layer which is concerned with the physical aspects of transmitting the multiplexed bit stream on the line.

The control protocol is proprietary with each vendor and determines the overall efficiency of the Stat Mux.

9.4 Bit Map Stat Mux Protocol

In the bit map Stat Mux protocol, the multiplexed data frame formed by the control sublayer consists of a map field and several data fields (Fig. 42). The map field has one bit for each sub-channel. It is two bytes long for the sixteen-port Stat Mux. If a bit is "1" in the map field, it indicates that the frame contains data field of the corresponding sub-channel. A "0" in the map field of a frame indicates that data field of the corresponding sub-channel is missing from this particular frame.

Note that the map field is present in all frames and has fixed length. The size of data field of a channel, if present, is fixed in the frame. It can be set to any value while configuring the Stat Mux. Fixed sizes of the data field enable the receiving Stat Mux to identify the boundaries of these

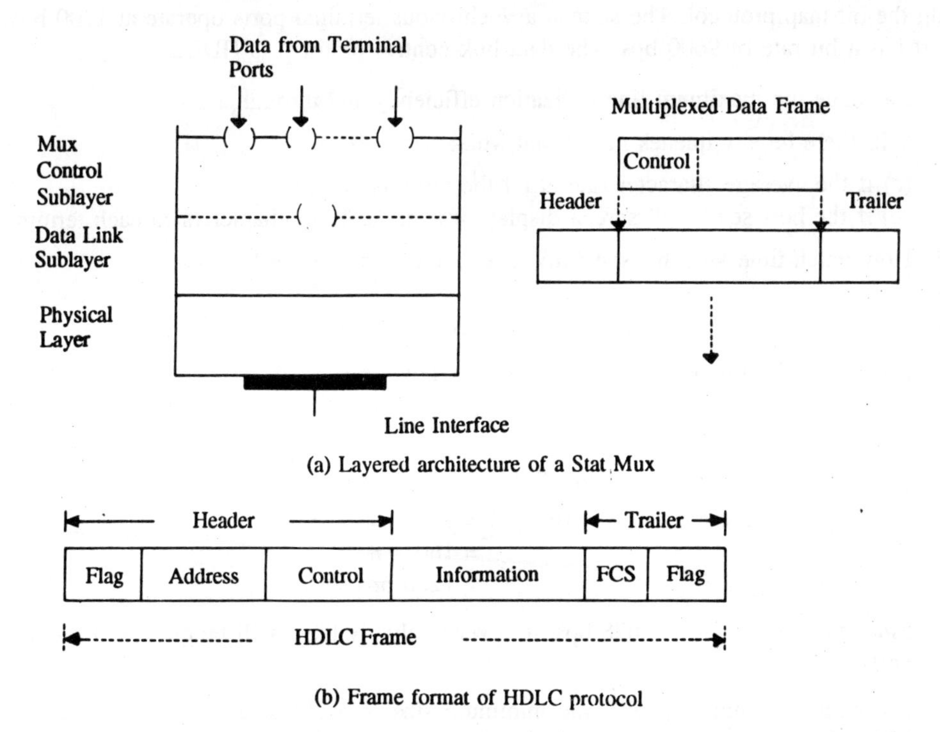

(a) Layered architecture of a Stat Mux

(b) Frame format of HDLC protocol

Fig. 41 Architecture of a Stat Mux.

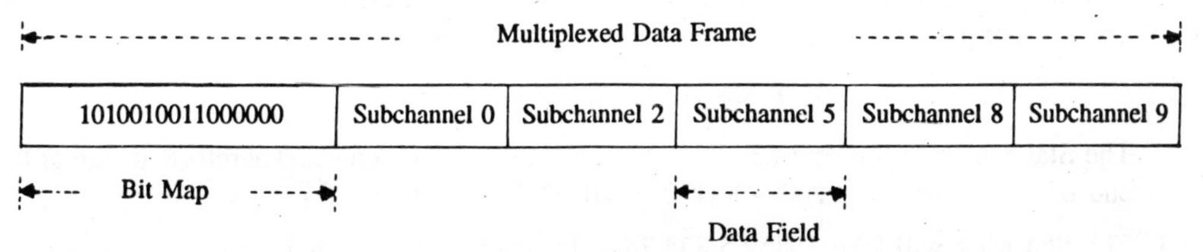

Fig. 42 Frame format of bit map Stat Mux protocol.

fields. For asynchronous terminal ports, the data field size is usually set to one character. The start-stop bits are stripped before multiplexing and reinserted after demultiplexing.

The HDLC frame transmitted on the line contains seven overhead bytes† (Flag-1, address-1, control-1, FCS-2, bit map-2) which reduce effective line utilization. If there are N bytes in the data fields of the control frame, the maximum line utilization efficiency E can be estimated by

$$E = \frac{N}{N + 7}$$

EXAMPLE 4

A host is connected to 16 asynchronous terminals through a pair of statistical time division multiplexers

†For maximum line utilization efficiency, we assume concatenation of HDLC frames so that the trailing flag also acts as leading flag for the next frame.

utilizing the bit map protocol. The sixteen asynchronous terminal ports operate at 1200 bps. The line port has a bit rate of 9600 bps. The data link control protocol is HDLC.

1. Calculate the maximum line utilization efficiency and throughput.
2. Will there be any queues in the Stat Mux
   (a) if the average character rate at all the ports is 10 cps?
   (b) if the host sends full screen display of average 1200 characters to each terminal?
3. How much time will the Stat Mux take to clear the queues?

Solution

1. As $N = 16$, the line utilization efficiency is given by

$$E = 16/(7+16) = 0.696$$

Throughput $T = E \times 9600 = 0.696 \times 9600 = 6678$ bps

2. (a) Aggregate average input $= 16 \times 10 = 160$ cps
$= 160 \times 8$
$= 1280$ bps

Since the throughput is 6678 bps, it is very unlikely there will be queues at the terminal ports.

(b) With start and stop bits, the minimum size of a character is 10 bits. Therefore, at 1200 bps, the host will take 10 seconds to transfer 1200 characters of one screen of a terminal. The Stat Mux will get $1200 \times 16 = 19200$ characters in 10 seconds from the host. The throughput is

$$6678 \text{ bps} = 6678/8 = 834.75 \text{ cps}$$

The Stat Mux will transmit 834.75×10 characters in 10 seconds. Therefore, queue at the end of 10 seconds $= 19200 - 8347.5 = 10852.5$ characters

3. The Stat Mux will take $10852.5/834.75 = 13$ additional seconds to clear the queue.

9.5 Multiple-Character Stat Mux Protocol

The bit map Stat Mux protocol has one limitation. The number of bytes in the data field of a sub-channel cannot be varied from frame to frame. Multiple-character Stat Mux protocol overcomes this limitation by including additional fields in the frame for indicating the sizes of the various data fields. The frame format of this protocol is shown in (Fig. 43).

The data field of each sub-channel which is present in a frame is identified by the sub-channel identifier of four bits. Thus, there can be a maximum of 16 sub-channels. The identifier field is followed by a four-bit sub-channel control field for management purposes.

The control field is followed by a length field which indicates the number of bytes in the data field of the sub-channel. The length field is also one byte long and, therefore, there can be maximum 256 bytes per sub-channel per frame. The data field follows immediately after the length field. The format is repeated for each sub-channel in the frame.

If the data link protocol is HDLC, total overhead bytes will be $5 + 2N$ per HDLC frame,

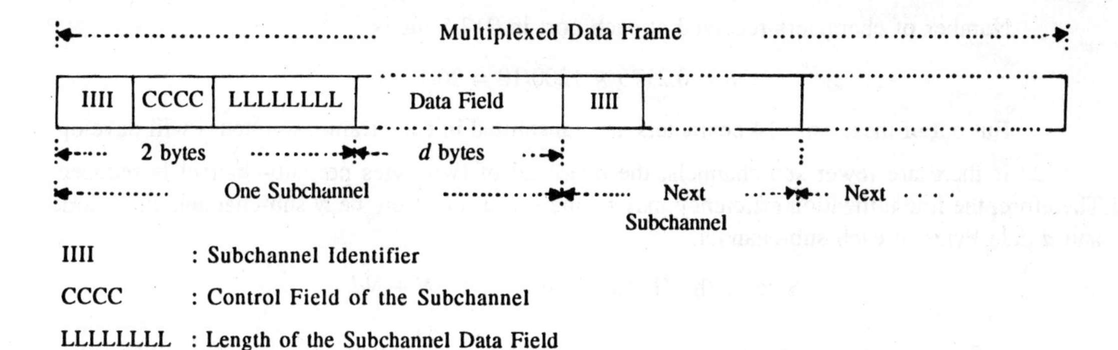

Fig. 43 Frame format of multiple-character Stat Mux protocol.

where N is the number of sub-channels present in a frame. Therefore, the line utilization efficiency E is given by

$$E = \frac{\sum_N d_i}{5 + 2N + \sum_N d_i}$$

where d_i is the number of data bytes in ith sub-channel.

EXAMPLE 5

A host is connected to 16 asynchronous terminals through a pair of statistical time division multiplexers utilizing the multiple-character protocol described above. The sixteen asynchronous terminal ports operate at 1200 bps. The line port has a bit rate of 9600 bps. The data link control protocol is HDLC and the maximum size of the HDLC frame is 261 bytes.

1. Calculate the line utilization efficiency when all the ports generate their maximum traffic. Will queues develop for this load?

2. What is the maximum line utilization efficiency without having the queues?

3. If the host sends full screen display of average 1200 characters to each terminal, will there be any queue?

   If so, how much time will the Stat Mux take to clear the queue.

Solution

1. If all the 16 users simultaneously generate a burst of data, each HDLC frame will contain all the sub-channels. As the HDLC frame size is 261 bytes, each sub-channel will occupy (261 – 5)/16 = 16 bytes. The data field of each channel will be 16 – 2 = 14 bytes. Therefore,

$$E = \frac{16 \times 14}{261} = 0.8582$$

$$\text{Time to transmit one frame } t_0 = \frac{261 \times 8}{9600} = 217.5 \text{ ms}$$

Number of characters received at each port in 217.5 ms is

$$n = 0.2175 \times 1200/10 = 26.1$$

But out of these only 14 characters are transmitted in each frame; so queues will develop.

2. If there are fewer sub-channels, the overhead of two bytes per sub-channel is reduced. Therefore, the line utilization efficiency may be increased. Let there be N sub-channels in a frame and d data bytes in each sub-channel.

$$\text{Size of the HDLC frame} = 5 + 2N + Nd$$

$$\text{Time to transmit the frame on the line } t_0 = \frac{(5 + 2N + Nd) \times 8}{9600}$$

Time taken by the terminals to generate d characters is $10d/1200$. If there are no queues, then

$$\frac{10d}{1200} = \frac{(5 + 2N + Nd) \times 8}{9600}$$

Simplifying, we get

$$d = (5 + 2N)/(10 - N), \quad N \neq 10$$

We need to solve the above equation for integer values of d and N. Line utilization efficiency is given by

$$E = \frac{Nd}{5 + 2N + Nd}$$

Substituting the value of d, we get

$$E = N/10$$

As $N \neq 10$, maximum line utilization efficiency is obtained when $N = 9$. Therefore,

$$E = 0.9, \quad N = 9, \quad d = 23$$

3. Time required by the host to transfer one screen = 1200 × 10/1200 = 10 s.

Number of characters to be transferred in 10 seconds = 16 × 1200 = 19,200

At the line rate of 9600 bps, time taken to transmit one HDLC frame is given by

$$t_0 = \frac{261 \times 8}{9600}$$

Assuming all the sub-channels are present in the frame, the data character transfer rate per HDLC frame is 224 characters/frame. Therefore, number of data characters transferred in 10 seconds is

$$\frac{224 \times 10}{t_0} = 10298.85 \text{ characters}$$

Additional time required to clear the queues

$$\frac{(19200 - 10298.85) \times 10}{10298.85} = 8.64 \text{ s}$$

10 COMPARISON OF DATA MULTIPLEXING TECHNIQUES

When compared with other types of data multiplexers, Stat Mux offers many advantages. Table 15 gives a general comparison of the data multiplexing techniques. The parameters used for comparison are:

Line Utilization Efficiency. It gives the potential to effectively utilize the line capacity.

Channel Capacity. It gives the aggregate capacity of all the sub-channels.

High Speed Channels. This parameter compares the ability to support high speed data sub-channels.

Flexibility. This parameter compares the ability to change speed of sub-channels.

Error Control. This parameter compares the ability to detect and correct transmission errors.

Multidrop Capability. This parameter compares the ability to use multidrop techniques on a sub-channel.

Transmission Delay. This parameter compares the additional transmission delays introduced by the multiplexer, over and above the propagation delay.

Table 15 Comparison of Data Multiplexer Techniques

Parameter	FDM	TDM	Stat Mux
Line efficiency	Poor	Good	Excellent
Channel capacity	Poor	Good	Excellent
High speed sub-channel	Very poor	Poor	Excellent
Flexibility	Very poor	Good	Excellent
Error control	None	None	Possible
Multidrop capability	Good	Difficult	Possible
Cost	High	Low	Medium
Transmission delay	None	Low	Random

11 SUMMARY

Transmission of digital signals using the limited bandwidth of the speech channel of the telephone network necessitates use of digital modulation methods, namely, Frequency Shift Keying (FSK), differential Phase Shift Keying (PSK) and Quadrature Amplitude Modulation (QAM). FSK is used in the low speed modems. PSK and QAM are used in the medium and high speed modems.

A modem has two interfaces, a digital interface which is connected to the Data Terminal Equipment (DTE) and a line interface which is connected to the transmission line. It comprises several functional blocks besides a modulator and a demodulator. Encoding, scrambling, equalizing and timing extraction are some of the additional functions carried out in a modem. CCITT recommendations for modems are summarized below. The number within brackets is the speed of the modem in bits per second. Half duplex modems are indicated by the letters "HD".

2-Wire-Asynchronous Modems: V.21 (300).

2-Wire-Synchronous Modems: V.22 (1200), V.22*bis* (2400), V.26*bis* (2400 HD), V.26*ter* (2400), V.27*ter* (4800), V.32 (9600).

4-Wire-Synchronous Modems: V.23 (1200), V.26 (2400), V.27 (4800), V.27*bis* (4800), V.29 (9600), V.33 (14400), V.36 (72 k), V.37 (144 k).

Limited distance modems, baseband modems and line drivers are designed for copper cable connection between the modems. These modems require the wider bandwidth of the cable and cannot work within the 300–3400 Hz band of the speech channel.

Data multiplexers are used to economize on lines and modems. Frequency division and time division data multiplexers offer limited capabilities and do not make optimum use of the channel capacity. Statistical time division multiplexers offer a very high potential utilization of channel capacity. They also offer high flexibility of configuring terminal port speeds.

PROBLEMS

1. Tick the right answer.

(a) Full duplex mode of transmission is possible in
- (i) 4-wire modems only,
- (ii) 2-wire modems only,
- (iii) 2-wire or 4-wire modems.

(b) The secondary channel in a modem uses the following:
- (i) ASK
- (ii) FSK
- (iii) PSK

(c) One of the functions of the training sequence is to
- (i) test the modems,
- (ii) synchronize the descrambler, and
- (iii) test the DTE,

(d) High-speed modems for switched telephone connections are equipped with
- (i) adaptive equalizers,
- (ii) fixed equalizers, and
- (iii) manually adjustable equalizers.

(e) Echo canceller is required in
- (i) 4-wire full duplex high speed modems,
- (ii) 2-wire half duplex high speed modems, and
- (iii) 2-wire full duplex high speed modems.

2. In a differential BPSK modulator, binary "0" results in a phase change of π and binary "1" does not cause any phase change. Write the phase states of the carrier for the following binary sequence.

1 0 1 1 0 0 0 1

3. Draw the phasor diagram of the output if the carrier is given a phase shift of $-\pi/2$ instead of $\pi/2$ in Fig. 8.

4. If each element of the QAM signal shown in Fig. 14 has a time duration of 1 μs, what are the baud and bit rates?

5. For the PSK modulator shown below, analog outputs of *D/A* converters are indicated in the table. Draw the phasor diagram of the output. If the bit rate is 4800 bps, what is the baud rate?

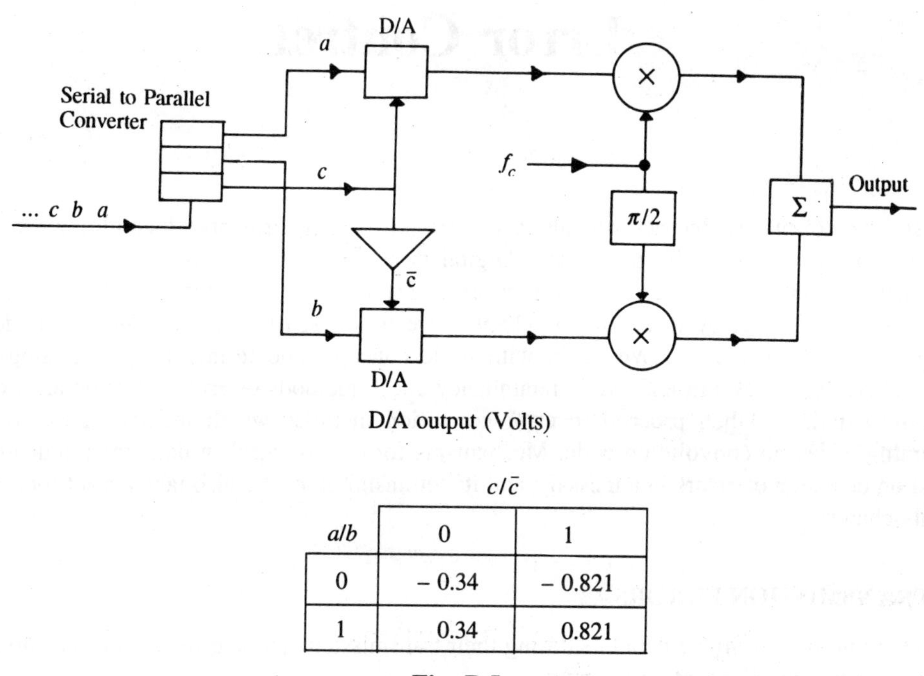

D/A output (Volts)

	$c/\bar{c}$	
a/b	0	1
0	– 0.34	– 0.821
1	0.34	0.821

Fig. P.5

6. Solve the bit map Stat Mux problem given in the chapter when the line speed is 19.2 kbps and the average number of characters in the full screen is 1920.

7. A Stat Mux based on multiple-character multiplexing protocol described in the chapter is connected to 16 asynchronous 1200 bps terminals. The line speed is 9600 bps. At data link layer, the Stat Mux utilizes HDLC protocol with 261 bytes frame length. If only one user is active, what is the maximum line utilization efficiency?

CHAPTER 4

Error Control

Transmission of bits as electrical signals suffers from many impairments which ultimately result in introduction of errors in the bit stream. Digital systems are very sensitive to errors and may malfunction if error rate is above a certain level. Therefore, error control mechanisms are built into almost all digital systems. In this chapter, we will discuss some common error detection and correction mechanisms. We begin with basic concepts and terminology of coding theory; parity checking, checksum and cyclic redundancy check methods of error detection are examined in some detail. We then proceed to error correction methods which include block codes, the Hamming code and convolution code. Mechanisms for error control in data communications are based on detection of errors in a message and its retransmission. We also take a brief look of these in this chapter.

1 TRANSMISSION ERRORS

Errors are introduced in the data bits during their transmission. These errors can be categorized as: content errors, and flow integrity errors.

Content errors are errors in the content of a message, e.g., a "1" may be received as a "0". Such errors creep in due to impairment of the electrical signal in the transmission medium.

Flow integrity errors refer to missing blocks of data. For example, a data block may be lost in the network due to its having been delivered to a wrong destination.

In voice communication, the listener can tolerate a good deal of signal distortion and make sense of the received signal but digital systems are very sensitive to errors. Measures are, therefore, built into a data communication system to counteract the effect of errors. These measures include the following:

- Introduction of additional check bits in the data bits to detect content errors
- Correction of the errors
- Establishment of procedures of data exchange which enable detection of missing blocks of data
- Recovery of the corrupted messages.

Before we examine these measures in detail, let us first understand some important principles of coding theory and its terminology.

2 CODING FOR ERROR DETECTION AND CORRECTION

For error detection and correction, we need to add some check bits to a block of data bits. The check bits are also called redundant bits because they do not carry any user information. Check bits are so chosen that the resulting bit sequence has a unique "characteristic" which enables error detection. Coding is the process of adding the check bits. Some of the terms relating to coding theory are explained below:

- The block of data bits to which check bits are added is called a data word.
- The bigger block containing check bits is called the code word.
- Hamming distance or simply distance between two code words is the number of disagreements between them. For example, the distance between the two words given below is 3 (Fig. 1).
- The weight of a code word is the number of "1"s in the code word e.g., 11001100 has a weight of 4.
- A code set consists of all valid code words. All the valid code words have a built in "characteristic" of the code set.

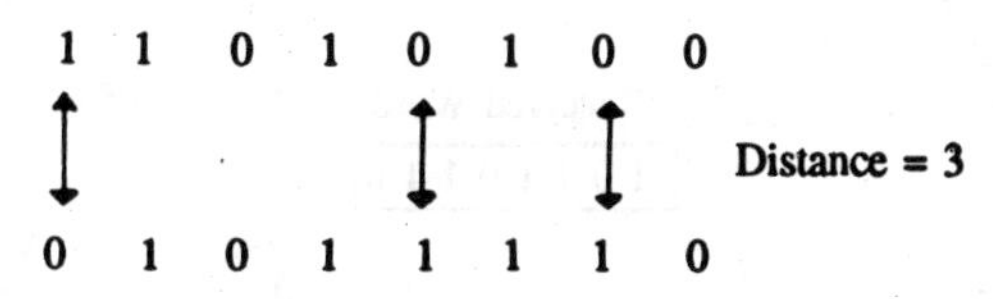

Fig. 1 Hamming distance.

2.1 Error Detection

When a code word is transmitted, one or more of its bits may be reversed due to signal impairment. The receiver can detect these errors if the received code word is not one of the valid code words of the code set.

When errors occur, the distance between the transmitted and received code words becomes equal to the number of erroneous bits (Fig. 2).

Transmitted Code Word	Received Code Word	Number of Errors	Distance
11001100	11001110	1	1
10010010	00011010	2	2
10101010	10100100	3	3

Fig. 2 Hamming distance between transmitted and received code words.

In other words, the valid code words must be separated by a distance of more than 1; otherwise, even a single bit error will generate another valid code word and the error will not be detected. The number of errors which can be detected depends on the distance between any two valid code words. For example, if the valid code words are separated by a distance of 4, up to three errors in a code word can be detected. By adding a certain number of check bits and properly choosing the algorithm for generating them, we ensure some minimum distance between any two valid code words of a code set.

2.2 Error Correction

After an error is detected, there are two approaches to correction of errors:

1. Reverse Error Correction (REC)
2. Forward Error Correction (FEC).

In the first approach, the receiver requests for retransmission of the code word whenever it detects an error. In the second approach, the code set is so designed that it is possible for the receiver to detect and correct the errors as well. The receiver locates the errors by analysing the received code word and reverses the erroneous bits.

An alternative way of forward error correction is to search for the most likely correct code word. When an error is detected, the distances of all the valid code words from the received invalid code word are measured. The nearest valid code word is the most likely correct version of the received word (Fig. 3).

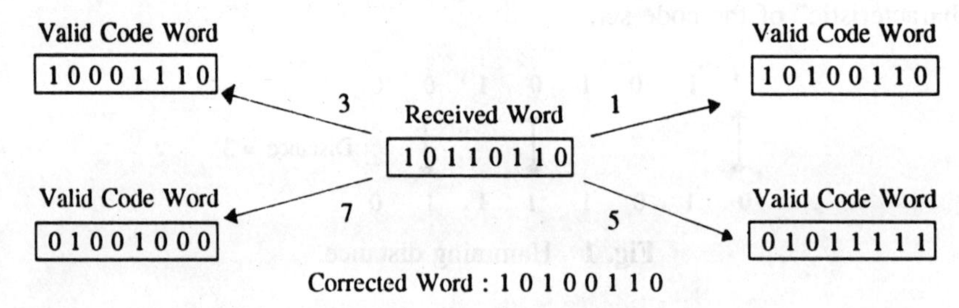

Fig. 3 Error correction approach based on the least Hamming distance.

If the minimum distance between valid code words is D, upto $D/2-1$ errors can be corrected. More than $D/2-1$ errors will cause the received code word to be nearer to the wrong valid code word.

2.3 Bit Error Rate (BER)

In analog transmission, signal quality is specified in terms of signal-to-noise ratio (S/N) which is usually expressed in decibels. In digital transmission, the quality of received digital signal is expressed in terms of Bit Error Rate (BER) which is the number of errors in a fixed number of transmitted bits. A typical error rate on a high quality leased telephone line is as low as 1 error in 10^6 bits or simply 1×10^{-6}.

EXAMPLE 1

If the average BER is 1 in 10^5, what is the probability of having

1. single bit error,
2. single bit correct,
3. at least one error in an eight-bit byte?

Solution

Probability of having single bit error $1/10^5 = 0.00001$

Probability of having single bit correct $1 - 0.00001 = 0.99999$

Probability of having one or more errors in an 8-bit byte $1 - (0.99999)^8 = 0.00008$

Just like BER, Character Error Rate (CER) and Frame Error Rate (FER) can be defined. CER is the average number of characters received with at least one error in a large sample of transmitted characters. The probability of having at least one error in a byte calculated in the above example gives a CER of 8 in 10^5 characters. FER, likewise, refers to the average number of frames received with at least one error in a large sample of transmitted frames. It can also be calculated on the same lines as CER. For low values of BER, the CER and FER can be calculated from BER as below:

$$\text{CER} = b \times \text{BER}$$
$$\text{FER} = f \times \text{BER}$$

where b is the number of bits per character and f is the number of bits per frame.

Whatever be the methods of error control, errors cannot be completely eliminated. There is always some residual error which goes undetected. Residual Error Rate (RER) refers to the error rate in the data bits after error control has been performed.

3 ERROR DETECTION METHODS

Some of the popular error detection methods are:

- Parity checking
- Checksum error detection
- Cyclic Redundancy Check (CRC).

Each of the above methods has its advantages and limitations as we shall see in the following sections.

3.1 Parity Checking

In parity checking methods, an additional bit called a "parity" bit is added to each data word. The additional bit is so chosen that the weight of the code word so formed is either even (even parity) or odd (odd parity) (Fig. 4). All the code words of a code set have the same parity (either odd or even) which is decided in advance.

Even Parity

P	Data Word
0	1 0 0 1 0 1 1
1	0 0 1 0 1 1 0

Odd Parity

P	Data Word
1	1 0 0 1 0 1 1
0	0 0 1 0 1 1 0

P: Parity Bit

Fig. 4 Even and odd parity bits.

When a single error or an odd number of errors occurs during transmission, the parity of the code word changes (Fig. 5). Parity of the code word is checked at the receiving end and violation of the parity rule indicates errors somewhere in the code word.

Transmitted Code	1 0 0 1 0 1 1 0	Even Parity
Received Code (Single error)	0 0 0 1 0 1 1 0	Odd Parity (Error is detected)
Received Code (Double error)	0 0 0 1 1 1 1 0	Even Parity (Error is not detected)

Fig. 5 Error detection by change in parity.

Note that double or any even number of errors will go undetected because the resulting parity of the code word will not change. Thus, a simple parity checking method has its limitations. It is not suitable for multiple errors. To keep the possibility of occurrence of multiple errors low, the size of the data word is usually restricted to a single byte.

Parity checking does not reveal the location of the erroneous bit. Also, the received code word with an error is always at equal distance from two valid code words. Therefore, errors cannot be corrected by the parity checking method.

EXAMPLE 2

Write the ASCII code of the word "HELLO" using even parity.

Solution

Bit Positions	8 7 6 5 4 3 2 1
H	0 1 0 0 1 0 0 0
E	1 1 0 0 0 1 0 1
L	1 1 0 0 1 1 0 0
L	1 1 0 0 1 1 0 0
O	1 1 0 0 1 1 1 1

3.2 Burst Errors

There is a strong tendency for the errors to occur in bursts. An electrical interference like lightning lasts for several bit times and, therefore, it corrupts a block of several bits. The parity checking method fails completely in such situations. Checksum and cyclic redundancy check are the two methods which can take care of burst errors.

3.3 Checksum Error Detection

In checksum error detection method, a checksum is transmitted along with every block of data bytes. Eight-bit bytes of a block of data are added in an eight-bit accumulator. Checksum is the resulting sum in the accumulator. Being an eight-bit accumulator, the carries of the most significant bits are ignored.

EXAMPLE 3

Find the checksum of the following message. The MSB is on the left-hand side of each byte.

10100101 00100110 11100010 01010101 10101010 11001100 00100100

Solution

```
1 1         1
    1   1   1   1   1       1        } Carries
---------------------------------
    1   0   1   0   0   1   0   1
    0   0   1   0   0   1   1   0
    1   1   1   0   0   0   1   0
    0   1   0   1   0   1   0   1    } Data Bytes
    1   0   1   0   1   0   1   0
    1   1   0   0   1   1   0   0
    0   0   1   0   0   1   0   0
---------------------------------
    1   0   0   1   1   1   0   0      Checksum Byte
```

After transmitting the data bytes, the checksum is also transmitted. The checksum is regenerated at the receiving end and errors show up as a different checksum. Further simplification is possible by transmitting the 2's complement of the checksum in place of the checksum itself. The receiver in this case accumulates all the bytes including the 2's complement of the checksum. If there is no error, the contents of the accumulator should be zero after accumulation of the 2's complement of the checksum byte.

The advantage of this approach over simple parity checking is that 8-bit addition "mixes up" bits and the checksum is representative of the overall block. Unlike simple parity where even number of errors may not be detected, in checksum there is 255 to 1 chance of detecting random errors.

3.4 Transport Protocol Checksum

It may be noted in Example 3 that the checksum will remain same even if the bits interchange their position within a column. If such errors occur during transmission, the received checksum will fail to reflect any error. A more powerful method of generating checksum is used in the Transport Protocol standardized by ISO and CCITT in IS 8072 and X.224. In this method, two checksum bytes are generated instead of the usual one. The procedure for generating these bytes involves the following steps.

1. To start with, assume checksum bytes X and Y are 00000000.

2. Define variables P_i & Q_i such that

$$P_i = P_{i-1} + B_i \qquad B_i = i\text{th byte}$$

$$Q_i = Q_{i-1} + P_i \qquad P_0 = Q_0 = 0$$

3. P_i and Q_i are calculated for all the data bytes in the message and for the initial values of the checksum bytes X and Y.

4. From the resulting values of P_i and Q_i variables, calculate the final values of checksum bytes X and Y as below:

$$X = P_l - Q_l \qquad l = \text{No. of data bytes} + 2$$

$$Y = Q_l - 2P_l$$

These checksum bytes are transmitted along with the message bytes.

At the receiving end, steps 2 and 3 are again carried out and variables P_l and Q_l are generated. If there are no errors in the received data block, the values of these variables will be zero.

EXAMPLE 4

Generate the checksum bytes as per the transport protocol for the following data bytes.

10100101 00100110 11100010 01010101

Regenerate variables P_l and Q_l at the receiving end and show that they are zero if there are no errors.

Solution

Transmitting End

Data byte					
	P_0	0 0 0 0 0 0 0 0			
1 0 1 0 0 1 0 1	B_1	1 0 1 0 0 1 0 1	Q_0		0 0 0 0 0 0 0 0
	P_1	1 0 1 0 0 1 0 1		$\longrightarrow$	1 0 1 0 0 1 0 1
0 0 1 0 0 1 1 0	B_2	0 0 1 0 0 1 1 0	Q_1		1 0 1 0 0 1 0 1
	P_2	1 1 0 0 1 0 1 1			1 1 0 0 1 0 1 1
1 1 1 0 0 0 1 0	B_3	1 1 1 0 0 0 1 0	Q_2		0 1 1 1 0 0 0 0
	P_3	1 0 1 0 1 1 0 1		$\longrightarrow$	1 0 1 0 1 1 0 1
0 1 0 1 0 1 0 1	B_4	0 1 0 1 0 1 0 1	Q_3		0 0 0 1 1 1 0 1
	P_4	0 0 0 0 0 0 1 0		$\longrightarrow$	0 0 0 0 0 0 1 0
0 0 0 0 0 0 0 0	X	0 0 0 0 0 0 0 0	Q_4		0 0 0 1 1 1 1 1
	P_5	0 0 0 0 0 0 1 0		$\longrightarrow$	0 0 0 0 0 0 1 0
0 0 0 0 0 0 0 0	Y	0 0 0 0 0 0 0 0	Q_5		0 0 1 0 0 0 0 1
	P_6	0 0 0 0 0 0 1 0		$\longrightarrow$	0 0 0 0 0 0 1 0
			Q_6		0 0 1 0 0 0 1 1

$X = P_6 - Q_6$		0 0 0 0 0 0 1 0
		0 0 1 0 0 0 1 1
	X	1 1 0 1 1 1 1 1
$Y = Q_6 - 2P_6$		0 0 1 0 0 0 1 1
		0 0 0 0 0 1 0 0
	Y	0 0 0 1 1 1 1 1

Receiving End

Summation upto P_4 and Q_4 will be same as above. Rest of the steps are shown below:

Data byte					
	P_4	0 0 0 0 0 0 1 0			
1 1 0 1 1 1 1 1	X	1 1 0 1 1 1 1 1	Q_4		0 0 0 1 1 1 1 1
	P_5	1 1 1 0 0 0 0 1		$\longrightarrow$	1 1 1 0 0 0 0 1
0 0 0 1 1 1 1 1	Y	0 0 0 1 1 1 1 1	Q_5		0 0 0 0 0 0 0 0
	P_6	0 0 0 0 0 0 0 0		$\longrightarrow$	0 0 0 0 0 0 0 0
			Q_6		0 0 0 0 0 0 0 0

Thus P_6 and Q_6 are zero indicating no errors.

In the next example, we consider errors in the third bit position of the second and fourth bytes and try to detect the errors using the method just described. These errors cannot be detected by the usual checksum method as the summation over the third column will not change.

EXAMPLE 5

Check whether the following bytes have any error. The last two bytes are the transport protocol checksum bytes.

10100101 00000110 11100010 01110101 11011111 00011111

Solution

	P_0	0 0 0 0 0 0 0 0		
1 0 1 0 0 1 0 1	B_1	1 0 1 0 0 1 0 1	Q_0	0 0 0 0 0 0 0 0
	P_1	1 0 1 0 0 1 0 1	⟶	1 0 1 0 0 1 0 1
0 0 0 0 0 1 1 0	B_2	0 0 0 0 0 1 1 0	Q_1	1 0 1 0 0 1 0 1
	P_2	1 0 1 0 1 0 1 1	⟶	1 0 1 0 1 0 1 1
1 1 1 0 0 0 1 0	B_3	1 1 1 0 0 0 1 0	Q_2	0 1 0 1 0 0 0 0
	P_3	1 0 0 0 1 1 0 1	⟶	1 0 0 0 1 1 0 1
0 1 1 1 0 1 0 1	B_4	0 1 1 1 0 1 0 1	Q_3	1 1 0 1 1 1 0 1
	P_4	0 0 0 0 0 0 1 0	⟶	0 0 0 0 0 0 1 0
1 1 0 1 1 1 1 1	X	1 1 0 1 1 1 1 1	Q_4	1 1 0 1 1 1 1 1
	P_5	1 1 1 0 0 0 0 1	⟶	1 1 1 0 0 0 0 1
0 0 0 1 1 1 1 1	Y	0 0 0 1 1 1 1 1	Q_5	1 1 0 0 0 0 0 0
	P_6	0 0 0 0 0 0 0 0	⟶	0 0 0 0 0 0 0 0
			Q_6	1 1 0 0 0 0 0 0

Since Q_6 is not zero, there are errors in the received bytes.

3.5 Cyclic Redundancy Check

Cyclic Redundancy Check (CRC) codes are very powerful and are now almost universally employed. These codes provide a better measure of protection at the lower level of redundancy and can be fairly easily implemented using shift registers or software.

A CRC code word of length N with m-bit data word is referred to as (N, m) cyclic code and contains $(N-m)$ check bits. These check bits are generated by modulo-2 division. The dividend is the data word followed by $n = N-m$ zeros and the divisor is a special binary word of length $n + 1$. The CRC code word is formed by modulo-2 addition of the remainder so obtained and the dividend.

EXAMPLE 6

Generate CRC code for the data word 110101010 using the divisor 10101.

Solution

Data Word	1 1 0 1 0 1 0 1 0
Divisor	1 0 1 0 1

```
                111000111          Quotient
10101 ) 1101010100000          Dividend
        10101
         11111
         10101
          10100
          10101
              11000
              10101
               11010
               10101
                11110
                10101
                 1011          Remainder
         1101010100000
                  1011
Code Word 1101010101011
```

In the above example, note that the CRC code word consists of the data word followed by the remainder. The code word so generated is completely divisible by the divisor because it is the difference of the dividend and the remainder (Modulo-2 addition and subtraction are equivalent). Thus, when the code word is again divided by the same divisor at the receiving end, a non-zero remainder after so dividing will indicate errors in transmission of the code word.

EXAMPLE 7

The code word of Example 6 be received as 1100100101011. Check if there are errors in the code word.

Solution

Dividing the code word by 10101, we get

```
         111110001
10101 ) 1100100101011
        10101
         11000
         10101
          11010
          10101
           11111
           10101
            10100
            10101
                11011
                10101
                 1110     Remainder
```

Non-zero remainder indicates that there are errors in the received code word.

3.6 Undetected Errors in CRC

Not all the types of errors can be detected by CRC code. The probability of error detection and the types of errors which can be detected depends on the choice of the divisor. If the number of check bits in CRC code is n, the probabilities of error detection for various types of errors are as given below:

- Single errors 100%
- Two bit errors 100%
- Odd number of bits in error 100%
- Error bursts of length $< n + 1$ 100%
- Error bursts of length $= n + 1$ $1 - (1/2)^{n-1}$
- Error bursts of length $> n + 1$ $1 - (1/2)^n$

3.7 Algebraic Representation of Binary Code Words

For the purpose of analysis, the binary codes are represented using algebraic polynomials. In a polynomial of variable x, coefficients of the powers of x are the bits of the code, the most significant bit being the coefficient of the highest power of x. The data word of Example 6 can be represented by a polynomial $M(x)$ as:

$$M(x) = 1x^8 + 1x^7 + 0x^6 + 1x^5 + 0x^4 + 1x^3 + 0x^2 + 1x^1 + 0x^0$$

or

$$M(x) = x^8 + x^7 + x^5 + x^3 + x$$

The polynomial corresponding to the divisor is called the generating polynomial $G(x)$. $G(x)$ corresponding to divisor used in last example would be

$$G(x) = 1x^4 + 0x^3 + 1x^2 + 0x^1 + 1x^0$$

or

$$G(x) = x^4 + x^2 + 1$$

The polynomial $D(x)$ corresponding to the dividend (1101010100000) is

$$D(x) = x^{12} + x^{11} + x^9 + x^7 + x^5 = x^4 \cdot M(x)$$

If $Q(x)$ is the quotient and $R(x)$ is the remainder when $D(x)$ is divided by $G(x)$,

$$D(x) = Q(x) \cdot G(x) + R(x)$$

$$D(x) + R(x) = Q(x) \cdot G(x) + R(x) + R(x)$$

$$D(x) + R(x) = Q(x) \cdot G(x)$$

Thus, the CRC code $D(x) + R(x)$ is completely divisible by $G(x)$. This characteristic of the code is used for detecting errors.

Some of the common generating polynomials and their applications are:

- CCITT V.41 $x^{16} + x^{12} + x^5 + 1$

  It is used in HDLC/SDLC/ADCCP protocols.

- CRC-12 $x^{12} + x^{11} + x^3 + x^2 + x + 1$

  It is employed in BISYNC protocol with 6-bit characters.

- CRC-16 $x^{16} + x^{15} + x^2 + 1$

  It is used in BISYNC protocol with 8-bit characters.

- CRC-32 $X^{32} + x^{26} + x^{23} + x^{22} + x^{16} + x^{12} + x^{11} + x^{10} + x^8 + x^7 + x^5 + x^4 + x^2 + x + 1$

  It is used with 8-bit characters when very high probability of error detection is required.

4 FORWARD ERROR CORRECTION METHODS

To locate and correct errors require a bigger overhead in terms of number of check bits in the code word. Some of the important error-correction codes which find application in data transmission devices are:

- Block parity
- Hamming code
- Convolutional code.

4.1 Block Parity

The concept of parity checking can be extended to detect and correct single errors. The data block is arranged in a rectangular matrix form as shown in Fig. 8 and two sets of parity bits are generated, namely,

1. Longitudinal Redundancy Check (LRC)
2. Vertical Redundancy Check (VRC).

VRC is the parity bit associated with the character code and LRC is generated over the rows of bits. LRC is appended to the end of a data block. The bit 8 of the LRC represents the VRC of the other 7 bits of the LRC. In Fig. 6, even parity is used for the LRC and the VRC.

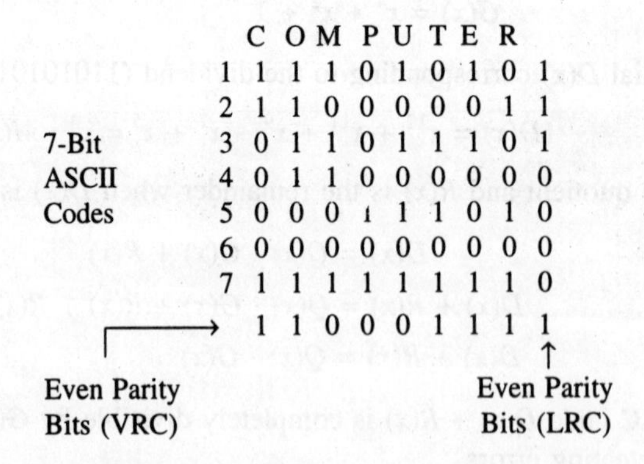

Bit Transmission Sequence

11000011 11110011 10110010 00001010 10101010 00101011 10100011 01001011 11100001

Fig. 6 Vertical and longitudinal parity check bits.

Even a single error in any bit results in failure of longitudinal redundancy check in one of the rows and vertical redundancy check in one of the columns. The bit which is common to the row and column is the bit in error.

Multiple errors in rows and columns can be detected but cannot be corrected as the bits which are in error cannot be located.

EXAMPLE 8

The following bit stream is encoded using VRC, LRC and even parity. Correct the error, if any.

11000011 11110011 10110010 00001010 10111010 00101011 10100011
01001011 11100001

Solution

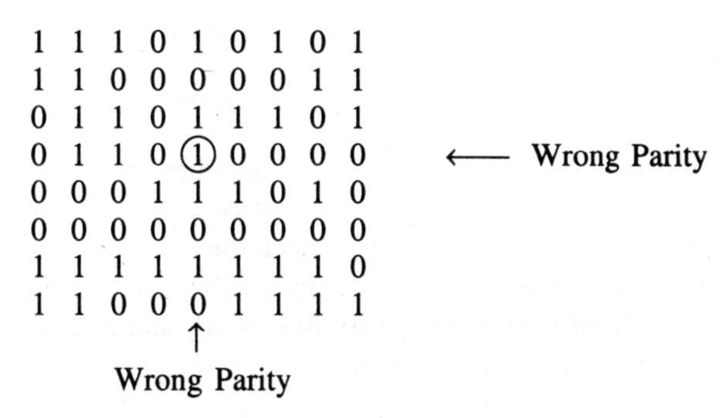

1 1 1 0 1 0 1 0 1
1 1 0 0 0 0 0 1 1
0 1 1 0 1 1 1 0 1
0 1 1 0 1 0 0 0 0
0 0 0 1 1 1 0 1 0
0 0 0 0 0 0 0 0 0
1 1 1 1 1 1 1 1 0
1 1 0 0 0 1 1 1 1
Wrong Parity
Wrong Parity

Fourth bit of the fifth byte is in error. It should be "0".

4.2 Hamming Code

It is the single error correcting code devised by Hamming. In this code, there are multiple parity bits in a code word. Bit positions 1, 2, 4, 8 ... etc. of the code word are reserved for the parity bits. The other bit positions are for the data bits (Fig. 7). The number of parity bits required for

1	2	3	4	5	6	7	8	9	10	11	......
P_1	P_2	D	P_4	D	D	D	P_8	D	D	D	......

P: Parity Bit D: Data Bit

Fig. 7 Location of parity bits in Hamming code.

correcting single bit errors depends on the length of the code word. A code word of length n contains m parity bits, where m is the smallest integer satisfying the condition:

$$2^m \geq n + 1$$

The MSB of the data word is on the right-hand side and its position is third in Fig. 7. As usual, the LSB is transmitted first.

Each data bit is checked by a number of parity bits. Data bit position expressed as sum of the powers of 2 determines parity bit positions which check the data bit. For example, a data bit in position 6 is checked by parity bits P_4 and P_2 ($6 = 2^2 + 2^1$). Similarly, data bit in position 11 is checked by parity bits P_8, P_2 and P_1 ($11 = 2^3 + 2^1 + 2^0$). Table 1 gives the parity bit positions which check the various data bit positions.

Each parity bit is determined by the data bits it checks. Even or odd parity can be used. For

Table 1 Data Bit Positions Checked by the Parity Bits

Data bit positions	Parity bit positions			
	P_1	P_2	P_4	P_8
3	×	×		
5	×		×	
6		×	×	
7	×	×	×	
9	×			×
10		×		×
11	×	×		×
12			×	×

example, if even parity is used, P_2 is such that the number of "1"s in 2nd, 3rd, 6th, 7th, 10th and 11th positions is even. The logic behind this way of generating the parity bits is that when a code word suffers an error, all the parity bits which check the erroneous bit will indicate violation of the parity rule and the sum of these parity bit positions will indicate the position of the erroneous bit. For example, if the 11th bit is in error, parity bits P_8, P_2 and P_1 will indicate error and 8 + 2 + 1 = 11 will immediately point to the 11th bit.

EXAMPLE 9

Generate the code word for ASCII character "K" = 1001011. Assume even parity for the Hamming code. No character parity is used.

Solution

	Bit positions											
	1	2	3	4	5	6	7	8	9	10	11	
	P_1	P_2	1	P_4	0	0	1	P_8	0	1	1	
First parity bit	P_1		1		0		1		0		1	$P_1 = 1$
Second parity bit		P_2	1			0	1			1	1	$P_2 = 0$
Third parity bit				P_4	0	0	1					$P_4 = 1$
Fourth parity bit								P_8	0	1	1	$P_8 = 0$
Code Word	1	0	1	1	0	0	1	0	0	1	1	

EXAMPLE 10

Detect and correct the single error in the received Hamming code word 10110010111. Assume even parity.

Solution

	Bit positions											Parity	Check	
	1	2	3	4	5	6	7	8	9	10	11			
	P_1	P_2	D	P_4	D	D	D	P_8	D	D	D			
Code word	1	0	1	1	0	0	1	0	1	1	1			
Flirst check (P_1, 3, 5, 7, 9, 11)	1		1		0		1		1		1	Odd	Fail	1
Second check (P_2, 3, 6, 7, 10, 11)		0	1			0	1			1	1	Even	Pass	
Third check (P_4, 5, 6, 7)				1	0	0	1					Even	Pass	
Fourth check (P_8, 9, 10, 11)								0	1	1	1	Odd	Fail	8
														9

Thus, the 9th bit position is in error. Correct code word is 10110010011.

4.3 Convolutional Codes

Unlike block codes in which the check bits are computed for a block of data, convolutional codes are generated over a "span" of data bits, e.g., a convolutional code of constraint length 3 is generated bit by bit always using the "last 3 data bits".

Figure 8 shows a simple convolutional encoder consisting of a shift register having three stages and EXOR gates which generate two output bits for each input bit. It is called a rate 1/2 convolutional encoder.

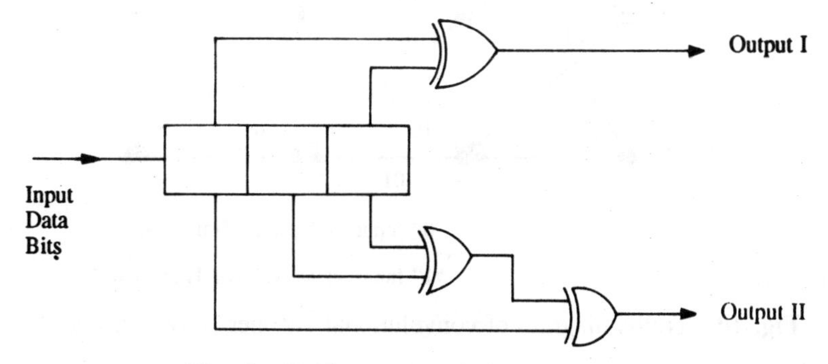

Fig. 8 Half-rate convolutional encoder.

State transition diagram of this encoder is shown in Fig. 9. Each circle in the diagram represents a state of the encoder, which is the content of two leftmost stages of the shift register. There are four possible states 00, 01, 10, 11. The arrows represent the state transitions for the input bit which can be 0 or 1. The label on each arrow shows the input data bit by which the transition is caused and the corresponding output bits. As an example, suppose the initial state of the encoder is 00 and the input data sequence is 1011. The corresponding output sequence of the encoder will then be 11010010.

Trellis Diagram. An alternative way of representing the states is by using the trellis diagram (Fig. 10). Here the four states 00, 01, 11, 10 are represented as four levels. The arrows represent

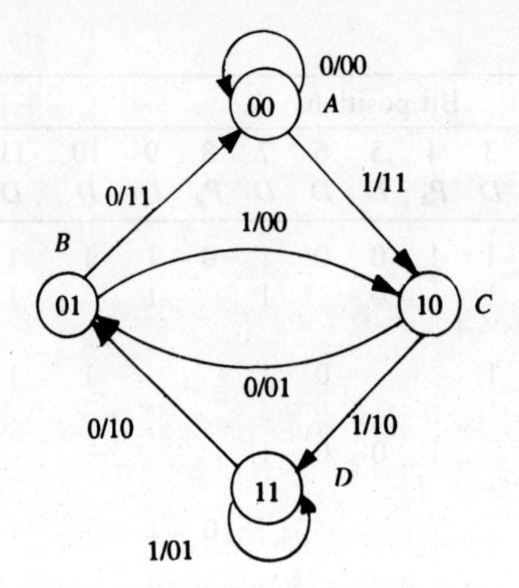

Fig. 9 State transition diagram of convolutional encoder shown in Fig. 8.

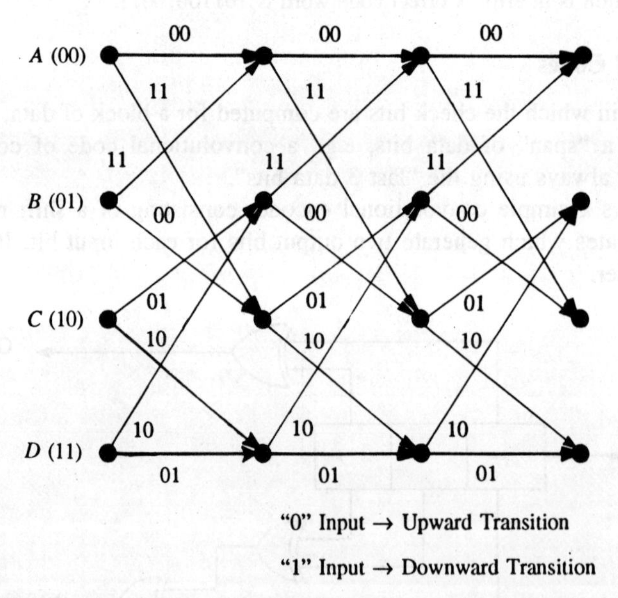

Fig. 10 Trellis diagram of convolutional encoder shown in Fig. 8.

state transitions as in the state transition diagram. The labels on the arrows indicate the output. By convention, a "0" input is always represented as an upward transition and a "1" input as a downward transition. The trellis diagram can be obtained from the state transition diagram.

EXAMPLE 11

Generate the convolutional code using the trellis diagram of Fig. 10 for the input bit sequence 0101 assuming the encoder is in state *A* to start with.

Solution

Starting from state *A* at top left corner in Fig. 10 and tracing the path through the trellis for the input sequence 0101, we get

Present state	Input bit	Next state	Output bits
A	0	*A*	0 0
A	1	*C*	1 1
C	0	*B*	0 1
B	1	*C*	0 0

Output bit sequence 0 0 1 1 0 1 0 0

Decoding Algorithm. Decoder for the convolutional code is based on the maximum likelihood principle called the Viterbi algorithm. Knowing the encoder behaviour and the received sequence of bits, we can find the most likely transmitted sequence by analysing all the possible paths through the trellis. The path which results in the output sequence which is nearest to the received sequence is chosen and the corresponding input bits are the decoded data bits.

Let the data bit sequence be 1011 which is encoded as 11010010 using the encoder shown in Fig. 8. The received sequence is 11110010 having an error in the third bit position.

Now we need to analyse all possible paths through the trellis and select the path which results in an output sequence nearest the received sequence. We will do it in two steps. After the first step we will be in a position to exclude further analysis of some of the paths.

Step 1 Let us first analyse the first three pairs of bits, that is, 111100. If we start from state *A* and trace all possible paths through the trellis shown in Fig. 10, we get the output bit sequences, and their distances from the received sequence 111100 as given in Table 2.

Table 2 Alternative Paths through the Trellis

Step 1				Step 2			
Data bits	Path	Output sequence	Distnace from 111100	Next data bit	Next state	Output sequence	Distance from 11110010
000	*AAAA*	000000	4				
100	*ACBA*	110111	3†	0	*A*	11011100	4
				1	*C*	11011111	4
110	*ACDB*	111010	2†	0	*A*	11101011	3
				1	*C*	11101000	3
010	*AACB*	001101	3				
001	*AAAC*	000011	6				
101	*ACBC*	110100	1†	0	*B*	11010001	3
				1	*D*	11010010	1†
111	*ACDD*	111001	2†	0	*B*	11100110	2
				1	*D*	11100101	4
011	*AACD*	001110	3				

†Chosen paths having smaller distance.

Note that a pair of paths terminate on each state, e.g., state *A* can be reached via *AAAA* or *ACBA*. But path *AAAA* results in output sequence 000000 which is at a distance of 4 from the first six bits of the received sequence. In the case of the other path *ACBA*, this distance is only 3. Because we are looking for a sequence with the smallest distance, we need not consider the first path for further analysis. We can drop some more paths in similar manner.

Step 2 Having considered the first three pairs of bits, let us move further. Transitions from the last state arrived at in the first step, will result in two potential states depending on the next input bit. Distances of the resulting bit sequences from the received sequence are given in Table 2. Note that we have computed the distances for only the selected paths of the first step. The minimum distance is for the path *ACBCD* which corresponds to the correct data bit sequence 1011.

EXAMPLE 12

What is the message sequence if the received rate 1/2 encoded bit sequence is 00010100? Use the trellis diagram given in Fig. 10.

Solution

Step 1				Step 2			
Data bits	Path	Output sequence	Distance from 000101	Next data bit	Next state	Output sequence	Distance from 00010100
000	*AAAA*	000000	2	0	*A*	00000000	2
				1	*C*	00000011	4
100	*ACBA*	110111	3				
110	*ACDB*	111010	6				
010	*AACB*	001101	1	0	*A*	00110111	3
				1	*C*	00110100	1
001	*AAAC*	000011	2	0	*B*	00001101	3
				1	*D*	00001110	2
101	*ACBC*	110100	3				
111	*ACDD*	111001	4				
011	*AACD*	001110	3	0	*B*	00111010	4
				1	*D*	00111001	4

From the above table, it can be seen that minimum distance is for the path *AACBC* which corresponds to the message bit sequence 0101.

5 REVERSE ERROR CORRECTION

We have seen some of the methods of forward error correction but reverse error correction is more economical than forward error correction in terms of the number of check bits. Therefore, usually error detection methods are implemented with an error correction mechanism which requires the receiver to request the sender for retransmission of the code word received with errors. There are three basic mechanisms of reverse error correction:

1. Stop and wait
2. Go-back-N
3. Selective retransmission.

5.1 Stop and Wait

In this scheme, the sending end transmits one block of data at a time and then waits for acknowledgement from the receiver. If the receiver detects any error in the data block, it sends a request for retransmission in the form of negative acknowledgement. If there is no error, the receiver sends a positive acknowledgement in which case the sending end transmits the next block of data. Figure 11 illustrates the mechanism.

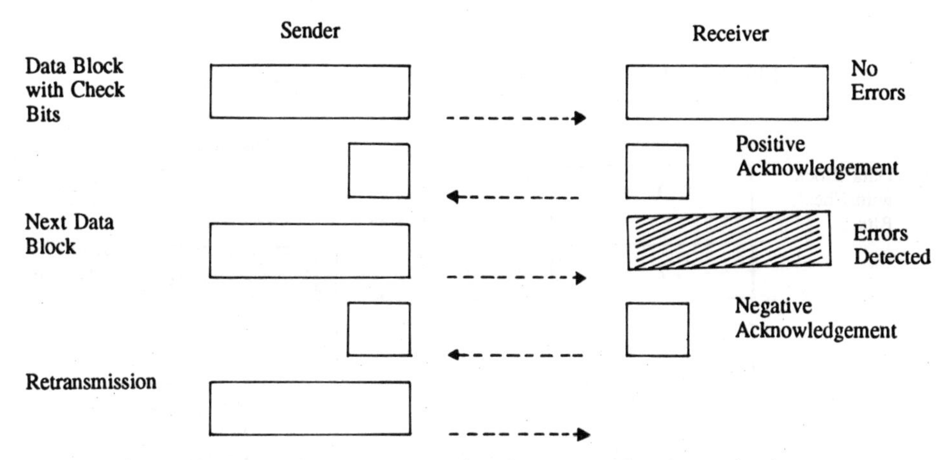

Fig. 11 Reverse error correction by stop-and-wait mechanism.

5.2 Go-Back-N

In this mechanism all the data blocks are numbered and the sending end keeps transmitting the data blocks with check bits. Whenever the receiver detects error in a block, it sends a retransmission request indicating the sequence number of the data block received with errors. The sending end then starts retransmission of all the data blocks from the requested data block onwards (Fig. 12).

5.3 Selective Retransmission

If the receiver is equipped with the capability to resequence the data blocks, it requests for selective retransmission of the data block containing errors. On receipt of the request, the sending end retransmits the data block but skips the following data blocks already transmitted and continues with the next data block (Fig. 13).

In data communications, we use reverse error correction using one of the mechanisms described above. We shall describe these mechanisms in detail in Chapter 7.

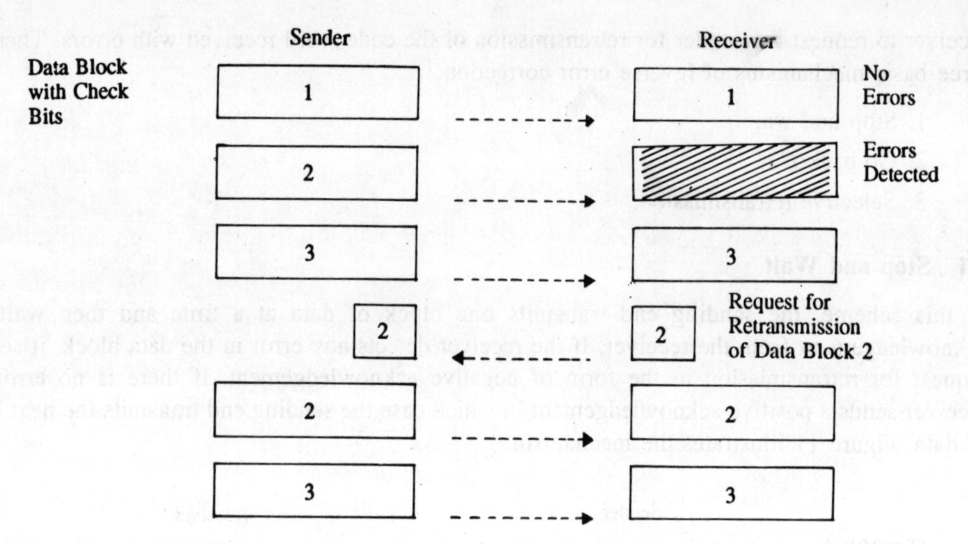

Fig. 12 Reverse error correction by go-back-*N* mechanism.

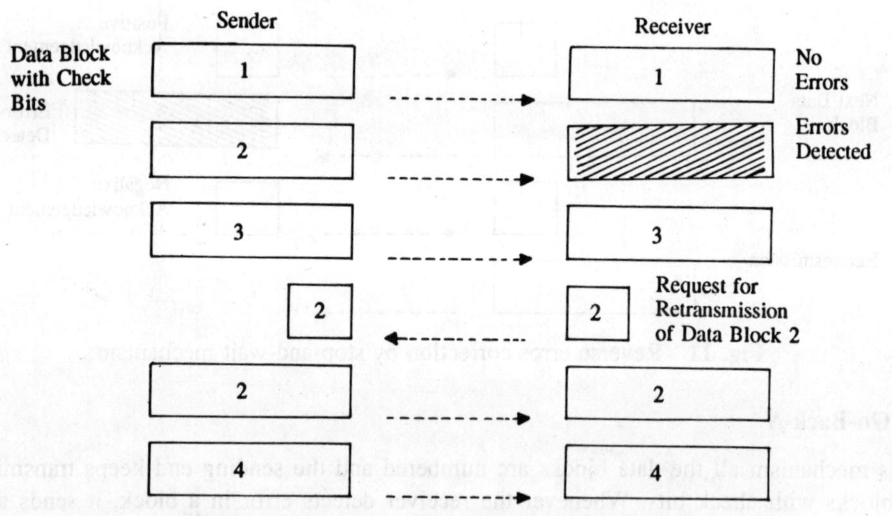

Fig. 13 Reverse error correction by selective retransmission mechanism.

6 SUMMARY

Errors are introduced due to imperfections in the transmission media. For error control, we need to detect the errors and then take corrective action. Parity bits, checksum and cyclic redundancy check (CRC) are some of the error detection methods. Out of the three, CRC is the most powerful and widely implemented.

Error correction methods include forward error correction or reverse error correction. Forward error correction requires additional check bits which enable the receiver to correct the errors as well. However, reverse error correction mechanisms, namely, stop and wait, go-back-*N* or selective retransmission are more common. In these mechanisms the receiver requests for retransmission of the data blocks received with errors.

PROBLEMS

1. There are 200 bytes in a data block, each byte being 8 bits. If the error rate is 1×10^{-5}, what is the probability of the block being received in error?

2. (a) Write the bit sequence corresponding to the message "DECODER". Assume ASCII code with odd parity bit.

(b) What is the text of the message corresponding to the following transmitted bit sequence? Assume ASCII code and even parity:

0010001011110011001100110011001110000010010010111

(c) Detect errors if any in the following message assuming it is in ASCII code with odd parity:

00001011100000110100111000101010

3. Write the bit transmission stream for the message "START BIT" which is coded in ASCII with VRC and LRC for error detection. Assume odd parity.

4. The following bit stream is encoded using VRC and LRC. Correct the error if any. What is the transmitted message? Assume ASCII code with even parity in VRC and LRC.

11001010000101111000011001100110001011100000010110001101100010111

5. The following bit stream is encoded using VRC and LRC. Detect the errors present in the message. Can the errors be corrected? Assume even parity.

11001010000101101001011001100111001011100000010110001101100010111

6. (a) Find simple checksum of the following bytes using modulo-256 addition. The MSBs are on the left of each byte.

10101010 10000001 11011011 01101100 10010101

(b) The above bit stream is transmitted with 2's complement of the checksum as the last byte. Check whether there are errors in the following received sequence:

10101010 10010001 11011011 01101110 10010101 11111001

(c) Repeat (b) for the following received sequence:

10001010 10000001 11011011 01101100 10110101 11111001

Why are the transmission errors not detected?

7. (a) Find the two checksum bytes as per the transport protocol for the following data bytes. The MSBs are on the left of each byte.

10101101 11101110 11111011 00111100 10001001

(b) Verify that the checksum is correct by performing the receiver check on the data bytes and the checksum bytes.

8. Generate CRC code for the data word 1010001011 using the divisor 11101.

9. If the CRC code is 10100010111100 and the generating polynomial is $x^4 + x^3 + x^2 + 1$, check if there is any error in the code word.

10. Received Hamming code word is 11110000101. Even parity is used. Locate and correct the bit in error.

11. Generate Hamming code for the following characters using even or odd parity as indicated. The character parity bit is not used in the ASCII code set.

(a) "U" ASCII Parity Odd
(b) "?" ASCII Parity Even
(c) "M" EBCDIC Parity Even

12. What are the characters corresponding to the received Hamming codes given below. The parity used is indicated in parentheses.

(a) 00011111110 ASCII (Odd)
(b) 11100011010 ASCII (Even)
(c) 100010011100 EBCDIC (Odd)

13. Generate convolutional code for message bits 1101 using rate 1/2 encoder shown in Fig. 8.

14. What are the message bits, if the received rate 1/2 code word is 11100010? Use the trellis diagram given in Fig. 10.

CHAPTER 5

Reference Model for Open System Interconnection

Communication is based on transfer of information but has a wider scope. In this chapter, we examine a man-to-man communication analogy to identify the functional requirements for meaningful communication. In the context of these requirements, we develop a model for communication in a computer network. The model is based on the concept of layered architecture. Then we proceed to the concepts of Open Systems and discuss the Reference Model for Open System Interconnection (OSI). Understanding the basic concept of the OSI model is essential for a systematic grasp over the computer networking issues. This chapter, therefore, is prerequisite to an understanding of the remaining chapters of this book.

1 TOPOLOGY OF A COMPUTER NETWORK

A computer network consists of end systems which are sources and sinks of information, and which communicate through a transit system interconnecting them (Fig. 1). The transit system is also called an interconnection subsystem or simply a subnetwork.

An end system comprises computers, terminals, software and peripherals forming an autonomous whole capable of performing information processing.

Each end system has an interaction point through which it is physically connected to the transmission media. The interaction point has an address by which the end system is identified.

Each end system hosts one or more application entities. It is due to these application entities that communication takes place between the end systems. They determine the subject and duration of their communication.

The subnetwork performs all transmission and switching activities required for transporting messages between the end systems. It is without an application entity and may consist of one or more types of transmission and switching equipments.

Transmission media connect end systems and the subnetwork and carry electrical signals.

2 ELEMENTS OF MEANINGFUL COMMUNICATION

The purpose of communication between the application entities is not served just by exchanging bits. The communication needs to be meaningful. Meaningful communication is always done with a purpose and aims at enlarging common understanding between communicating entities. Physical transfer of bits is transmission not communication.

There are some basic constituents of the communication process which must be present in any communication to make it meaningful. This is true for any type of communication, man-to-man or

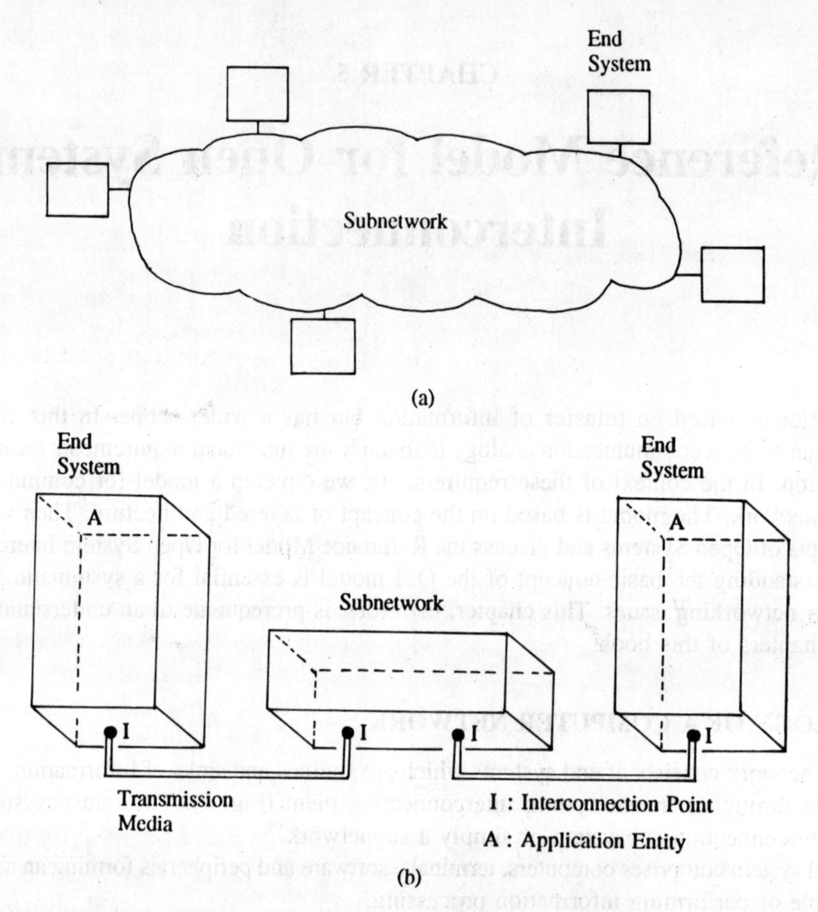

Fig. 1 Topology of a computer network.

computer-to-computer. We shall take a man-to-man communication analogy to understand the elements of the communication process. From this analogy we shall try to derive functional requirements for meaningful communication between two end systems.

Consider that a young Indian, Ravi, would like to tell his fiancee Mary who lives with her mother in Britain, about her visit to India during her vacations. Mary is learning Hindi for her forthcoming visit. Ravi decides to call Mary over the telephone and dials her telephone number. Mary's mother picks up the receiver. The conversation which ensues could be as follows:

Ravi : *Hello, Ravi speaking.*
Mother : *Hello Ravi, mama here.*
Ravi : *Mama, could I speak to Mary?*
Mother : *Please wait. I'll call her.*
.
.
.
Mary : *Hello Ravi, Mary here.*

Authentication
Identification of communicating entities.

Ravi : *Hello, I have made plans for your visit to India.*
Mary : *Thanks.*

Common Theme
Agreement on the common theme.

Ravi : *Are your Hindi lessons continuing? May I speak in Hindi?*
Mary : *No. I am still not at ease with Hindi. Please continue in English.*

Common Language
Agreement on the common language.

Ravi : *O.K. ...*
(Ravi tells her the programme.)

.
.
.

Marry : *Yes, fine.*

Synchronization (Forward)
Point of common understanding and indication of willingness to proceed.

.
.
.

(There is some disturbance on the line)
Mary : *Please repeat the dates. I did not hear you clearly.*
Ravi : *... (Ravi repeats) ...*
Mary : *Yes.*

Error Recovery
Recovery of the lost messages.

Ravi : *.....*
Mary : *Please speak slowly. Let me note it down.*
Ravi : *... (Ravi slows down) ...*

Flow Control
Control of the flow of messages.

.
.
.

Mary : *I could not follow after the visit to Agra.*
Ravi : *... (Ravi repeats and continues)...*

Synchronization (Backward)
Loss of the point of common understanding. Going back to the last point of common understanding.

.
.
.

Ravi : *Good bye, Mary.*
Mary : *Bye, Ravi.*

The above communication involved two communicating entities—Ravi and Mary. After identification of the communicating entities, there was need to decide the common theme and common language for communication. This was followed by an orderly dialogue session which required flow control, error control, and synchronization. Common theme, common language and an orderly session are the essential elements of any type of meaningful communication, though they may not be explicitly spelled out in every incidence of man-to-man communication.

The elements of meaningful communication discussed above are also applicable to a distributed computing system where application entities residing in different end systems communicate. Authentication (user password), login, code and data format conversion, and orderly exchange of messages with markers for dialogue synchronization are built into an end system. For each communication, these elements must be decided and agreed upon explicitly for the communication to be meaningful.

3 TRANSPORT ORIENTED FUNCTIONS

Communication results in generation of messages which are to be truthfully transported between the communicating end systems. The subnetwork provides means of transporting these messages but some additional functions, as discussed below, must also be built into the end systems for transportation of the messages through the subnetwork without any error.

3.1 Interaction with the Subnetwork

In the man-to-man communication situation described above, the communicating entities interacted with the telephone network by way of lifting the handset of the telephone instrument, waiting for the dial tone and dialling the telephone number. A computer network is somewhat analogous in this respect. The end systems need to interact with the subnetwork for transporting the messages to the destination. This interaction is in the form of specifying the address of the destination, answering an incoming call and releasing the connection.

3.2 Quality of Transport Service

The decision to use the telephone network and not the other available means (e.g. telegram, post) was taken by Ravi on considerations of delivery delay, cost and reliability. In a computer network, the end systems need to set up an appropriate transport connection of the required quality of service which is specified in terms of error rate, transit delay of message delivery, throughput, and of course, the cost.

3.3 Conversion of Signals

The messages generated by the communicating entities in the above example were speech signals which were suitably converted to electrical signals for transmission by the telephone instrument. In digital devices, messages are in the form of bits and the bits need to be converted by the end systems into electrical signals having suitable voltage levels and impedance for the transmission media.

3.4 Error Control

Speech signals have so much built in redundancy that even if there is some corruption of the signal, the communicating entities are usually able to make sense of the received signal. But computer communication is very sensitive to the errors which get introduced due to noise and distortion of the electrical signals during transmission. Some mechanism in the end systems to control these errors is required.

4 MEANINGFUL COMMUNICATION IN A DISTRIBUTED COMPUTING SYSTEM

From the above analogy of man-to-man communication, certain functional capabilities which must be built into an end system for meaningful communication can be derived. These capabilities are listed below. This list is not exhaustive but it does indicate broad categorization of the required capabilities.

- Authentication and login
- Code and format conversion
- Establishment of an orderly exchange of messages with markers
- Providing transport connection of required quality
- Interacting with the subnetwork
- Error control
- Conversion of bits into electrical signals and vice versa.

5 COMPONENTS OF A COMPUTER NETWORK

The above communication functions are implemented and controlled by using many hardware (physical) and software (logical) components in a computer network.

- Physical components
  - — Computer hardware
  - — Front end processor
  - — Terminals
  - — Modems, concentrators, multiplexers
  - — Transmission media
  - — Data switching equipment etc.
- Logical components
  - — Operating system
  - — File management system
  - — Communication software
  - — Application software etc.

All these components of a computer network function in a coordinated fashion to realise the functional requirements of meaningful communication between the end systems. Design and implementation of such a system is one of the most complex tasks that man has ever tried.

6 ARCHITECTURE OF A COMPUTER NETWORK

The architecture of a system, whether it is a building, organization or a computing system, describes how the system has been assembled using various components of the system. It defines the specifications of the components and their interrelationships.

The architecture of a computer network, or simply network architecture, specifies a complete set of rules for the connections and interactions of its physical and logical components.

7 NETWORK ARCHITECTURE MODELS

Designing the architecture of a complex system requires a model. A model helps in visualising and understanding the complex structure of the system. Standardization can then follow. The approach usually adopted for modelling a complex system is to partition it into meaningful functional pieces. After identifying these functional pieces, their interrelationships, interfaces, services and functionality are so defined that on integration they form a complete model.

7.1 Benefits of Partitioning

Partitioning is beneficial in many ways. A complex problem is broken down into smaller tasks. Each smaller task can be attended to by a specialist team. Partitioning can be so done that some of the tasks which already have an acceptable solution are not attempted again, thus reducing the developmental effort. Partitioning also results in a modular structure which permits flexibility of upgradation and reconfiguration.

7.2 Features of a Partitioned Structure

Partitioning is not a new concept. It is built into every system either intentionally or otherwise, e.g., an organization can have several offices and each office can have several functional levels. For the functionality of the overall organization, procedures for interactions are defined at various levels. Figure 2 shows the sequence of events which take place when a manager communicates with another manager in an organization having geographically distributed offices.

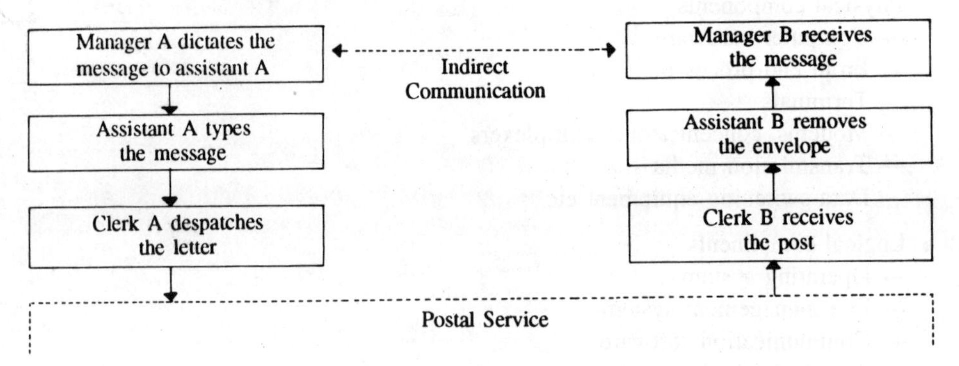

Fig. 2 Partitioned model of communications in an organization.

In the above example, the organizational structure has a vertical partition and several horizontal partitions. Some of the important features of this partitioned structure are:

- Different functions in the organization are separated and each function is distinctly implemented, e.g., typing is restricted to the second layer only.
- There is hierarchy of functions, i.e., to carry out functions assigned to one layer, services of the next lower layer are also needed. For example, manager A needs typing services of the assistant to send the message.
- To provide service to the higher layer, the lower peer layers may coordinate, e.g., to trace a lost letter the clerks will interact.
- The services are transparent, i.e., the lower layer does not restrict the higher layer in any way. It communicates whatever it receives from the higher layer, e.g., clerk A despatches the letter. He is in no way concerned about its contents.
- There is no bypassing of layers. The interaction is between the adjacent layers only, e.g., the managers do not directly interact with the clerks.
- There is indirect interaction between the peer layers. The managers interact with each other through the services provided by their assistants.

8 LAYERED ARCHITECTURE OF A COMPUTER NETWORK

Decomposition of the organization into offices and each office into hierarchical functional levels and the interaction procedures define the overall organization architecture. A computer network is also partitioned into end systems interconnected using a subnetwork and the communication process is decomposed into hierarchical functional layers (Fig. 3).

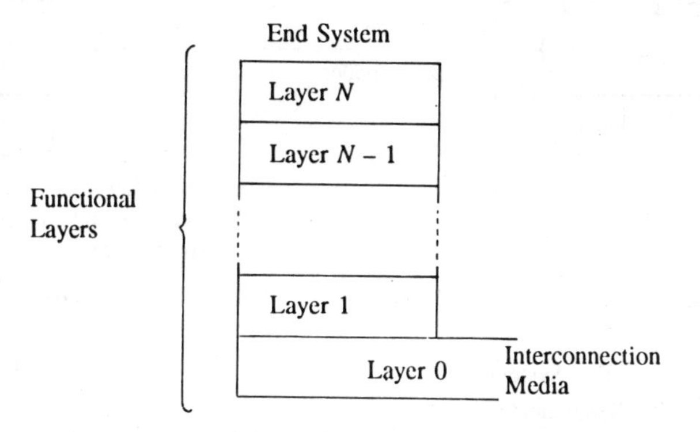

Fig. 3 Layered architecture of an end system.

Just like in an office, each layer has a distinct identity and a specific set of functions assigned to it. Each layer has an active element, a piece of hardware or software, which carries out the layer functions. It is called layer entity.

The general criteria for defining the boundaries of a layer are:

- Each function is distinctly identified and implemented precisely in one layer.
- Sequentiality of the functions is ensured by proper design of the hierarchy.
- Number of layers should be minimum.
- Boundaries of a layer are defined taking into consideration the existing acceptable implementation.
- The implementation details of a function in a layer are hidden so that any change in the implementation does not affect other layers.

8.1 Functionality of the Layered Architecture

The layered architecture emphasizes that there is hierarchy of functions. Each layer provides certain services to the next higher layer which uses these services to carry out its assigned functions (Fig. 4).

Each layer also needs to interact with the peer layer of another end system or the subnetwork to carry out its functions. Since there is no direct path between peer layers, they interact using the services of the lower layers. Therefore, two types of communication take place in the layered architecture to make it work properly (Fig. 5):

1. Hierarchical communication
2. Peer-to-peer communication.

Hierarchical Communication. Hierarchical communication between adjacent layers of a system is for requesting and receiving services from the lower layer. The rules and procedures for hierarchical

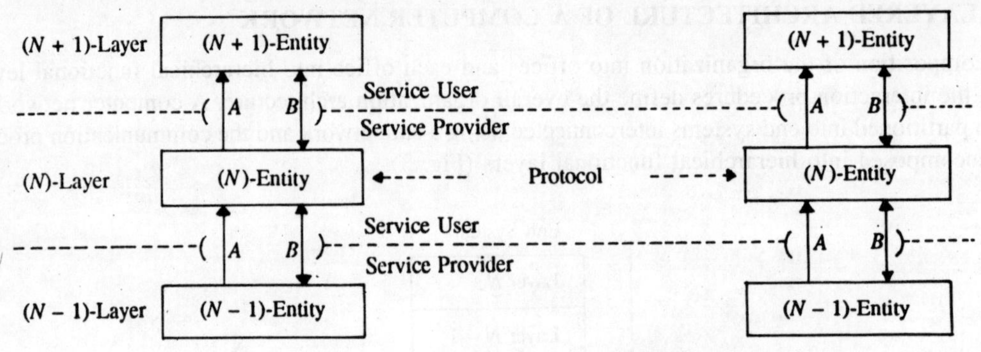

Fig. 4 Service providers and users in the layered architecture.

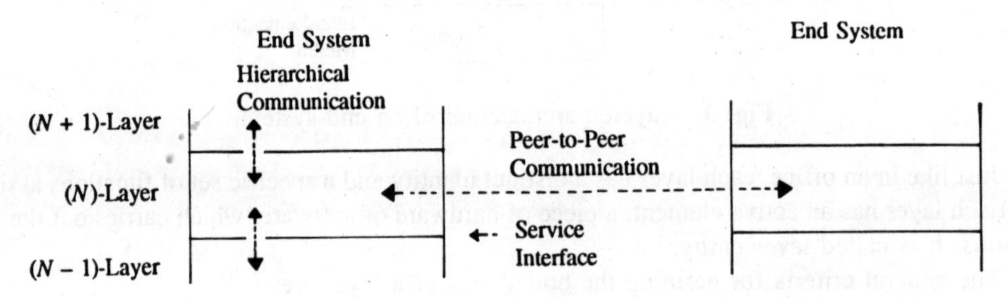

Fig. 5 Hierarchical and peer-to-peer communications in the functional layers.

communication are specified by the service interface definition which consists of: description of the services provided by the lower layer; and rules, procedures and parameters for requesting and utilizing the services.

The messages exchanged between the adjacent layers during hierarchical communication are called Interface Control Information (ICI), see Fig. 6.

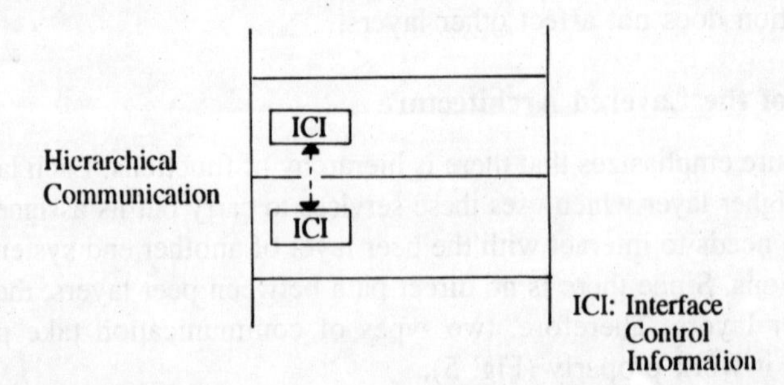

Fig. 6 Interface control information for hierarchical communication.

Peer-to-Peer Communication. Peer-to-peer communication is between the peer layers for carrying out an assigned set of functions. Rules and procedures for peer-to-peer communication are called protocol. The messages which are exchanged between the peer layers are called Protocol Control Information, PCI (Fig. 7).

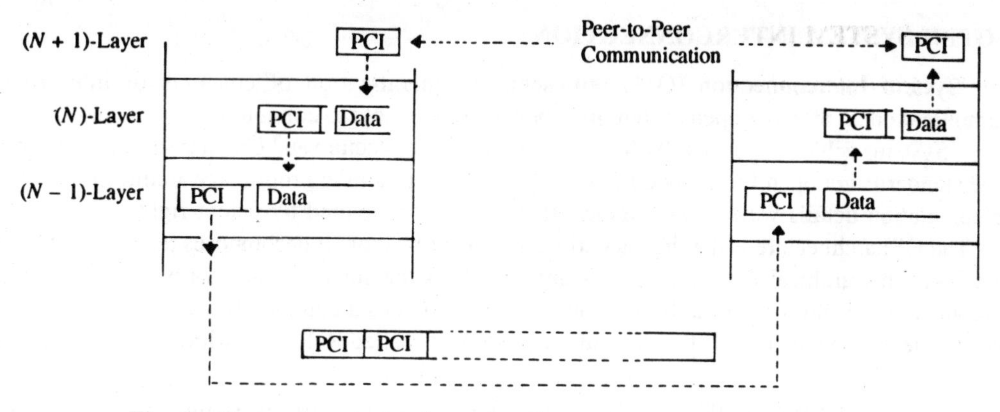

Fig. 7 Protocol control information for peer-to-peer communication.

Since there is no direct path between the peer layers, protocol control information is exchanged using the services provided by the lower layer. The mechanism commonly utilized is shown in Fig. 7.

- Whatever a layer receives from the layer above is treated as "data".
- PCI is added as header to the "data" and handed over to the layer below. The PCI may consist of one or several fields.
- The process is repeated at each layer.
- At the other end, each layer strips off the PCI of its peer layer from the received block of data and hands over the remaining block to the layer above.

For the messages to be meaningful to the peer layers, format, contents and sequence of the messages contained in the PCI fields should be known to peer layers in advance. The protocol definition of each layer specifies these attributes.

8.2 Need for Standardization of Network Architecture

The layered architecture concept was built into many systems but different vendors defined proprietary protocols and interfaces. The layer partitioning also did not match. As a result, there was total integration incompatibility of architectures developed by different vendors. Standardization of network architecture can solve many problems and save a lot of effort required for developing interfaces for networking different architectures. This was later realised and efforts were made in this direction.

Today, we have several network architectures developed by manufacturers and by standardization organizations. Some of the important network architectures are:

- IBM's System Network Architecture (SNA)
- Digital's Digital Network Architecture (DNA)
- Open System Interconnection (OSI) Reference Model developed by ISO (International Organization for Standardization) and CCITT.

SNA and DNA are vendor-specific layered architectures while the OSI Model has been accepted as an international standard. It is covered in detail in the following sections.

9 OPEN SYSTEM INTERCONNECTION

Open System Interconnection (OSI) represents a generalization of concepts of interprocess communication so that any open system may be technically able to communicate with another open system. Systems achieve openness by following a certain architecture and obeying standard protocols. These standards are open for anybody to use and implement unlike proprietary architectures whose implementation details were always either trade secrets or covered by patent rights.

The OSI architecture is the first step towards standardization. It decomposes the communication process into hierarchical functional layers and identifies the standards necessary for open system interconnection. It does not specify the standards but provides a common basis for coordination of standards development. The OSI architecture is, therefore, called Reference Model for Open System Interconnection.

This model was developed primarily by ISO and was approved as international standard IS 7498 in 1983. CCITT did parallel work and their Recommendation X.200 is in complete alignment with IS 7498.

10 LAYERED ARCHITECTURE OF THE OSI REFERENCE

In the OSI Reference Model, the communication functions are divided into a hierarchy of seven layers as shown in Fig. 8. It is also referred to as the 7-layer model. The transmission medium is not included in the seven layers and, therefore, it can be regarded as the 0th layer.

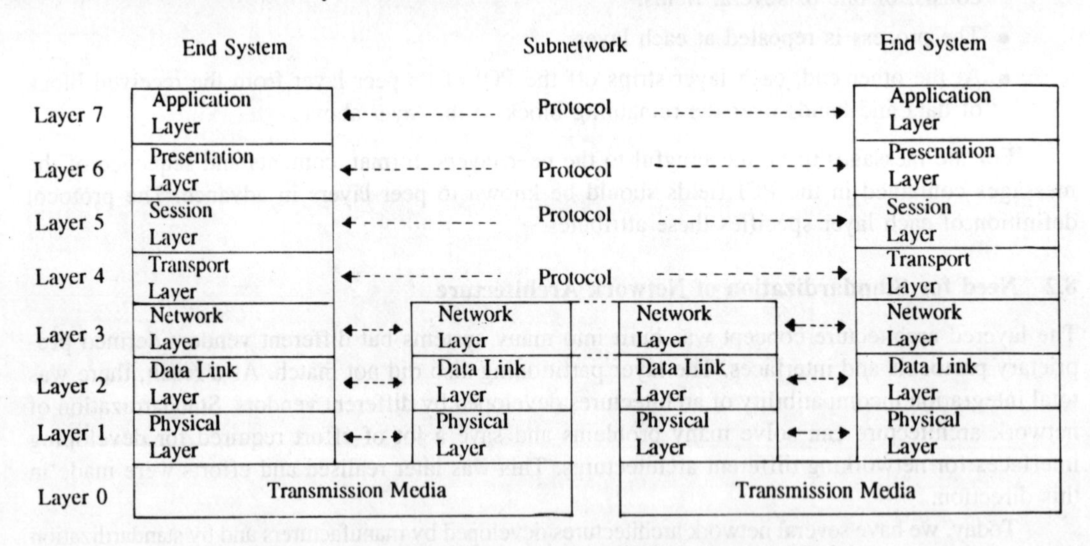

Fig. 8 Layered architecture of the OSI Reference Model.

The subnetwork has at most three layers as shown. These three layers interface with the corresponding peer layers of the end systems to carry out functions relating to the transport of messages from one end system to the other.

Table 1 gives a summary of the functions and services provided by the layers of the OSI model.

We will take up these layers individually and describe them in detail later in the book. A brief description of their functions and services is given below.

Table 1 Summary of Functions and Services Provided by Various Layers of the OSI Reference Model

Level	Layer	Primary functions	Services provided to next higher layer
7	Application	Support the end user, LOGIN, Pass word, File transfer	This is the highest layer and provides user-oriented services.
6	Presentation	Code and Format conversion	Freedom from compatibility problems
5	Session	Session management, Synchronization	Dialogue management
4	Transport	Optimum utilization of the network resources	End-to-end transport connection of the required quality in a cost-effective manner
3	Network	Interaction with the subnetwork; Routing and relaying	Network connection linking the end systems
2	Data link	Error control; Flow control	Reliable transfer of bits across the physical connection
1	Physical	Conversion of bits into electrical signal of suitable characteristics	Transmission of bits
0	Media	Transmission of electrical signals	Transmission of electrical signals

10.1 Application Layer

As the highest layer in the OSI model, the Application layer provides services to the users of the OSI environment. The purpose of the Application layer is to serve as a window between the communicating entities. LOGIN, password checking, file transfer, etc. are some of the functions of the Application layer.

10.2 Presentation Layer

The purpose of the Presentation layer is to present the information to the communicating application entities in a way that preserves the meaning while resolving the syntax (code and data format) differences. There are three syntactic versions of data being transferred, the syntax used by the application entity of the originator of the data, the syntax used by the recipient of the data, and the "transfer" syntax used to transfer the data between presentation entities (Fig. 9).

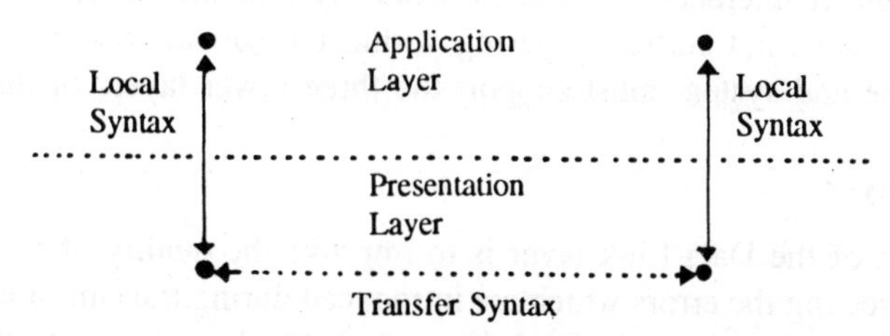

Fig. 9 Local and transfer syntaxes of the Presentation layer.

These syntaxes may be same or different. When they are not same, the Presentation layer contains functions necessary to transform the transfer syntax to the required syntaxes used by the Application entities preserving the meaning. There is no fixed transfer syntax, and is to be negotiated by the Presentation entities.

10.3 Session Layer

The purpose of the Session layer is to provide the means necessary for the cooperating Presentation entities to organize and synchronize their data exchange. It provides functions which are necessary for opening a communication relationship called a session, for carrying it out in an orderly fashion and for terminating it. At the time of session termination, the Session entities ensure that there is no data loss unless there is a request from the Presentation entities to abort the session.

The Session layer provides two-way simultaneous, two-way alternate and one-way communication services. The Session synchronization service enables the Presentation entities to mark and acknowledge identifiable ordered synchronization points. During a session, the Session entities enable resynchronization if the Transport service is disrupted and reestablished.

10.4 Transport Layer

The overall function of the Transport layer is to provide Transport service of the quality required by the Session entities in a cost-effective manner. The quality of the Transport connection is specified in terms of residual error rate, delay, throughput and other quality determining parameters.

For optimum utilization of network resources and for achieving the quality of service, the Transport layer may do multiplexing, splitting, blocking or segmenting. These functions are described in the next section.

It may be seen from Fig. 8 that the Transport layers provides end-to-end connectivity. Depending on the quality of the network service, the Transport layer may be required to carry out sequencing of the messages and exercise end-to-end error control also to ensure quality of the Transport service provided to the Session entities.

Figure 10 shows a hypothetical situation—two end systems having access to several subnetworks. These could be PSPDN (Packet Switched Public Data Network), PSTN (Public Switched Telephone Network), point-to-point dedicated circuit and CSPDN (Circuit Switched Public Data Network). Assuming each subnetwork provides different quality of service at a cost, the Transport layer carries out functions necessary to meet the quality of service in a cost-effective manner by choosing an appropriate subnetwork.

10.5 Network Layer

The Network layer provides the means to access the subnetwork for routing the messages to the destination end system. It interacts with the Network layer of the subnetwork for this purpose.

The OSI model does not address layering within the subnetwork, but an access node of a subnetwork facing the end system must support the three lower layers of the OSI model.

10.6 Data Link Layer

The primary function of the Data Link layer is to improve the quality of service provided by the Physical layer by correcting the errors which are introduced during transmission of electrical signals. It appends error detection bits to a block of data before handing it over to the Physical layer for

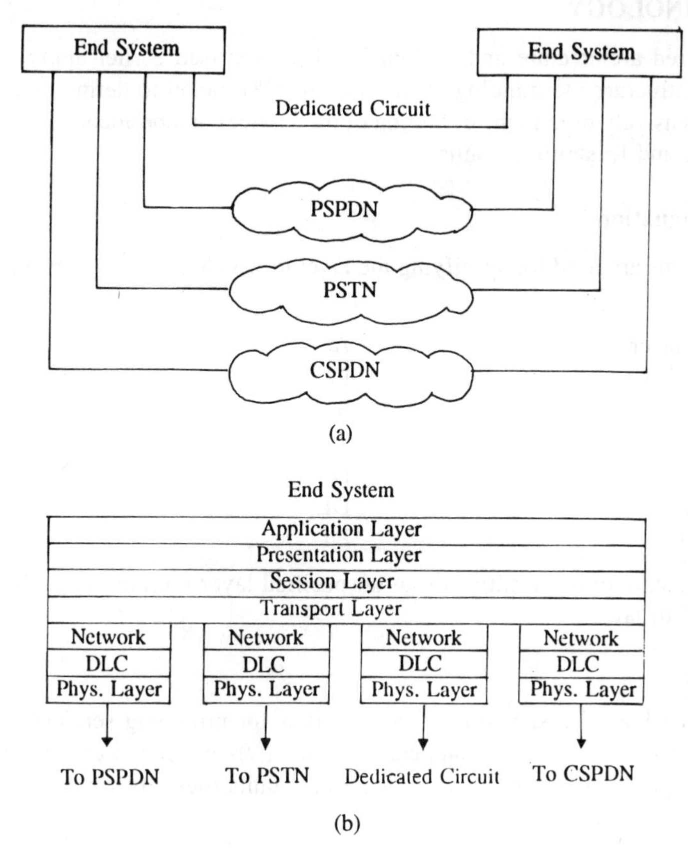

Fig. 10 Selection of an appropriate Network connection by the Transport layer.

transmission. These bits are used for detecting the errors if any in the data blocks received by the Data Link layer at the other end. Usually, retransmission mechanisms are employed to correct the errors.

It is necessary that the receiving end be provided with some control to regulate the flow of the incoming frames. Therefore, flow control mechanisms are also an integral part of the error control mechanism in the Data Link layer.

10.7 Physical Layer

The Physical layer is primarily concerned with transmission of bits across the interconnection media. To this end, the Physical layer carries out the following functions:

- Conversion of the bits into electrical signals having characteristics suitable for transmission over the media
- Signal encoding if required
- Relaying of the digital signals using intermediary devices like modems.

The Physical layer does not have capability to detect and correct errors which are introduced due to noise and distortion of electrical signals during transmission.

11 OSI TERMINOLOGY

Principles of layered architecture and its functionality discussed earlier apply to the OSI model also. A very well structured terminology is used in the OSI model to define its functionality. This terminology is extensively used in the definition of the services and protocols. If it is not understood, a lot of confusion and frustration results.

11.1 Layer Designations

The following initials are used for specifying the layer to which an entity, a data unit or a primitive belongs:

Application layer	A
Presentation layer	P
Session layer	S
Transport layer	T
Network layer	N
Data Link layer	DL
Physical layer	Ph

Any entity, data unit or primitive of an unspecified layer prefixed with (*N*), indicates that it belongs to the "*N*"th layer.

11.2 Connection

A connection is the logical association of peer entities for providing services to the next higher layer. In Fig. 11 (*N* + 1)-entities communicate over the (*N*)-connection established by (*N*)-entities on the former's request. They send their protocol data units (described below) transparently over the connection.

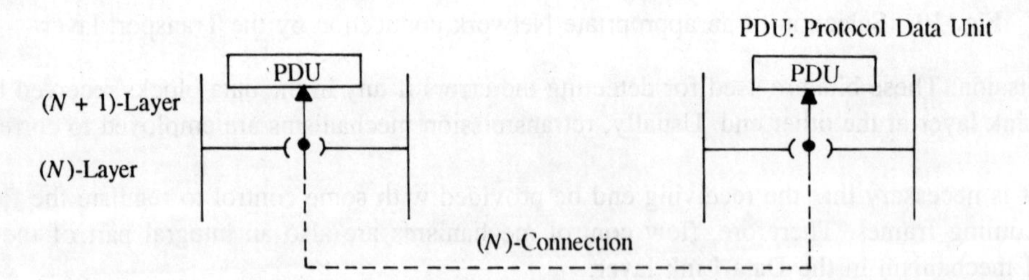

Fig. 11 (*N*)-connection for providing services to (*N* + 1)th layer.

11.3 Service Access Point (SAP)

For hierarchical communication, the adjacent layer entities interact through a Service Access Point (SAP) which is at the interface between the layers (Fig. 12). Each service access point supports one communication path.

11.4 Service Access Point Address

(*N*)-service-access-point-address or (*N*)-address for short, identifies the service access point located between (*N* + 1)-layer and (*N*)-layer. The (*N* + 1)-entity is accessible through the (*N*)-address and, therefore, it is also the address of the (*N* + 1)-entity (Fig. 12).

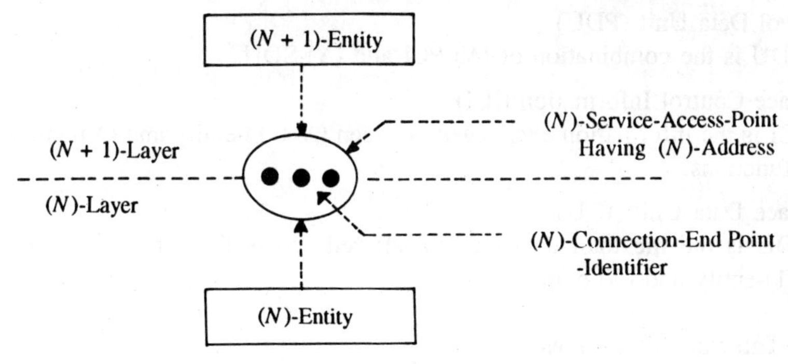

Fig. 12 (*N*)-service access point.

11.5 Connection End Point Identifier

A service access point path can support multiple connections on its communication path. (*N*)-connection-endpoint-identifier uniquely identifies a connection (Fig. 12). It consists of two parts:

1. (*N*)-address of the (*N*)-service-access-point
2. A suffix which is unique within the scope of the (*N*)-service access point.

11.6 Data Units

There are several types of data units which are exchanged between the adjacent and the peer layers (Fig. 13):

- Protocol Control Information (PCI)
  (*N*)-PCI is the protocol control information exchanged between the (*N*)-entities to coordinate their functions.
- Service Data Unit (SDU)
  (*N*)-SDU is the data unit which is transferred between the ends of a (*N*)-connection and whose identity is preserved during the transfer.

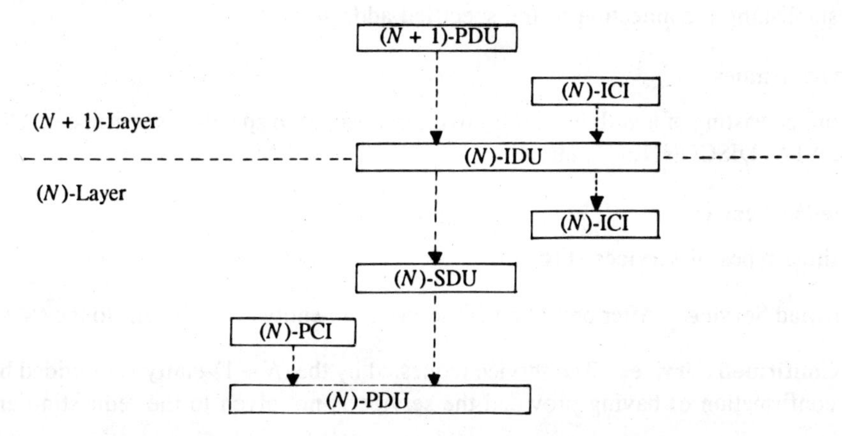

Fig. 13 Data units in the OSI Reference Model.

- Protocol Data Unit (PDU)
  (*N*)-PDU is the combination of (*N*)-PCI and (*N*)-SDU.
- Interface Control Information (ICI)
  (*N*)-ICI is the information exchanged between (*N* + 1)-entity and (*N*)-entity to coordinate their functions.
- Interface Data Unit (IDU)
  (*N*)-IDU is the the total data unit transferred across the service access point between (*N* + 1)-entity and (*N*)-entity.

11.7 Service Interface Primitives

Services are provided across a service interface between (*N* + 1)-layer and (*N*)-layer. Primitives used for providing and using the services are as follows (Fig. 14):

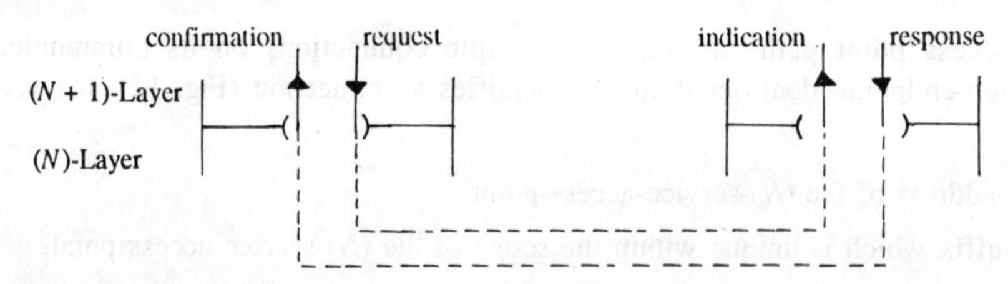

Fig. 14 Service interface primitives.

Request. It is used by the (*N* + 1)-entity to request (*N*)-entity to provide a particular service.

Indication. Indication of providing a service is given by (*N*)-entity to (*N* + 1)-entity.

Response. It is given by the (*N* + 1)-entity to (*N*)-entity in reply to the indication primitive.

Confirmation. This primitive is used by the (*N*)-entity to indicate the completion of the service as requested by the (*N* + 1)-entity.

These primitives are associated with the name of a service and parameters. For example, T-CONNECT request (parameter such as address) is a request by the Transport layer to the Network layer for establishing a connection to the specified address.

11.8 Service Names

A single word consisting of a verb in its infinitive form is used to specify a service, e.g., CONNECT, ABORT, DATA, DISCONNECT, etc.

11.9 Types of Services

There are three types of services (Fig. 15):

(i) **Confirmed Service.** After providing the service, (*N*)-entity confirms this to the (*N* + 1)-entity.

(ii) **Non-Confirmed Service.** The service requested by the (*N* + 1)-entity is provided by the (*N*)-entity but confirmation of having provided the service is not given to the requesting entity.

(iii) **Provider Initiated Service.** In this case the (*N*)-entity initiates and provides the service.

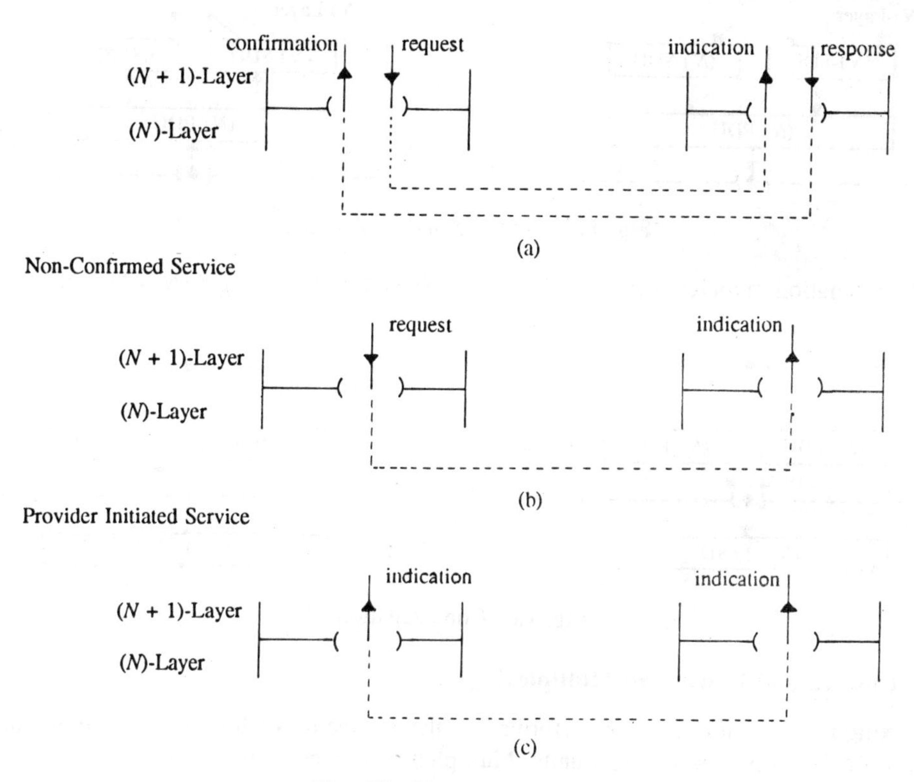

Fig. 15 Types of service.

11.10 Segmenting, Blocking and Concatenation

Segmenting, blocking and concatenation are carried out by the layer entities to accommodate incompatible sizes of the data units. (*N*)-entity may segment an (*N*)-SDU into several (*N*)-PDUs within an (*N*)-connection (Fig. 16). At the other end of the connection, the (*N*)-PDUs are reassembled into one (*N*)-SDU. An SDU always preserves its identity between the ends of a connection.

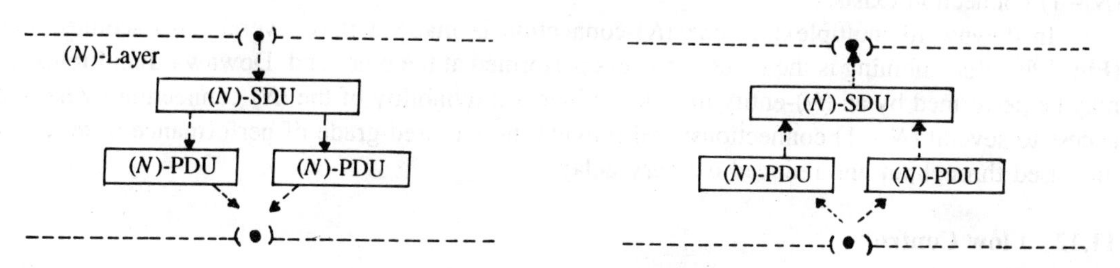

Fig. 16 Segmentation and reassembly.

In blocking, several (*N*)-SDUs with their (*N*)-PCIs are mapped into one (*N*)-PDU. Deblocking is the reverse process carried out at the other end of the connection to get back the (*N*)-SDUs (Fig. 17).

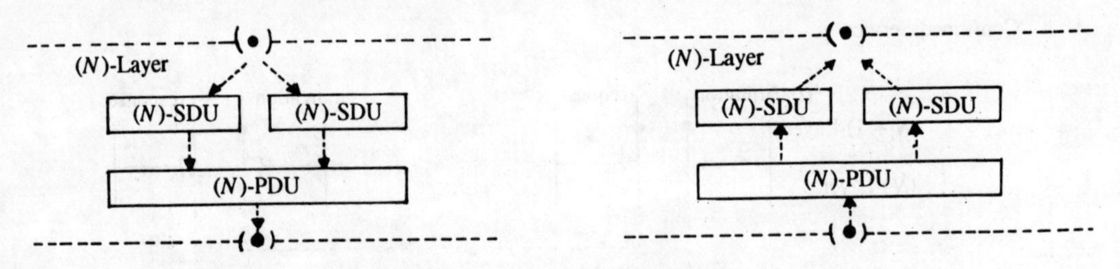

Fig. 17 Blocking and deblocking.

Concatenation involves mapping of several (*N*)-PDUs into a single (*N* – 1)-SDU (Fig. 18).

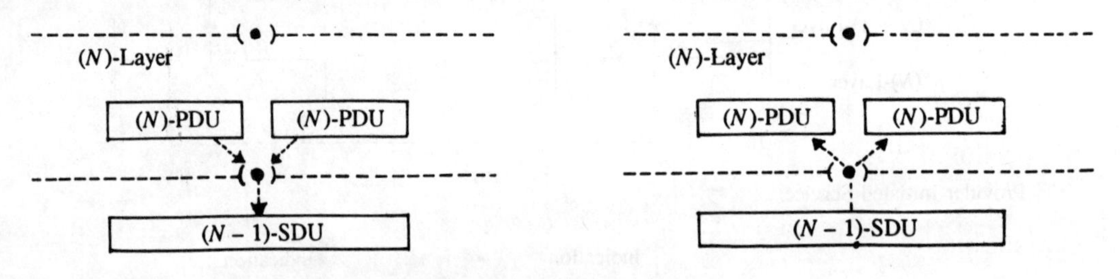

Fig. 18 Concatenation.

11.11 Upward and Downward Multiplexing

Multiplexing is a mapping function performed on the connections while segmenting and blocking functions are performed on the data units. Multiplexing can be of two types:

1. Upward multiplexing
2. Downward multiplexing.

In upward multiplexing, several (*N*)-connections are mapped into one (*N* – 1)-connection (Fig. 19a). Demultiplexing is the reverse process performed at the other end. Upward multiplexing may be performed by an (*N*)-entity in order to make more efficient and economic use of the (*N* – 1)-connection; and provide several (*N*)-connections in an environment where only one (*N* – 1)-connection exists.

In downward multiplexing, one (*N*)-connection is mapped into several (*N*–1)-connections (Fig. 19b). Recombining is the reverse process performed at the other end. Downward multiplexing may be performed by an (*N*)-entity in order to increase reliability of the (*N*)-connection by having access to several (*N* – 1) connections; and provide the required grade of performance in terms of increased throughput and reduced delivery delay.

11.12 Flow Control

Flow control refers to regulating the flow of incoming data units by the layer entities so that they are not overwhelmed with data units and may process the data units already received. Two types of flow are provided (Fig. 20):

1. Peer flow control which regulates the flow of (*N*)-PDUs between (*N*)-entities.

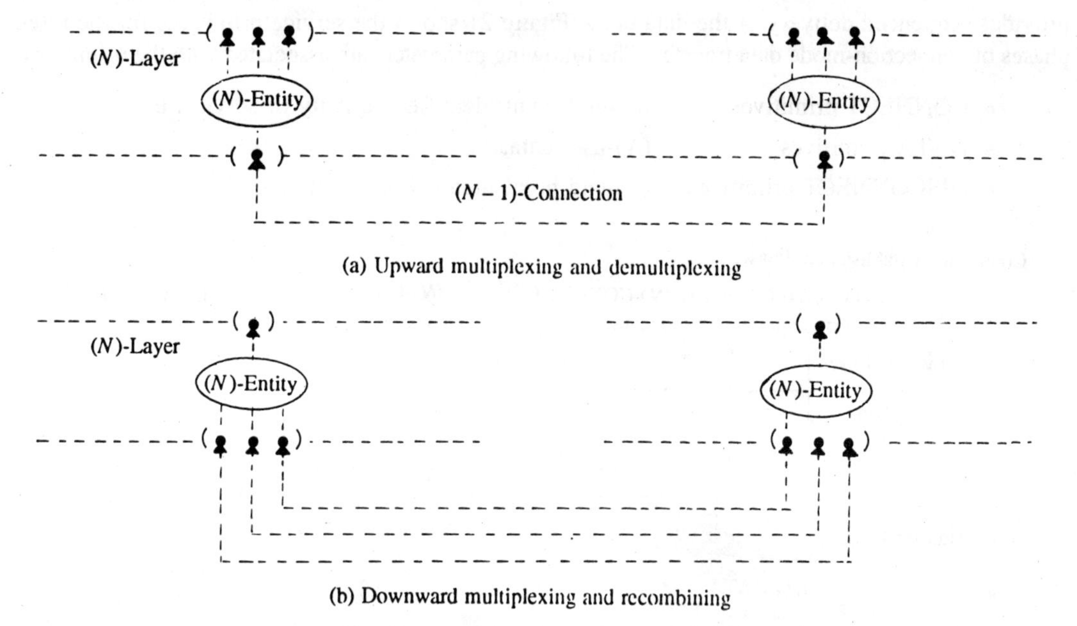

Fig. 19 Multiplexing of connections.

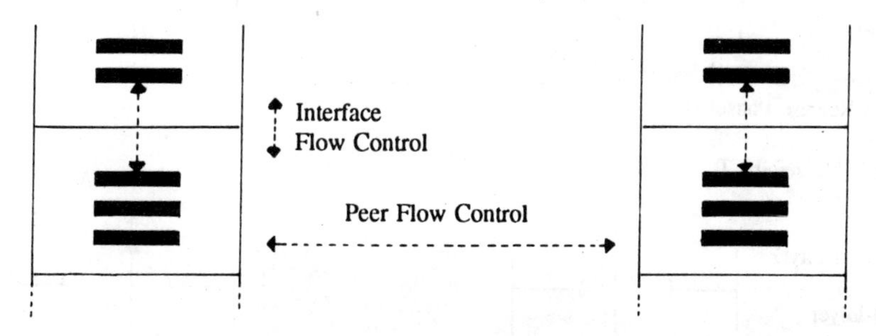

Fig. 20 Types of flow control.

2. Interface flow control which regulates the flow of (*N*)-IDUs between the (*N*)-entity and (*N* + 1)-entity.

11.13 Connection-Mode Data Transfer

An OSI entity can transfer data to the peer entity using the connection-mode data transfer which involves three phases:

1. Connection establishment phase
2. Data transfer phase
3. Connection release phase.

Connection is established between the communicating (*N* + 1)-entities using the service provided by the (*N*)-entities. Establishment of a connection involves negotiation and agreement between the service user entities and the service provider entities. The connection once established provides sequenced delivery of the data units. Figure 21 shows the service primitives for the three phases of connection-mode data transfer. The following parameters are associated with the primitives:

- CONNECT primitives (*N*)-end point identifiers, quality of service, etc.
- DATA primitives (*N*)-user data.
- DISCONNECT primitives (*N*)-end point identifiers, reason, etc.

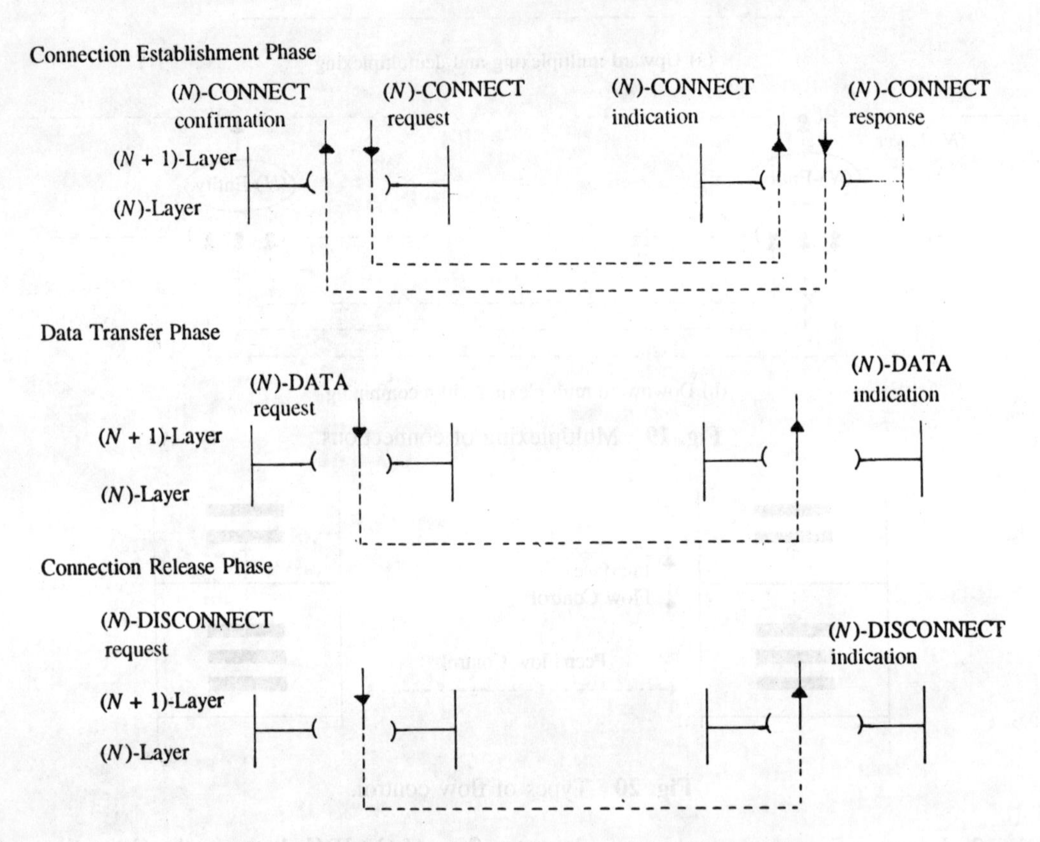

Fig. 21 Primitives for connection-mode data transfer.

11.14 Connectionless Data Transfer

Connectionless data transfer is a single self-contained action without establishing, maintaining and releasing a connection. Connectionless-mode service provides transmission of one (*N*)-SDU from a source (*N*)-SAP to one or more (*N*)-SAPs. There is no prior negotiation between the service users and the service providers. There is no assurance of delivery of the data unit. Figure 22 shows the service primitives for connectionless service. The parameters associated with the primitives are source (*N*)-SAP, destination (*N*)-SAP, quality of service parameters, (*N*)-user data.

12 ROLE OF OSI REFERENCE MODEL IN STANDARDS DEVELOPMENT

The OSI model provides complete reference for the totality of standards necessary for open system interconnection. It enables the developers of standards to keep existing standards in perspective,

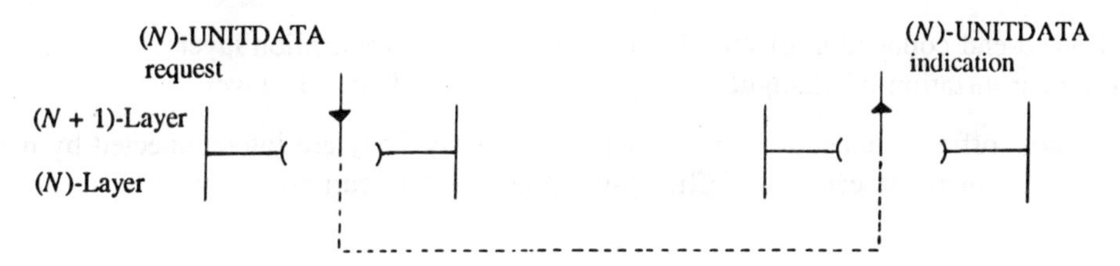

Fig. 22 Primitives for connectionless data transfer.

identify areas in existing standards requiring additional development and areas where new standards should be developed.

Tremendous amount of activity is presently underway in standards development. Many standards have been developed, many are still in draft stage and development of some is yet to be attempted. We will examine some of the important standards for services and protocols in the remaining chapters. Development of standards has been taken up by several organizations. Some of the standards-making organizations which are important to us are:

- American National Standards Institute (ANSI)
- Electronic Industries Association (EIA)
- European Computer Manufacturers Association (ECMA)
- Institute of Electrical and Electronics Engineers (IEEE)
- International Organisation for Standardization (ISO)
- International Telegraph and Telephone Consultative Committee (CCITT)
- National Bureau of Standards.

13 SUMMARY

A standard network architecture is required for meaningful communication between end systems. The reference model for Open System Interconnection specifies a seven-layered architecture and provides a common basis for development of standards.

Functionality of layered architecture is based on a service-provider and service-user concept. Each layer uses the service provided by the lower layer to carry out the assigned functions and in turn, provides services to the next higher layer. Protocols are the rules and procedures of interaction between peer layers of different systems.

The OSI model is a significant step towards standardization of the network architecture and will facilitate interconnection between systems from different vendors.

PROBLEMS

1. Match the following:

(a) Data encryption	(i) Session layer
(b) Bit synchronization	(ii) Network layer
(c) Parity bits	(iii) Presentation layer
(d) Interacting with the subnetwork	(iv) Data Link layer
(e) Login	(v) Physical layer

(f) End-to-end connection of required quality (vi) Application layer
(g) Synchronization of dialogue (vii) Transport layer

2. Three post offices cooperate to provide postal service. They are interconnected by mail vans which carry the mail. At each post office three processes are carried out in order to provide the postal service.

(a) Collection of mail from service users.
(b) Sorting of mail.
(c) Packing of mail for the other post offices.

Draw a layered model of the postal department. Define the functions of each layer. Define the service provided by each layer. Define possible peer layer interactions. Build required security error control measures in the system. Assume that the mail packing department hands over the mail to the mail van. How will a letter posted within the same postal area be processed?

3. An intermediate OSI system carries out the relaying function between two or more end systems. Relaying function is the responsibility of the Network layer and the intermediate OSI system has the first three layers. Draw a layered model of such a system which is connected to two end systems through two of its physical ports.

4. Two end systems are interconnected using a pair of modems. A modem has only the Physical layer and two ports, one towards the end system and the other towards the second modem. Draw the layer model of the configuration.

5 In Fig. 8, error control is carried out by the four Data Link layers and the two Transport layers. Justify the need for error control at so many stages and levels.

6. In Fig. 8, a user at one of the end system wishes to access a database at the other end system. Write the primitives which are exchanged at various interfaces of the OSI reference model for establishing the connection. Assume confirmed service at each interface.

CHAPTER 6

The Physical Layer

Transmission of digital information from one device to another is the basic function for the devices to be able to communicate. This chapter describes the first layer of the OSI model, the Physical layer, which carries out this function. After examining the services it provides to the Data Link layer, functions of the Physical layer are discussed. Relaying through the use of modems is a very important data transmission function carried out at the Physical layer level. Various protocols and interfaces which pertain to the relaying functions are put into perspective. We then proceed to examine EIA-232-D, a very important interface of the Physical layer. We discuss its applications and limitations. Before closing the chapter, we take a look at two other less popular interfaces, namely, RS-449 and X.21.

1 THE PHYSICAL LAYER

Let us consider a simple data communication situation shown in Fig. 1, where two digital devices A and B need to exchange data bits.

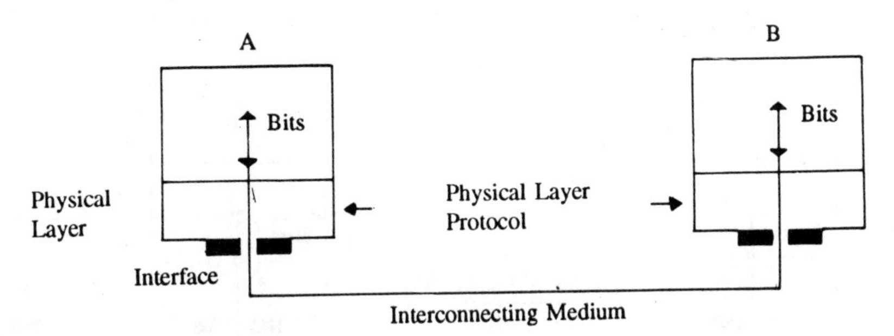

Fig. 1 Transmission of bits by the Physical layer.

The basic requirements for the devices to be able to exchange bits are the following:

1. There should be a physical interconnecting medium which can carry electrical signals between the two devices.

2. The bits need to be converted into electrical signals and vice versa.

3. The electrical signal should have characteristics (voltage, current, impedance, rise time etc.) suitable for transmission over the medium.

4. The devices should be prepared to exchange the electrical signals.

These requirements, which are related purely to the physical aspects of transmission of bits, are met out by the Physical layer. The rules and procedures for interaction between the Physical layers are called Physical layer protocols (Fig. 1).

The Physical layer provides its service to the Data Link layer which is the next higher layer and uses this service. It receives service of the physical interconnection medium for transmitting the electrical signals.

1.1 Physical Connection

The Physical layer receives the bits to be transmitted from the Data Link layer (Fig. 2). At the receiving end, the Physical layer hands over these bits to the Data Link layer. Thus, the Physical layers at the two ends provide a transport service from one Data Link layer to the other over a "Physical connection" activated by them. A Physical connection is different from a physical transmission path in the sense that it is at the bit level while the transmission path is at the electrical signal level.

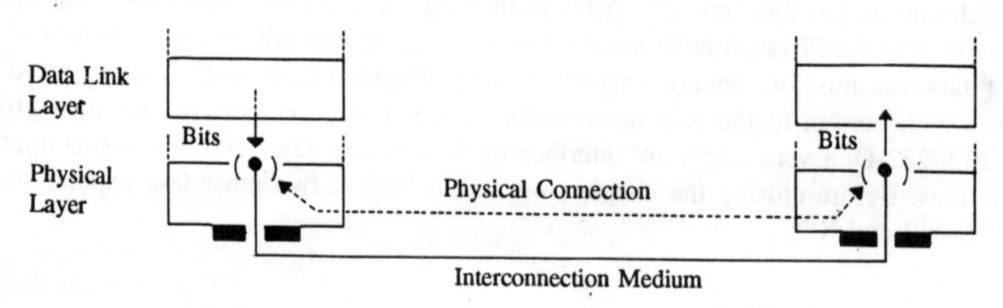

Fig. 2 Physical connection.

The Physical connection shown in Fig. 2 is point-to-point. Point-to-multipoint Physical connection is also possible as shown in Fig. 3.

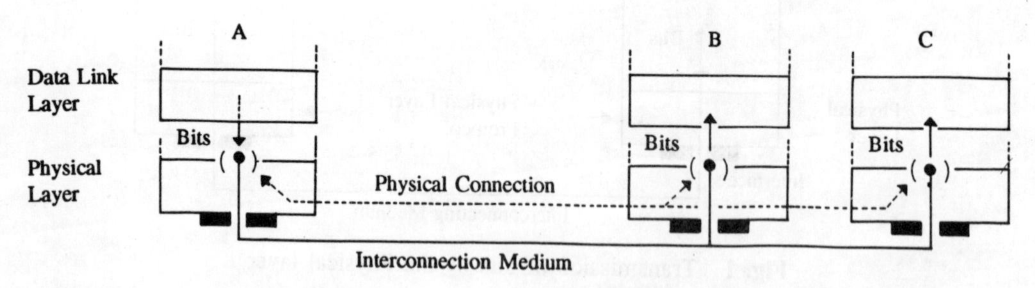

Fig. 3 Point-to-multipoint Physical connection.

1.2 Basic Service Provided to the Data Link Layer

The basic service provided by the Physical layer to the Data Link layer is the bits transmission service over the Physical connection. The Physical layer service is specified in ISO 10022 and CCITT X.211 documents. Some of the features of this service are now described.

Activation/Deactivation of the Physical Connection. The Physical layer, when requested by the Data Link layer, activates and deactivates a Physical connection for transmission of bits. Activation ensures that if one user initiates transmission of bits, the receiver at the other end is ready to receive them. The activation and deactivation service is non-confirmed, i.e., the user activating or deactivating a connection is not given any feedback of the action having been carried out by the Physical layer.

A Physical connection may allow full duplex or half duplex transmission of the bits. In half duplex transmission, the users themselves decide which of the two users may transmit. It is not done by the Physical layer protocol.

Transparency. The Physical layer provides transparent transmission of the bit stream between the Data Link entities over the Physical connection. Transparency implies that any bit sequence can be transmitted without any restriction imposed by the Physical layer.

Physical Service Data Units (Ph-SDU). Ph-SDU received from the Data Link layer consists of one bit in serial transmission and of "*n*" bits in parallel transmission.

Sequenced Delivery. The Physical layer tries to deliver the bits in the same sequence as they were received from the Data Link layer but it does not carry out any error control. Therefore, it is likely that some of the bits are altered, some are not delivered at all, and some are duplicated.

Fault Condition Notification. Data Link entities are notified in case of any fault detected in the Physical connection.

Service Primitives. The Physical layer provides a non-confirmed service to the Data Link layer. The service names and primitives for activation of the Physical connection, data transfer and for deactivation of the Physical connection are shown in Fig. 4.

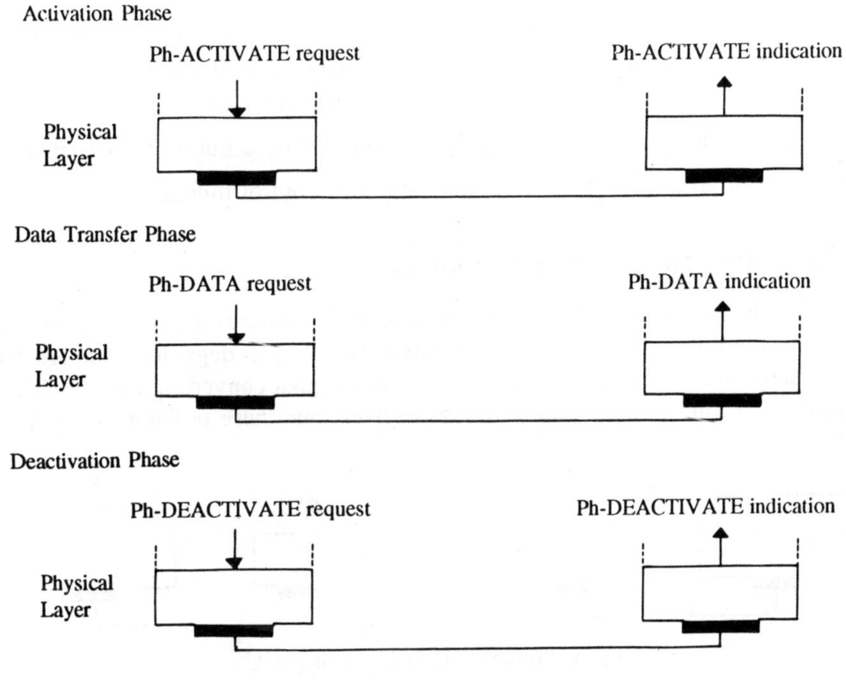

Fig. 4 Service primitives of the Physical layer.

The parameters associated with connection activation primitives have not been defined. The parameters associated with the data transfer primitives are the user data. There are no parameters associated with deactivation primitives.

2 FUNCTIONS WITHIN THE PHYSICAL LAYER

To provide the services as listed above to the Data Link layer, the Physical layer carries out the following functions:

1. It activates and deactivates the Physical connection at the request of the Data Link layer entity. These functions involve interaction of the Physical layer entities. Thus, the Physical layer exchanges control signals with the peer entity.

2. A Physical connection may necessitate the use of a relay at an intermediate point to regenerate the electrical signals (Fig. 5). Activation and deactivation of the relay is carried out by the Physical layer. This function is explained in detail in the next section.

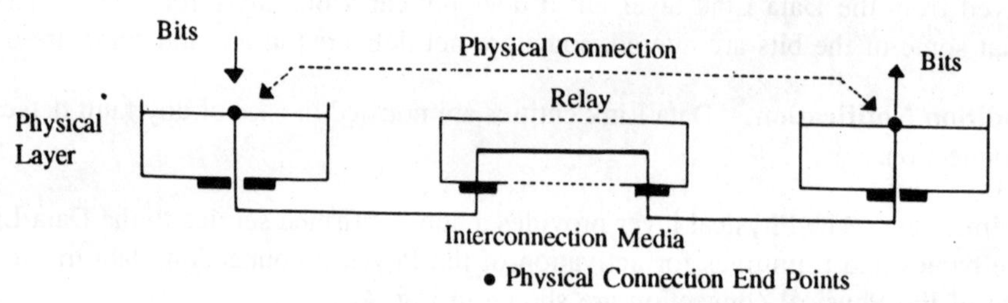

Fig. 5 Relaying function of the Physical layer.

3. The Physical transmission of the bits may be synchronous or asynchronous. The Physical layer provides synchronization signals necessary for transmission of the bits. Character level or frame level synchronization is the responsibility of the Data Link layer.

4. If the signal encoding is required, this function is carried out by the Physical layer.

5. The Physical layer does not incorporate any error control function.

3 RELAYING FUNCTION IN THE PHYSICAL LAYER

It may not always be practical to directly connect two digital devices using a cable if the distance between them is very long. The quality of the received signals gets degraded by noise, attenuation and phase characteristics of the interconnecting medium. Signal converting units (SCUs) are used in the physical interconnecting medium as relays to overcome these problems (Fig. 6).

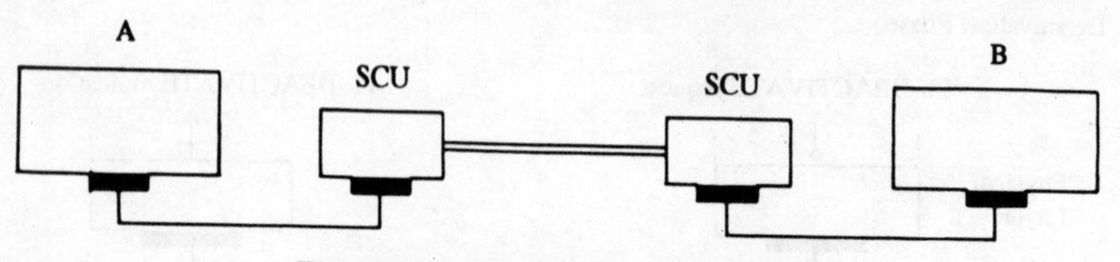

Fig. 6 Signal converting unit (SCU).

SCUs employ one or more of the following methods to ensure acceptable quality of the signal received at the distant end:

- Amplification
- Regeneration
- Equalization of media characteristics
- Modulation.

Examples of SCUs which carry out these functions are: modems, LDMs (Limited Distance Modems), line drivers, digital service unit, and optical transceiver.

A pair of these devices is always required, one at each end. These two devices together act as a relay. They receive electrical signals representing data bits at one end and deliver the same signals at the other end.

The digital end devices face the SCUs and interact with the SCUs at the Physical layer level. This is shown in detail in Fig. 7. Notice that a number of protocols and interfaces at Physical layer level are involved when SCUs are used as relay units.

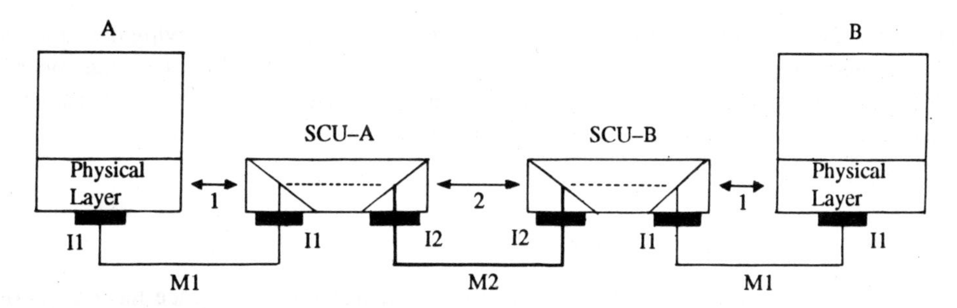

M1 Transmission Medium between End Device and SCU
M2 Transmission Medium between the Two SCUs
I1 Physical Medium Interface between End Device and SCU
I2 Physical Medium Interface between Two SCUs
1. Physical Layer Protocol between End Device and SCU
2. Physical Layer Protocol between the Two SCUs

Fig. 7 Interfaces and protocols in a Physical connection involving signal converting units.

In the above example, the media M1 and M2 are usually different. M1 consists of a bunch of copper wires, each carrying data or a control signal. M2, on the other hand, can be a telephony channel or even optical fibre. Physical medium interfaces I1 and I2 depend on the type of medium used.

As regards the Physical layer protocols, note that the Physical layer of device A no longer interacts with the Physical layer of device B. It interacts with the Physical layer of SCU-A to carry out the Physical layer functions. The two SCUs have a different set of Physical layer protocols between them.

4 PHYSICAL MEDIUM INTERFACE

The Physical layers need to exchange protocol control information between them. Unlike the other layers which send the protocol control information as a separate field, the Physical layers use the interconnecting medium for sending the protocol control signals. These signals are sent on separate wires as shown in Fig. 8. Note that the control signals originate and terminate in the Physical layers. They have no functional significance beyond the Physical layer. This is in conformity with the principles of the layered architecture.

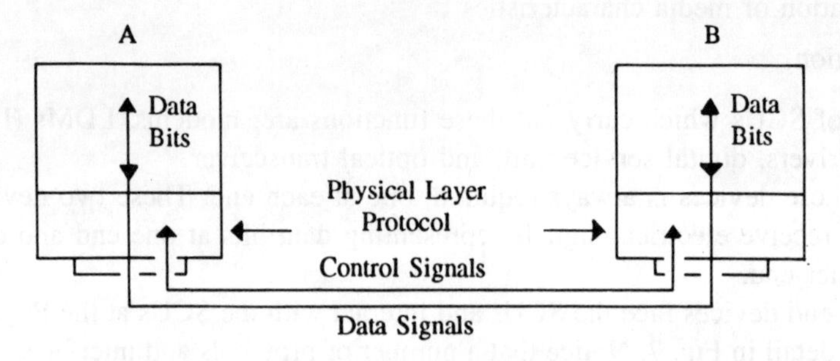

Fig. 8 Transmission of control signals of the Physical layer.

The physical interconnecting medium consists of a number of wires carrying data and control signals. It is essential to specify which wire carries which signal. Moreover, the mechanical specifications of the connector, type of the connector (male or female) and the electrical characteristics of the signals need to be specified. Definition of the physical medium interface includes all these specifications.

5 PHYSICAL LAYER STANDARDS

Historically, the specifications and standards of the physical medium interface have also covered the Physical layer protocols. But these specifications have not identified the Physical layer protocols as such.

Physical layer specifications can be divided into the following four components (Fig. 9):

1. Mechanical specification
2. Electrical specification
3. Functional specification
4. Procedural specification.

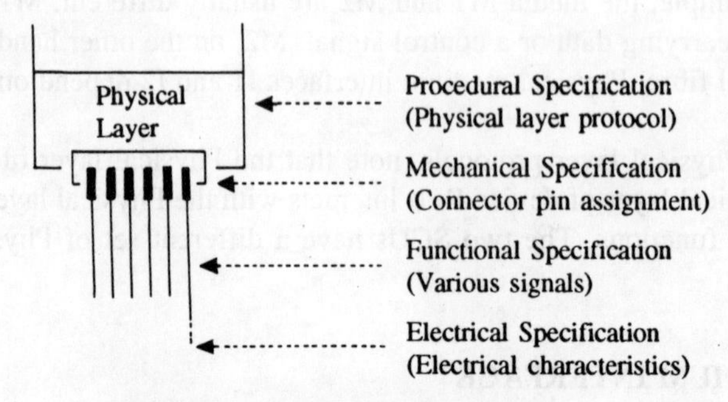

Fig. 9 Physical layer specifications.

The procedural specification is the Physical layer protocol definition and the other three specifications constitute the physical medium interface specifications.

- The mechanical specification gives details of the mechanical dimensions and the type of connectors to be used on the device and the medium. Pin assignments of the connector are also specified.
- The electrical specification defines the permissible limits of the electrical signals appearing at the interface in terms of voltages, currents, impedances, rise time, etc. The required electrical characteristics of the medium are also specified.
- The functional specification indicates the functions of various control signals.
- The procedural specification indicates the sequence in which the control signals are exchanged between the Physical layers for carrying out their functions.

Although there are many standards of the Physical layer, only a few are of wide significance. Some examples of Physical layer standards are given below.

EIA:	EIA-232-D
	RS-449, RS-422-A, RS-423-A
CCITT:	X.20, X.20*bis*
	X.21, X.21*bis*
	V.35, V.24, V.28
ISO:	ISO 2110

Out of the above, the EIA-232-D interface is the most common and is found in almost all computers. We will examine EIA-232-D in detail in the following sections. Other less important Physical layer standards will also be discussed in brief.

6 EIA-232-D DIGITAL INTERFACE

The EIA-232-D digital interface of Electronics Industries Association (EIA) is the most widely used physical medium interface. RS-232-C is the older and more familiar version of EIA-232-D. It was published in 1969 as RS-232 interface and the current version was finalised in 1987. EIA-232-D is applicable to the following modes of transmission:

- Serial transmission of data
- Synchronous and asynchronous transmission
- Point-to-point and point-to-multipoint working
- Half duplex and full duplex transmission.

6.1 DTE/DCE Interface

EIA-232-D is applicable to the interface between a Data Terminal Equipment (DTE) and a Data Circuit Terminating Equipment (DCE) (Fig. 10). The terminal devices are usually called Data Terminal Equipment (DTE). The DTEs are interconnected using two intermediary devices which carry out the relay function. The intermediary devices are categorized as Data Circuit-terminating Equipment (DCE). They are so called because standing at the Physical layer of a DTE and facing the data circuit, one finds oneself looking at an intermediary device which terminates the data circuit.

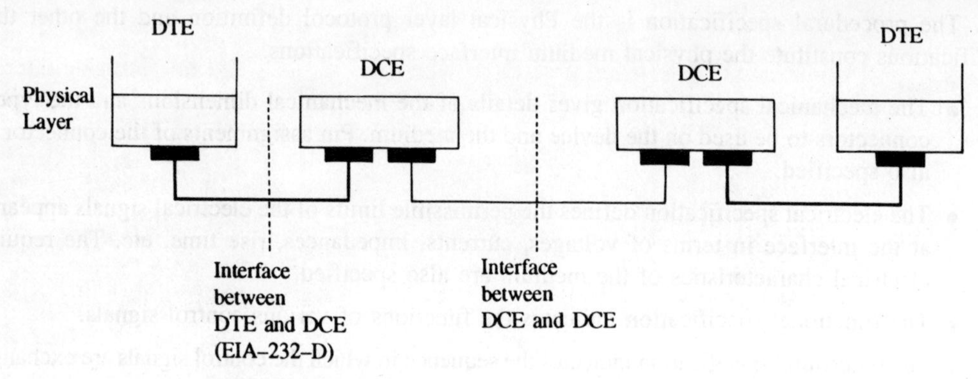

Fig. 10 DTE/DCE interfaces at the Physical layer.

Two types of Physical layer interfaces are involved in the above configuration:

1. Interface between a DTE and a DCE
2. Interface between the DCEs.

EIA-232-D defines the interface between a DTE and a DCE. There are other standards for DCE-to-DCE interface.

The physical media between the DTE and the DCE consist of several circuits carrying data, control and timing signals. Each circuit carries one specific signal, either from the DTE or from the DCE. These circuits are called interchange circuits.

6.2 DTE and DCE Ports

Over the years, use of the terms DTE and DCE for classifying two kinds of devices on the basis of their functions has declined. We can have today a terminal equipment which looks like a DCE at its transmission port. The situation becomes even more confusing when we come across a device which has multiple ports of different types (Fig. 11).

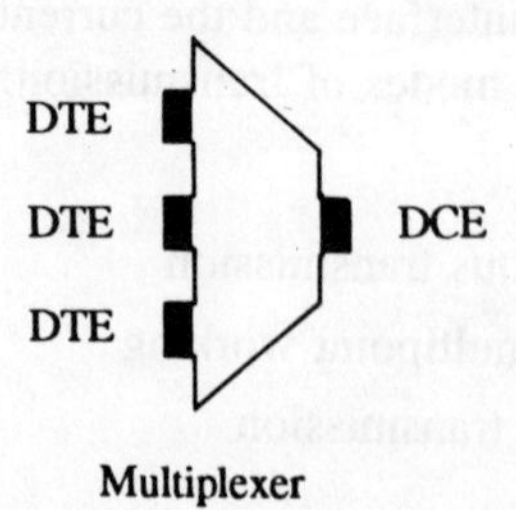

Fig. 11 Multiplexer with DTE and DCE ports.

Therefore, the use of the terms DTE and DCE these days at the Physical layer level refers to the description of the transmission port of a device rather than the device itself.

EIA-232-D, however, was designed with modem as DCE, and the terminology which has been used to specify its signals and functions also refers to modem as DCE. We shall, therefore, describe the interface with modem as the DCE.

6.3 DCE-DCE Connection

A DCE has two interfaces, DTE-side interface which is EIA-232-D, and the line-side interface which interconnects the two DCEs through the transmission medium. There can be several forms of connection and modes of transmission between the DCEs as shown in Fig. 12.

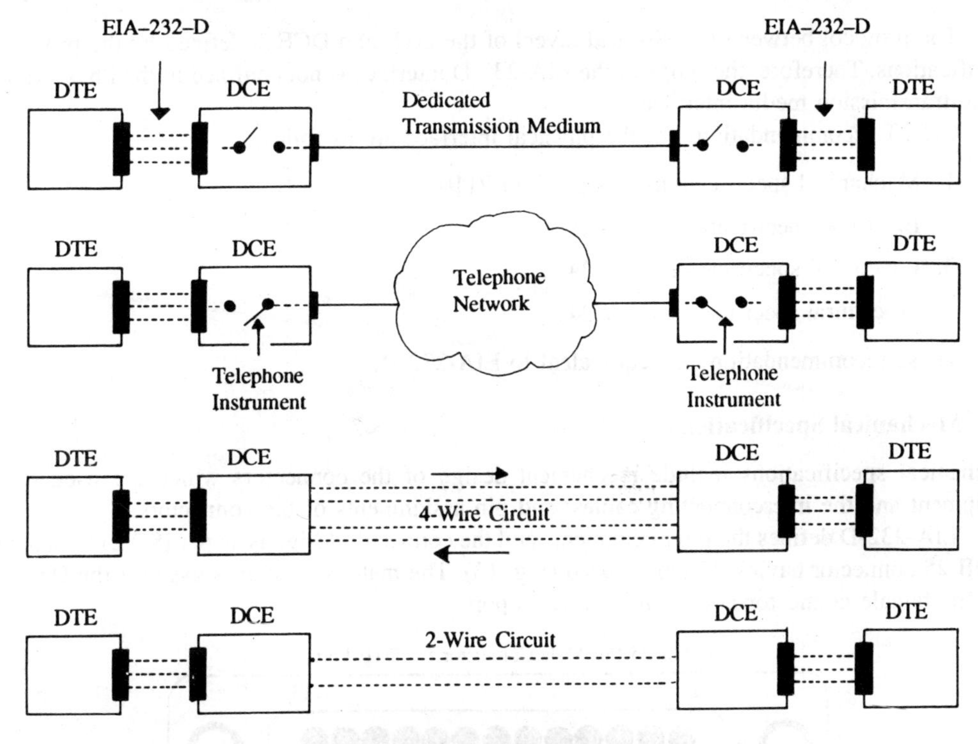

Fig. 12 Transmission alternatives between two DCEs.

1. The two DCEs may be connected directly through a dedicated transmission medium.
2. The two DCEs may be connected to PSTN (Public Switched Telephone Network).
3. The connection may be on a 2-wire transmission circuit or on a 4-wire transmission circuit.
4. The mode of transmission between the DCEs may be either full duplex or half duplex.

Full duplex mode of transmission is easily implemented on a 4-wire circuit. Two wires are used for transmission in one direction and the other two in the opposite direction. Full duplex operation on a 2-wire circuit requires two communication channels which are provided at different frequencies on the same medium.

PSTN provides a 2-wire circuit between the DCEs and the circuit needs to be established and released using a standard telephone interface.

Note that electronics of the DCE may not be directly connected to the interconnecting transmission circuit. This connection is made on request from the DTE as we shall see later.

7 EIA-232-D INTERFACE SPECIFICATIONS

EIA-232-D interface defines four sets of specifications for the interface between a DTE and a DCE:

1. Mechanical specifications
2. Electrical specifications
3. Functional specifications
4. Procedural specifications.

The protocol between the Physical layers of the DTE and DCE is defined by the procedural specifications. Therefore, the scope of the EIA-232-D interface is not confined to the Physical layer to the transmission media interface.

CCITT recommendations for the physical interface are as follows:

1. Mechanical specifications as per ISO 2110
2. Electrical specifications V.28
3. Functional specifications V.24
4. Procedural specifications V.24

These recommendations are equivalent to EIA-232-D.

7.1 Mechanical Specifications

Mechanical specifications include mechanical design of the connectors which are used on the equipment and the interconnecting cables; and pin assignments of the connectors.

EIA-232-D defines the pin assignments and the connector design is as per ISO 2110 standard. A DB-25 connector having 25 pins is used (Fig. 13). The male connector is used for the DTE port and the female connector is used for the DCE port.

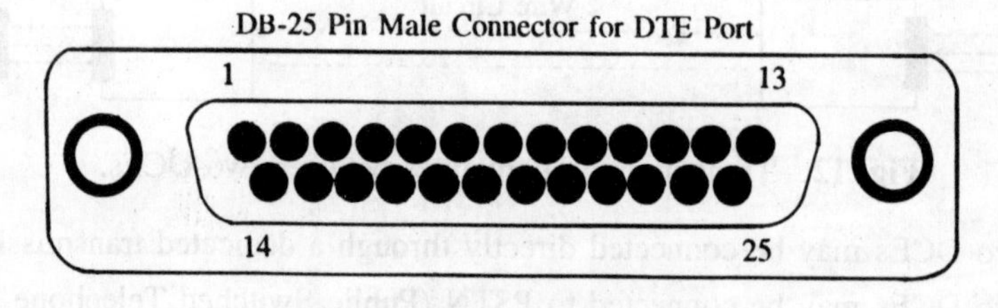

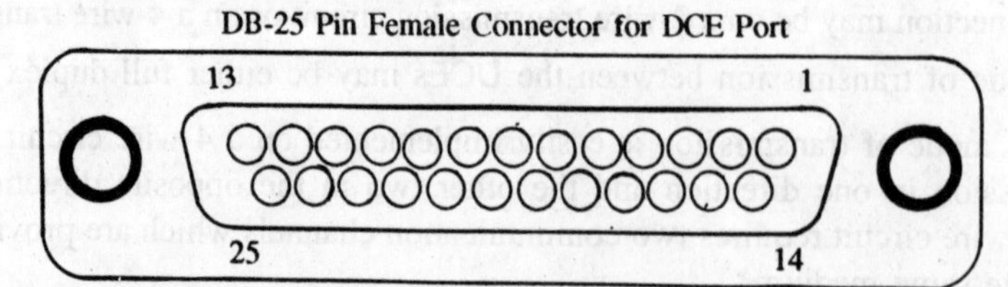

Fig. 13 25-pin connector of EIA-232-D interface.

7.2 Electrical Specifications

The electrical specifications of the EIA-232-D interface specify characteristics of the electrical signals. EIA-232-D is a voltage interface. Positive and negative voltages within the limits as shown in Fig. 14 are assigned to the two logical states of a binary digital signal.

Limit	....	+ 25 Volts
	Logic 0, On, Space	
Nominal	....	+ 12 Volts
	....	+ 3 Volts
0 Volt	....	
	....	– 3 Volts
Nominal	....	– 12 Volts
	Logic 1, Off, Mark	
Limit	....	– 25 Volts

Fig. 14 Electrical specifications of EIA-232-D interface.

All the voltages are measured with respect to the common ground. The 25-volts limit is the open circuit or no-load voltage. The range from – 3 to + 3 volts is the transition region and is not assigned any state.

DC resistance of the load impedance is specified to be between 3000 to 7000 ohms with a shunt capacitance less than 2500 pF. The cable interconnecting a DTE and a DCE usually has a capacitance of the order of 150 pF per metre which limits its maximum length to about 16 metres. EIA-232-D specifies the maximum length of the cable as 50 feet (15.3 metres) at the maximum data rate of 20 kbps.

7.3 Functional Specifications

Functional specifications describe the various signals which appear on different pins of the EIA-232-D interface. Table 1 lists these signals which are divided into five categories:

1. Ground or common return
2. Data circuits
3. Control circuits
4. Timing circuits
5. Secondary channel circuits.

A circuit implies the wire carrying a particular signal. The return path for all the circuits in both directions (from DTE to DCE and from DCE to DTE) is common. It is provided on pin 7 of the interface. EIA has used a two- or three-letter designation for each circuit. CCITT, on the other hand, has given a three digit number to each circuit. In day-to-day use, however, acronyms based on the function of individual circuits are more common.

Not all the circuits are always wired between a DTE and a DCE. Depending on configuration and application, only essential circuits are wired. Functions of the commonly used circuits are now described.

Signal Ground (AB). It is the common earth return for all data and control circuits in both directions. This is one circuit that is always required whatever be the configuration.

Data Terminal Ready (CD), DTE ⟶ DCE. The ON condition of the signal on this circuit informs the DCE that the DTE is ready to operate and the DCE should also connect itself to the transmission medium.

Table 1 EIA-232-D Interchange Circuits

Pin	To DTE	To DCE	Circuit names	CCITT	EIA
1	Common		Shield	101	—
7	Common		Signal ground	107	AB
2		⟶	Transmitted data	103	BA
3	⟵		Received data	104	BB
4		⟶	Request to send	105	CA
5	⟵		Clear to send	106	CB
6	⟵		DCE ready	107	CC
20		⟶	Data terminal ready	108.2	CD
22	⟵		Ring indicator	125	CE
8	⟵		Received line signal detector	109	CF
21	⟵		Signal quality detector	110	CG
23		⟶	Data rate selector (DTE)	111	CH
23*	⟵		Data rate selector (DCE)	112	CI
24		⟶	Transmitter signal element timing (DTE)	113	DA
15	⟵		Transmitter signal element timing (DCE)	114	DB
17	⟵		Receiver signal element timing (DCE)	115	DD
14		⟶	Secondary transmitted data	118	SBA
16	⟵		Secondary received data	119	SBB
19		⟶	Secondary request to send	120	SCA
13	⟵		Secondary clear to send	121	SCB
12*	⟵		Secondary received line signal detector	122	SCF
18		⟶	Local loopback	141	LL
21		⟶	Remote loopback	140	RL
25	⟵		Test mode	142	TM

* If SCF is not used then CI is on pin 23.

DCE Ready (CC), DTE ⟵ DCE. This circuit is usually turned ON in response to CD and indicates ready status of the DCE. When this signal is ON, it means that power of the DCE is switched on and it is connected to the transmission medium.

If the DCE-to-DCE connection is through PSTN, ON status of the CC implies that the call has been established.

Request to Send (CA), DTE ⟶ DCE. Transition from OFF to ON on the CA triggers the local DCE to perform such set-up actions as are necessary to transmit data. These set-up activities include sending a carrier to the remote DCE so that it may further alert the remote DTE and get ready to receive data.

Transition of the CA from ON to OFF instructs the DCE to complete transmission of all data and then withdraw the carrier.

Clear to Send (CB), DTE ⟵ DCE. Clear to Send signal indicates that the DCE is ready to receive data from the DTE on Transmitted Data (BA) circuit. This control signal is changed to the ON state in response to the Request to Send (CA) from the DTE after a predefined delay. This delay is provided to give sufficient time to the remote DCE and DTE to get ready for receiving data. Figure 15 illustrates how the Request to Send (CA) signal works with Clear to Send (CB) signal to coordinate data transmission between a DTE and a DCE.

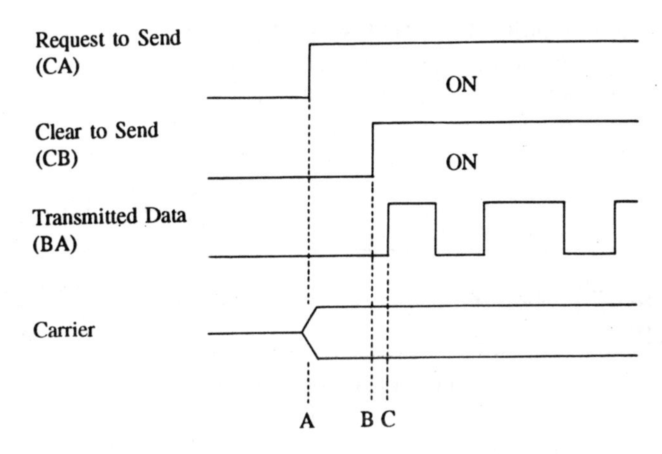

A : DTE switches CA "ON" indicating its wish to transmit data ; DCE sends carrier on the transmission media

B : DCE accepts to receive data by switching CB "ON"

C : DCE receives data from DTE on BA

Fig. 15 Time sequence of Request to Send and Clear to Send circuits.

Transmitted Data (BA), DTE ⟶ DCE. Data from DTE to DCE is transmitted on this circuit. When no data is being transmitted, the DTE keeps the signal on this circuit in "1" state.

Data can be transmitted on this circuit only when the following control signals are ON:

1. Request to Send (CA)
2. Clear to Send (CB)
3. DCE Ready (CC)
4. Data Terminal Ready (CD).

The ON state of these signals ensures that the local DCE is in readiness to transmit data and sufficient opportunity has been given to the remote DCE and DTE to get ready for receiving data.

Received Data (BB), DTE ⟵ DCE. Data from DCE to DTE is received on this circuit. DCE maintains the signal on this circuit in "1" state when no data is being received.

Received Line Signal Detector (CF), DTE ⟵ DCE. When a DTE asserts CA, the local DCE sends a carrier to the remote DCE so that it may get ready to receive data. When the remote DCE detects the carrier on the line, it alerts the DTE to get ready to receive data by turning the CF circuit ON.

Transmitter Signal Element Timing (DA), DTE ⟶ DCE. When operating in the synchronous mode of transmission, the DTE clock is made available to the DCE on this circuit.

Transmitter Signal Element Timing (DB), DTE ⟵ DCE. When operating in synchronous mode of transmission, the DCE clock is made available to the DTE on this circuit. One of the two clocks, DA or DB, is used as timing reference.

Receiver Signal Element Timing (DD), DTE ⟵ DCE. At the receiving end, the circuit DD provides the receive clock from the DCE to the DTE. This clock is extracted from the received signal by the DCE and is used by the DTE to store the data bits in a shift register. Figure 16 shows two typical methods of configuring the timing circuits.

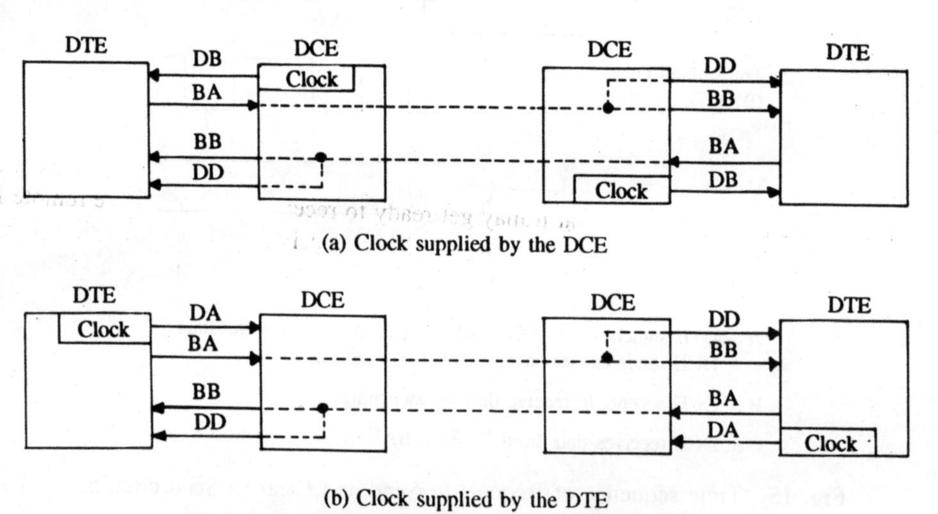

Fig. 16 Clock supply alternatives in synchronous transmission.

In the first alternative, the DCE supplies clock to the DTE on circuit DB for the transmitted data. At each clock transition, one data bit is pushed out of the DTE. At the remote end, the clock is extracted from the received data and supplied to the DTE on circuit DD for the received data.

In the second alternative, the DTE supplies clock to the DCE on circuit DA. For the received data, the DCE extracts the clock from data and supplies it to the DTE as before.

Ring Indicator (CE), DTE ⟵ DCE. The ON state of this circuit indicates to the DTE that there is an incoming call and the DCE is receiving a ringing signal. On receipt of this signal the DTE is expected to get ready and indicate this to the DCE by turning its Data Terminal Ready signal ON.

Local Loopback (LL), DTE ⟶ DCE. The ON condition of this circuit causes a local loopback at the DCE line output so that the data transmitted on the circuit BA is made available on the received data circuit BB for conducting local tests.

Remote Loopback (RL), DTE ⟶ DCE. The ON condition of this circuit causes loopback at the remote DCE so that the local DCE line and the remote DCE could be tested.

Test Mode (TM), DTE ⟵ DCE. After establishing the loopback condition, the DCE indicates its loopback status to the local DTE by the ON condition of the TM circuit.

Secondary Channel Circuits (SBA, SBB, SCA, SCB, SCF). These circuits are used when a secondary channel is provided by a DCE. The secondary channel operates at a lower data signalling rate (typically 75 bits/s) than the data channel and is intended to be used for return of supervisory control signals. The control circuits for the secondary channel, SCA and SCB, are functionally the same as CA and CB except that they are associated with the secondary channel rather than the data channel.

7.4 Procedural Specifications

Procedural specifications lay down the procedures for the exchange of control signals between a DTE and a DCE. The sequence of events which comprise the complete procedure for data transmission can be divided into the following four phases:

1. Equipment readiness phase
2. Circuit assurance phase
3. Data transfer phase
4. Disconnect phase.

Equipment Readiness Phase. The following functions are carried out during the equipment readiness phase:

1. The DTE and DCE are energized.
2. Physical connection between the DCEs is established if they are connected to PSTN.
3. The transmission medium is connected to the DCE electronics.
4. The DTE and DCE exchange signals which indicate their ready state.

We shall consider two simple configurations of connection between the DCEs:

1. The DCEs having dedicated transmission medium between them
2. The DCEs having a switched connection through PSTN between them.

Dedicated Transmission Connection: A DTE which wants to transmit, asserts the Data Terminal Ready signal (CD) which connects the DCE electronics to the transmission medium. If the DCE is energized, it replies with the DCE Ready signal (CC) as shown in Fig. 17.

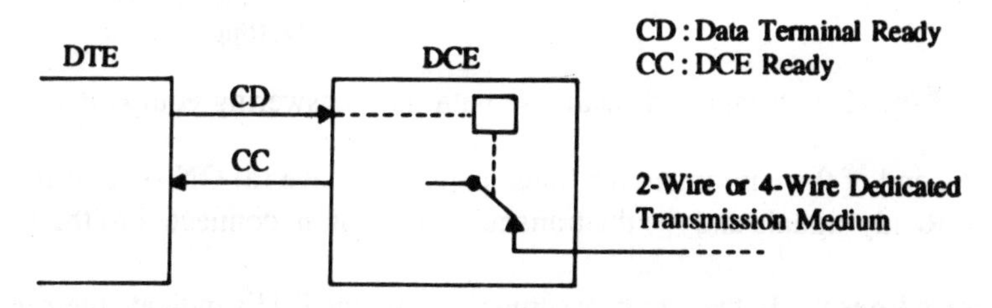

Fig. 17 Equipment readiness phase for dedicated transmission media.

Switched Connection: In this case, the physical connection of DCEs needs to be established through a switched telephone network. This is done either manually by the operators at both ends or automatically through using automatic calling and answering equipment.

In the manual operation, the DCEs are fitted with a telephone instrument. The operator wishing to establish the connection dials the distant end telephone number and indicates his intent to the distant end operator. The operators then press appropriate switches on their respective DTEs to send the Data Terminal Ready signals (CD). The Data Terminal Ready signal causes the transmission medium to changeover from the telephone instrument to the DCE at both ends (Fig. 18).

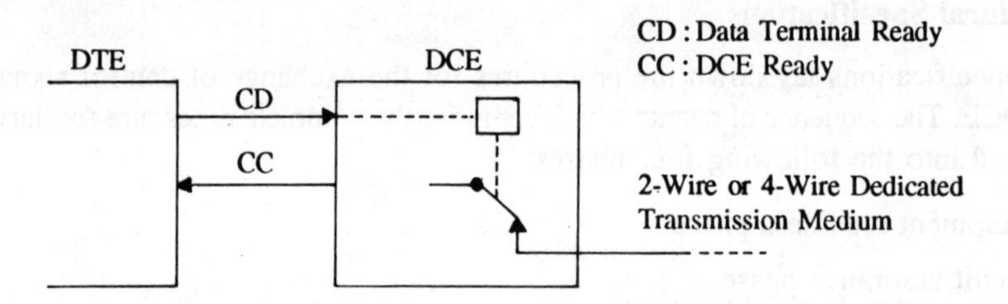

Fig. 18 Equipment readiness phase for transmission on switched media.

If automatic answering equipment is used, the incoming call is detected by the DCE and indicated to the DTE by the Ring indicator signal (CE). If the DTE is in energized condition, it sends the Data Terminal Ready signal (CD) which causes connection to the transmission medium. The DCE indicates its readiness status simultaneously to the DTE on the DCE Ready circuit (CC) (Fig. 19).

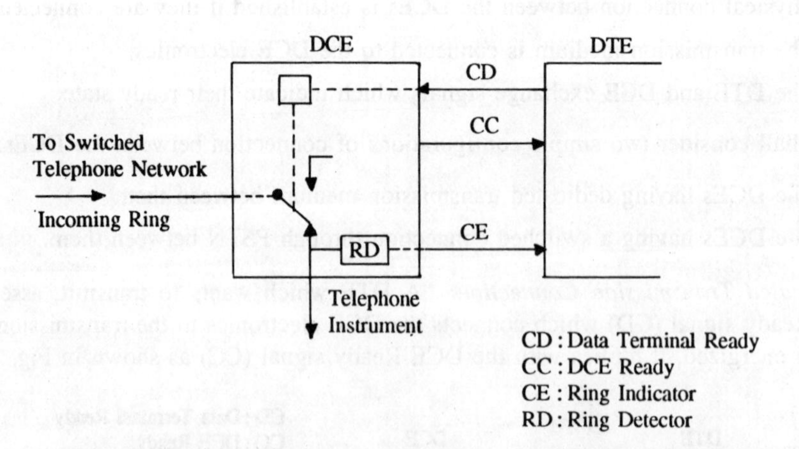

Fig. 19 Distant end readiness with auto answering equipment.

Thus, at the end of the equipment readiness phase, we have (a) ON state of the Data Terminal Ready and DCE Ready signals and (b) the transmission medium connected to the DCE electronics.

Circuit Assurance Phase. In the circuit assurance phase, the DTEs indicate their intent to transmit data to the respective DCEs and the end-to-end (DTE to DTE) data circuit is activated. If the transmission mode is half duplex, only one of the two directions of transmission of the data circuit is activated.

Half Duplex Mode of Transmission: A DTE indicates its intent to transmit data by asserting the Request to Send signal (CA) which activates the transmitter of the DCE and a carrier is sent to the distant end DCE (Fig. 20). The Request to Send signal also inhibits the receiver of the DCE.

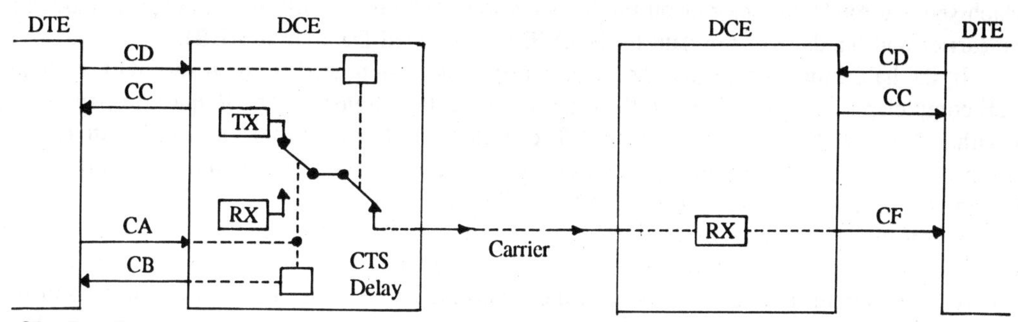

CD : Data Terminal Ready
CC : DCE Ready
CA : Request to Send
CB : Clear to Send
CF : Received Line Signal Detector
TX : Transmitter
RX : Receiver

Fig. 20 Circuit assurance phase in half duplex mode of transmission.

After a short interval of time equal to the propagation delay, the carrier appears at the input of the distant end DCE. The DCE detects the incoming carrier and gets ready to demodulate data from the carrier. It also alerts the DTE using the Received Line Signal Detector circuit (CF) as shown in the Fig. 20.

After activating the circuit, the sending end DCE signals the DTE to proceed with data transmission by returning the Clear to Send signal (CB) after a fixed delay. This delay ensures that sufficient opportunity is given to the distant end to get ready to receive data. With the Clear to Send signal, the equipment readiness and end-to-end data circuit readiness are assured and the sending end DTE can initiate data transmission.

In half duplex operation, the Clear to Send signal is given in response to Request to Send only if the local Received Line Signal Detector circuit is OFF.

Full Duplex Operation: In full duplex operation, there are separate communication channels for each direction of data transmission so that both the DTEs may transmit and receive simultaneously. The circuit assurance phase is exactly the same in half duplex transmission mode except that both the DTEs can independently assert Request to Transmit. In this case, the receivers always remain connected to the receive side of the communication channel.

Data Transfer Phase. Once the circuit assurance phase is over, data exchange between DTEs can start. The following circuits are in ON state during this phase:

Transmitting End	*Receiving End*
Data Terminal Ready	Data Terminal Ready
DCE Ready	DCE Ready
Request to Send	Received Line Signal Detector
Clear to Send	

At the transmitting end, the DTE sends data on Transmitted Data circuit (BA) to the DCE which sends a modulated carrier on the transmission medium. The distant end DCE demodulates the carrier and hands over the data to the DTE on Received Data circuit (BB).

In the half duplex operation, the direction of transmission needs to be reversed every time a DTE completes its transmission and the other DTE wants to transmit. The Request to Send signal is withdrawn after the transmitting end DTE completes its transmission. The DCE withdraws its carrier and switches the communication channel to its receiver. The DCE also inhibits further flow of data from the local DTE by turning off the Clear to Send signal.

When the distant end DCE notices the carrier disappear, it withdraws the Received Line Signal Detector circuit. Noticing that the transmission medium is free, the distant end DTE performs actions of the circuit assurance phase and then transmits data. Thus, a DTE wanting to transmit, checks each time if the channel is free by sensing Received Line Signal Detector circuit and if it is OFF, it asserts the Request to Send.

Disconnect Phase. After the data transfer phase, disconnection of the transmission media is initiated by a DTE. It withdraws Data Terminal Ready signal. The DCE disconnects from the transmission media and turns off the DCE Ready signal.

8 COMMON CONFIGURATIONS OF EIA-232-D INTERFACE

Not all the circuits defined in EIA-232-D specifications are always implemented. Depending on application and communication configuration only a subset of the circuits is implemented. Figure 21 shows the circuits commonly implemented in a standard full duplex configuration.

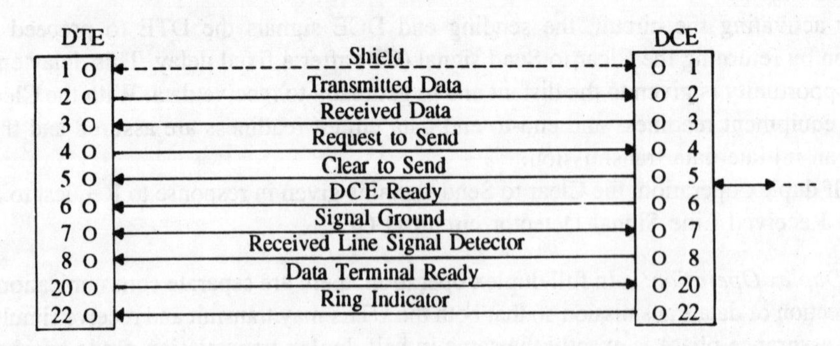

Fig. 21 Commonly implemented circuits in a standard full duplex configuration.

Standard full duplex configuration implementation as shown above is required for communication involving modems and telephone network. In practice, however, the following non-standard configurations are also quite often used.

Three-Wire Interconnection. Figure 22 depicts a three-wire interconnection which is quite adequate for many interfacing configurations. This interconnection provides a bare minimum number of circuits necessary for full duplex communication. The circuits present are Transmitted Data, Received Data and Signal Ground.

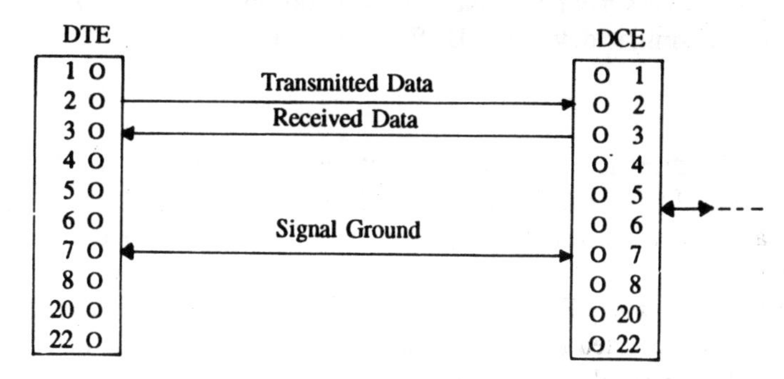

Fig. 22 Three-wire interconnection for full duplex operation.

Three-Wire Interconnection with Loopback. If Request to Send and Clear to Send circuits are implemented in a DTE port, the three-wire interconnection shown in Fig. 22 does not work because the DTE will not transmit data unless it receives the Clear to Send signal. A three-wire interconnection with loopback overcomes this problem (Fig. 23) by locally generating the signals required for initiating the transmission. The following jumpers are provided.

- Request to Send circuit is jumpered to Clear to Send and Received Line Signal Detector circuits
- Data Terminal Ready circuit is jumpered to DCE Ready circuit.

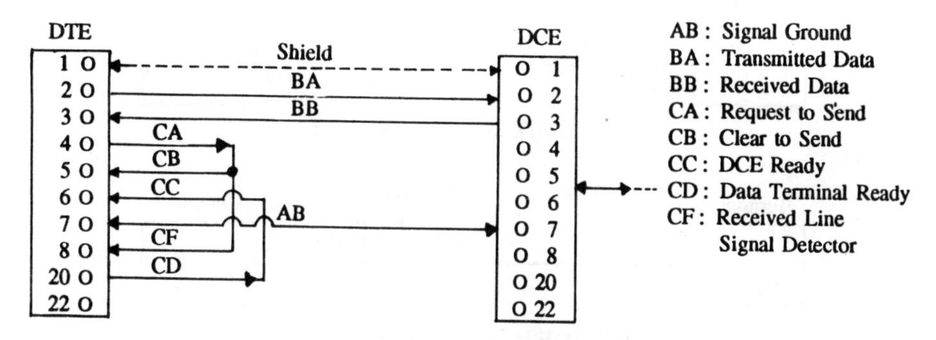

Fig. 23 Three-wire interconnection with loopbacks.

By jumpering the Data Terminal Ready circuit to DCE Ready circuit, the equipment readiness phase is completed as soon as the DTE asserts the Data Terminal Ready signal. Quite often, this occurs when power is applied to the DTE.

When the DTE asserts the Request to Send signal, the circuit assurance phase is immediately completed because it receives immediately the Clear to Send and Received Line Signal Detector signals.

By providing the loopbacks, the number of interconnecting wires is reduced but it should be kept in mind that certain features of EIA-232-D interface have also been omitted. There are many other configurations each tailored to a particular requirement and with its own merits and limitations. In the following section we shall discuss the special class of interface configurations associated with interconnection of devices having similar interface ports even though EIA-232-D was designed to work between two dissimilar devices, a DTE and a DCE.

8.1 Null Modem

If we view the EIA-232-D interface by standing between the DTE and the DCE, it is seen that a signal which comes out of a particular pin of the DTE port goes towards the DCE on the same pin. In other words, in any pair of corresponding pins of the DTE and DCE ports, one is output pin and the other is input pin.

Therefore, in order to apply EIA-232-D to interconnect any two devices, it is necessary that a DTE thinks that it is connected to a DCE, whether the other device is actually a DCE or not. Thus, a computer and a terminal can be directly interconnected using EIA-232-D interface if one of them has a DCE port and the other a DTE port (Fig. 24a).

On the other hand, if both the devices which are to be interconnected have DTE ports, one of the devices needs to be suitably modified to look like a DCE (Fig. 24b). A null modem carries out this job externally by converting a DTE port to a DCE port and vice versa (Fig. 24c).

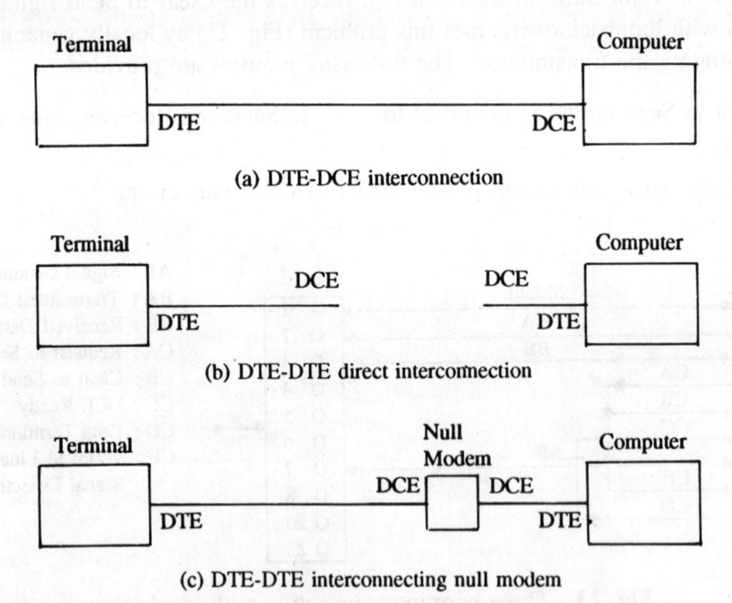

Fig. 24 Need for null modem.

Null Modem with Loopback. Figure 25 shows a three-wire null modem used for interconnecting two DTEs. Notice that null modem is a cable with DCE connectors (female connectors) at the ends. The Transmitted/Received Data wires are crossed so that data transmitted by one DTE may be received by the other at its appropriate pin. The loopback jumpers for three-wire interconnections explained earlier are also provided.

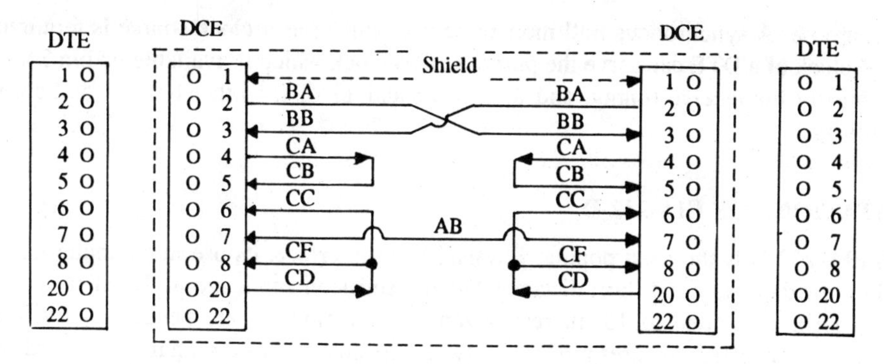

Fig. 25 Internal configuration of a null modem.

Null Modem with Loopback and Multiple Crossovers. Figure 26 shows another variation of null modem cable. The following jumpers and crossovers are provided.

- Jumpers from
  - — Request to Send to Clear to Send
  - — Ring Indicator to DCE Ready.
- Crossovers between
  - — Transmitted Data and Received Data
  - — Request to Send and Received Line Signal Detector
  - — Data Terminal Ready and Ring Indicator.

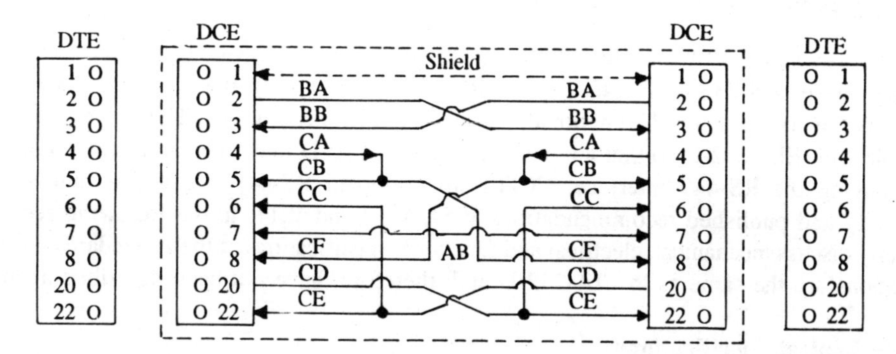

Fig. 26 Null modem with loopbacks and multiple crossovers.

When a DTE asserts a Data Terminal Ready signal, the other DTE is immediately given a stimulus, the Ring indicator, to believe that it has an incoming call. It responds with its Data Terminal Ready which results in the DCE Ready signal at the calling DTE. Thus, the equipment readiness phase is complete. Before transmitting data, the calling DTE asserts the Request to Send which raises the Received Line Signal Detector at the other DTE. The Request to Send signal is looped back at the calling DTE as Clear to Send. Therefore, the circuit assurance phase is also immediately completed and data transmission can begin.

The above discussion applies to the asynchronous mode of operation because we have not considered the clock. If the terminal devices require external clock, the null modem cable will not serve the purpose. A synchronous null modem device which has a clock source is required. Else, the internal clock of a DTE can serve the purpose. This clock which is available on pin 24, is wired to pin 17 locally for receive timing, and to pins 15 and 17 of the other device for transmit and receive timings.

9 LIMITATIONS OF EIA-232-D

Although EIA-232-D is the most popular Physical layer interface, its use in computer networking is limited to low data rates and short distance data transmission applications. The distance between a DTE and DCE is limited to 15 metres, beyond which modems are necessary. Even a small industrial plant or an office requires modems between the host and its terminals. As regards the data rate, the EIA-232-D interface meets the local transmission requirements which are usually below 9600 bps but higher data rates of 48 kbps and above are required for computer networking. The upper limit of 20 kbps of EIA-232-D is not sufficient for these applications.

The above limitations of the EIA-232-D interface are due to the following two reasons:

1. Unbalanced transmission mode of its signals
2. Shared common ground for all signals flowing in both the directions.

Raised ground potential, crosstalk and noise due to these factors result in introduction of errors at high bit rates and for longer separation between the DTE and the DCE. These limitations of the EIA-232-D have been overcome in the interface standards developed subsequently.

10 RS-449 INTERFACE

In the early 1970s, the EIA introduced RS-422-A, RS-423-A and RS-449 interfaces to overcome the limitations of RS-232-C. RS-422-A and RS-423-A cover only the electrical specifications, and RS-449 covers mechanical, functional and procedural specifications. These specifications are compatible with EIA-232-D so that a device having EIA-232-D interface can be interconnected to another having the RS-449 interface. CCITT also adopted RS-449, RS-422-A and RS-423-A subsequently and published recommendations V.54, V.10 and V.11. In the following sections we shall briefly discuss mechanical, electrical and functional specifications of these standards. Procedural specifications are the same as in EIA-232-D and, therefore, have not been described again.

10.1 Mechanical Specifications

RS-449 gives detailed mechanical specifications of the interface. Since RS-449 incorporates more than 25 signals, two connectors, one with 37 pins and the other with 9 pins have been specified. Mechanical designs of the connectors are as per ISO 4902 standard. All signals associated with the basic operation of the interface appear on the 37-pin connector. The secondary channel circuits are grouped on the 9-pin connector. Table 2 gives a list of the signals present in the RS-449 interface with their pin assignments. For purposes of comparison, we have included the signals which are present in the EIA-232-D interface also in the table.

Mechanical compatibility between EIA-232-D and RS-449 is accomplished at connector level using an adapter as shown in Fig. 27.

The RS-449 standard also specifies the maximum cable length and the corresponding data rate supported by the cable. Figure 28 shows this relationship graphically.

Table 2 RS-449 Interface Circuits

A. 37 Pin Connector

	RS-449 Circuit name	Pin No.	To DTE	To DCE		EIA-232-D Circuit name
	Shield	1	←	→		Shield
SG	Signal Ground	19	←	→	AB	Signal ground
SC	Send Common	37		→		—
RC	Receive Common	20	←			—
TS	Terminal in Service	28		→		—
IC	Incoming Call	15	←		CE	Ring Indicator
TR	Terminal Ready	12, 30		→	CD	Data Terminal ready
DM	Data Mode	11, 29	←		CC	DCE Ready
SD	Send Data	4, 22		→	BA	Transmitted Data
RD	Receive Data	6, 24	←		BB	Received Data
TT	Terminal Timing	17, 35		→	DA	Transmitter Signal Element Timing (DTE)
ST	Send Timing	5, 23	←		DB	Transmitter Signal Element Timing (DCE)
RT	Receive Timing	8, 26	←		DD	Receiver Signal Element Timing (DCE)
RS	Request to Send	7, 25		→	CA	Request to Send
CS	Clear to Send	9, 27	←		CB	Clear to Send
RR	Receive Ready	13, 31	←		CF	Received Line Signal Detector
SQ	Signal Quality	33	←		CG	Signal Quality Detect
NS	New Signal	34		→		—
SF	Select Frequency	16		→		—
SR	Signal Rate Selector	16		→	CH	Data Signal Rate Selector (DTE)
SI	Signal Rate Indication	2	←		CI	Data Signal Rate Selector (DCE)
LL	Local Loop-back	10		→	LL	Local Loop-back
RL	Remote Loop-back	14		→	RL	Remote Loop-back
TM	Test Mode	18	←		TM	Test Mode
SS	Select Standby	32		→		—
SB	Standby indicator	36				—
	spare	3, 21				—

Table 2 RS-449 Interface Circuits (Cont.)

B. 9 Pin Connector

	RS-449 Circuit name	Pin No.	To DTE	To DCE		EIA-232-D Circuit name
	Shield	1	←	→		—
SG	Signal Ground	5	←	→	AB	Signal ground
SC	Send Common	9		→		—
RC	Receive Common	6	←			—
SSD	Secondary Send Data	3		→	SBA	Secondary Transmitted Data
SRD	Secondary Received Data	4	←		SCB	Secondary Received Data
SRS	Secondary Request to Send	7		→	SCA	Secondary Request to Send
SCS	Secondary Clear to Send	8	←		SCB	Secondary Clear to Send
SRR	Secondary Receiver Ready	6	←		SCF	Secondary Received Line Signal Detector

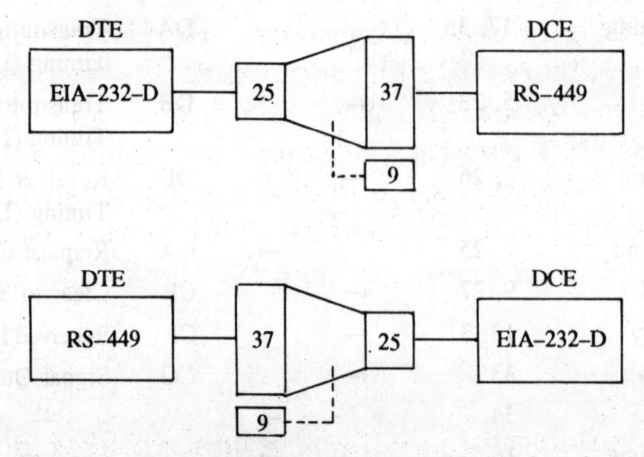

Fig. 27 Adapter for EIA-232-D and RS-449 interfaces.

10.2 Electrical Specifications

To ensure electrical compatibility with EIA-232-D, both balanced and unbalanced transmissions can be used. RS-422-A specifies electrical characteristics of the balanced circuits while RS-423-A specifies electrical characteristicts of the unbalanced circuits. Circuits of RS-449 are divided into two categories. Category I circuits are as follows:

- Send Data (SD)
- Receive Data (RD)
- Terminal Timing (TT)
- Send Timing (ST)
- Receive Timing (RT)

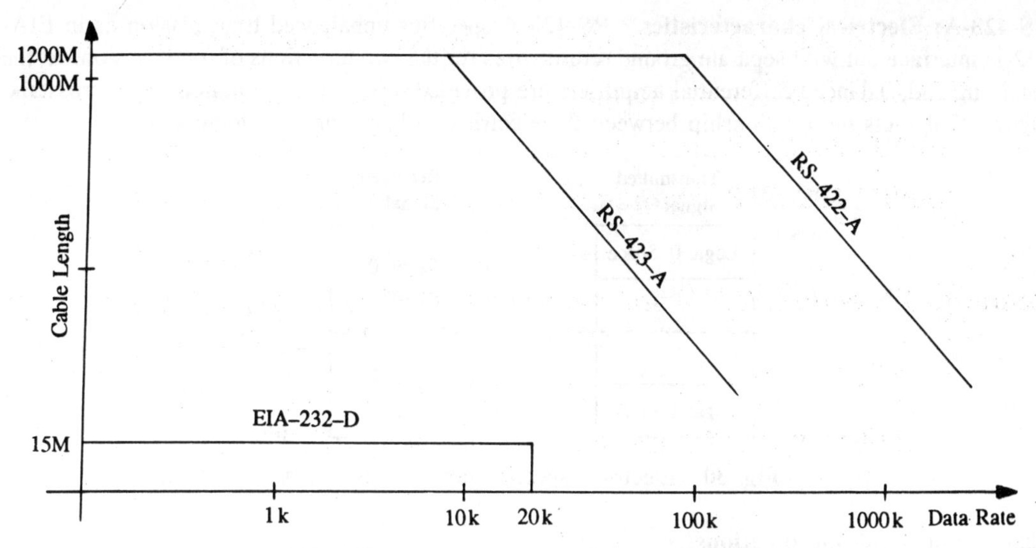

Fig. 28 Data rates supported by RS-449.

- Request to Send (RS)
- Clear to Send (CS)
- Receive Ready (RR)
- Terminal Ready (TR)
- Data Mode (DM).

The rest of the circuits belong to Category II.

For data rates of less than 20 kbps (upper limit for EIA-232-D circuits), Category I circuits may be implemented using either RS-422-A or RS-423-A electrical characteristics. For data rates over 20 kbps, balanced RS-422-A electrical characteristics must be used. Circuits belonging to Category II are always implemented using RS-423-A characteristics.

RS-422-A: Electrical characteristics. RS-422-A specifies electrical characteristics of the balanced circuits. Figure 29 shows the voltage levels corresponding to the two logic states of the electrical signals. Note that the transition region on the received side is much narrower because of the better crosstalk performance and elimination of earth potential problems.

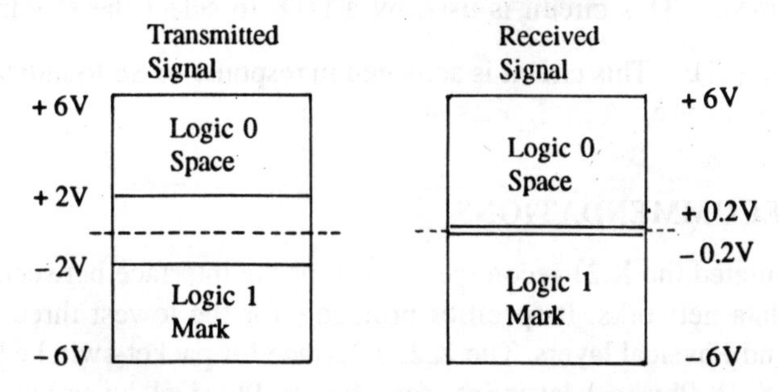

Fig. 29 Electrical specifications of RS-422-A.

RS-423-A: Electrical characteristics. RS-423-A specifies unbalanced transmission as in EIA-232-D interface but with separate ground return wires for the two directions of transmission. At the receiving end, balanced differential amplifiers are provided to reject the common mode crosstalk. Figure 30 depicts the relationship between the electrical voltages and the logic states.

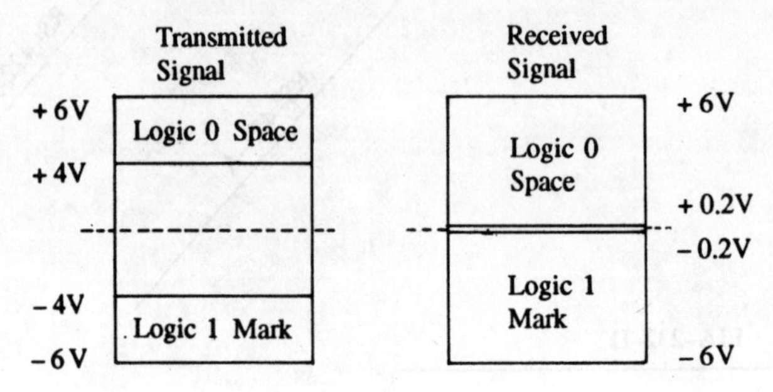

Fig. 30 Electrical specifications of RS-423-A.

10.3 Functional Specifications

Table 2 gives complete listing of the RS-449 circuits and corresponding EIA-232-D circuits where applicable. There are ten new signals defined in the RS-449 standard. Their functions are briefly described below:

Send Common (SC): This is the common return path for all unbalanced circuits from DTE to DCE.

Receive Common (RC): This is the common return path for all unbalanced circuits from DCE to DTE.

Terminal in Service (TS): This circuit indicates to the DCE whether or not the DTE is operational.

New Signal (NS): In multipoint polling applications, the NS circuit is used by the control DTE to indicate, to its DCE, the end of a message from one of the remote DTEs and to alert it to get ready for receiving message from a new DTE.

Select Frequency (SF): A DTE selects the transmit frequency of a DCE on the communication channel by signalling on the SF circuit.

Select Standby (SS): This circuit is used by a DTE to select the standby equipment.

Standby Indicator (SI): This circuit is activated in response to SS to indicate whether standby or normal equipment is in use.

11 CCITT X.21 RECOMMENDATIONS

In 1976, CCITT formulated the X.21 recommendation for the interface between a DTE and a DCE for circuit-switched data networks. It specifies protocols for the lowest three layers, namely, the Network, Data Link and Physical layers. The X.25 interface for packet-switched data networks also specifies use of the X.21 Physical layer interface for its Physical layer but there are very few circuit-switched data networks and terminals which support the X.21 interface.

The Physical layer of X.21 specifies the protocol and media interface for synchronous transmission. Its mechanical, electrical, functional and procedural specifications are given below in brief.

11.1 Mechanical Specifications

X.21 specifies a 15-pin connector. The mechanical design and the pin assignments are as per the ISO 4903 standard.

11.2 Electrical Specifications

The electrical specifications are as per X.26 for unbalanced transmission for rates 9600 bps and below; and X.27 for balanced transmission for rates above 9600 bps.

The interface can operate at data rates from 600 to 48000 bps.

11.3 Functional Specifications

Definitions of the interchange circuits used in X.21 are:

Transmit (T): This circuit is used for DTE to DCE data signals and call control signals sent during call establishment and call clearing phases.

Receive (R): This circuit is used for DCE to DTE data signals and call control signals during call establishment and clearing phases.

Signal Element Timing (S): This circuit is used by the DCE to provide the clock to the DTE.

DTE Signal Timing Element (X): This circuit is used by the DTE to transmit the DTE clock to the DCE.

Control (C): This circuit is used by the DTE to indicate condition of the Transmit circuit. It is ON when data is being transmitted.

Indication (I): This circuit is used by the DCE to indicate condition of the Receive circuit. It is ON when data is being transmitted.

Byte Timing (B): This signal is transmitted by the DCE for byte synchronization at the DTE.

Frame Start Identification (F): This circuit is used by the DCE to mark the start of the frame being transmitted on the Receive circuit.

Signal Ground (G): This is the common ground for the signals when the transmission is as per V.28 recommendation.

DTE Common Return (Ga): This is the common return for circuits from the DTE and is used by the receivers in the DCE when the transmission is unbalanced as per X.26.

DCE Common Return (Gb): This is the common return for circuits from the DCE and is used by the receivers in the DTE when the transmission is unbalanced as per X.26.

11.4 Procedural Specifications

Unlike the EIA-232-D interface where each control signal has a separate circuit, X.21 defines a sequence of signal combinations and states of the interface for transfer of data. A typical procedure for establishing and clearing an X.21 connection is summarized in Table 3.

Table 3 Typical X.21 Procedure

DTE		Description of interface states	DCE	
T	C		R	I
1	OFF	Ready	1	OFF
0	ON	Call request	1	OFF
0	ON	Proceed to select	+++...	OFF
Address	ON	DTE transmits address	+++...	OFF
1	ON	Connection establishment in progress	Call progress signals	
1	ON	Connection established, ready for data	1	ON
Data	ON	Data transfer	Data	ON
0	OFF	Clear request by DTE	X	X
0	OFF	Clear confirmation by DCE	0	OFF

11.5 X.21*bis* Recommendation

As most of the commercially available terminal devices do not conform to the X.21 interface, X.21*bis* was defined by CCITT to connect existing DTEs having the V.24 interface. X.21*bis* is a V.24 compatible interface with some additional signalling procedures to obtain connections through a switched data network. The DTE can have the following mechanical and electrical interfaces together with the V.24 interface.

- V.28 with 25-pin connector and pin allocation as per ISO 2110.
- X.26 with 37-pin connector and pin allocation as per ISO 4902.
- X.27 with 37-pin connector and pin allocation as per ISO 4902.

The X.26 and X.27 are the same as V.10 and V.11 respectively.

12 SUMMARY

The Physical layer is responsible for transport of bits from one device to the other on the Physical connection. It converts the bits into electrical signals having characteristics suitable for transmission over the physical medium. It also supports the relaying function in the transmission medium. The physical medium interface has been defined in terms of its mechanical, electrical, functional and procedural specifications. The procedural specification is the definition of the Physical layer protocol. EIA-232-D is the most popular physical medium interface but it has limitations of data rate and distance. Other interfaces which overcome these limitations have been defined.

The standards for the physical medium interface are summarized as follows:

ISO 2110 25-pin connector applicable to EIA-232-D, V.24

ISO 4902	37/9-pin connector applicable to RS-449, V.10, V.11
ISO 4903	15-pin connector for X.21
V.28	Electrical specifications of the low data rate unbalanced circuits; compatible with EIA-232-D.
V.10/X.26	Electrical specifications of the high data rate unbalanced circuits; compatible with RS-423-A.
V.11/X.27	Electrical specifications of the high data rate balanced circuits; compatible with RS-422-A.
V.24	Functional and procedural specifications; compatible with EIA-232-D, RS-449.
X.21	Procedures for synchronous operation over public data network.
X.21*bis*	Procedures for synchronous operation over public data networks for DTEs designed to interface with V series modems, compatible with EIA-232-D and RS-449.

PROBLEMS

1. The Physical service is a non-confirmed service. If some data bits are lost during transmission over the interconnecting media, which layer will detect their loss and take recovery action?

2. Trellis-coded modem carries out forward error correction and improves the error performance of the transmission media. This function is carried out at the Physical layer in the modems. Show by a layered model how the end systems are not involved in the error correction process.

3. In Fig. 3, when point-to-multipoint Physical connection is activated, there is possibility of intermixing of electrical signals transmitted by different end systems in the shared media. Can the Physical layer avoid such mixing of signals?

4. In the following diagram, statistical multiplexers have the first two layers of the OSI reference model. Draw the layered model showing the Physical connections which exist in this configuration.

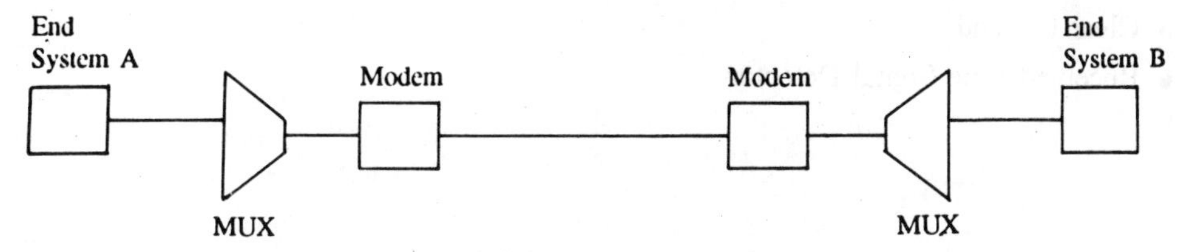

Fig. P.4

5. In Fig. 1 of this chapter, devices A and B have half duplex Physical connection between them. Indicate the service primitives when:

- A activates the connection,
- A sends one data unit,
- A deactivates the connection,
- B activates the connection,
- B sends one data unit,
- B deactivates the connection.

6. State whether true or false:
(a) In half duplex transmission, the Physical layers decide who will transmit. (True/False)
(b) Activation and deactivation of the Physical connection is a confirmed service. (True/False)
(c) The protocol control information (PCI) at the Physical layer is usually sent on seperate wires. (True/False)
(d) If modems are used for relaying electrical signals, the Physical layer protocol is between the end system and the modem. (True/False)
(e) The Physical layer just converts bits into electrical signals and vice versa. Clock and signal encoding are Data Link layer functions. (True/False)

7. If the Received Line Signal Detector circuit becomes ON, which circuit at the other end has been raised?

8. If a modem is OFF, which circuit will not be ON on the EIA-232-D interface?

9. A DTE using the EIA-232-D interface sends the ASCII character "K" with odd parity. The mode of transmission is asynchronous. Sketch the Transmitted Data signal assuming 15 V logic circuits and stop pulse of 1-bit duration.

10. At times, the DTE and DCE ports do not have the specified gender of the connector. It becomes impossible to identify the type of port by its physical attributes. Which are the pins likely to indicate the type of port when the device is energized?

11. If DTE-A is transmitting data to DTE-B over a half duplex transmission line with modems at either end, list the events which takes place when B starts transmitting. The interface between the DTEs and modems is EIA-232-D.

12. If two DTEs having EIA-232-D interface are connected through modems and have full duplex transmission between them, indicate the ON/OFF status of the following circuits at both the ends:

- Data Terminal Ready
- DCE Ready
- Request to Send
- Clear to Send
- Received Line Signal Detector.

CHAPTER 7

The Data Link Layer

The basic service provided by the Physical layer is transportation of bits over the Physical connection. This service is unreliable in the sense that disturbed line conditions of the media may introduce errors which are not taken care of by the Physical layer. Error control and other associated functions are carried out by the second or Data Link layer of the OSI model.

We begin this chapter, with a description of the purpose and functions of the Data Link layer. After a brief discussion on the Data Link service, we move over to frame design considerations. Error control and flow control functions using stop-and-wait and sliding window mechanisms are discussed at length.

1 NEED FOR DATA LINK CONTROL

Let us consider a situation, as shown in Fig. 1, where two digital devices A and B need to exchange information. These devices could be computers, concentrators or other data terminal equipment.

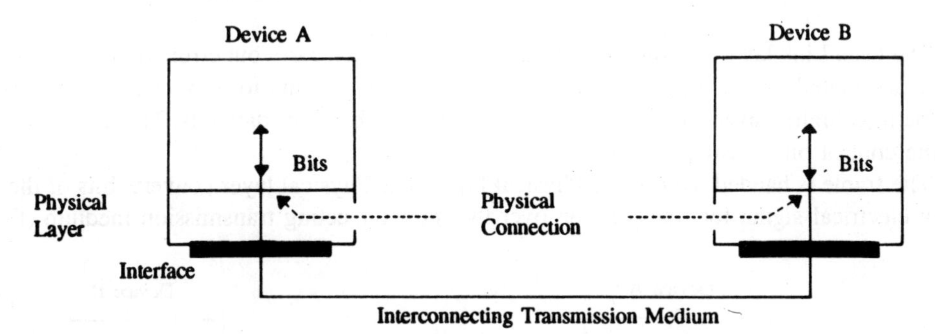

Fig. 1 Exchange of bits over Physical connection.

To exchange digital information between devices A and B we require an interconnecting transmission medium to carry the eletrical signals (e.g., copper wire); a standard interface (e.g. EIA-232-D), and the Physical layer to convert bits into electrical signals and vice versa.

Together with the physical medium, Physical layers of the devices provide the capability for transparent exchange of bits over the Physical connection but this capability has certain limitations:

- If the electrical signal gets impaired due to the noise encountered during transmission or due to the medium characteristics, errors may be introduced in the data bits. There is need to establish mechanisms to control transmission errors.

- Errors can also be introduced if the receiving device is not ready for the incoming bits and some of the bits are lost. Therefore, a data flow control mechanism also needs to be implemented.

The Physical layer does not meet these requirements. Error and flow control functions are implemented in the Data Link layer which ensures error-free transfer of bits from one device to the other.

2 THE DATA LINK LAYER

The Data Link layer constitutes the second layer of the hierarchical OSI model. The Data Link layers together with Physical layers and the interconnecting medium provide a Data Link connection for reliable transfer of data bits over an imperfect Physical connection (Fig. 2).

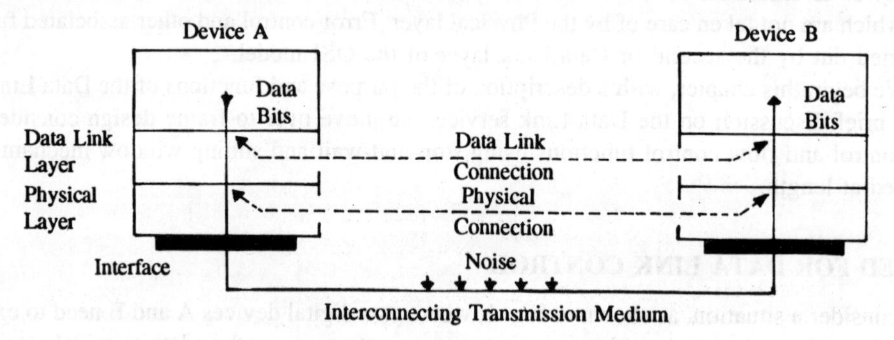

Fig. 2 Reliable transfer of bits over Data Link connection.

The Data Link layer incorporates certain processes which carry out error control, flow control, and the associated link management functions. It receives the data to be sent to the other device from the next higher layer and adds some control bits to a block of data bits. The data block along with the control bits is called a *frame*.

The frame is handed over to the Physical layer. The Physical layer converts bits of the frame into an electrical signal for transmission over the interconnecting transmission medium (Fig. 3).

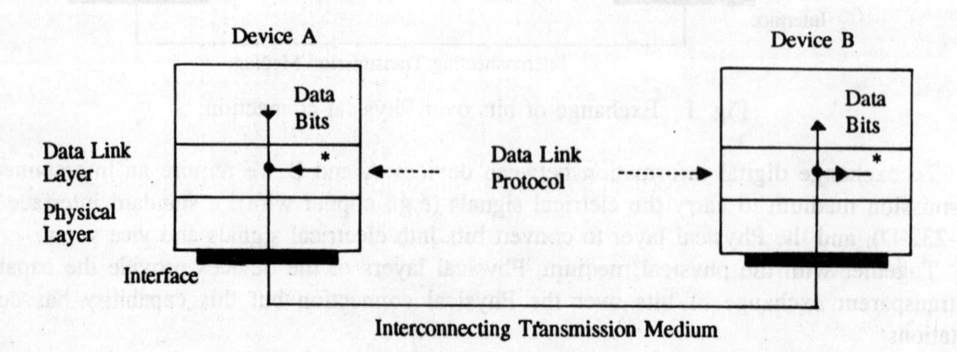

Fig. 3 Formation of frames in the Data Link layer.

At the receiving end, the incoming electrical signal is converted back to bits by the Physical layer and the frame is handed over to the Data Link layer. The Data Link layer removes the control bits and checks for errors. If there is no error, it hands over the received data bits to the next layer.

The control bits include error-check bits, addresses, sequence numbers etc. These additional bits usually constitute more than one field in a frame and enable error control, flow control and link management.

2.1 Service Provided by the Data Link Layer

The Data Link layer receives service from the Physical layer and provides service to the Network layer which is the user of these services. In OSI terminology, the user data block received from the Network layer is called a Data Link Service Data Unit (DL-SDU) and the frame formed by adding protocol control information bits to the DL-SDU is called a Data Link Protocol Data Unit (DL-PDU) as shown in Fig. 4.

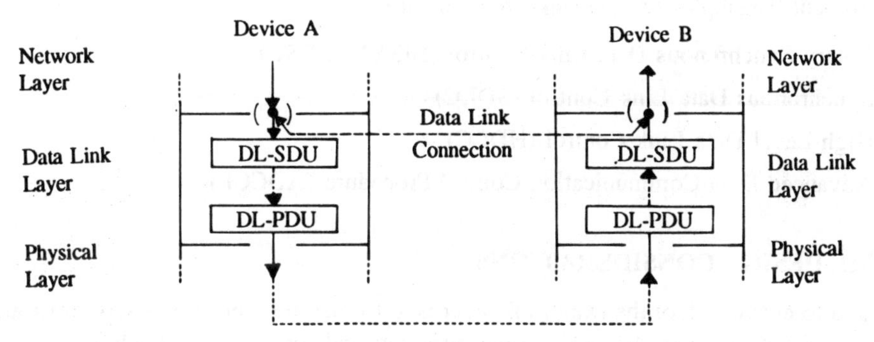

Fig. 4 Data Link service data units and protocol data units.

The basic service provided by the Data Link layer to the Network layer is reliable transfer of DL-SDUs over the Data Link connection which is established, maintained and released by the Data Link layer. This basic service has the following associated features.

Sequencing: The sequence integrity of the DL-SDUs is maintained.

Error Notification: If the Data Link layer detects an unrecoverable error, it notifies the Network layer.

Flow Control: The Network layer can control the rate at which it receives the DL-SDUs from the Data Link layer. This control may be reflected in the rate at which the Data Link layer will accept DL-SDUs at the other end.

Quality of Service Parameters: The Data Link layer provides selectable quality of service parameters which include residual error rate, transit delay, throughput, etc. The selected quality of service is maintained during Data Link connection.

The Data Link service, described above, is connection-mode service but it can be connectionless-mode service also. The service primitives for connection-mode and connectionless-mode are given in Appendix A at the end of the chapter. Connectionless-mode Data Link service is described in Chapter 11. The complete definition of Data Link service is given in ISO 8886. The corresponding CCITT Recommendation is X.212.

2.2 Data Link Protocols

It is essential that the structure of the frame is known to both the Data Link layers so that control bits can be identified. The Data Link layers should also agree on the set of procedures to be adopted for exchange of control information. The specified set of rules and procedures for carrying out data link control functions is called data link protocol. A data link protocol specifies the following:

- Format of the frame, i.e., locations and sizes of the various fields
- Contents of these fields
- Sequence of messages to be exchanged to carry out the error control, flow control and link management functions.

There are many data link protocols developed by various manufacturers and organizations. While all the protocols broadly satisfy the basic functional requirements of the Data Link layer, the services offered are different. Frame formats and contents of various fields are also very specific to each protocol. Examples of data link protocols are:

- Binary Synchronous Data Link Control (BISYNC, BSC)
- Synchronous Data Link Control (SDLC)
- High-Level Data Link Control (HDLC)
- Advanced Data Communication Control Procedure (ADCCP).

3 FRAME DESIGN CONSIDERATIONS

The first and foremost task of the Data Link layer is to format the user data as series of frames each having a predefined structure. A frame contains user data and control fields. Each frame is processed as one entity for error and flow control, i.e., if an error is detected, the whole frame is retransmitted.

The format of a frame, in general, consists of three components as shown in Fig. 5a.

1. Header
2. Data
3. Trailer.

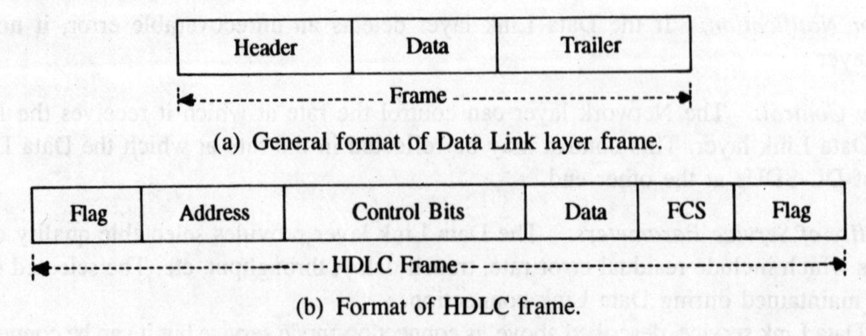

(a) General format of Data Link layer frame.

(b) Format of HDLC frame.

Fig. 5 Data Link layer frame formats.

The data field contains the user data bits to be transmitted. The header and the trailer consist of one or more fields containing data link protocol control information. A typical example of the composition of a frame used for carrying user data in HDLC protocol is shown in Fig. 5. The frame starts with a flag to identify the start of a frame. The flag is followed by an address field. The control field contains the sequence number of the frame, acknowledgement of the receipt of a good frame or other control information. The data field contains the user data received from the Network layer. The Frame Check Sequence (FCS) contains error detection bits. Finally, there is a flag indicating the end of the frame.

3.1 Types of Frame Formats

The frame format is so designed that the receiver is always able to locate the beginning of a frame and its various fields, and is able to separate the data field.

To identify a frame and its various fields, field identifiers/delimiters are incorporated in it. These are unique symbols which indicate by their presence the beginning and end of a frame or a specific field. For example, the flags in an HDLC frame indicate the start and end of the frame.

The requirement of field identifiers and delimiters is determined by the frame structure. A data link protocol may adopt fixed or variable formats of the frames. In fixed format, all the fields are always present in all the frames. In variable format, the presence of any field is optional. The length of a frame may also be fixed or variable. Thus, there are several possibilities:

1. Variable format-variable length
2. Fixed format-fixed length
3. Fixed format-variable length.

A variable format-fixed length frame is not possible and is never used. Let us consider a simple case of a frame having only three fields—header, data, trailer and try to determine the requirement of the delimiters and identifiers for the above-mentioned options.

Variable Format-Variable Length. Figure 6a shows the frame format. All the fields are optional and if a field is present, its size is variable. In all, five delimiters/identifiers are required. X is the frame start identifier. The presence of each field is indicated by a field identifier which also acts

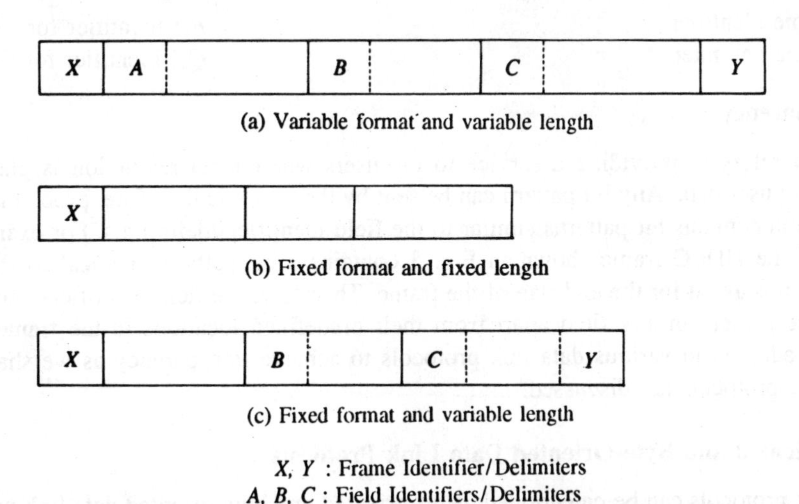

Fig. 6 Types of frame structures.

as delimiter for the previous field. As the size of frame is variable, an end delimiter *Y* is required to indicate the end of the frame.

Fixed Format-Fixed Length. In this case the format of the frame is decided once for all and the field sizes are also fixed in all the frames. The frame format is shown in Fig. 6b. Only one identifier is required at the beginning of the frame. On receipt of the identifier, the receiver is able to identify all the fields as the format of the frame and the sizes of the fields are known to the receiver in advance.

Fixed Format-Variable Length. In this case frame start and end identifier/delimiters are required (Fig. 6c). The identifier for the first field is not required as the frame identifier also identifies the field. As regards other fields, field delimiters are required for each field.

EXAMPLE 1

Consider a frame consisting of four fields, *A*, *B*, *C* and *D*. Field *A* is always present and is of fixed length. Fields *B* and *C* are of variable length and optional. Field *D* is fixed and always present. Give one possible frame structure.

Solution

1. Frame identifier (*X*) is always required.
2. Field *A* is fixed and always present, therefore can be located without an identifier/delimiter.
3. Fields *B* and *C* require identifiers and delimiters. They can be delimited by the following identifier.
4. Field *D* can be located by the frame delimiter (*Y*) by counting back the bits.

X	A	P	B	Q	C	D	Y

X : Frame identifier
Y : Frame delimiter
P : Identifier for field *B*
Q : Identifier for field *C*

3.2 Transparency

Transparency refers to providing a service to the users wherein no restriction is placed on the contents of the user data. Any bit pattern can be sent by the user and therefore, problems may arise if the data field contains bit patterns similar to the field identifiers/delimiters. For example, if the data field of the HDLC frame shown in Fig. 5 contains a bit pattern identical to the flag, the receiver may mistake it for the end flag of the frame. Therefore, the field identifiers and delimiters should not be present in any field apart from their predefined locations in the frame. Different methods are adopted in various data link protocols to achieve transparency as we shall see later when specific protocols are discussed.

3.3 Bit-Oriented and Byte-Oriented Data Link Protocols

The data link protocols can be categorized as bit-oriented and byte-oriented data link protocols. In a bit-oriented data link protocol, control information is coded at bit level and the length of the data field may not be a multiple of bytes. Bit level implies that a control symbol need not be one full byte. For example, in HDLC protocol which is a bit-oriented protocol, the first bit of the control field of the HDLC frame indicates the type of frame.

Byte-oriented data link protocols define all control symbols which are at least one byte long. The size of data field is also a multiple of bytes. BISYNC is a byte-oriented protocol.

4 FLOW CONTROL

Flow control mechanisms are incorporated to ensure that the Data Link layer at the sending end does not send more frames containing data than what the Data Link layer at receiving end is capable of handling. Therefore, the receiver is provided with a control to regulate the flow of the incoming frames. This control is in the form of an acknowledgement (ACK) which is sent by the receiver. The acknowledgement serves two purposes:

1. It clears the sending end to transmit the next data frame.
2. It acknowledges receipt of all previous frames.

Two commonly used flow control mechanisms are stop-and-wait and sliding window.

4.1 Stop-and-Wait Flow Control

In the stop-and-wait flow control mechanism, the sending end sends one data frame at a time and waits for an acknowledgement from the receiver. The receiver can temporarily stop flow of data frames by witholding acknowledgement. Alternatively, it can request for temporary suspension of transmission of the data frames by sending ACK and WAIT (WACK). On receipt of WACK, the sending end has to wait for ACK to commence transmission of the next data frame. The sending end is equipped with a timer so that it may challenge the receiver after time out if no acknowledgement is received. Figure 7 illustrates the mechanism.

Link Utilization. In the stop-and-wait flow control mechanism, only one data frame is sent at a time and each frame is individually acknowledged. The data frame and the acknowledgement take a certain amount of propagation time to travel across the link. Propagation time can be as large as 270 milliseconds for a satellite communication link or a few milliseconds for a terrestrial link. Large propagation time makes the stop-and-wait mechanism very inefficient from the point of view of link utilization. Let us calculate the link utilization efficiency for the stop-and-wait flow control mechanism.

Figure 8 shows the sequence of events and the associated time instants of their occurrence. If the frame size is N bits and the data rate is R, device A shall complete transmission of one frame in time t_f given by

$$t_f = \frac{N}{R}$$

If t_p is the propagation time, the frame is completely received by device B after time $t_f + t_p$. To calculate the best possible utilization of the link, let us assume that

1. B sends back acknowledgement immediately on receipt of a data frame.
2. The size of the acknowledgement frame is very small.

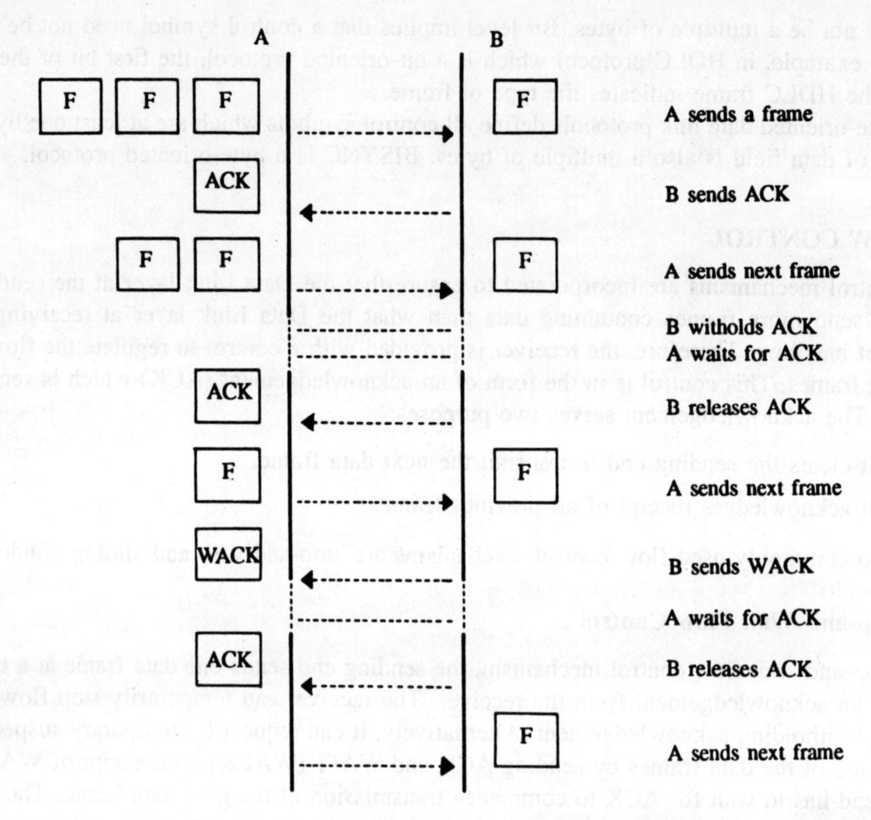

Fig. 7 Stop-and-wait flow control mechanism.

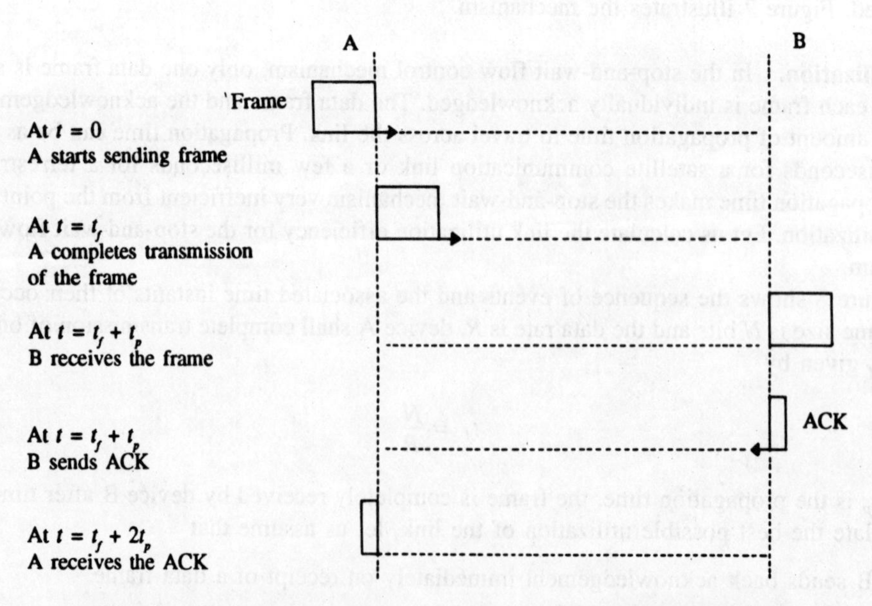

Fig. 8 Time analysis of stop-and-wait flow control mechanism.

A will receive the acknowledgement after time $t_f + t_p + t_p$ and can send the next data frame immediately thereafter. From time $t_f + 2t_p$, A has utilized the link only for time t_f. Therefore, link utilization efficiency U is given by

$$U = \frac{t_f}{t_f + 2t_p} = \frac{1}{1 + 2A}, \qquad A = \frac{t_p}{t_f}$$

EXAMPLE 2

Calculate the maximum link utilization efficiency for stop-and-wait flow control mechanism if the frame size is 2400 bits, bit rate is 4800 bps and distance between the devices is 2000 km. Speed of propagation over the transmission media can be taken as 200,000 km/s.

Solution

$$\begin{aligned} \text{Frame transmission time } 2400/4800 &= 0.5 \text{ s.} \\ \text{Propagation time } 2000/200000 &= 0.01 \text{ s.} \\ A = 0.01/0.5 &= 0.02 \\ U = 1/(1+ 0.04) &= 96\% \end{aligned}$$

Table 1 gives link utilization efficiencies for some typical cases. Speed of propagation in cable media is assumed to be 200,000 km/s. Note that link utilization is very poor in case of a satellite link. The link utilization can be improved by keeping a large frame size and thus reducing A. But it may not be always possible. It is more likely that transmission errors will occur in a data frame of large size and the effective throughput will be reduced. Therefore, the stop-and-wait mechanism is suitable only when the propagation time is less than or comparable to the frame transmission time. The sliding window flow control mechanism overcomes this limitation and is universally adopted.

Table 1 Link Utilization in Stop-and-Wait Mechanism

Type	$t_f^\dagger$	t_p	A	U
Local cable link (10 km)	0.1	0.00005	~ 0.0	~ 100%
Coaxial cable link (1000 km)	0.1	0.005	0.05	~ 91%
Satellite link	0.1	0.270	2.70	~ 15.6%

$^\dagger N = 960$ bits, $R = 9600$ bits/s

4.2 Sliding Window Flow Control

In sliding window flow control mechanism, multiple data frames can be transmitted without waiting for acknowledgements of individual data frames. To understand its operation, let us examine its basic features:

- Each data frame carries a sequence number for its identification.
- The sending end maintains a window containing a fixed number of data frames ready for transmission. These frames can be sent without waiting for any acknowledgement. But a copy of each transmitted frame is retained in the window till it is acknowledged.
- The number of frames in a window is called its size. Its typical value is seven.
- The receiver acknowledges receipt of one or more data frames by sending back a numbered acknowledgement (Receive Ready, RR-*N*) frame. *N* is the sequence number of the next frame it expects to receive.
- All previous data frames are assumed acknowledged on receipt of an acknowledgement. For example, by sending RR-5, the receiver is acknowledging receipt of frames bearing numbers 4, 3, 2, etc. Note that all data frames are still acknowledged but not individually.
- When an acknowledgement is received by the sending end, it slides the window deleting the copies of acknowledged data frames and inserting the same number of new frames from the queue of data frames awaiting transmission.
- To stop the transmission temporarily, the receiving end can send another type of acknowledgement, Receive Not Ready (RNR-*N*). RNR-*N* is acknowledgement upto frame *N*-1 and a request to stop further transmission temporarily. Transmission can be resumed when RR-*N* is released by the receiving end.
- A timer is provided so that the sending end can request for acknowledgement after time out.

RR and RNR in sliding window flow control are equivalent to ACK and WACK in the stop-and-wait flow control. Figure 9 illustrates operation of the mechanism. Device A is sending frames to B. Let us assume that the window size is seven and the window is initially located on frames F1 to F7.

A initiates the transmission with its first frame F1 followed by frames F2, F3, etc. A can send upto F7 without waiting for an acknowledgement from B.

While A is in the process of sending F4, it receives an acknowledgement RR-3 from B indicating to A that F1 and F2 have been received by B. A slides the window by two frames deleting F1 and F2 from the window. F3 to F9 now occupy the window and frames upto F9 can be sent without waiting for further acknowledgement.

To temporarily stop the flow of frames, B sends RNR-9. Transmission is resumed when B sends back RR-9.

Sequence Numbering of the Frames. In the sliding window mechanism, all data frames are given a sequence number which is a binary number of a fixed number of bits. Any numbering scheme of fixed size has a finite count sequence after which it must start all over again from the beginning. If there are "*n*" bits, the length of the count sequence would be 2^n. The two-bit numbering scheme counts from 0 to 3 and then starts again from 0. Similarly, the three-bit numbering scheme counts from 0 to 7 and then starts again from 0. Since no sequence number may be repeated within the window, the length of the count sequence must be at least equal to the size of the window. But window size equal to the count sequence results in some ambiguous situations as illustrated below.

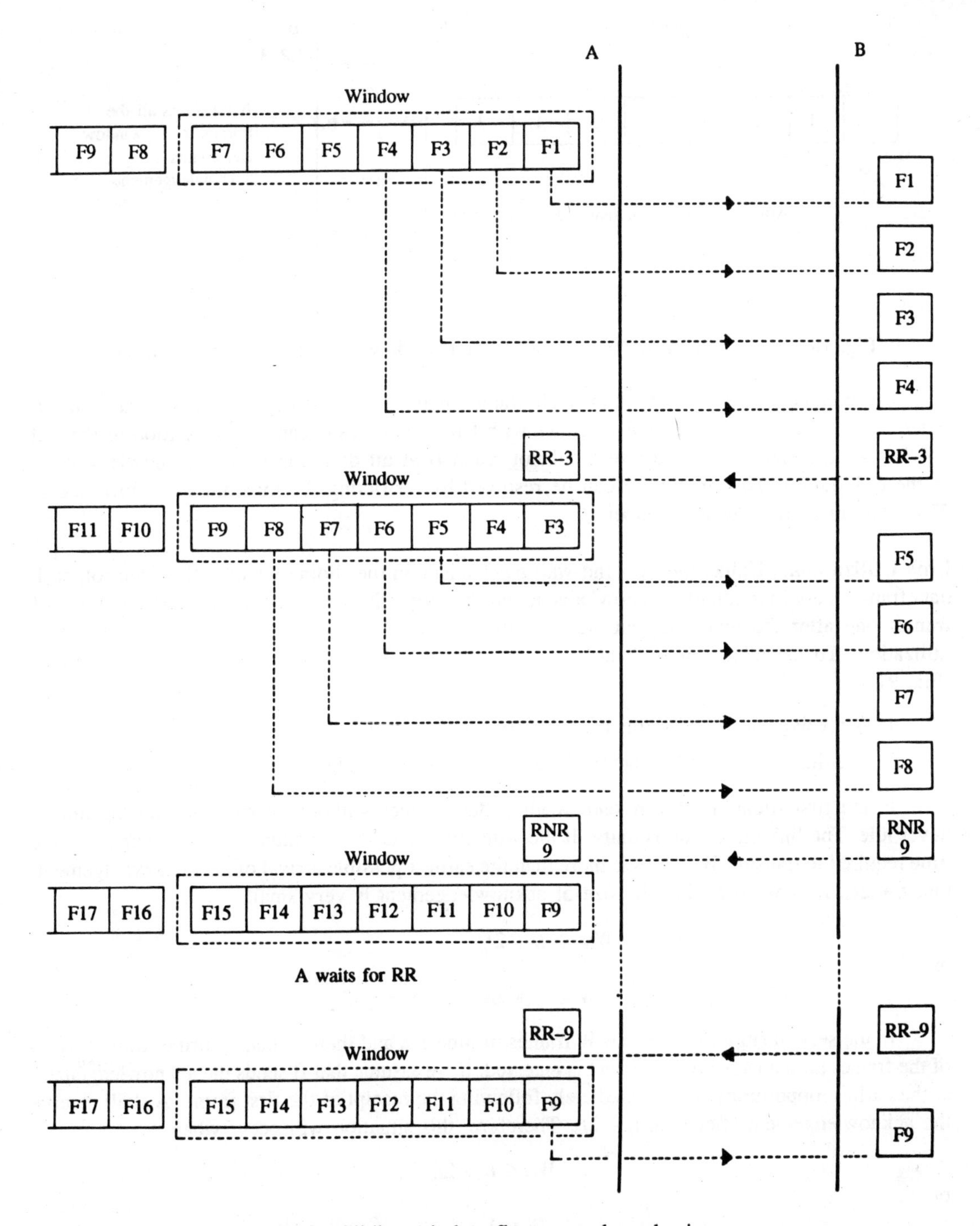

Fig. 9 Sliding window flow control mechanism.

Let us suppose window size is eight and the three-bit numbering scheme is used. The status of window on receipt of RR-3 is shown in Fig. 10.

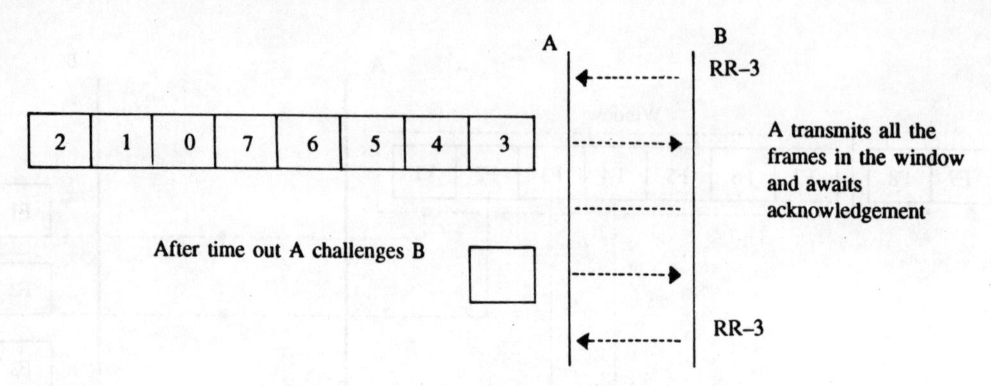

Fig. 10 Ambiguous acknowledgement when window size equals count sequence.

On receipt of RR-3, A sends all the eight frames in the window and waits for acknowledgement. After time out it challenges B which replies with RR-3. This RR-3 can be a repetition of the last RR-3 if all the eight frames just sent did not reach B at all or it can be the acknowledgement of these 8 frames. This ambiguity can be resolved by restricting the maximum window size to $2^n - 1$ for an n-bit sequence number.

Link Utilization. Unlike the stop-and-wait mechanism, in the sliding window flow control, each data frame is not individually acknowledged, and therefore, the sending end can send a number of frames one after the other without waiting for acknowledgement, which results in better link utilization. To calculate link utilization, let us consider the following two possible situations (Fig. 9):

1. A receives an acknowledgement before it exhausts the window.
2. A exhausts the window before it receives an acknowledgement.

In the first situation, A can keep sending data frames without interruption and the link is never idle. The link utilization is unity. If the window size is W, this situation will occur when the time required to transmit W frames is more than the earliest possible arrival of an acknowledgement, i.e., $t_f + 2t_p$. It is assumed that the size of acknowledgement is very small.

$$Wt_f \geq t_f + 2t_p$$

or

$$W \geq 1 + 2A, \qquad A = t_p/t_f$$

In the second situation, A sends W frames in time Wt_f and then suspends further transmission of the frames until an acknowledgement is received. If we assume that B sends the acknowledgement at the earliest opportunity, i.e., immediately following the receipt of the first frame, A shall receive the acknowledgement after time $t_f + 2t_p$. Therefore, this situation will occur when

$$Wt_f \leq t_f + 2t_p$$

or

$$W \leq 1 + 2A$$

In this case the link has been engaged for time $t_f + 2t_p$ while A has utilized it for time Wt_f. Therefore, link utilization is given by

$$U = \frac{Wt_f}{t_f + 2t_p} = \frac{W}{1 + 2A}$$

Figure 11 shows link utilization efficiency as a function of A. Curves for three window sizes 1, 7, and 127 have been shown. When window size is one, the sliding window mechanism degenerates to simple stop-and-wait mechanism. Note that the sliding window flow control mechanism permits use of higher values of A without any impairment of link utilization efficiency. Of course, the window size needs to be appropriately chosen. For satellite links, a large window of 127 frames is used.

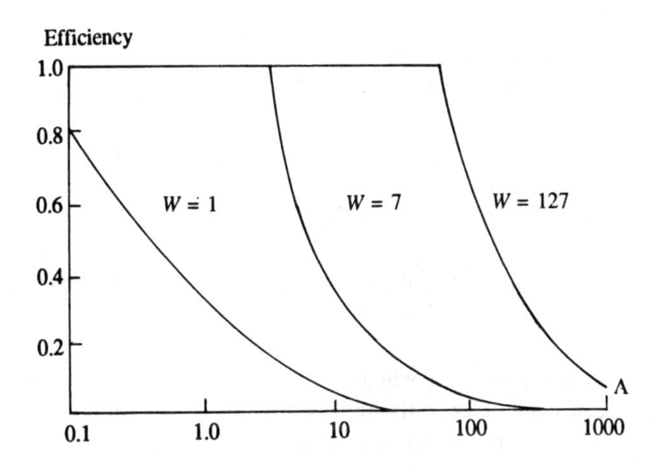

Fig. 11 Link utilization efficiency.

EXAMPLE 3

Calculate link utilization efficiency if

1. Bit rate = 19.2 kbps, frame size = 960 bits, window size = 3, propagation time = 0.06 s.
2. Bit rate = 19.2 kbps, frame size = 960 bits, window size = 7, propagation time = 0.06 s.

What is the minimum window size for 100 per cent link utilization?

Solution

1. $t_f = 960/19200 = 0.05$ s, $\quad A = 0.06/0.05 = 1.2$

   $2A + 1 = 3.4, \quad W < 2A + 1, \quad U = 3/3.4 = 88\%$

2. $W = 7 > 2A + 1, \quad U = 100\%$

Minimum size of the window for $U = 100\%$ is given by

$$W = 2A + 1 = 3.4$$

So window size should be at least equal to 4 for 100 per cent efficiency.

5 DATA LINK ERROR CONTROL

Two types of errors can occur during transmission of frames from one device to the other:

1. Content errors
2. Flow integrity errors.

Errors contained in a received frame are termed content errors. These errors are detected using check bits. Flow integrity errors refer to the lost/duplicate data frames and acknowledgements. Data link error control takes care of both the types of errors. It involves three phases:

1. Error detection
2. Error correction
3. Recovery.

Content errors are detected using parity check or cyclic redundancy check bits. The check bits are added as the trailer in a frame at the sending end. Their span of check usually cover all bits from the frame identifier onwards.

The most common method of content error correction is retransmission of the frame. The receiver informs the sending end of the error and the sending end retransmits the frame. Note that it is essential for the sending end to retain a copy of the transmitted frame until it is acknowledged by the receiver.

For flow integrity errors, the data link protocols specify the procedures to be adopted to detect and recover the missing frames and acknowledgements. These procedures are built into the flow control mechanisms.

Recovery refers to the manner in which the system gets back into normal operating mode after a correction has been made. Generally, the sending end treats each retransmission as the first transmission and continues with the next frame. It keeps a historical record to detect if the link has become noisy, resulting in very frequent retransmissions. In such an eventuality, it may initiate recovery procedures which may involve re-establishment of the link.

It must be remembered that no error control method is 100 per cent effective. There will always be some undetected content and flow integrity errors. Residual Error Rate (RER) refers to the errors that still exist in the data stream after all error control procedures have been completed.

5.1 Error Control in Stop-and-Wait Mechanism

In the stop-and-wait mechanism, the receiver sends a positive acknowledgement (ACK) if there is no content error in the received data frame; else it responds with a negative acknowledgement (NAK). The sending end continues with the next data frame if it receives an ACK or repeats the previous frame if it receives a NAK. Figure 12 illustrates the mechanism.

To deal with flow integrity errors, the sending end is equipped with a timer. After sending a frame, the sending end waits for an acknowledgement (ACK or NAK) for a specified time period. When the time expires, it challenges the other end by sending an enquiry (ENQ). The receiver responds with ACK or NAK of the data frame received last.

The algorithm described above does not cover all contingencies. If it is a simple case of lost acknowledgement, the receiving end repeats the acknowledgement when it is challenged after time out, but if a data frame is lost during the transmission, the receiving end will have no knowledge of the frame and, therefore, when it is challenged by the sending end, it will respond with acknowledgement of the data frame received last. The sending end will obviously misinterpret it as acknowledgement for the lost data frame. So, a data frame will be completely lost without the sender or the receiver being aware of the loss (Fig. 12).

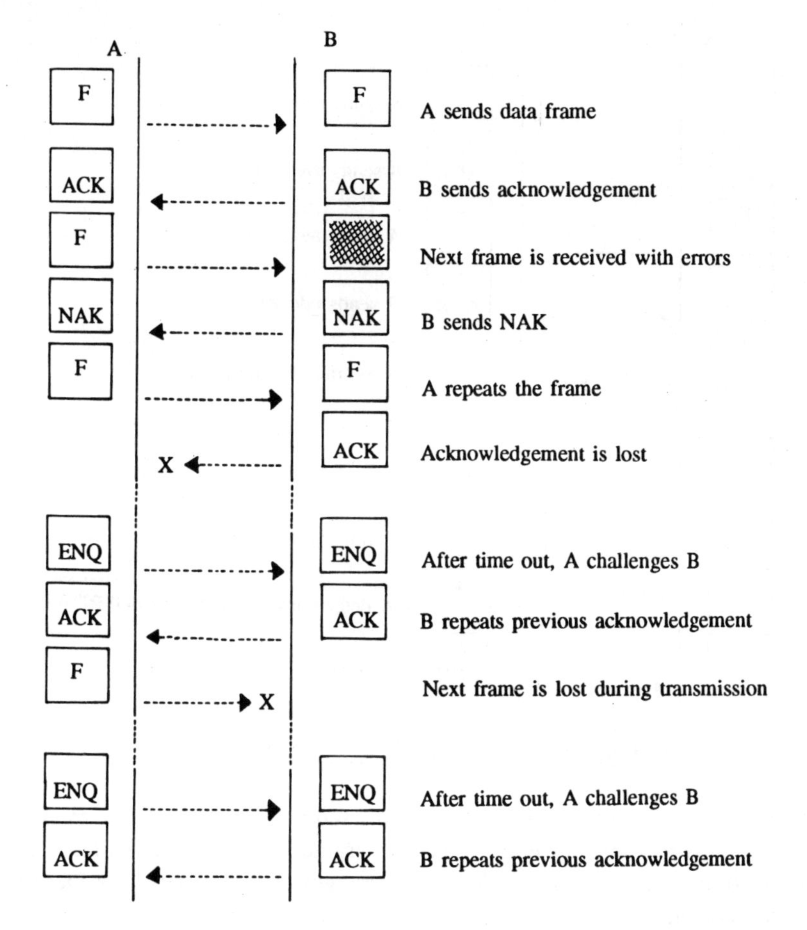

Fig. 12 Error control in stop-and-wait mechanism.

An alternative scheme overcomes this problem by distinguishing between the acknowledgements of consecutive data frames. Two types of acknowledgements are used—one for even data frames and the other for odd data frames. The data frames do not carry any number. The sender and receiver keep track of the even/odd designation of each frame. The mechanism is illustrated in Fig. 13.

The advantage of the stop-and-wait mechanism is its simplicity, but as discussed earlier, it is inefficient from the link utilization point of view.

5.2 Error Control in Sliding Window Mechanism

In the sliding window mechanism, each frame is assigned a sequence number and, therefore, a more elaborate error control scheme is feasible. The receiver keeps track of the sequence numbers of the incoming data frames. If any out-of-sequence frame is received, immediately, a request for retransmission of the missing frame is sent. There are two choices for the retransmission mechanism: selective retransmission or go-back-N.

Selective Retransmission: The receiver requests retransmission of the missing data frame

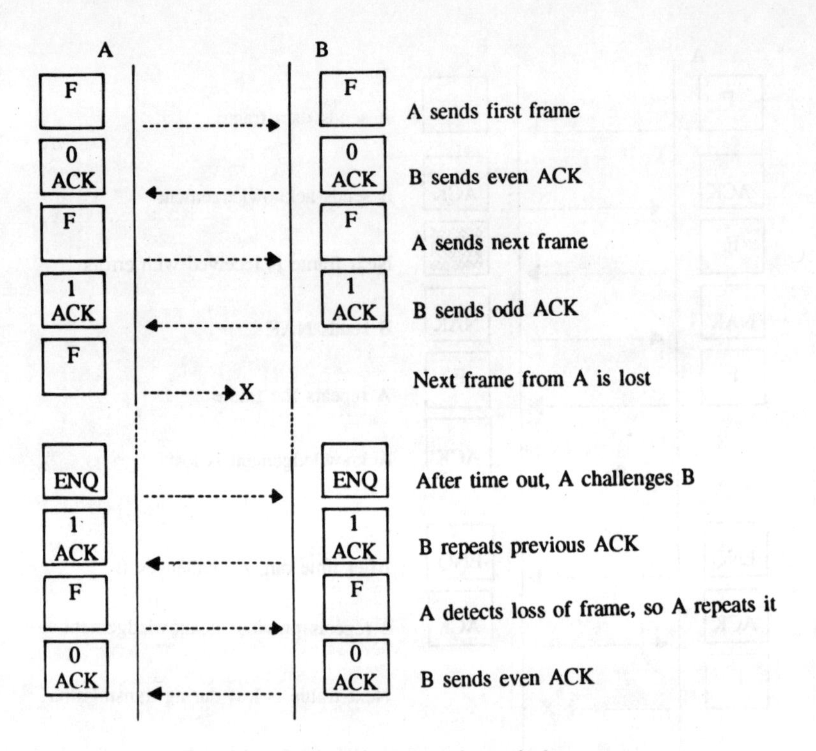

Fig. 13 Odd and even acknowledgements.

only by sending a Selective Reject (SREJ-*N*). *N* is the sequence number of the missing data frame. On receipt of SREJ-*N*, the sending end retransmits frame *N* only. Meanwhile, if the succeeding frames are received, the receiver accepts them and arranges all the frames in proper sequence when frame *N* is also received.

Go-Back-N: The receiver requests retransmission of the missing data frame and all the following frames by sending a Reject (REJ-*N*). REJ-*N* indicates request for retransmission of the data frames starting with the frame *N*.

REJ or SREJ are not immediately sent on detection of content error in a data frame as the error could be in the sequence number of the frame itself. So the receiver waits for the next correct frame and then declares the missing frame. Figure 14 illustrates the error control mechanism.

Both SREJ-*N* and REJ-*N* also acknowledge receipt of data frames upto *N*-1. Although SREJ is more efficient, the receiver requires enhanced capability for arranging the frames in proper sequence.

Acknowledgement Transmission. We have so far restricted ourselves to transmission of data frames from one device and the acknowledgements from the other device. In general, both the devices will need to exchange information and, therefore, will send data frames and acknowledgements as shown in Fig. 15. In sliding window flow control mechanism, the acknowledgements can be sent through special acknowledgement frames or, alternatively, they can be piggybacked on a data frame. Thus, a data frame will have two sequence numbers, one for the frame and the other for the acknowledgement.

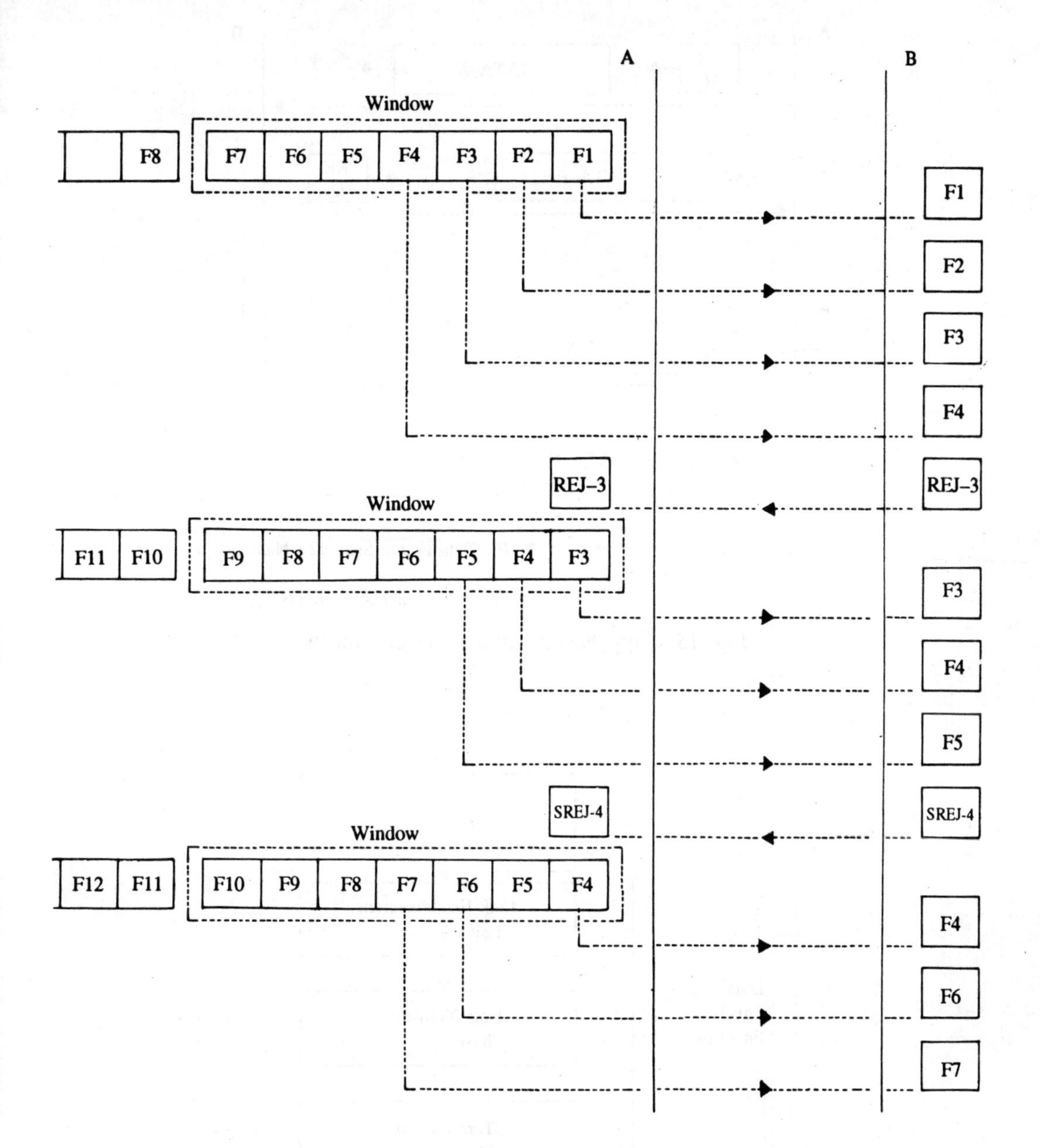

Fig. 14 Error control in sliding window flow control mechanism.

Only RR can be piggybacked on a data frame. If a data frame does not have any new RR to report, it repeats the last RR sequence number. REJ, SREJ and RNR are always sent as separate frames.

6 DATA LINK MANAGEMENT

The data transfer process between two devices can be viewed as consisting of the following five phases (Fig. 16):

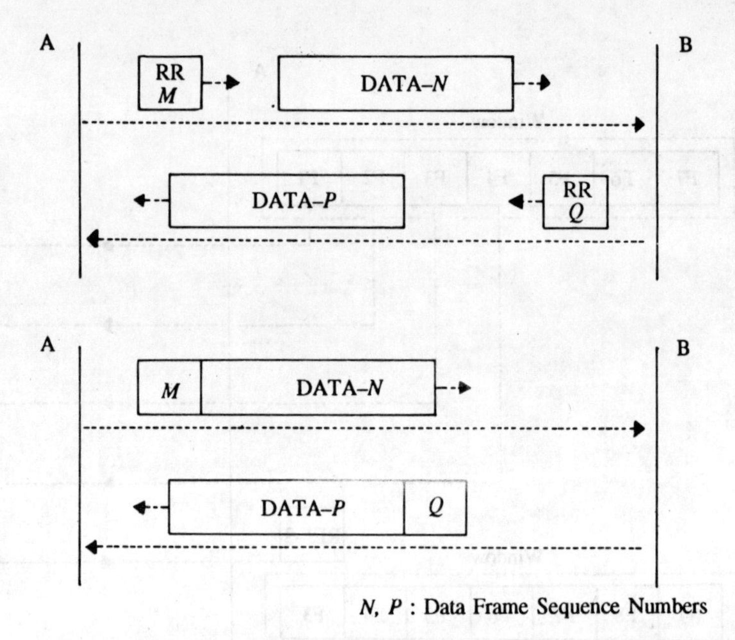

Fig. 15 Piggybacking of acknowledgements.

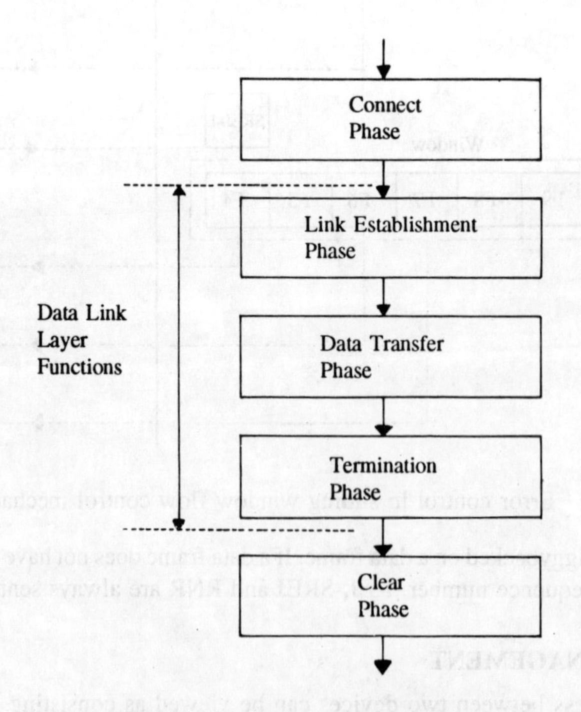

Fig. 16 Link management phases.

1. Connect phase
2. Link establishment phase
3. Data transfer phase
4. Termination phase
5. Clear phase.

The connect and clear phases consist of functions associated with establishing and releasing a physical connection between the two devices. These functions are part of the Physical layer protocol.

The span of data link protocols covers link establishment, information transfer and termination phases only. The data link management function involves execution of these phases.

Link establishment is the first phase in the span of data link control protocols. It includes processes required to initialize the data link, call/poll the other end and set the mode of data transfer (synchronous/asynchronous, TWA/TWS).

The data transfer phase involves exchange of data and acknowledgements. During data transfer, recovery procedures are initiated when some abnormal situation arises, which cannot be corrected by usual error control procedures. The link may be required to be re-established.

The termination phase consists of processes associated with disconnecting the link.

Each data link protocol defines a set of control symbols and procedures to execute the above-mentioned functions. These symbols and procedures are specific to a data link protocol and, therefore, cannot be generalized. BISYNC and HDLC are the commonly used protocols today. In the next chapter we will discuss HDLC protocol in detail as it is an international standard. An overview of the BISYNC protocol is given in Appendix B of this chapter.

7 SUMMARY

Data Link layer is the second layer of the OSI reference model. It improves the bit transport service of the Physical layer by controlling the errors. It provides sequenced and transparent delivery of DL-SDUs over the Data Link connection.

Data link protocols can be bit oriented or byte oriented. Flow control, error control and link management are the three basic functions of the Data Link layer. Either stop-and-wait or sliding window flow control mechanism is used. The stop-and-wait mechanism is very inefficient from the link utilization point of view.

Errors can be in the content or flow of the frames. Acknowledgements and time outs enable the detection of errors and retransmission of the frames.

The link management function involves establishment, maintenance and disconnection of the data link connection. It is specific to a data link protocol.

APPENDIX A

Data Link Service Primitives

Data Link service primitives for connection-mode and connectionless-mode of data transfer are given in Table A1 below.

Table A1 Data Link Service Primitives

Service	Primitive
Connection establishment	DL-CONNECT request DL-CONNECT indication DL-CONNECT response DL-CONNECT confirm
Connection release	DL-DISCONNECT request DL-DISCONNECT indication
Normal data transfer	DL-DATA request DL-DATA indication
Expedited data transfer	DL-EXPEDITED-DATA request DL-EXPEDITED DATA indication
Connection reset	DL-RESET request DL-RESET indication DL-RESET response DL-RESET confirm
Error reporting	DL-ERROR REPORT indication
Data transfer (Connectionless-Mode)	DL-UNITDATA request DL-UNITDATA indication

Connection establishment and reset are confirmed services. Expedited data allows transfer of a data unit which is not flow controlled, along with other data units.

APPENDIX B

B1 Binary Synchronous Communication Data Link Protocol (BISYNC)

Binary Synchronous Communication, BISYNC or BSC in short, is a data link layer protocol used for communication between IBM computers and terminals. Related ISO standards are ISO 1745, ISO 2111, ISO 2628 and ISO 2629. The basic features of BISYNC protocol are:

- It is a byte-oriented protocol.
- It supports three code sets—ASCII, EBCDIC and Transcode.
- It supports synchronous two-way alternate communication.
- It is applicable for point-to-point and point-to-multipoint communication.

B1.1 Types of Stations. There is a master-slave relationship between two communicating stations. The station which is to send a message is designated as the master and the station which receives messages and sends acknowledgements is designated as the slave.

B1.2 Point-to-Point Communication. Point-to-point communication is between two hosts. These two stations contend for master status whenever they want to transmit a message. Alternatively, one of the stations can be designated as control station which delegates master or slave status to the other depending on which station is to transmit messages.

B1.3 Point-to-Multipoint Communication. In point-to-multipoint communication, there is one host and several tributary stations (Fig. B1). The host decides who will send or receive messages. All the messages are sent by or to the host.

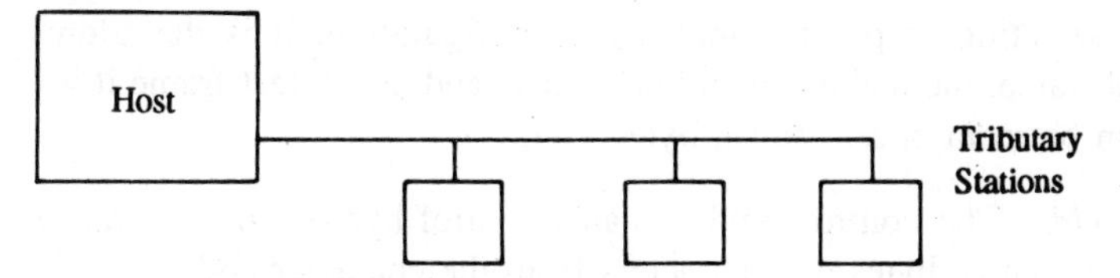

Fig. B1 Point-to-multipoint configuration.

Polling and Selecting. In point-to-multipoint communication, the host invites a station to transmit data by polling. The polled station becomes the master station and controls further communication. After satisfactory completion of transmission it returns the control to the host.

If the host wants to transmit data to a station, it alerts the station to receive messages. The process is called selecting. The selected station takes over the status of slave station for this communication.

B2 Transmission Frame

In IBM terminology, BISYNC frames are called blocks but we will stick to the terminology being used in this book. BISYNC utilizes two categories of frame types—supervisory and data frames (Fig. B2). Supervisory frames are used for sending control information and are not protected against content errors. Data frames contain user data and contain error detection bytes.

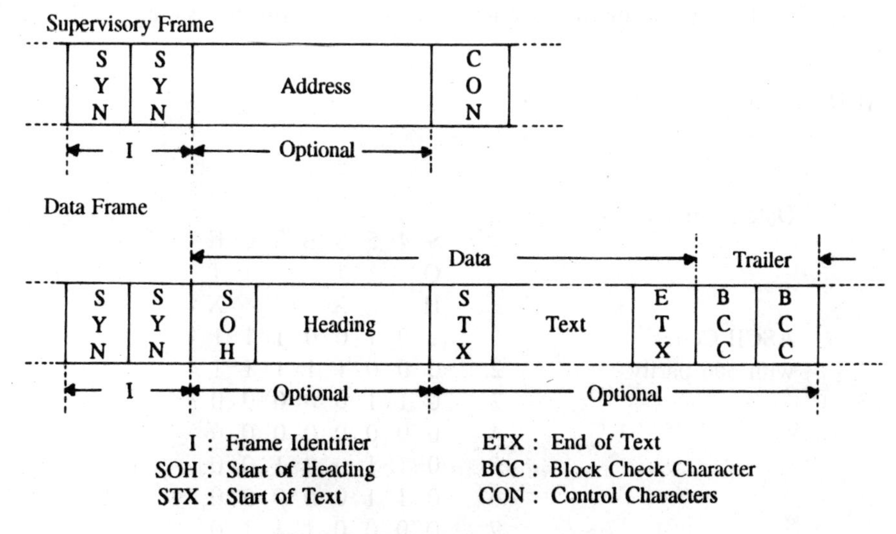

Fig. B2 BISYNC frame formats.

B2.1 Frame Format. Being a byte-oriented protocol, all the fields in BISYNC frames are of multiple bytes. BISYNC employs variable format and variable size frames. Therefore, several field identifiers/delimiters are required to indicate the presence of a field and to mark the end of a field.

Frame Identifier. Frame identifier is two-character long and consists of synchronizing characters SYN SYN. It is always present in all types of frames.

Address Field. The address field is optional and is present in the supervisory frames sent for polling and selecting in point-to-multipoint configuration. It is the address of the tributary station. In a poll frame, the address is in upper case; and in a select frame it is in lower case. The address field can be from one to seven bytes long.

Control Field. The control field contains control byte(s) for link management, error and flow control. The control bytes are characters from the character code set being used.

Heading Field. The heading field contains information for higher layer functions such as message identification, routing, device control and priority. It is optional and its presence is indicated by the field identifier SOH (Start of Heading). Length of the heading is variable and is delimited by the field identifier of the following field.

Text Field. The text field contains user data bytes. It is of variable size and is optional. A field identifier STX (Start of Text) indicates its presence. A field delimiter ETX (End of Text) or ETB (End of Transmission Block) is provided to mark end of the field.

Block Check Characters (BCC). BCC field is present in the data frames and is used for content-error detection. For ASCII code, it is one character long. For EBCDIC and Transcode, it is two characters long.

EXAMPLE B1

Show the data segment comprising SOH, heading, STX, Text and ETX for the following heading and text fields. Give the bit transmission sequence of the data segment assuming ASCII code with odd parity.

HEADING : 45 TEXT : BSC

Solution

Data Segment

		SOH	4	5	STX	B	S	C	ETX
ASCII Codes with odd parity	1	1	0	1	0	0	1	1	1
	2	0	0	0	1	1	1	1	1
	3	0	1	1	0	0	0	0	0
	4	0	0	0	0	0	0	0	0
	5	0	1	1	0	0	1	0	0
	6	0	1	1	0	0	0	0	0
	7	0	0	0	0	1	1	1	0
VRC	8	0	0	1	0	1	1	0	1

Bit transmission sequence

1000000000101100101011010100000001000011110010111100001011000001

B3 Control Characters

BISYNC utilizes ten characters from the character code set for link management, acknowledgements and framing. Some two-character control sequences are also defined as they are not readily available in the code set. Functions of some of the important control characters are now described.

SYN (Synchronous Idle). Two SYN characters are used as frame identifier as mentioned earlier. This character is also used to fill inter-frame idle time.

SOH (Start of Heading). It acts as field identifier for the heading field.

STX (Start of Text). This control character is used as field identifier for the text field.

ETB (End of Transmission Block). This control character is used as the field delimiter for text field.

ETX (End of Text). This control character is used as the field delimiter for the text field in the last frame of a message which may have been transmitted over several frames.

ENQ (Enquiry). ENQ is used during the link establishment phase to activate the other end. The control station polls or selects a station by sending ENQ. For communication between two hosts, ENQ is used for gaining master station status. It is also used during the data transfer phase to challenge the other station after time out if no acknowledgement is received.

EOT (End of Transmission). This control character signifies the end of a transmission. It results in relinquishment of the data link. It is also a negative response to a poll call.

ACK0/ACK1 (Positive Acknowledgements). ACK is a positive acknowledgement of the received frame. On receipt of ACK, the sending end can despatch the next frame. In BISYNC, ACK0 and ACK1 are used to acknowledge even and odd frames respectively.

NAK (Negative Acknowledgement). When a frame is received with errors, NAK is sent back as request for retransmission of the frame.

DLE (Data Line Escape). This character is used to change the meaning of the following contiguous characters. For example, when DLE is combined with STX as DLE-STX, it indicates a transparent data sequence.

WACK (Wait and Acknowledge). WACK is optional and is positive acknowledgement with a request for temporary suspension of further transmission.

B4 Error and Flow Control

BISYNC uses the stop-and-wait mechanism of flow control. Wait-and-acknowledge (WACK) is used to indicate inability of the receiving end to accept more data frames. For error control, BISYNC uses alternating acknowledgements ACK0 and ACK1. Timers are provided for error recovery. Depending on the character code set being used, BISYNC employs block parity check or cyclic redundancy check for error detection. With ASCII code set, BISYNC uses a VRC bit in each byte and an 8-bit LRC in the BCC field. With EBCDIC, a 16-bit CRC code based on CRC-16 polynomial is used in the BCC field. With 6-bit Transcode also the CRC code is used for detection of block errors. CRC-12 polynomial is used for generating the two 6-bit BCC bytes.

Accumulation of bytes which are covered by the BCC field for error detection, starts at the first SOH or STX character in a frame and all bytes upto (and including) ETX and ETB are covered (Fig. B3).

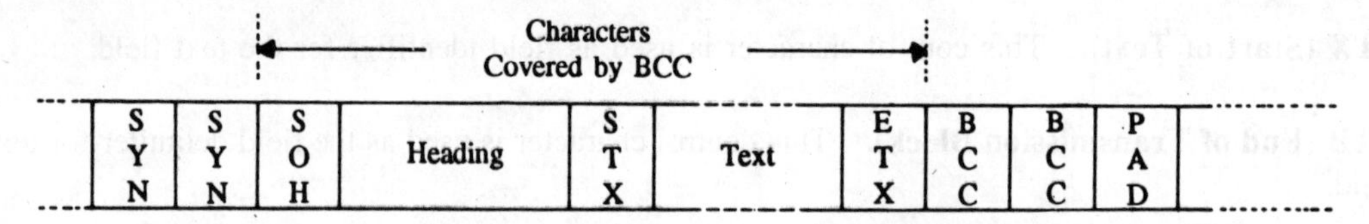

Fig. B3 Span of error detection by BCC field.

EXAMPLE B2

For the data segment of Example B1, determine the BCC character. Show the complete bit structure of the frame in order of transmission of bits.

Solution

		SYN	SYN	SOH	4	5	STX	B	S	C	ETX	BCC LRC
(LSB)	1	0	0	1	0	1	0	0	1	1	1	0
	2	1	1	0	0	0	1	1	1	1	1	0
	3	1	1	0	1	1	0	0	0	0	0	1
	4	0	0	0	0	0	0	0	0	0	0	1
	5	1	1	0	1	1	0	0	1	0	0	0
	6	0	0	0	1	1	0	0	0	0	0	1
	7	0	0	0	0	0	0	1	1	1	0	0
VRC	8	0	0	0	0	1	0	1	1	0	1	0

01101000011010001000000000101100101011010100000000100001111001011
1100001011000000100110100

B5 Transparency

Transparency is achieved in BISYNC by inserting the DLE control character before the text field identifier STX. The DLE STX sequence effectively instructs the receiving end to treat all characters in the text field as data bytes even if they are control characters. The transparent mode of the text field is terminated by DLE ETX (or DLE ETB) as shown in Fig. B4.

SYN	SYN	SOH	Heading	DLE	STX	Text (Transparent)	DLE	ETX	BCC	BCC	PAD

Fig. B4 Transparency in BISYNC.

If a byte representing DLE itself appears in the text field, the sending end stuffs another DLE into the bit stream to indicate to the receiving end that this is not the control character DLE which appears before ETX (or ETB) in the transparent mode of the text field. When the receiver detects two consecutive DLEs, it discards one and considers the other as part of the text field.

EXAMPLE B3

Show the structure of a data frame containing user message which is

1. ETB character;
2. DLE character.

Solution

1.

SYN	SYN	DLE	STX	ETB	DLE	ETX	BCC	BCC

2.

SYN	SYN	DLE	STX	DLE	DLE	DLE	ETX	BCC	BCC

B6 Protocol Operation

The protocol operation comprises data link establishment, information transfer and data link termination phases.

B6.1 Data Link Establishment Phase. Establishment of a data link involves exchange of certain supervisory frames between the data link entities to ensure their readiness to transmit and receive frames containing data. During this phase, master or slave status of a station is also decided.

Point-to-Point Communication. Usually, contention protocol is used for point-to-point communication between the host stations. Figure B5 shows the supervisory frames exchanged during link establishment phase.

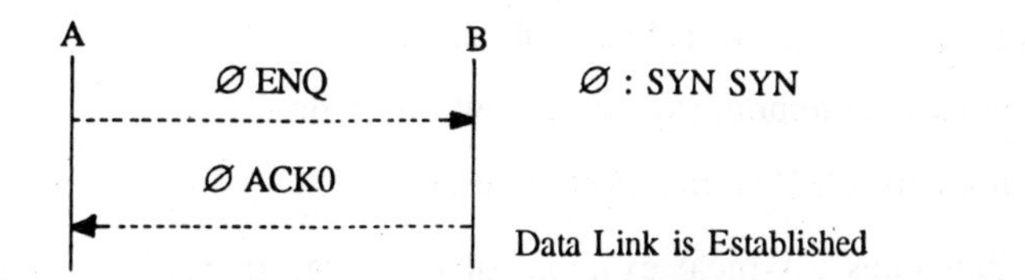

Fig. B5 Link establishment in point-to-point communication.

- A which wants to send a message bids for master status by sending SYN SYN ENQ.
- B replies SYN SYN ACK0 if it is ready to receive. If it is not ready to receive, it replies SYN SYN NAK in which case A needs to retry.
- If an invalid reply is received or if there is no reply, A transmits SYN SYN EOT and terminates the first attempt.
- Since it is possible that both the stations may bid together, contention is resolved by keeping different time outs for retry.

Point-to-Multipoint Communication. In point-to-multipoint link, the host activates one of the tributary stations at a time by sending a poll or a select. Figure B6 shows the frames which are exchanged during polling and selecting.

- The host sends SYN SYN address ENQ to the tributary station.
- Poll and select are differentiated by the address. Address is in upper case for polling and in lower case for selecting.
- If the polled tributary station B has data to send, it replies with a data frame and assumes master status. Else, it terminates the link by sending SYN SYN EOT.
- In the case of select call, reply of the tributary station B is SYN SYN ACK0 if it is ready to receive a data frame, or SYN SYN NAK if it is not ready to receive any data frame.

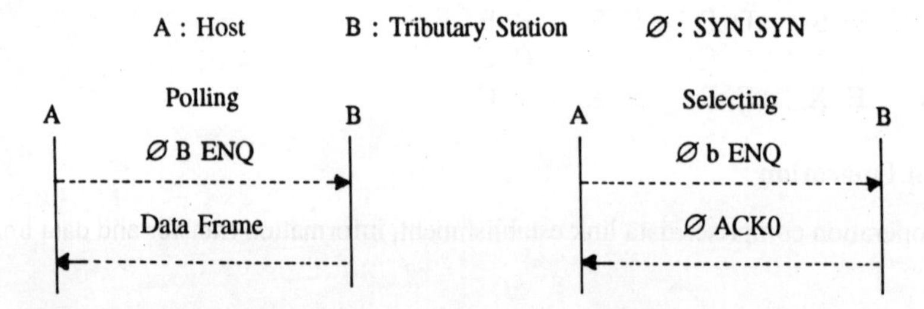

Fig. B6 Polling and selecting in point-to-multipoint communication.

B6.2 Data Transfer Phase. After establishing the data link, the master station sends the data frames containing user data bytes. The slave station sends acknowledgements using the supervisory frames.

- Alternating positive acknowledgements ACK0 and ACK1 are used for even and odd frames respectively.
- NAK is used for negative acknowledgement.
- WACK is used to temporarily stop flow of frames.
- ENQ is used for challenging after time out.

Figure B7 illustrates a typical example of exchange of frames during data transfer phase between stations A and B.

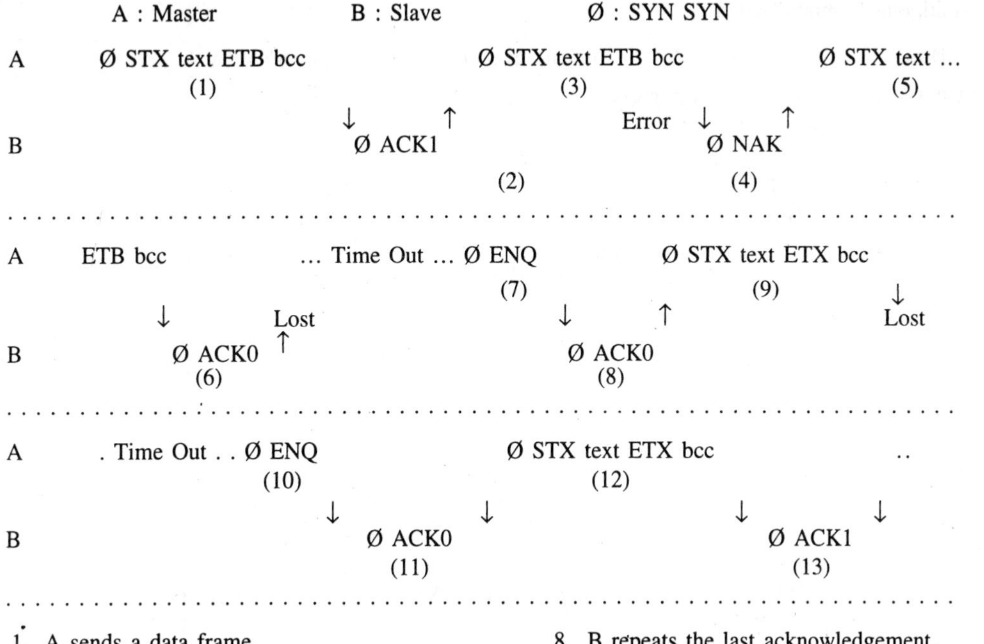

1. A sends a data frame.
2. B sends odd acknowledgement.
3. A sends next data frame.
4. B detects errors, sends negative acknowledgement.
5. A retransmits the data frame.
6. B sends even acknowledgement.
7. Acknowledgement is lost. After time out A challenges B.
8. B repeats the last acknowledgement.
9. A sends last data frame of the message.
10. The data frame is lost. After time out A challenges B.
11. B repeats the last acknowledgement.
12. A detects loss of frame, repeats the last data frame.
13. B sends odd acknowledgement.

Fig. B7 Example of data transfer in point-to-point communication

B6.3 Termination Phase. Termination is initiated by the master station following reception of a positive acknowledgement to the last data frame. It is effected by sending the supervisory frame SYN SYN EOT. Upon termination of the link, the master station loses its status and control of the link returns to the host. In point-to-point communication, the link becomes available to both the stations to contend for.

Figure B8 shows an example of data link dialogue on a point-to-multipoint link. In this example, all the three phases of data link operation are illustrated.

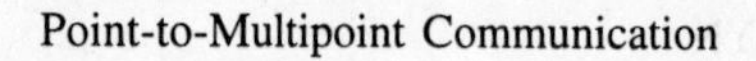

Point-to-Multipoint Communication

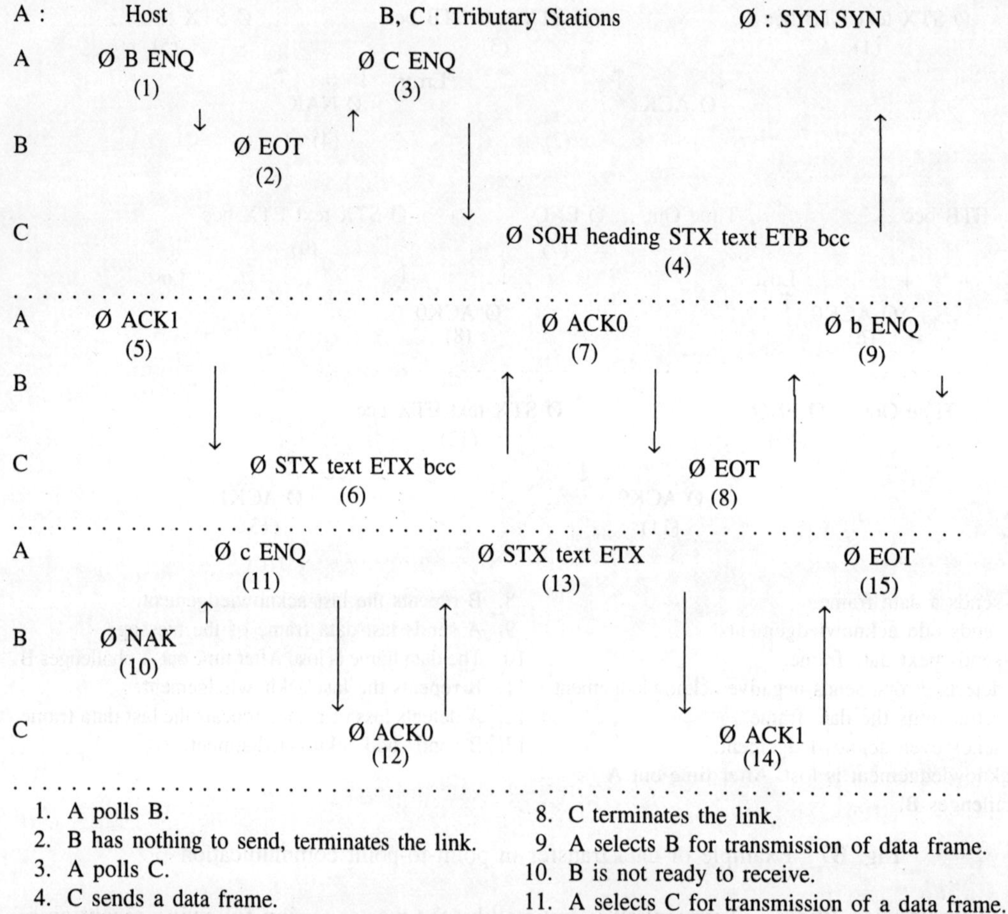

1. A polls B.
2. B has nothing to send, terminates the link.
3. A polls C.
4. C sends a data frame.
5. A acknowledges.
6. C sends next data frame indicating end of message.
7. A acknowledges.
8. C terminates the link.
9. A selects B for transmission of data frame.
10. B is not ready to receive.
11. A selects C for transmission of a data frame.
12. C acknowledges.
13. A sends a data frame.
14. C acknowledges.
15. A terminates the link.

Fig. B8 Example of data transfer in point-to-multipoint communication.

EXAMPLE B4

Shown below is an example of BISYNC point-to-point communication without any errors. Instead of showing complete frames only the important fields of the frames have been shown. Rewrite the exchange of frames had the second frame failed the error check.

ENQ		STX-ETB		STX-ETB		EOT
	ACK 0		ACK 1		ACK 0	

Solution

ENQ		STX-ETB		STX-ETB		STX-ETB		EOT
	ACK 0		ACK 1		NAK		ACK 0	

EXAMPLE B5

In Example B4, rewrite the exchange of frames if the second frame had been lost during transit.

Solution

ENQ		STX-ETB		STX-ETB	Time Out	ENQ		STX-ETB		EOT
	ACK 0		ACK 1				ACK 1		ACK 0	

B7 Limitations of BISYNC Protocol

Layered architecture concept is based on independence of functions and their distinct implementation. Since BISYNC was not originally designed with hierarchical functional layers in mind, it is awkward at places. For example, the heading field is a higher layer function but the field is defined at the Data Link layer. As regards Data Link layer functions, it meets the basic requirements but suffers from inherent limitations:

- Supervisory frames are not protected against errors.
- As a result of using characters from the code set, natural transparency is impossible. Therefore, transparency is achieved only as a special case by invoking transparent text mode.
- Communication is always two-way alternate even if a full duplex line is used. Its effect pervades all hierarchical layers.
- Link utilization is poor due to inherent limitations of the stop and wait flow control mechanism.

PROBLEMS

1. A channel is operating at 4800 bps and the propagation delay is 20 ms. What should be the minimum frame size for stop-and-wait flow control to get 50 per cent link utilization efficiency?

2. If the frame size is 960 bits on a satellite channel operating at 960 kbps, what is the maximum link utilization for the following:

(a) Stop-and-wait flow control mechanism?
(b) Sliding window flow control with window size 7?
(c) Sliding window flow control with window size 127?
(d) Sliding window flow control with window size 255?

Assume that the propagation delay is 270 ms.

3. In stop-and-wait mechanism, if the probability of receiving a data frame with errors is P, show that effective link utilization is given by

$$U = \frac{1 - P}{1 + 2A}$$

where $A = t_p/t_f$. Assume that acknowledgements are never received with errors.

4. Two stations A and B exchange frames using stop-and-wait flow control with odd/even acknowledgements for flow integrity errors and negative acknowledgement for content errors. Indicate the frames exchanged for the following actions:

- A sends a data frame.
- B detects content error and indicates so to A.
- A retransmits the frame.
- B receives the frame but asks A to wait.
- B asks A to go ahead with the next frame.
- A sends a frame but the frame is lost.
- A challenges B to respond.
- B responds.
- A retransmits the frame.
- B acknowledges.

5. If the window size is 7, and modulo 8 counting is used in sliding window flow control, show the exchange of frames and the frames in the window at each step. Assume A needs to transmit 10 frames to B and TWA mode of communication is used.

- A sends frames 0, 1, 2 and 3 to B.
- B acknowledges frames 0 and 1.
- A sends frames 4, 5 and 6.
- B rejects frame 3 and asks A to retransmit from frame 3 onwards.
- A transmits frame 3, 4, 5, 6 and 7.
- B acknowledges frames 3, 4, 5, 6 and 7.
- A transmits frames 0 and 1.
- B acknowledges frames 0 and 1.

6. Stations A and B exchange frames using sliding window flow control. The acknowledgements are piggybacked on the data frames. Each station has five data frames to transmit. The window size is 3 and modulo 4 counting scheme is used. Each data frame has two sequence numbers, first being the frame number and the second the acknowledgement number. Assume the previous acknowledgement number is repeated if no new data frame is received. Fill in the sequence numbers below. Assume there are no errors.

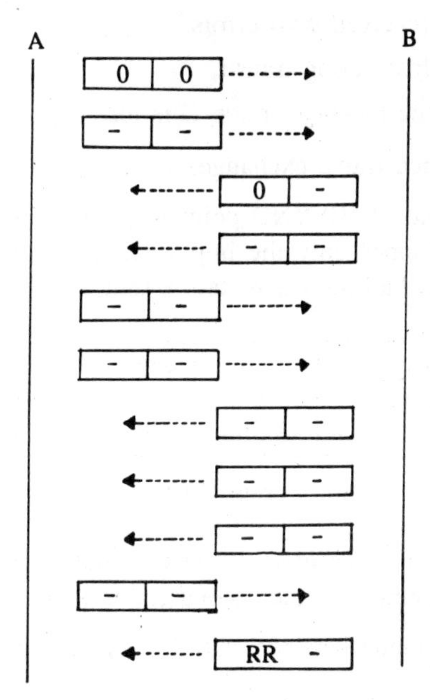

Fig. P.6

7. Show the data segment comprising SOH, heading, STX, Text and ETX for the following heading and text fields. Give the bit transmission sequence of the data segment assuming ASCII code with even parity.

HEADING : LQ
TEXT : D3

8. Show the frame format for the following heading and text fields:

(a) HEADING : REPORT
TEXT : PAYMENT *ETX* MONTH *DLE* JUNE

ETX and *DLE* control characters are part of the text field.

9. Find the BCC for the following frame. All the bytes are coded using ASCII and even parity is used for VRC and LRC.

S	S	S	9	0	%	E	B
Y	Y	T				T	C
N	N	X				B	C

10. A is a host which is connected to three tributary stations B, C, D. The polling sequence is B, C, D. The select sequence is C, B, D. A wants to send a message to each of the tributary station. Each tributary station has a message to send to A. Priority is given to the ouput messages from the host. Show the complete sequence of frames exchanged between A and the tributary stations. Use abbreviated frame notation as in Examples B4 and B5.

11. In Problem 10, the following frames are lost during transmission or are received with errors as indicated:

- Frame from A to D is received with errors,
- Frame from B to A is lost during transit,
- Acknowledgement for the message frame C is lost.

Show the complete sequence frame exchange.

12. Shown below is an example of BISYNC point to point communication without any errors. Instead of showing complete frames only the important fields of the frames have been shown. Rewrite the exchange of frames had the slave station sent a WACK for the first data frame.

ENQ		STX–ETB		STX-ETB		EOT
	ACK0		ACK1		ACK0	

Fig. P.12

13. For the data exchange shown in Problem 12, rewrite the frames if the first frame is received with errors and the acknowledgement of the second frame is lost.

14. Translate the following dialogue between stations A and B into BISYNC protocol:

A : Ready to receive?
B : Yes.
A : Here is data frame.
B : Received correctly.
A : Here is next data frame.
(The frame is lost. There is no response from B.)
A : Did you receive my last data frame?
B : The one I received last was correct.
A : Here is the data frame again.
B : Received correctly.
A : I am terminating transmission.
B : Ready to receive?
A : Yes.
B : Here is data frame.
A : Received correctly but wait.
B : Ready to receive?
A : O.K. Go ahead now.
B : Here is data frame.
A : There are errors in your data frame.
B : Here is the data frame again.
A : Received correctly.
B : I am terminating the transmission.

CHAPTER 8

HDLC—High-Level Data Link Control

High-Level Data Link Control (HDLC) protocol was developed by ISO and has become the most widely accepted data link protocol. It offers a high level of flexibility, adaptability, reliability and efficiency of operation for today's as well as tomorrow's synchronous data communication needs. ADCCP of ANSI (X3.66-1979) is almost similar to HDLC. IBM's SDLC (GA27-3093-2) is a proper subset of HDLC. LAP-B (Level 2, X.25) of CCITT is a permissible option of HDLC.

In this chapter, we first examine basic features, modes of operation, and frame structure of the HDLC protocol. The protocol operation is illustrated with the help of some typical examples of data communication situations. These examples include two-way alternate and two-way simultaneous communication in asynchronous and synchronous modes. We next discuss LAP-B CCITT. As it is similar to HDLC protocol, its special features including Multi-link Procedure are presented. Certain liberties have been taken in the level of completeness of description of the protocols so that the overall picture is not clouded with too many details.

1 GENERAL FEATURES

HDLC is a bit-oriented data link control protocol which satisfies a wide variety of data link control requirements including

- point-to-point and point-to-multipoint communication,
- two-way simultaneous communication over full duplex circuits,
- two-way alternate communication over half duplex or full duplex circuits,
- synchronous and asynchronous communication,
- communication between equal stations and between host and remote stations, and
- full data transparency.

The ISO standards for HDLC are ISO 3309, ISO 4335, ISO 6159 and ISO 6256.

2 TYPES OF STATIONS

To make HDLC protocol applicable to various possible network configurations, three types of stations have been defined:

1. Primary station
2. Secondary station
3. Combined station.

A primary station has the responsibility of data link management. A secondary station operates under the control of a primary station. A combined station can act both as a primary and a secondary station.

When communication is between a primary station and a secondary station, the primary station has the responsibility of activating, maintaining and disconnecting the data link. All the frames sent by a primary station are called commands and the frames sent by a secondary station are called responses (Fig. 1).

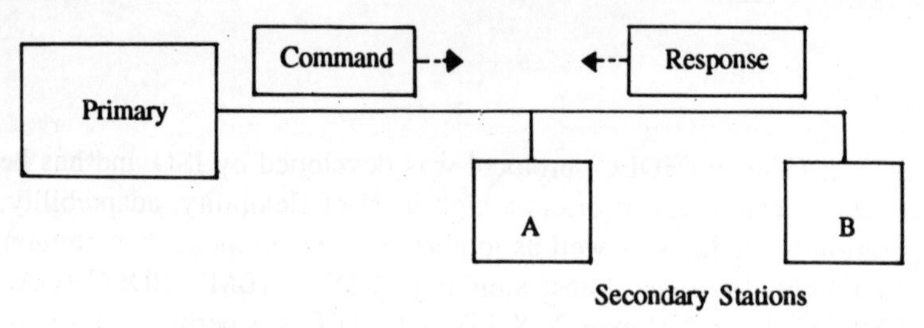

Fig. 1 Primary and secondary stations.

Communication can be between two logical equal status computers also, in which case they are designated as combined stations and can send and receive both, commands and response (Fig. 2). When a combined station sends a command, the other responds and vice-versa.

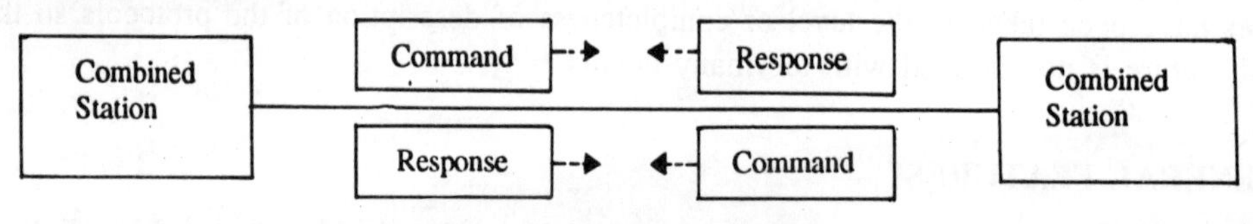

Fig. 2 Combined stations.

It must be kept in mind that the communicating entities in the above figure and in the discussion which follows are the Data Link layer entities of the stations. Thus, when we use the term "station", we mean the "Data Link layer of the station".

3 MODES OF OPERATION FOR DATA TRANSFER

HDLC permits both synchronous and asynchronous modes of communication. Synchronous and asynchronous terms do not imply "start-stop" bits or "clock" which refer to the Physical layer. Three modes of operation for data transfer are possible:

1. Normal Response Mode (NRM)
2. Asynchronous Reponse Mode (ARM)
3. Asynchronous Balanced Mode (ABM).

The normal and asynchronous response modes of operation provide an unbalanced type of data transfer capability between logical unequal stations—one primary and other secondary stations. The asynchronous balanced mode of operation is for logical equal or combined stations.

3.1 Normal Response Mode (NRM)

In the normal response mode the primary station controls the overall link management function. A secondary station can send a frame only as a result of receiving explicit permission to do so from the primary station. It is a synchronous mode of communication.

The normal response mode is applicable to point-to-point and point-to-multipoint configurations. It is suited for polled multipoint operation where ordered interaction between a host and a number of outlying computers/terminals is required.

3.2 Asynchronous Response Mode (ARM)

It is an asynchronous mode of communication between a primary and a secondary station. The secondary station can send a frame without any explicit permission from the primary station. ARM operation, therefore, is less disciplined than NRM operation. The link management function is the responsibility of the primary station.

The asynchronous response mode is applicable to both point-to-point and point-to-multipoint configurations. In multipoint environment, however, only one secondary station can be active at a time and other secondary stations must be kept in disconnected mode.

3.3 Asynchronous Balanced Mode (ABM)

Asynchronous balanced mode is applicable to point-to-point communication between two combined stations. Both the stations are capable of link management function when required. They can issue commands and responses. One station can force a response from the other station if required. Being in asynchronous communication mode, a station can send a frame without any explicit permission from the other station.

4 OTHER MODES OF OPERATION

In addition to the three data transfer modes explained above, there are three other modes of operation. These modes refer to the states before and after the data transfer phase.

1. Normal Disconnected Mode (NDM)
2. Asynchronous Disconnected Mode (ADM)
3. Initialization Mode (IM).

4.1 Normal Disconnected Mode (NDM)

In the disconnected modes, the stations are logically disconnected. They need to exchange mode-setting commands to come out of the disconnected mode. When in normal disconnected mode, a secondary station is activated by a mode-setting command for normal response mode from the primary station.

4.2 Asynchronous Disconnected Mode (ADM)

When in the asynchronous disconnected mode, the stations enter asynchronous response mode or asynchronous balanced mode when the corresponding mode-setting command is exchanged. A secondary station in asynchronous disconnected mode can request for a mode-setting command from the primary station in order to establish data transfer mode.

4.3 Initialization Mode (IM)

In the initialization mode, operational parameters are exchanged. It is invoked when a primary station concludes that the secondary station is operating abnormally and needs its operational parameters corrected. Also, a secondary station can request the primary station for IM if it is unable to function properly.

4.4 Mode Transition

Transition from one mode to another is effected by the primary station by giving mode-setting commands. We will study these commands shortly. Figure 3 shows how transition of the logical modes takes place when appropriate commands are given.

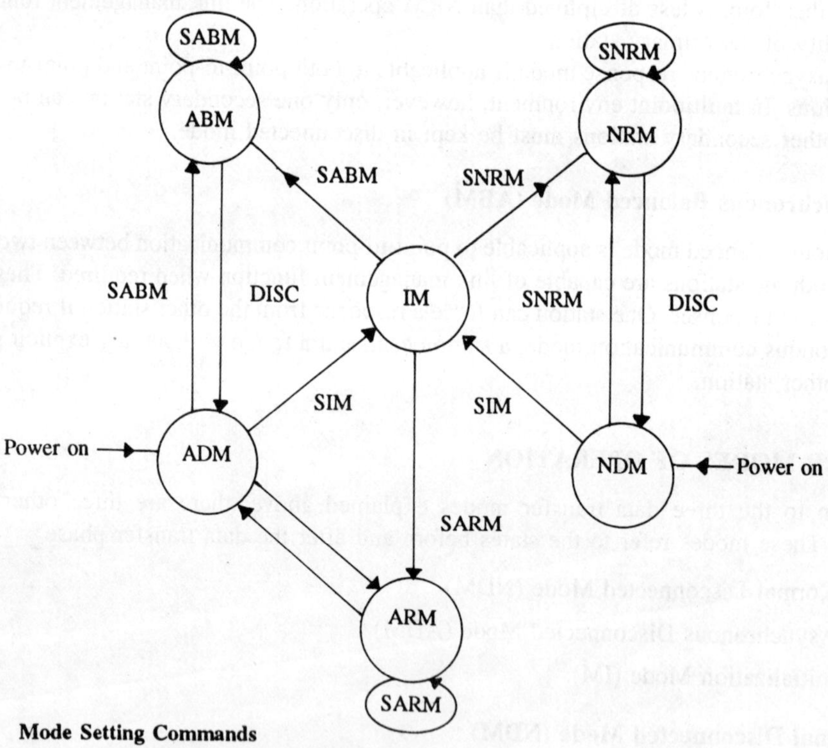

Fig. 3 Mode transition diagram.

Note that, when switched on, a station is in the disconnected mode. When it receives a mode-setting command from a primary station, it enters the mode corresponding to the command.

5 FLOW CONTROL

HDLC utilizes the sliding window flow control mechanism. The window size can be either 7 or 127. All the data frames and acknowledgements are numbered as required for the sliding window flow control mechanism. The receiving end sends acknowledgement in the form of RR–N (Receive Ready for frame *N*, frames upto *N*-1 acknowledged). When the receiver is not ready to receive more data frames, it sends RNR-*N* (Receive Not Ready for frame *N*, frames upto *N*-1 acknowledged).

6 ERROR CONTROL

Error control is based on retransmission of frames received with errors. Retransmission is requested by sending a Reject (REJ-*N*) or a Selective Reject (SREJ-*N*). Error detection is carried out using a 16-bit cyclic redundancy check (CRC) code generated using CCITT V.41 polynomial $x^{16} + x^{12} + x^5 + 1$ (10001000000100001).

For recovery purposes, the following parameters are specified:

- *T*1 : Time out for retransmission of a frame. Its typical value is 3 seconds.
- *T*3 : Time out for completion of link initialization. Its typical value is 90 seconds.
- *N*1 : Maximum number of transmissions and retransmissions before declaring link down. Its typical value is 20.

7 FRAMING

HDLC utilizes two types of frame formats as shown in Fig. 4. The frame format is fixed. Except the information field, all other fields have fixed sizes. Two frame identifier/delimiters are required, one at the start of the frame and the other at the end. The frame is transmitted from left to right and the low-order bit is transmitted first.

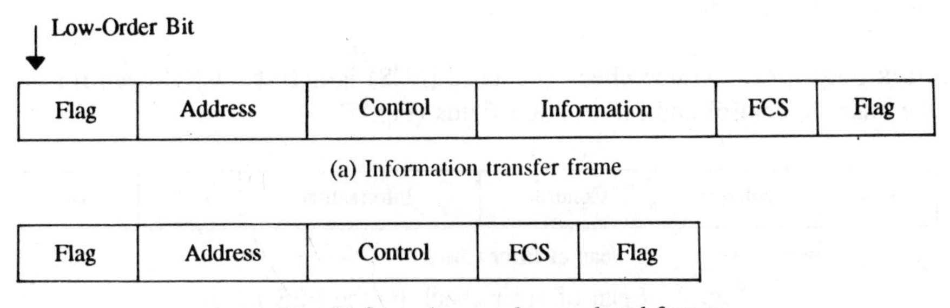

Fig. 4 Structure of HDLC frames.

Flag. The flag is a unique 8-bit word pattern (01111110) which identifies the start and end of each frame. It is also used for filling the idle time between consecutive frames (Fig. 5).

Flag	Flag	Flag	Flag	Other Fields	Flag	Flag	Flag

|←---- Frame ----→|

Fig. 5 Flag transmission to fill idle time between consecutive frames.

Address Field. The address field always contains the address of the secondary station whether a frame is being transmitted by a primary or a secondary station. In Fig. 6, when the primary station sends a frame, it indicates the address of the secondary station so that the secondary station may

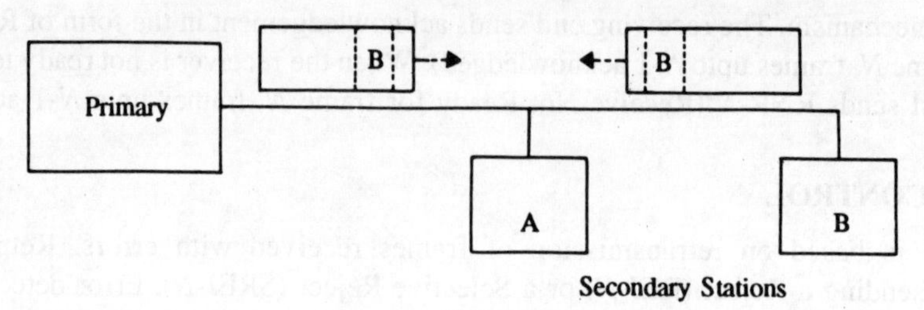

Fig. 6 Frames containing address of the secondary station.

receive the frame. However, when the primary station receives a frame, it would not be able to identify the sender unless the sending secondary station inserts its own address in the frame. Thus in command, a station gives the address of the destination, and in response, it gives its own address.

The address field consists of 8 bits, giving it a capability of 256 different addresses. Greater than 256 address capability is also possible as explained later. The all ones address is specified as global address. It addresses all the secondary stations simultaneously.

Control Field. The control field consists of 8 bits. It can be extended to 16 bits as explained later. It carries the sequence number of the frame, acknowledgements, request for retransmission and other control commands and responses.

Information Field. The information field has variable size and can consist of any number of bits. The maximum number of the bits in the information field is not specified. It contains the user data and is completely transparent.

Frame Check Sequence. Frame check sequence (FCS) is a 16-bit CRC code for detection of errors in the address, control and information fields (Fig. 7).

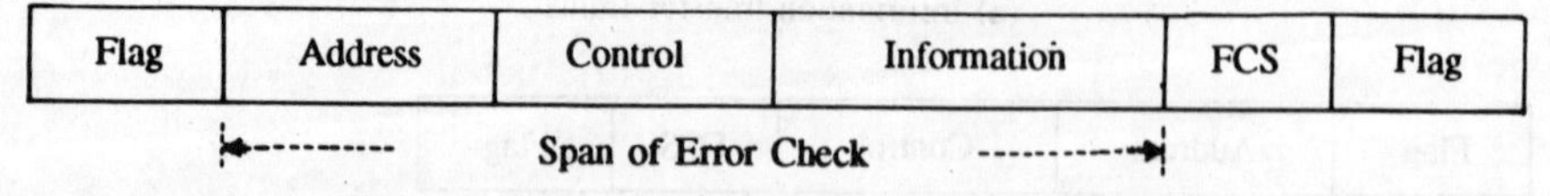

Fig. 7 Span of error check by the FCS field.

EXAMPLE 1

Shown below is a bit sequence containing an HDLC frame. Identify various fields of the frame.

01111110 01111110 10100011 01100010 01110000 11000011 01010101
10101011 01111110 01111110

Solution

1. Identify the start and end flags.
2. After the start flag, we have one octet each of the address and control fields.

3. Before the end flag, we have two octets of the FCS.
4. The remaining bits comprise the information field.

Flag	Start Flag	Address	Control	Information		FCS-1
01111110	01111110	10100011	01100010	01110000	11000011	01010101

10101011	01111110	01111110
FCS-2	End Flag	Flag

7.1 Types of HDLC Frames

There are three types of HDLC frames:

1. Information transfer frame (I-frame)
2. Supervisory frame (S-frame)
3. Unnumbered frame (U-frame).

The I-frame has a format as shown in Fig. 4a. The S-frame and U-frame have formats as shown in Fig. 4b.

Information Transfer Frame (I-Frame). The I-frame is used for transporting user data. It also carries acknowledgement of the received frames. The control field of the I-frame is as shown in Fig. 8. The first bit is "0" which identifies the frame as an I-frame. The next three bits are the sequence number *N*(*S*) of the frame.

The fifth bit is the Poll/Final (*P*/*F*) bit. Its use is explained later.

The last three bits are the sequence number *N*(*R*) of the acknowledgement (RR) which is piggybacked on the I-frame.

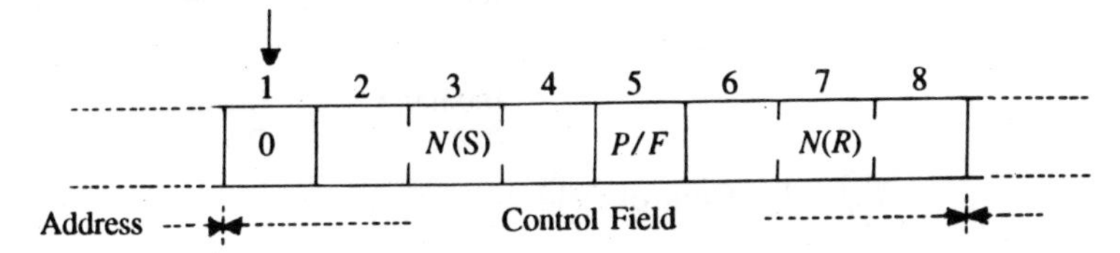

N (*S*) : Frame Sequence Number
N (*R*) : Acknowledgement Number
P/*F* : Poll / Final Bit

Fig. 8 Control field of an I-frame.

Supervisory Frame (S-Frame). The S-frame does not have a data field and is used to carry only acknowledgements and requests for retransmission. It is identified by the first two bits of the control field (Fig. 9). These two bits are "10" in an S-frame. The next two bits *SS* are used to indicate four supervisory states, Receive Ready (RR), Receive Not Ready (RNR), Reject (REJ), and Selective Reject (SREJ).

The fifth bit is the Poll/Final bit. The last three bits are the sequence number associated with the supervisory state RR, RNR, REJ or SREJ as indicated in the *SS* bits of the control field. As

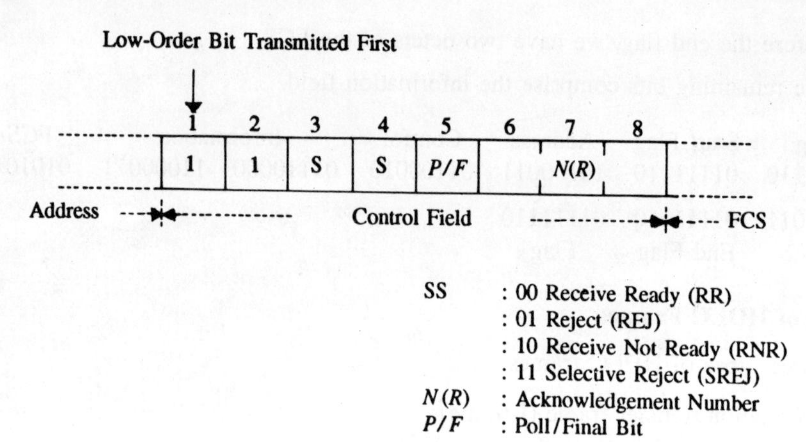

Fig. 9 Control field of a S-frame.

already explained, an acknowledgement (RR) can be sent either on a supervisory frame or piggybacked on an I-frame. On the other hand, RNR, REJ and SREJ are sent only through a supervisory frame.

Unnumbered Frame (U-Frame). A U-frame, as the name suggests, does not have any sequence number. It is used for link establishment, termination, mode setting and other control functions. The control field of a U-frame is shown in Fig. 10.

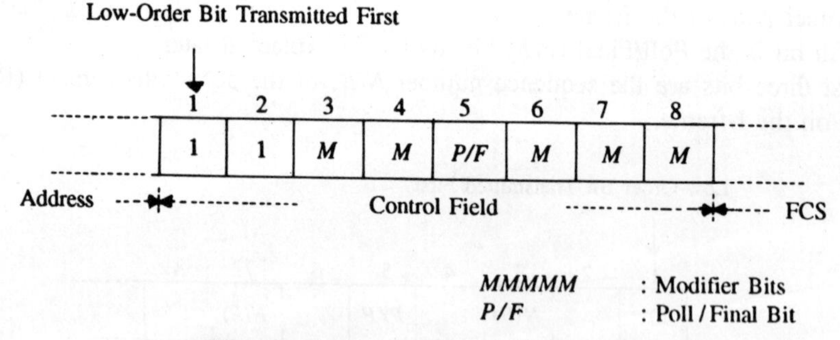

Fig. 10 Control field of an U-frame.

The first two bits of the control field are "11" which identify an unnumbered frame. The fifth bit is the Poll/Final bit. The remaining five bits are called modifier bits. They specify the control function. Table 1 gives the codes of the control field of a U-frame for the various commands and responses.

U-frame commands are sent by the primary station and the responses by the secondary station. These commands and responses are briefly described below:

Set-mode Commands (SNRM, SARM, SABM, SNRME, SARME, SABME): SNRM, SARM and SABM are the mode-setting commands. A secondary or a combined station sets itself in the mode corresponding to the received command. SNRME, SARME and SABME are used when an extended frame numbering format is used to accommodate a window size of 127.

Set Initialization Mode Command (IM): This command is used to establish the initialization mode of operation during which operational parameters are exchanged.

Table 1 Control Field of U-Frames

Control field bits								
1	2	3	4	5	6	7	8	
Low-order bit transmitted first ↓								**U-Frame Commands**
1	1	0	0	P	0	0	1	Set Normal Response Mode (SNRM)
1	1	1	1	P	0	0	0	Set Asynchronous Response Mode (SARM)
1	1	1	1	P	1	0	0	Set Asynchronous Balanced Mode (SABM)
1	1	1	1	P	0	1	1	Set NRM Extended (SNRME)
1	1	1	1	P	0	1	0	Set ARM Extended (SARME)
1	1	1	1	P	1	1	0	Set ABM Extended (SABME)
1	1	1	0	P	0	0	0	Set Initialization Mode (SIM)
1	1	0	0	P	0	1	0	Disconnect (DISC)
1	1	0	0	P	0	0	0	Unnumbered Information (UI)
1	1	0	0	P	1	0	0	Unnumbered Poll (UP)
1	1	1	1	P	0	0	1	Reset (RSET)
1	1	1	1	P	1	0	1	Exchange Identification (XID)
1	1	0	0	P	1	1	1	Test
1	1	0	1	P	0	*	*	Nonreserved commands
								U-Frame Responses
1	1	0	0	F	1	1	0	Unnumbered Acknowledgement (UA)
1	1	1	1	F	0	0	0	Disconnected Mode (DM)
1	1	1	0	F	0	0	0	Request Initialization Mode (RIM)
1	1	0	0	F	0	0	0	Unnumbered Information (UI)
1	1	1	0	F	0	0	1	Frame Reject (FRMR)
1	1	1	1	F	1	0	1	Exchange Identification (XID)
1	1	0	0	F	0	1	0	Request Disconnect (RD)
1	1	0	0	F	1	1	1	Test
1	1	0	1	F	0	*	*	Nonreserved Responses

* : '1' or '0'.

Disconnect Command (DISC): It is used to terminate a previously established link and to cause the stations to assume the disconnected mode.

Unnumbered Information Command/Response (UI): The unnumbered information frames are used to exchange miscellaneous information such as hourly reports, periodic time checks, etc. These frames are not acknowledged.

Unnumbered Poll Command (UP): The unnumbered poll command is used to solicit a response frame from a station without regard to sequencing.

Reset Command (RSET): Reset command is used for recovery purposes and sets sequence numbers $N(S)$ and $N(R)$ in one direction of transmission to zero.

Exchange Identification (XID): XID command and response are used to request and/or report identity of a station and optionally, its operational parameters.

Test Command/Response: Test command and response are used for testing the data link control.

Unnumbered Acknowledgement (UA): The UA response is used to acknowledge receipt and execution of a U-frame command.

Disconnected Mode Response (DM): It is sent to the primary station in response to a mode-setting command to indicate that mode-setting action has not been executed. It is also used as request to the primary station to send the mode-setting command.

Request Initialization Mode (RIM): This response is used for requesting the primary station to establish the initialization mode.

Frame Reject Response (FRMR): The FRMR response is used to report a condition which is not correctable by retransmission of frames e.g., receipt of invalid $N(R)$.

Request Disconnect (RD): The RD response is used for requesting the primary station to disconnect the link.

EXAMPLE 2

Identify the type of frame from the control field given below. Also identify the sub-fields within the control field. The low order bit is on the left hand side.

1. 01010111
2. 10111010
3. 11000000

Solution

1. I-frame, $N(S)$ = 101, P/F = 0, $N(R)$ = 111
2. S-frame, SREJ, P/F = 1, $N(R)$ = 010
3. U-frame, unnumbered information, P/F = 0

7.2 Poll/Final (P/F) Bit

Fifth bit of the control field of an HDLC frame is called Poll/Final (P/F) bit. It is called P bit when the frame is a command, i.e., the frame is being sent by a primary station. It is called F bit when the frame is a response, i.e., the frame is being sent by a secondary station.

A frame is identified as a command or response by the address field. If the address field contains the address of destination, the frame is a command. If it is the station's own address then the frame is a response.

In normal response mode, a primary station invites a secondary station to transmit a frame by setting the P of the control field to "1". Having received the invitation to transmit, the secondary station sends frames and finally returns the permission by explicitly marking its last frame. The secondary station utilizes the F bit of the control field in the last frame. It sets this bit to "1".

In the asynchronous modes of data transfer, ARM and ABM, the P bit is used to solicit response from a secondary/combined station. When a frame with P set to "1" is received, the receiving station responds with the F bit set to "1" at the earliest opportunity.

Once a command with P bit set to "1" is sent, the primary station awaits a response with F set to "1" and does not send another frame with the P bit set to "1" until it is established that such response will not be forthcoming. This may happen if either the command or the response is lost.

8 TRANSPARENCY

In HDLC, transparency is achieved by ensuring that the unique flag sequence (01111110) does not occur in the address, control, information and FCS fields. A technique called "zero stuffing" is used. At the sending end an extra "0" bit is inserted after five contiguous "1"s occurring anywhere after the opening flag and before the closing flag. At the receiving end, the extra "0" bit following five contiguous "1"s is deleted.

The steps involved in assembling an HDLC frame are given below. Note that zero stuffing is performed before the flags are appended to the rest of the frame. Therefore, any sequence of bits (including the flag sequence) can be transmitted in the address, control, information and FCS fields without affecting the data link control operation.

- Build address and control fields and append to the information field
  ↓
- Generate CRC
  ↓
- Carry out zero stuffing
  ↓
- Append flags

At the receiving end, the above steps are carried out in reverse order.

- Identify flags and delete them
  ↓
- Remove the stuffed zeros
  ↓
- Compute and check FCS
  ↓
- Check address and control fields

EXAMPLE 3

The following bit stream represents an HDLC frame from the address field to the FCS field. Do the zero stuffing for transparency and then construct the full HDLC frame.

1011111101110111110111111111111111000000

Solution

1. Inserting an extra zero after every five consecutive ones, we get

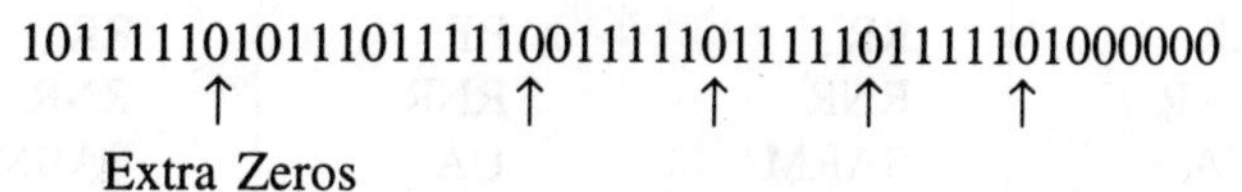

2. Constructing the complete frame by adding flags, we get

01111110101111101011101111100111110111110111110100000001111110

EXAMPLE 4

Identify various fields after destuffing extra zeros from the following bit stream of an HDLC frame.

01111110101111010011101111100111110111110011110100000001111110

Solution

1. Removing the flags, we get

   1011110100111011111001111101111100111101000000

2. Removing extra zeros after every five consecutive ones, we get

   1011110100111011111 011111 11111 0111101000000

   ↓ ↓ ↓

   0 0 0

3. Address field : 10111101
   Control field : 00111011
   FCS : 1110111101000000
   Information field : 1110111111111101

9 PROTOCOL OPERATION

Having reviewed the basic features of the HDLC protocol, let us now examine its operation. Typical data communication situations include point-to-point and point-to-multipoint links in various data transfer operating modes, namely, NRM, ARM and ABM. As HDLC protocol has been designed to serve a wide variety of applications, only a subset of its capabilities is required in a specific situation. There are many subsets and these subsets are categorized into three classes which correspond to the three modes of data transfer:

1. Unbalanced Normal Class (UNC)
2. Unbalanced Asynchronous Class (UAC)
3. Balanced Asynchronous Class (BAC).

Each class has a basic repertoire of commands and responses as shown in Table 2.

Table 2 Classes of HDLC Operation and Their Basic Repertoire of Commands and Responses

UNC		UAC		BAC	
Primary commands	Secondary responses	Primary commands	Secondary responses	Primary commands	Secondary responses
I	I	I	I	I	I
RR	RR	RR	RR	RR	RR
RNR	RNR	RNR	RNR	RNR	RNR
SNRM	UA	SARM	UA	SABM	UA
DISC	DM	DISC	DM	DISC	DM
	FRMR		FRMR		FRMR

The capabilities of each class can be modified by adding other optional commands and responses or deleting some from the basic repertoire as required for a specific application. Table 3 gives the optional functions. Each function is identified by the corresponding serial number.

Table 3 Optional Functions

	Optional functions		Command	Response
1.	For switched circuits	Add	XID	XID, RD
2.	For two-way simultaneous operation	Add	REJ	REJ
3.	For single frame retransmission	Add	SREJ	SREJ
4.	For unnumbered information	Add	UI	UI
5.	For initialization	Add	SIM	RIM
6.	For unnumbered polling	Add	UP	
7.	For multiple octet addressing			
8.	For command I-frames only	Delete		I
9.	For response I-frames only	Delete	I	
10.	For extended sequence numbering			
11.	For one-way reset	Add	RSET	
12.	For data link testing	Add	Test	Test
13.	For request disconnect	Add		RD

A subset of HDLC application is denoted by its class followed by the serial number of the additional function. For example, UNC 3, 4 denotes unbalanced normal class with basic repertoire of commands and responses supplemented with SREJ and UI.

10 EXAMPLES OF PROTOCOL OPERATION

Figures 10–17 illustrate some examples of the protocol operation for two-way alternate and two-way simultaneous links in these modes. These examples serve to illustrate the protocol operation in typical situations but it must be noted that these examples are not exhaustive and do not cover all possibilities.

In these examples, we consider the following three phases of operation:

1. Link establishment
2. Data transfer
3. Link disconnection

Link establishment is always initiated by the primary station by sending a mode-setting U-frame command with the P bit set to "1". The link is established when an unnumbered acknowledgement with the F bit set to "1" is received from the secondary station.

Link disconnection is also carried out in the same manner by the primary station. It sends an unnumbered disconnect command with P bit set to "1". The secondary station responds with an unnumbered acknowledgement having the F bit set to "1". It is ensured by the primary station that all I-frames have been acknowledged and all acknowledgements have reached the destination before the link disconnection is initiated.

In the asynchronous balanced mode, either of the two combined stations can establish and disconnect the link.

In these examples, the frames are represented as a code of five symbols *A B C D E*, where

A = Address of the secondary station

B = Type of frame—I, S, U

C = Sequence number of the I-frame
or
Acknowledgement in S-frame
or
Link management command, response of U-frame

D = Sequence number associated with the acknowledgement (I- and S-frames only)

E = Poll/Final bit

The poll/final bit is shown only when it is "1". In commands, it is shown as *P* and in responses as *F*. When it is not shown, it means that *P*/*F* bit is "0".

10.1 Normal Response Mode, Point-to-Point

Figures 11 to 13 illustrate the operation of the protocol in Normal Response Mode for point-to-point communication over two-way alternate and two-way simultaneous links.

Two-Way Alternate (TWA) Communication without Errors (Fig. 11)

- The primary station "A" sends mode-setting command SNRM with *P* bit set to "1".
- A mode-setting command is always acknowledged with UA response and with *F* bit set to "1".
- Being TWA operation, only one station transmits at a time.
- A secondary station can initiate transmission only after it receives explicit permission from the primary station in the form of *P* bit set to "1".
- The link is disconnected by sending DISC command which is acknowledged with UA response.

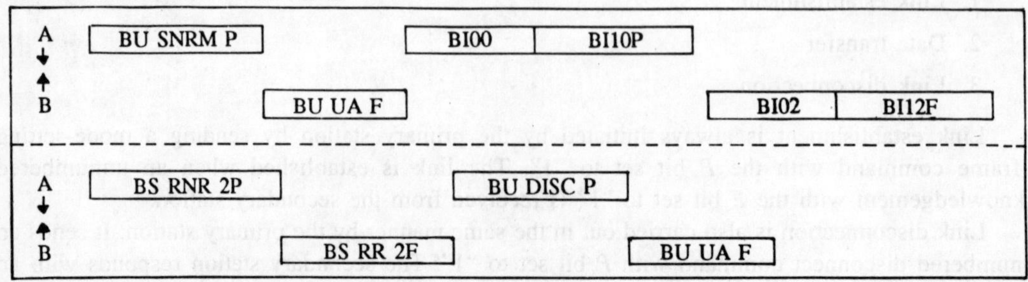

Fig. 11 Example of point-to-point two-way alternate communication in normal response mode (without errors).

- Before the link is disconnected, it is ensured that
  - — all frames have been acknowledged,
  - — all acknowledgements have been received, and
  - — secondary station is prevented from sending any I-frame.

In Fig. 11 station A sends RNR-2 to acknowledge I-frames of station B and to indicate its unreadiness to accept more I-frames. Station A ensures that this message reaches station B by sending the *P* bit set to "1". The only alternative left for B is to respond RR-2 with the *F* bit set to "1". On receipt of RR-2 from B, station A sends the disconnect command.

Two-way Alternate (TWA) Communication with Errors (Fig. 12)

- A mode setting command is retransmitted after time out if it is not acknowledged.
- Loss of a frame is detected when the next frame in sequence is received. A frame received with error is also considered as lost because the error may even be in its sequence number.
- Loss of a frame is communicated when its acknowledgement is not received in the frames sent subsequently e.g. BI00 of B could have been sent only after receipt of BI10P of A. As BI00 of B does not acknowledge BI00 of A, the frame must have been lost.
- Loss of a frame with *P* bit set to "1" is detected when there is no response. After time out,

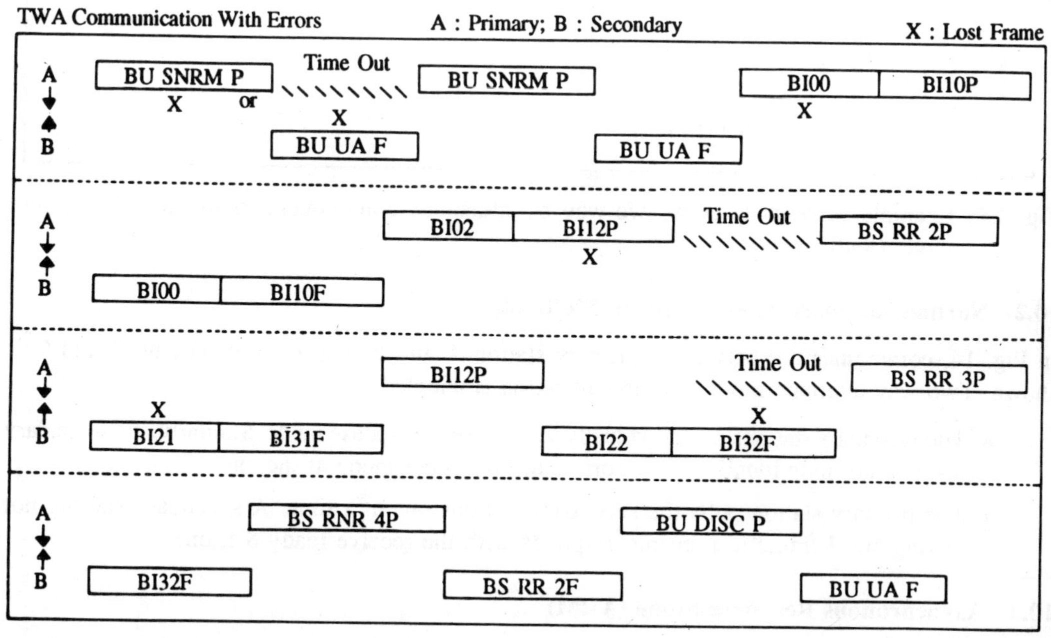

Fig. 12 Example of point-to-point two-way alternate communication in normal response mode (with errors).

the primary station challenges the secondary by sending another frame with P bit set to "1" e.g. BS RR 2P is sent when there is no response to BI12P.

- Loss of a frame with F bit set to "1" is detected by the primary when there is no activity on the link after it sends a frame with P bit set to "1". It challenges the secondary after time out by sending a frame P bit set to "1".

Two-Way Simultaneous (TWS) Communication with Errors (Fig. 13)

- Both the stations can transmit and receive simultaneously, but the secondary station can send a frame only after it receives permission to transmit from the primary station.
- When loss of a frame is detected, a supervisory frame with REJ is sent.
- When REJ is received, all the frames from the lost frame onward are retransmitted.

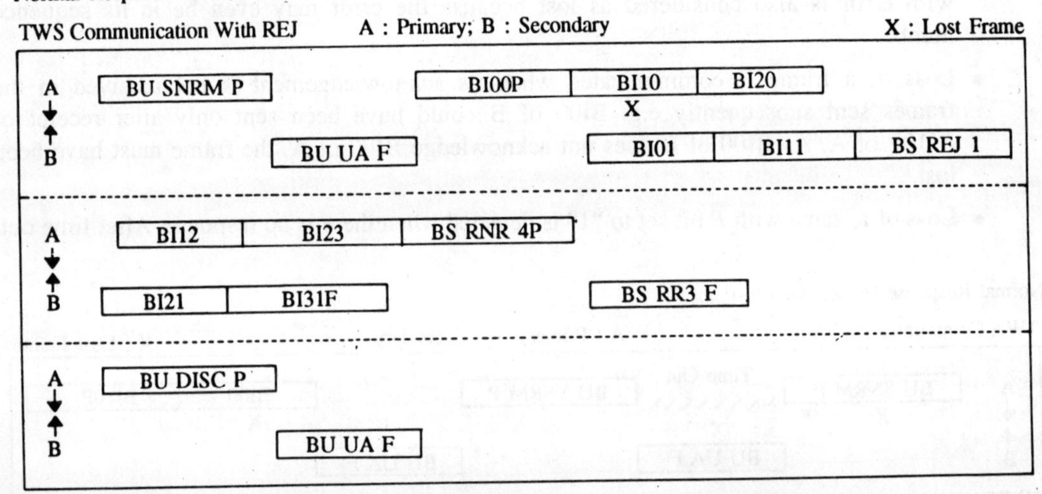

Fig. 13 Example of point-to-point two-way simultaneous communication in normal response mode (with errors).

10.2 Normal Response Mode, Point-to-Multipoint

In Fig. 14, communication between a primary station A and two secondary stations B and C is shown. Two-way alternate mode of communication is adopted.

- The secondary stations are individually set to normal response mode. Similarly, the secondary stations are individually set to normal disconnected mode at the end.
- The primary station polls the secondary stations one at a time. A secondary station, not having any I-frame to transmit, responds with the receive ready S-frame.

10.3 Asynchronous Response Mode (ARM)

Figures 15 to 17 illustrate the operation of the protocol in asynchronous response mode of communication over two-way alternate and two-way simultaneous links.

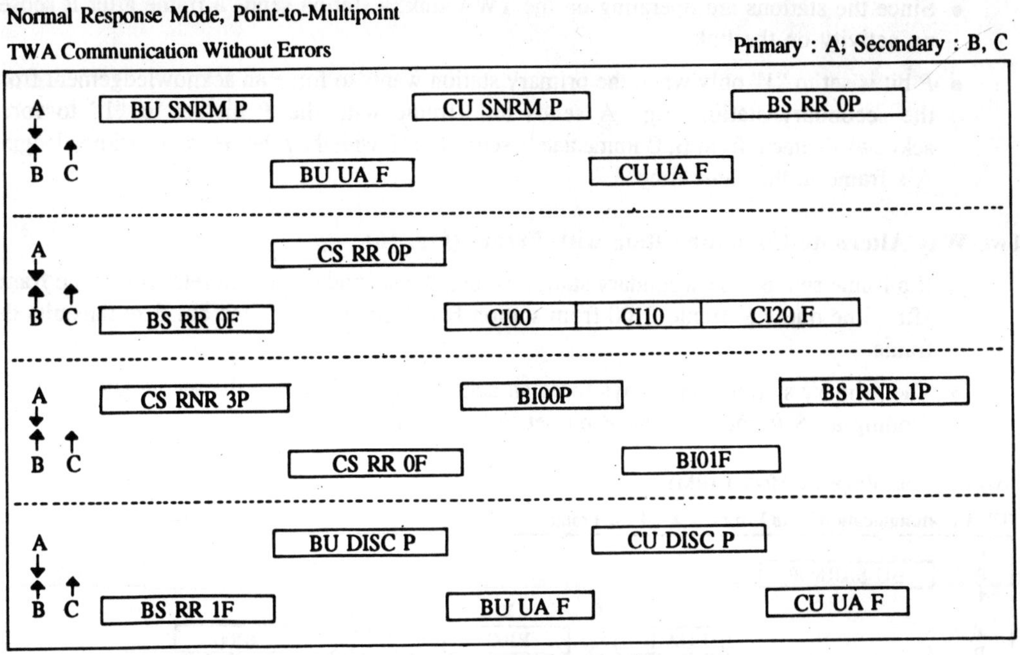

Normal Response Mode, Point-to-Multipoint

TWA Communication Without Errors

Primary : A; Secondary : B, C

A → B, C: BU SNRM P; BU UA F; CU SNRM P; CU UA F; BS RR 0P

CS RR 0P; BS RR 0F; CI00; CI10; CI20 F

CS RNR 3P; CS RR 0F; BI00P; BI01F; BS RNR 1P

BU DISC P; BS RR 1F; CU DISC P; BU UA F; CU UA F

Fig. 14 Example of point-to-multipoint two-way alternate communication in normal response mode.

Two-Way Alternate Communication without Errors (Fig. 15)

- The secondary station need not wait for the poll from the primary station to transmit, e.g., after receiving the mode-setting command from the primary station A, the secondary station B sends the BI00 and BI10 frames to A.

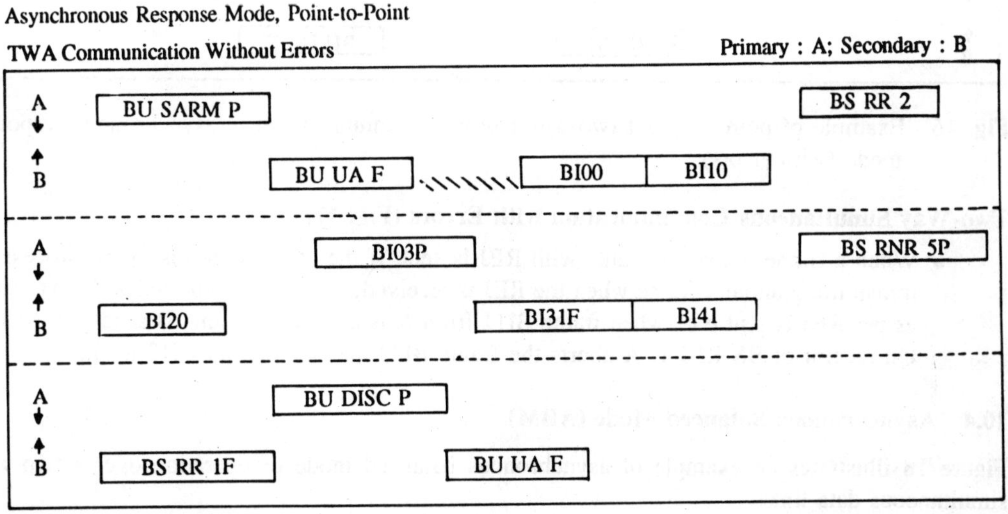

Asynchronous Response Mode, Point-to-Point

TWA Communication Without Errors

Primary : A; Secondary : B

A → B: BU SARM P; BU UA F; BI00; BI10; BS RR 2

BI03P; BI20; BI31F; BI41; BS RNR 5P

BU DISC P; BS RR 1F; BU UA F

Fig. 15 Example of point-to-point two-way alternate communication in asynchronous response mode (no errors).

- Since the stations are operating on the TWA link, a station sends a frame after it senses no activity on the link.
- *P* bit is set to "1" only when the primary station wants to force an acknowledgement from the secondary station, e.g., A sends an I-frame with the *P* bit set to "1" to force acknowledgement from B. B immediately sends BI31F with the *F* bit set to "1" acknowledging A's frame at the same time.

Two-Way Alternate Communication with Errors (Fig. 16)

- If a frame sent by the secondary station is lost, the secondary station retransmits the frame after time out, e.g. frame BI00 from station B is lost and after time-out B retransmits the frame.
- The primary station can always force an acknowledgement from the secondary station by sending an S-frame with the *P* bit set to "1" after time out.

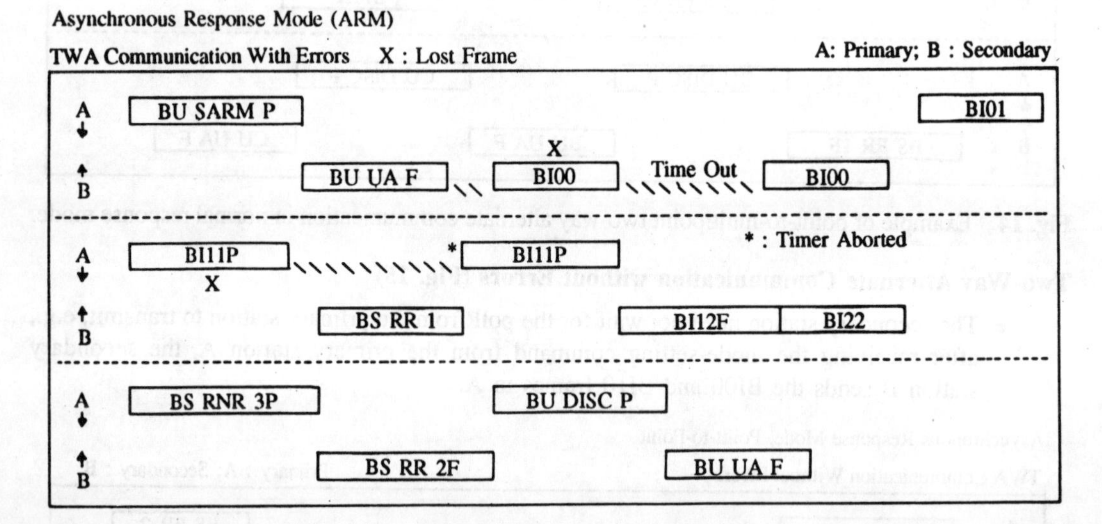

Fig. 16 Example of point-to-point two-way alternate communication in asynchronous response mode (with errors).

Two-Way Simultaneous Communication with Errors (Fig. 17)

- When a frame is lost, S-Frame with REJ is sent. If the other station is in the process of transmitting another frame when the REJ is received, the frame is aborted and next frame as per REJ is sent e.g., when frame BI11 from A is detected missing, B sends BS REJ-1. On receipt of BS REJ-1, A aborts the frame BI33 and starts from BI13 again.

10.4 Asynchronous Balanced Mode (ABM)

Figure 18 illustrates an example of asynchronous balanced mode of operation over a two-way simultaneous data link.

- Link can be set up by either of the two stations by sending SABM mode setting command.

Asynchronous Response Mode

TWA Communication With Errors X : Lost Frame A : Primary; B : Secondary

A
BU SARM P
BI00P
BI11
BI22
B I 3 *
X
B
BU UA F
BI00
BI10
BI21F
BS REJ 1

A
BI13
BI23
BI34P
BS RNR 4P
B
BI31
BS RR 4F

A
BU DISC P
B
BS RR 4F
BU UA F

* : Frame Aborted

Fig. 17 Example of point-to-point two-way simultaneous communication in asynchronous response mode (with errors).

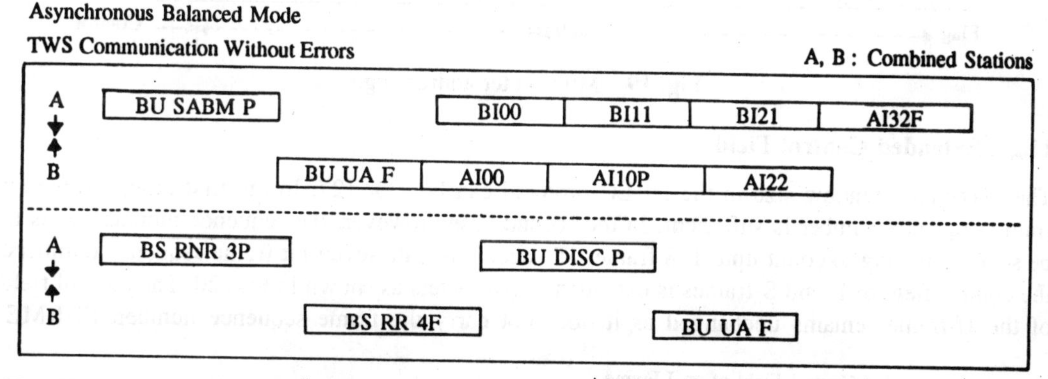

Fig. 18 Example of point-to-point two-way simultaneous communication in asynchronous balanced mode.

- A station may send an I-frame as a command or as a response. The command or response status of a frame is indicated by the address field.
- Both the stations can force a response by sending a command with the *P* bit set to "1".

11 ADDITIONAL FEATURES

HDLC protocol has provision for extension of the address and the control fields. Extended address field is required when more than 256 addresses are to be accommodated. Extended control field is required when the window size is greater than seven. The format of HDLC frame is modified to accommodate the extended fields as described below.

11.1 Extended Addressing

Two address field options are defined in the HDLC protocol; namely, single-octet addressing and multi-octet addressing. Single octet addressing provides for a maximum of 256 different addresses. Addressing capability can be enhanced using multiple octet addressing scheme. In case of multi-octet addressing, the address field is recursively extendable using the first bit of each octet to indicate the extended format of the address field (Fig. 19). The first bit of each octet is set to "0" indicating that the next octet is also to be considered as part of the address field. In the final octet, the first bit is set to "1". The addressing scheme, either single octet or multi-octet is chosen once and thereafter it cannot be dynamically changed.

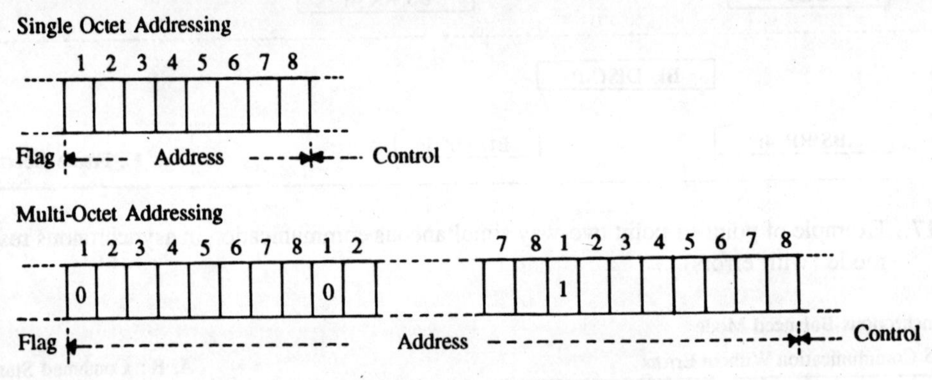

Fig. 19 Multi-octet addressing.

11.2 Extended Control Field

The maximum window size in the HDLC protocol can be 7 or 127. In the first case, a three-bit frame sequence number is sufficient. In the second case, however, the sequence number needs to be seven bits long to count upto 128 frames. To accommodate seven-bit frame sequence numbers, the control field of I- and S-frames is extended to two octets as shown in Fig. 20. The control field of the U-frame remains unchanged as it does not carry the frame sequence number. SNRME,

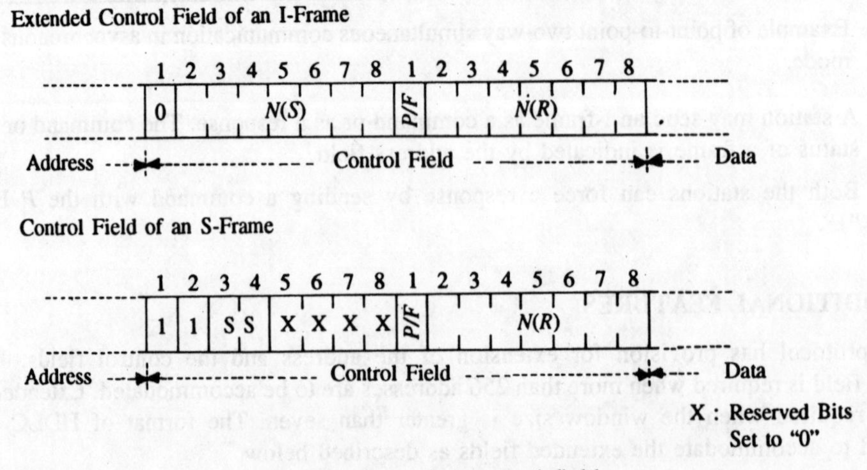

Fig. 20 Extended control field.

SARME and SABME mode-setting commands are used in place of corresponding SNRM, SARM and SABM commands, for the extended format of the control field.

12 COMPARISON OF BISYNC AND HDLC FEATURES

Table 4 gives comparison of the features of BISYNC and HDLC protocols. The Physical layer characteristics required for supporting these protocols are also indicated.

Table 4 Comparative Features of BISYNC and HDLC Protocols

Feature	Characteristic	BISYNC	HDLC
Transmission	Serial	Yes	Yes
	Asynchronous	No	No
	Synchronous	Yes	Yes
Communication mode	Asynchronous	No	Yes
	Synchronous	Yes	Yes
Directional mode	TWA	Yes	Yes
	TWS	No	Yes
Configuration	Point-to-point	Yes	Yes
	Point-to-multipoint	Yes	Yes
Flow control		Stop-and-wait	Sliding window
Error detection and correction	Content errors	LRC/CRC	CRC
	Flow integrity errors	ACK-0/ACK-1	Sequence number
Code set		ASCII/EBCDIC/ Transcode	Any
Control characters		Many	None
Framing	Frame identifier	SYN SYN	Flag
	Frame delimiter	ETB/ETX	Flag
	Information field	Multiple bytes	Multiple bits
	Transparency	DLE stuffing	Zero stuffing

13 LINK ACCESS PROCEDURE (BALANCED)

CCITT Recommendation X.25 defines interface between a DTE and a DCE operating in the packet mode. The DCE is an access node of a packet switched data subnetwork and the DTE is a terminal equipment owned by the subscriber. A packet is N-PDU generated by the Network layer of the DTE and is routed to the destination by the Network layer of the subnetwork. We will study packet switching and X.25 recommendation in considerable detail in Chapter 10.

X.25 defines the interface for the first three layers. The recommendation specifies X.21 and X.21*bis* interfaces for the Physical layer. For the Data Link layer, X.25 specifies the Link Access Procedure—Balanced (LAP-B).

LAP-B is a data link protocol between a packet mode DTE and the access node of the subnetwork which is referred to as DCE (Fig. 21). It is a permissible option of the HDLC protocol. Asynchronous balanced mode is used in the LAP-B protocol. The frame structures for I-frames, S-frames and U-frames and the basic protocol operation are the same as described earlier.

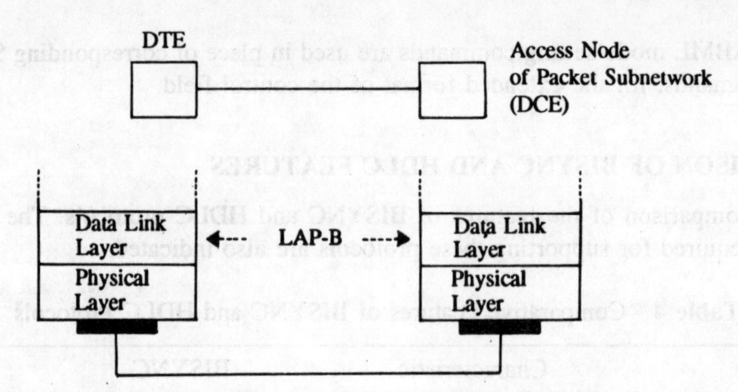

Fig. 21 LAP-B protocol for the Data Link layers of a DTE and an access node of packet switched data subnetwork.

Note that LAP-B is used for reliable transfer of data bits from the DTE to the DCE and vice versa over the physical connection established by the Physical layers. The Data Link layers of the DTE and the DCE are given addresses 00000011† and 00000001 respectively. These addresses enable identification of commands and responses as described earlier.

LAP-B specifies the following commands and responses for the U-frames:

- SABM Set to Asynchronous Balanced Mode command
- DISC Disconnect command
- DM Disconnected Mode response
- UA Unnumbered Acknowledgement response
- FRMR Frame Reject Response.

FRMR is sent by the DTE or by the DCE to report a non-recoverable error condition such as receipt of an invalid command or response, receipt of an I-frame whose information field exceeds the maximum established length, and receipt of invalid $N(R)$.

The S-frames use RR, RNR and REJ acknowledgements. The I-frames contain the packets received from the Network layer in their information field.

14 MULTILINK PROCEDURE (MLP)

CCITT Recommendation X.25 permits multiple physical connections between the packet mode DTE and the subnetwork access node, DCE (Fig. 22). Multiple physical connections provide increased

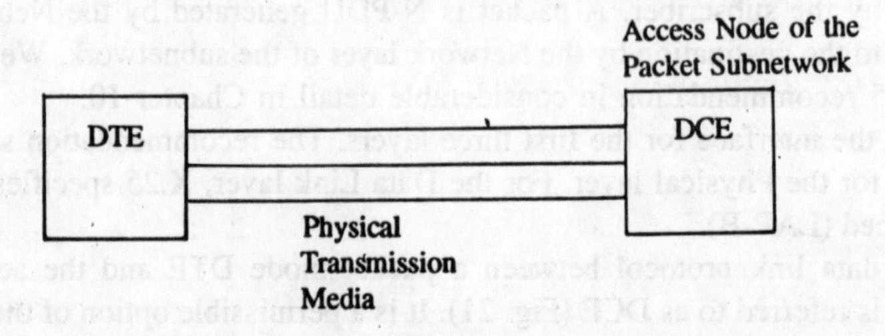

Fig. 22 Multiple physical media for increased reliability of transmission.

†Low-order bit has been indicated first.

reliability of operation. The protocol for utilizing the multiple connections is implemented in the Data Link layer.

The Data Link layer is divided into a Multilink Procedure (MLP) sublayer and Single Link Procedure (SLP) sublayer (Fig. 23). The SLP sublayer is LAP-B and, therefore, each single link connection operates as described in the last section.

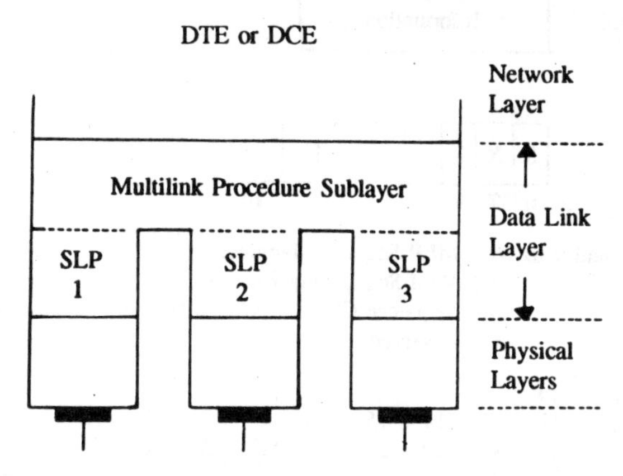

Fig. 23 Architecture of the Data Link layer for multilink procedure.

The multilink procedure forms a common sublayer above the SLP sublayers and makes all single links look like one logical link. It provides for optimum load sharing and resequencing of frames received from the different single links.

To distribute the data units over different SLP sublayers and to resequence the data units at the other end, the MLP sublayer adds a multilink control field to the data units (Fig. 24). The multilink frame so formed fits inside the information field of the LAP-B frame.

The MLP control field is two octets long and contains 12-bit multilink sequence number, void sequencing bit (*V* bit) and sequencing check bit (*S* bit). A *V* bit equal to "1" indicates resequencing is not required and *V* bit equal to "0" implies that the frames must be sequenced. The *S* bit is significant only when the *V* bit is equal to "1". An *S* bit equal to "1" indicates that no sequence number is assigned to the frame. When the *S* bit is "0", it indicates that a sequence number is assigned to the frame to check for duplicate frames.

The multilink sequence number is used only for resequencing the received frames and for detecting the duplicate frames. It is not used for acknowledgement. All error control functions are implemented at the SLP level.

15 SUMMARY

High-level Data Link Control (HDLC) is a bit-oriented protocol for layer 2 of the OSI reference model. It provides for two-way alternate or two-way simultaneous communication between primary and secondary stations or between combined stations. The mode of communication can be synchronous or asynchronous. It is applicable to point-to-point or point-to-multipoint configurations. It provides full data transparency. It utilizes sliding window mechanism for flow control. The usual window size is 7 frames. Error recovery is done using the 16-bit CRC code and timers. HDLC offers

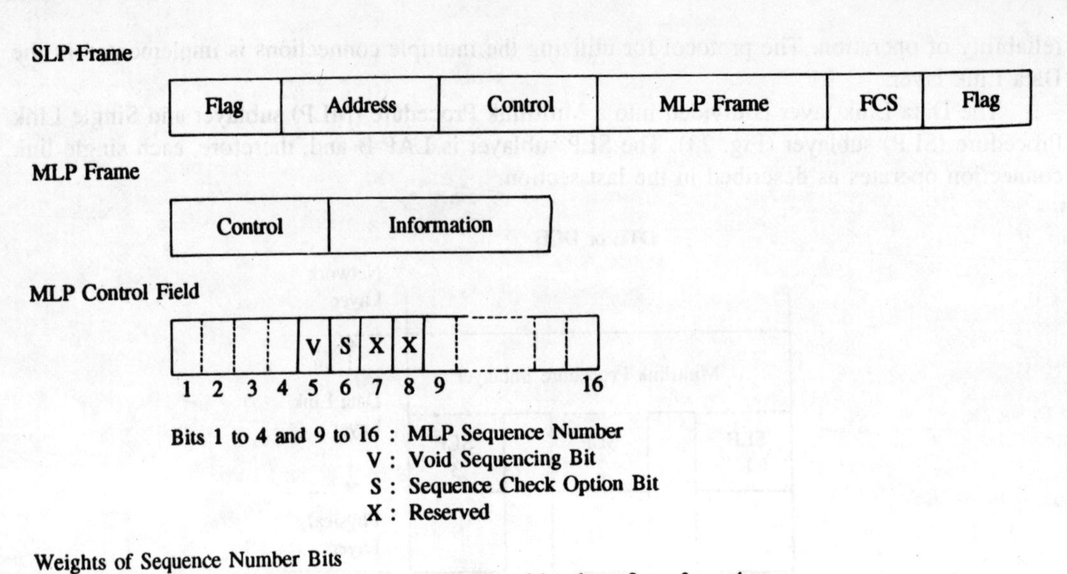

Fig. 24 Format of SLP frame.

flexibility and adaptability for its application in a variety of network configurations. LAP-B of CCITT is an option of HDLC and adopts asynchronous balanced mode of data transfer. CCITT has also specified a Multilink Procedure (MLP) which is implemented as a sublayer in the Data Link layer for increasing the reliability of the Data Link connection.

PROBLEMS

1. Locate an HDLC frame from the following bit streams and identify its various fields.

(a) 01111110011111101100110001110010000011100101010101001111100111110

(b) 0111111011011111001111001101010111110111010101111101111110.

2. Various fields of an HDLC frame are given below. The bits are shown in their transmission order. Compose the HDLC frame.

Address : 00011111

Control : 00110111

Information : 11111000011

FCS : 1000000101110001.

3. Write the control fields of the following frames. Low-order bits have been shown first:

(a) I-frame, $N(S)$ = 010, $N(R)$ = 101, command with P bit equal to "0".

(b) S-frame, RNR, $N(R)$ = 101, command with P bit equal to "1".

(c) U-frame, Unnumbered acknowledgement, response with F bit equal to "1".

4. Fill in the blanks. Frame representation as explained in the chapter has been used. A is the primary station and B the secondary station.

(a) Normal Response Mode, no errors, TWA communication.

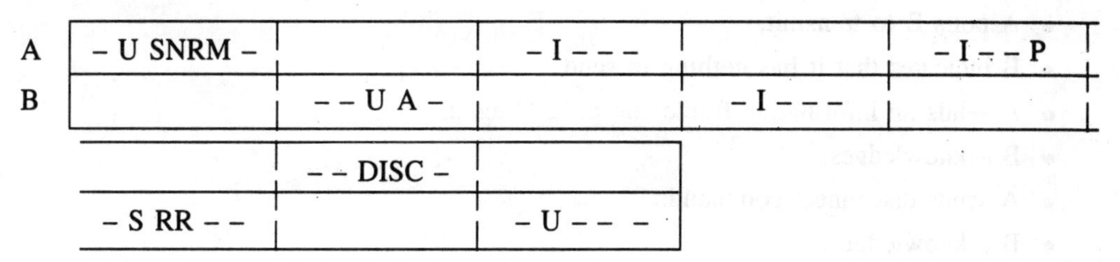

Fig. P.4a

(b) Normal Response Mode, with errors, TWA communication. X indicates that the frame is lost.

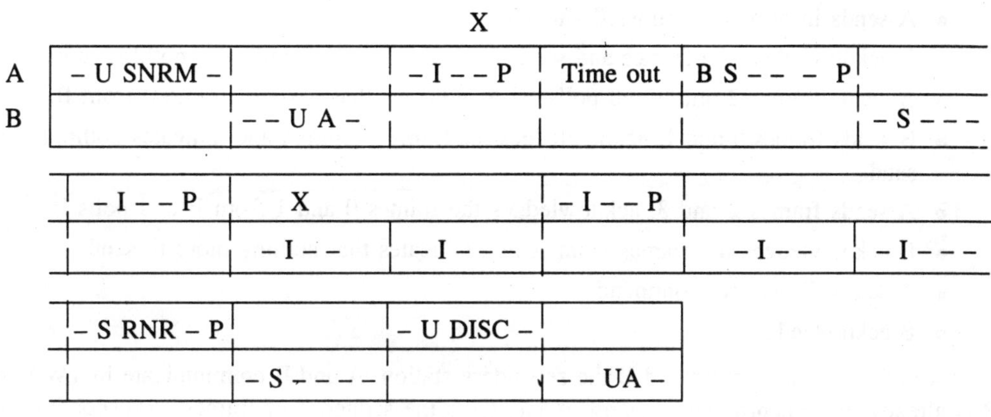

Fig. P.4b

(c) Normal Response Mode, with errors, TWS communication. X indicates that the frame is lost.

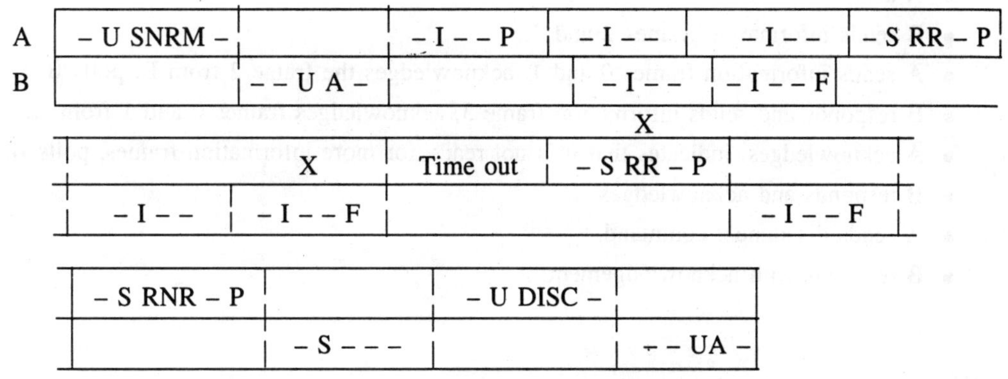

Fig. P.4c

5. A is the primary station and B the secondary station. A and B communicate in TWA mode. Give the sequence of HDLC frames, which corresponds to the following sequence of actions:

- A sends a command to set B in normal response mode.

- B acknowledges.
- A polls B to transmit.
- B indicates that it has nothing to send.
- A sends an information frame and polls B again.
- B acknowledges.
- A sends disconnect command.
- B acknowledges.

6. A is the primary station and B the secondary station. A and B communicate in TWA mode. B is already in normal response mode. Give the sequence of frames corresponding to the following actions:

- A sends information frames 0 and 1 and polls B.
- B acknowledges the frames and sends its frames 0 and 1, indicates nothing more to send.
- A sends frames 2 and 3 and polls B. A indicates that it wants frame 0 from B.
- B sends frames 0 and 1, acknowledges the frame 1 from A and indicates nothing more to send.
- A sends frames 2 and 3, acknowledges the frames 0 and 1 from B and polls B.
- B acknowledges the frames from A and indicates that nothing more to send.
- A sends disconnect command.
- B acknowledges.

7. A is the primary station and B the secondary station. A and B communicate in TWA mode. B is already in asynchronous response mode. Give the sequence of frames corresponding to the following actions:

- B sends information frames 0 and 1.
- A acknowledges.
- B sends information frames 2 and 3.
- A sends information frames 0 and 1, acknowledges the frame 2 from B, polls B.
- B responds and sends information frame 3, acknowledges frames 0 and 1 from A.
- A acknowledges, indicates that it is not ready for more information frames, polls B.
- B responds and acknowledges.
- A sends disconnect command.
- B responds with acknowledgement.

CHAPTER 9

The Network Layer

The Physical and Data Link layers are concerned primarily with error-free transport of bits between two adjacent digital devices. The transfer of error-free messages from one end-system to another across a subnetwork is the next function to be implemented in a computer network. In this chapter, we first examine the subnetwork which provides switched data transfer service. We discuss three data switching techniques, namely, circuit switching, message switching and packet switching. Packet switching is covered in considerable detail and two very important packet switching concepts—Datagram and Virtual Circuit Routing are introduced.

Having introduced the switching technologies used in subnetworks, we proceed to the Network layer of the OSI Reference model which is primarily responsible for the routing function of the subnetwork. After a brief description of the purpose, functions and services of the Network layer, we look at its internal architecture and apply these concepts to develop the layered architecture of the various switched data subnetworks. In this chapter, we lay the foundations for the very important X.25 interface which we will discuss in detail in the next chapter.

1 THE SUBNETWORK CONNECTIONS

Basic configuration of a distributed computing system consists of end-systems and a subnetwork which provides resources for interconnection of the end-systems (Fig. 1). The subnetwork may provide fixed or switched connections. Fixed connection may be in the form of dedicated physical media which link the end systems. These subnetwork resources are allocated permanently for use by the end systems. In the switched connections, on the other hand, the subnetwork resources are shared and allocated temporarily on request of the end systems. We will concentrate on the subnetworks which provide switched connections.

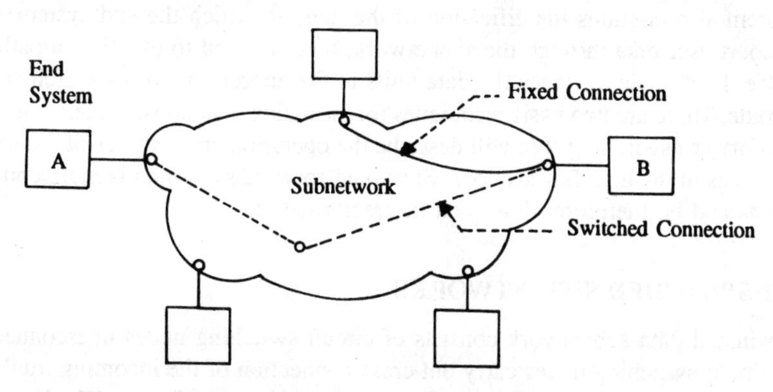

Fig. 1 Configuration of a distributed computing system.

1.1 Switched Data Subnetworks

Switching is the selection and establishment of a path from a source to a specific destination through the subnetwork. Switching is carried out on specific demand from the source. The motivation for using switched subnetworks arises from the following two major requirements:

Flexible topology: Switching enables delivery of information presented at one access point of the subnetwork to a variety of destinations which can be selected by the users. Thus, switching provides a flexible interconnection topology.

Resource sharing: The subnetwork resources are available to all users of the subnetwork. As and when a user requires the services of the subnetwork, the resources are allocated to it.

A switched data subnetwork consists of an interconnected collection of nodes. The node interconnecting links are called trunks (Fig. 2). Data units are transmitted from source to destination by being routed through these nodes. For example, data units from end system 4A intended for 6F

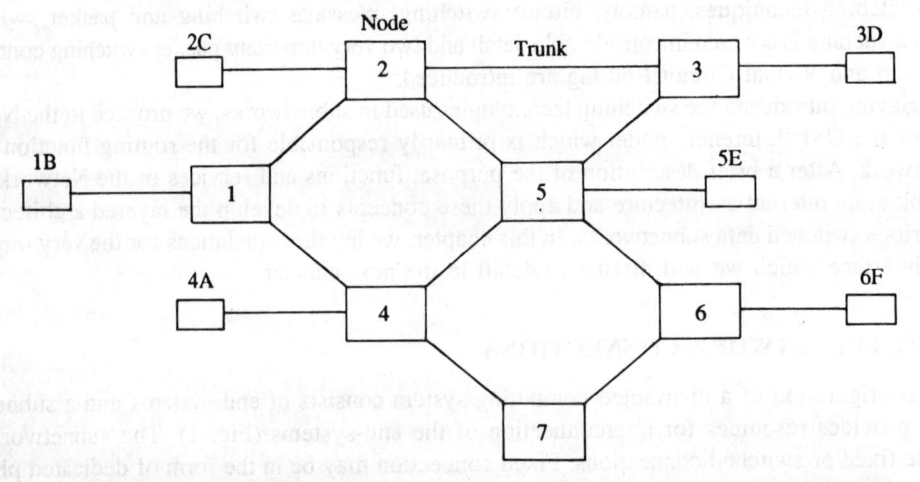

Fig. 2 Switched data subnetwork consisting of nodes and interconnecting trunks.

are sent to the entry node 4, switched to the transit node 5 and then to the exit node 6. Each end system is identified by a unique address to facilitate routing of the information. Usually the address of an end system also contains identification of the node to which the end system is attached.

To transport user data through the subnetwork, there is need to establish a path which leads to the exit node. Each node switches the data units to the appropriate trunk which connects to the next node enroute. There are two basic techniques for switching data units, namely, circuit switching, and store-and-forward switching. We will describe the operation and features of the service offered by these techniques in the next few sections. Store-and-forward switching is more common for data communications and is, therefore, described in greater detail.

2 CIRCUIT SWITCHED SUBNETWORKS

The circuit switched data subnetwork consists of circuit switching nodes interconnected by trunk circuits. The circuit switching nodes carry out cross connection of the incoming trunk circuits/user data circuits and outgoing trunk circuits to establish a through transmission path. The most common example of a circuit switched subnetwork is the telephone network which is primarily used for voice communication.

2.1 Phases of Operation for Data Transfer

Transfer of data units through a circuit switched subnetwork involves the following three phases of operation:

1. Connection establishment phase
2. Data transfer phase
3. Connection release phase.

Connection Establishment Phase. When a call request with the destination address is received from the originating end system, the entry node builds up a path by cross connecting one of the outgoing trunk circuits in the direction of the destination end system (Fig. 3). The address information is transferred to the next node where again a cross-connection between the incoming and outgoing

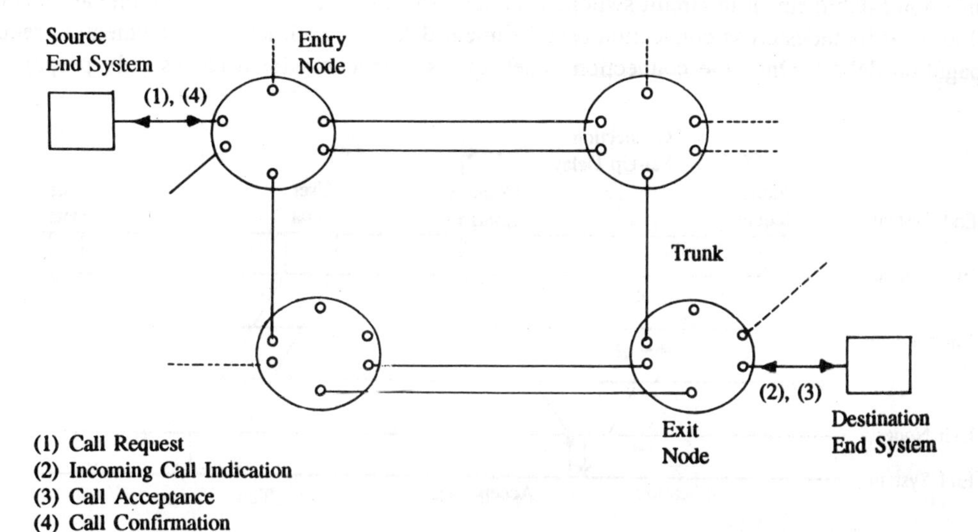

Fig. 3 Circuit switched data subnetwork.

trunk is made. This process is repeated at each intermediate node and at the exit node which serves the destination end system. The exit node sends in incoming call indication to the destination end system which returns a call acceptance. The subnetwork confirms establishment of connection to the originating end system.

The subnetwork resources, trunks and cross-connection switches allocated for the purpose of building up the connection are assigned for exclusive use of end systems for transporting their data. Trunks may be real (metallic pairs, FDM channels, PCM channels) or "virtual". If they are virtual, they must be immediately available to their user whenever information is to be transmitted.

Data Transfer Phase. After the connection confirmation is received, data transfer can begin on the connection. Some of the basic features of the data transfer service are given below.

- The same connection is used by both the end systems to communicate, i.e., the connection is bidirectional.
- The subnetwork nodes cannot store, even temporarily, the data bits. Therefore, the data rates at the source and destination and on the trunks are the same.
- Address of the destination is specified only once during call set up. All subsequent data blocks are transmitted on the path already established.
- The cross-connection at each node involves connection of physical channels or copper wires. Therefore, the nodes do not carry out any form of error control.

Connection Release Phase. The connection is released at the request of the end-system and after the release, the subnetwork resources which were engaged for setting up of the connection are also released.

2.2 Delays in Circuit Switched Subnetwork

Connection establishment in circuit switched subnetworks involves certain set up time as shown in the Fig. 4. It includes cross connection establishment delays at each node and connection request propagation delays. Once the connection is set up, user data transfer involves only propagation

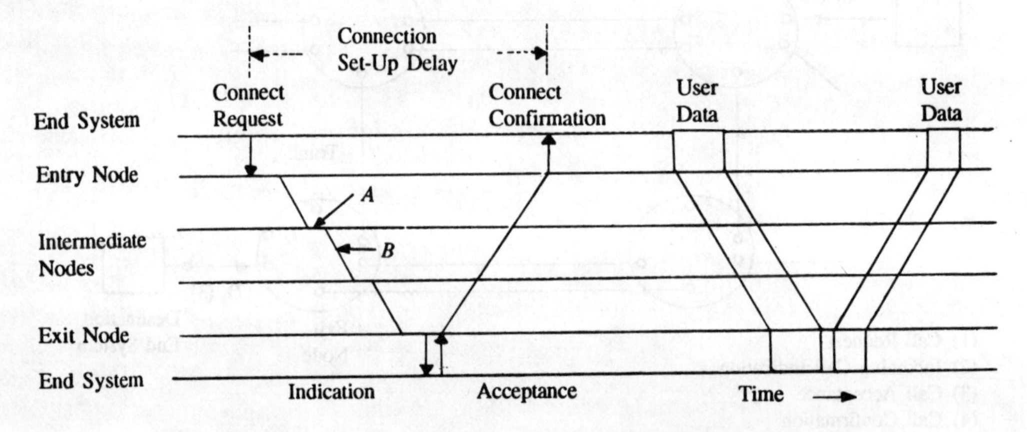

Fig. 4 Data transfer delays in circuit switched data subnetwork.

delay and it is constant. There is almost no delay at the nodes during the data transfer phase. Data blocks are transmitted immediately and incrementally as soon as they are presented to the subnetwork.

During the data transfer phase, the delivery delay from source to destination is constant as all the data blocks are transmitted on the same path through the subnetwork. Therefore, time relationships of data blocks and their sequence of transmission are maintained.

During peak traffic hours, connection set up delay may increase because the subnetwork resources may not be free but once the connection is established, there is no increase in delivery delay through the subnetwork as an unshared transmission path is always available to the end systems.

Table 1 summarizes the features of the service provided by the circuit switched data subnetworks.

Table 1 Service Features of Circuit Switched Data Subnetwork

- Call establishment and release phases
- Call set up delay which increases with traffic
- Destination address is specified only once
- Delivery delay is constant irrespective of the traffic
- Delivery delay is minimal
- Data rates at the source and destination are same
- Time relationship and order of the data blocks are maintained
- No error control within the subnetwork

3 STORE-AND-FORWARD DATA SUBNETWORKS

In store-and-forward switching, a data unit is accepted by the subnetwork node, stored, put in a queue and when its turn comes, forwarded to the next node. Subnetworks employing this basic technique can be of two types:

1. Message switched subnetworks
2. Packet switched subnetworks.

In message switched subnetworks, the complete message is switched at the subnetwork nodes. In packet switched subnetworks, the message is first divided into smaller packets of data and then these packets are switched through the subnetwork. Although both the approaches employ store-and-forward switching, their service features are quite different.

3.1 Message Switched Subnetworks

A message switched subnetwork consists of store-and-forward nodes interconnected by trunks. A single trunk is usually sufficient between a pair of nodes. Multiple trunks can be provided to increase reliability. Each node is equipped with a storage device wherein all incoming messages are temporarily stored for onward transmission. The basic operation of the store-and-forward service is similar to the telegram service. A message along with the destination address is sent from node to node till it reaches the destination.

Let us say end system A wants to send a message to end system B (Fig. 5). A sends its message along with the address of the destination and its own address to entry node 1. The addresses are included in the header of the message (Fig. 6).

Node 1 accepts the message and analyzes the destination address. A routing table is maintained at each node. It contains entries indicating destination nodes and the corresponding outgoing trunks from the node. There is a separate queue for each trunk. Since the destination node may be accessible via more than one route, decision to send the message to a particular next node depends on the expected delay in its queue. Let us say, the message from A is put in the queue for node 2.

The message received at node 2 is again put in a queue of messages awaiting transmission to node 4. When its turn comes, the message is sent to node 4 which delivers it to the destination. Some of the basic features of store-and-forward message switching are:

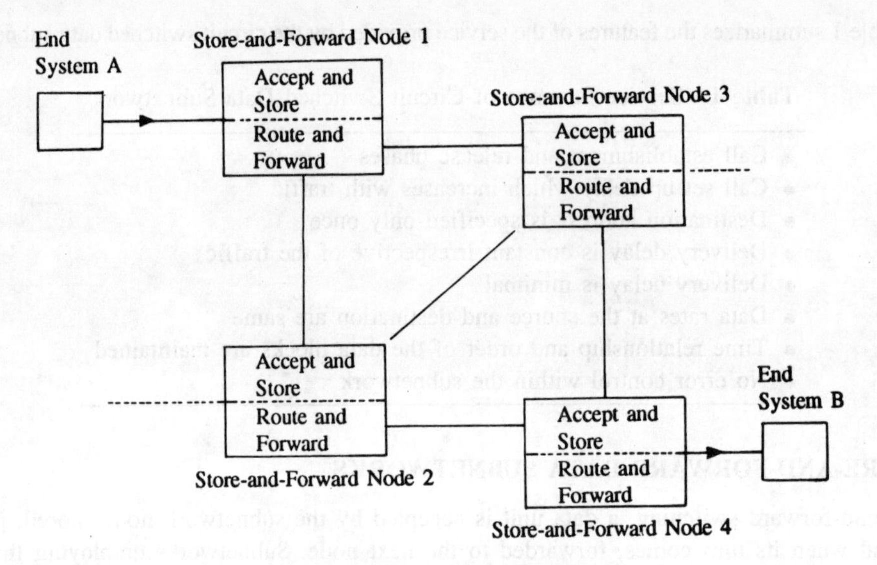

Fig. 5 Message switched data subnetwork.

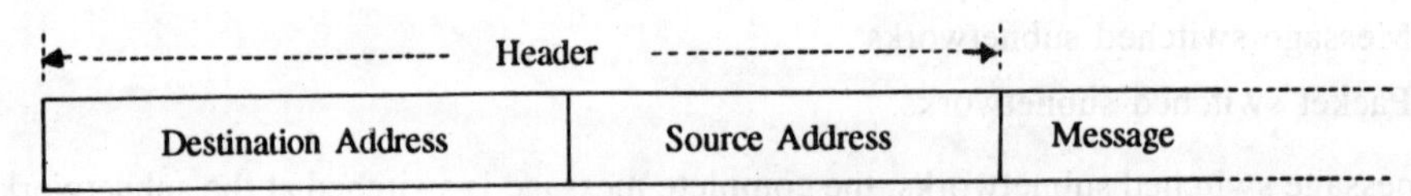

Fig. 6 Message format.

- The store-and-forward service is unidirectional. After delivery of the message, the subnetwork does not send back any confirmation to the source. If end system B is required to send an acknowledgement to the message received from A, the acknowledgement is treated like any other message by the subnetwork and carries the addresses of the destination and the source.
- For node-to-node transfer of the message, the subnetwork may employ some error control mechanism. The message may be appended with error-checking bits and if any error is detected by the receiving node, it may request the sending node for retransmission of the message. Therefore, the sending node is required to keep a copy of the message till an acknowledgement is received.
- Since the message is stored in a buffer at the node at each stage of transmission, each node-to-node transfer is an independent operation. The trunks can operate at different data rates. Even the source and destination end systems can operate at different data rates.
- In message switching every message is treated as an independent entity by the subnetwork and, therefore, destination and source addresses are repeated on each message.

Delay in Delivery. Figure 7 shows the timing diagram for routing a message through a message switched subnetwork. The message passes through the entry node, two transit nodes and finally through the exit node to arrive at the destination.

Message delivery time is the sum of the following components:

- Time required to send the message to the entry node

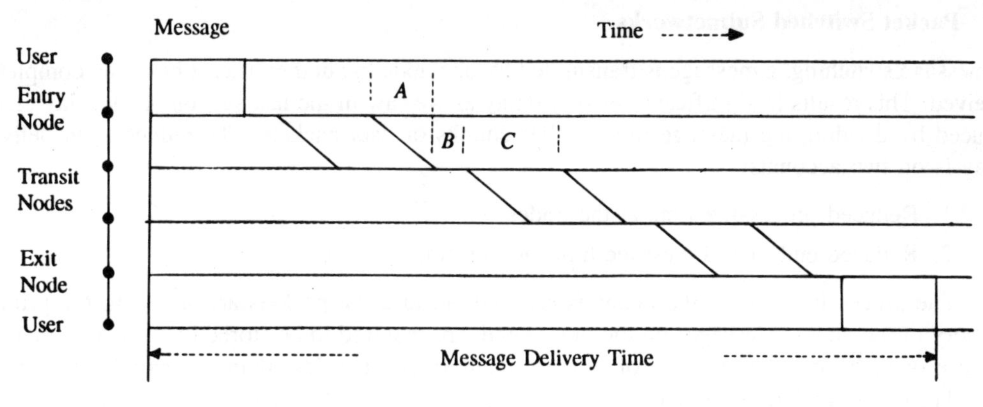

Fig. 7 End-to-end delay in message delivery.

- Node delay
- Transmission time at each node.

The time required to send a message to the entry node is determined by the transmission data rate and the message size. Propagation time to the entry node is usually negligible. Node delay is due to two factors:

1. Message processing at each node (time required for error checking, routing etc.)
2. Waiting time in the queues at each node.

The transmission time at each node is determined by the transmission data rate and propagation time for transmission across the trunk.

The total time required to deliver the message is the linear sum of all these components of time as they occur in a sequential manner. The delivery time varies from message to message because of random waiting times in queues and alternate routes between the same pair of entry and exit nodes. Therefore, time relationship of the messages and their sequence are not guaranteed in a message switched subnetwork. As traffic increases there is increase in message delivery time because the queues get longer and there may be congestion on the route.

Table 2 summarizes the features of the service provided by a message switched subnetwork.

Table 2 Service Features of Message Switched Data Subnetwork

- No call establishment and release phases
- Destination address is specified on each message
- Delivery delay is significant and random
- Delivery delay increases with the traffic
- Data rates at the source and destination need not be the same
- Time relationships and order are not maintained
- Some error control is possible within the subnetwork

3.2 Packet Switched Subnetworks

In message switching, a message is transmitted by one node to another after it has been completely received. This results in significant delivery delay as we saw in the last section. This delay can be reduced by dividing the message into smaller chunks of data packets. The reduction in delivery delay is on two accounts:

1. Reduced processing time at the nodes
2. Reduced end-to-end message transmission time.

The processing time at the nodes is reduced because the packets are stored in the primary memory of the nodes. Messages on the other hand, are required to be stored in secondary memory because of their size. Access time of primary memory is much less than of secondary memory.

The end-to-end message delivery time is reduced because the packets are transmitted as soon as they are available for transmission (Fig. 8). Note that total delivery time in this case is not the linear sum of all the components of delay as there is some overlapping. In message switching, the

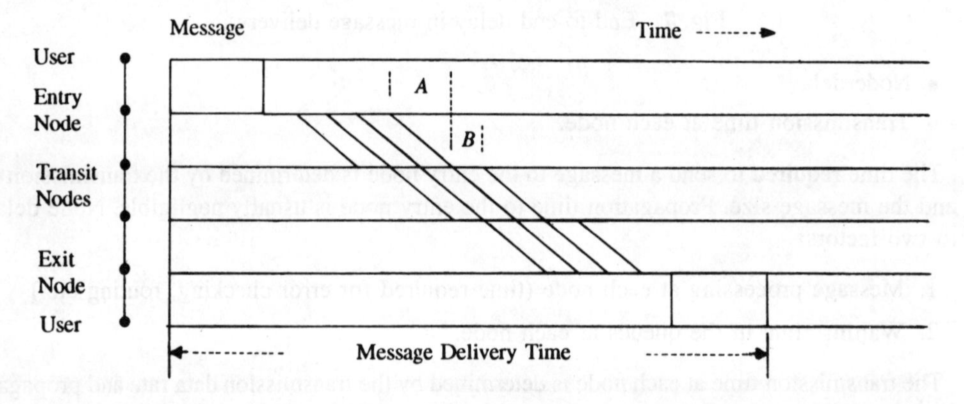

Fig. 8 Message delivery delay in packet switched data subnetwork.

whole message must be received before any step is taken to retransmit it and the total delivery time is the linear sum of all the time components. Conversion of messages into packets and vice versa may involve some additional processing time but it is insignificant. Other features of packet switched subnetworks are determined by the approach adopted for routing the packets through the subnetworks. We discuss the routing methods in the next section.

4 ROUTING OF DATA PACKETS

There are two approaches for routing the data packets through a subnetwork:

1. Datagram routing
2. Virtual circuit routing.

In its simplest form, a datagram is a packet of data with the complete address of a destination. Datagrams are sent out from one node to the other through the subnetwork and at each node, a routing decision is taken for each datagram. Datagrams of a message may take different routes through the subnetwork to reach the destination. In virtual circuit approach, packets are delivered to the destination over a fixed route which is established beforehand at the request of the users. Let us examine datagram routing and virtual routing approaches in some detail.

4.1 Datagram Routing

To send a datagram across the subnetwork, each node examines the destination address on the datagram and uses a routing algorithm to decide the next node. Some of the possible approaches for deciding the route of the datagram across the subnetwork are given below:

- The simplest approach could be to send the datagram on one of the trunk circuits at random. The datagram will eventually reach the destination.
- Another similar approach could be to send it on the trunk which has the shortest queue irrespective of its destination.
- A brute force approach could be to send the datagram on all the outgoing trunks except in the direction from which the datagram came. The datagram would reach the destination by the shortest path but large redundant traffic would be generated in the subnetwork. Also duplicate datagrams may reach the destination.
- A much better approach than those mentioned above is to set up a routing table at each node. Given an address, the node can look up the routing table and decide the next node.

The last approach appears to be very attractive, but we have not considered two very important aspects, namely, creation and updating of routing tables. The routing tables are defined using an algorithm before the subnetwork is made operational and then updated periodically. Let us examine a simple approach for creating routing tables.

Static Routing Algorithm. A routing table maintained at a node indicates the next node to which a data packet must be sent so that it eventually reaches the destination. Routes for all possible destinations are indicated in the routing table. The routing table also maintains alternate paths and associated cost parameters. The cost parameter could be in terms of the number of hops to the destination, transmission delay or a combination of these.

To understand the creation of a routing table, let us consider a simple subnetwork consisting of nodes *A*, *B*, *C*, *D* and *E* which are interconnected using trunks (Fig. 9). The cost parameter for

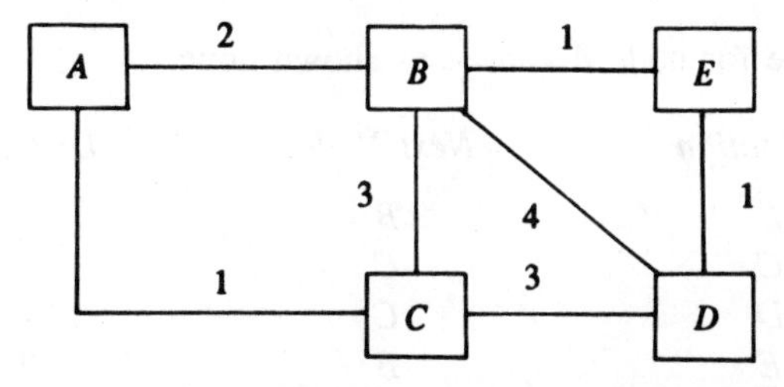

Fig. 9 Example of a subnetwork for generating routing tables.

each trunk has also been defined. The algorithm for generating the routing table is based on least cost and involves the following steps.

Step 1 Let $P(i, j)$ be the cost of transmission between node i and node j. If there is no direct connection between the two nodes, the cost is taken as infinite.

$$P(A, B) = 2, \quad P(B, A) = 2, \quad P(C, A) = 1, \quad P(D, A) = \infty, \quad P(E, A) = \infty$$
$$P(A, C) = 1, \quad P(B, C) = 3, \quad P(C, B) = 3, \quad P(D, B) = 4, \quad P(E, B) = 1$$
$$P(A, D) = \infty, \quad P(B, D) = 4, \quad P(C, D) = 3, \quad P(D, C) = 3, \quad P(E, C) = \infty$$
$$P(A, E) = \infty, \quad P(B, E) = 1, \quad P(C, E) = \infty, \quad P(D, E) = 1, \quad P(E, D) = 1$$

Step 2 To develop the routing table for node A, we write down the paths and the distances for each node from a set of nodes S. The distance (Q) between any two nodes is expressed as the sum of cost parameters along the path interconnecting the stations. Only direct paths are considered from the set S of nodes. To start with, the set contains only node A.

	Node B		Node C		Node D		Node E	
	Path	Q	Path	Q	Path	Q	Path	Q
$S\{A\}$	$A - B$	2	$A - C$	1		∞		∞

Step 3 From step 2, it is evident that node C is at the least distance from node A. So we include node C also in set S and find the distances from set S again.

	Node B		Node C		Node D		Node E	
	Path	Q	Path	Q	Path	Q	Path	Q
$S\{A, C\}$	$A - B$	2	$A - C$	1	$A - C - D$	4		∞

Step 4 Of the remaining nodes, node B is nearest to node A, so it is also included in the set S.

	Node B		Node C		Node D		Node E	
	Path	Q	Path	Q	Path	Q	Path	Q
$S\{A, B, C\}$	$A - B$	2	$A - C$	1	$A - C - D$	4	$A - B - E$	3

Step 5 Between D and E, node E is nearer to A, and so the next node to be included in the set S is E.

	Node B		Node C		Node D		Node E	
	Path	Q	Path	Q	Path	Q	Path	Q
$S\{A, B, C, E\}$	$A - B$	2	$A - C$	1	$A - C - D$	4	$A - B - E$	3

Step 6 The routing table for node A can be as shown now:

Destination	*Next Node*	*Distance*
B	B	2
C	C	1
D	C	4
E	B	3

We can express the above results graphically in the form of a routing diagram as shown in Fig. 10.

Similar routing tables are created at each node. Since a subnetwork may consist of a very

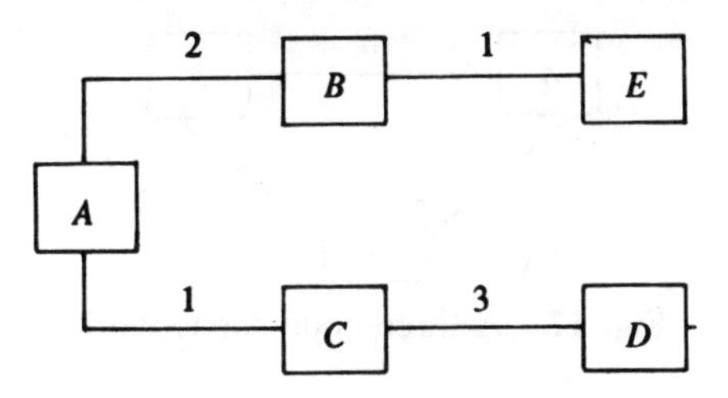

Fig. 10 Routing diagram.

large number of nodes, the size of routing tables will tend to be big. It is quite possible to divide the subnetwork into several regions so that the nodes belonging to any one region are required to maintain routing tables for the nodes within that region. Interregional traffic is routed through specified nodes. If a new node is added in a region, the routing tables of the nodes belonging to the region need only to be updated.

Dynamic Routing. The routing tables discussed so far are static and become outdated very soon. There are several reasons for this. These tables must be updated for every new node as and when it is added to the subnetwork. If there is failure of trunks between two nodes, or there is congestion in some part of the subnetwork, alternate routes for traffic are to be established. Dynamic routing addresses these routing issues. There are several algorithms for dynamic routing. These algorithms require subnetwork maintenance data packets. In the subnetworks with central control, all the nodes exchange maintenance packets with the central controller which updates their routing tables.

The routing control can be distributed also. In distributed control, the nodes exchange maintenance packets with the neighbouring nodes. Suppose in Fig. 11 nodes *A*, *B*, *C* and *D* are interconnected and have up to date routing tables, i.e., they have entries for nodes *A*, *B*, *C* and *D*.

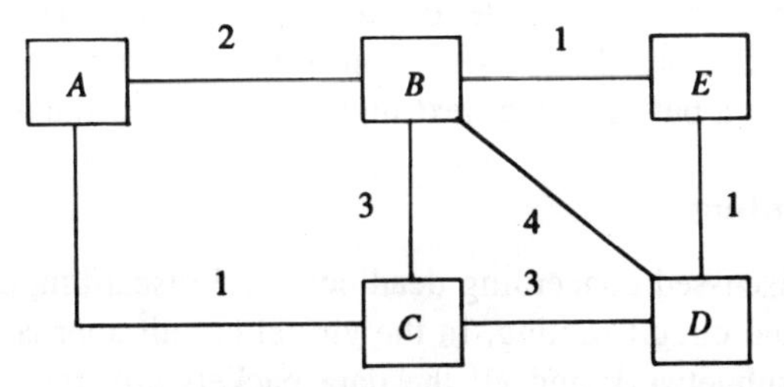

Fig. 11 Updating routing tables on creation of a new node.

Now imagine that a new node *E* is added to the subnetwork. As soon as node *E* becomes active, it informs nodes *B* and *D* of its distance from these nodes. Nodes *B* and *D* update their routing tables and also inform the other nodes they are connected to. Node *D* informs nodes *B* and *C*. Node *B* informs node *A*, *C* and *D*. Thus each node updates the routing table by adding a new path for the new node, *E*. Some of the nodes will receive this information from more than one source, e.g., node *B* receives information about node *E* from *E* and from *D*. The alternative routes are also thus established.

Congestion and Deadlock. Datagram routing is efficient so long as traffic is sufficiently low. In heavy traffic conditions, datagram movement across the subnetwork can stop altogether. To understand this, let us consider a subnetwork consisting of three nodes *A*, *B* and *C* (Fig. 12).

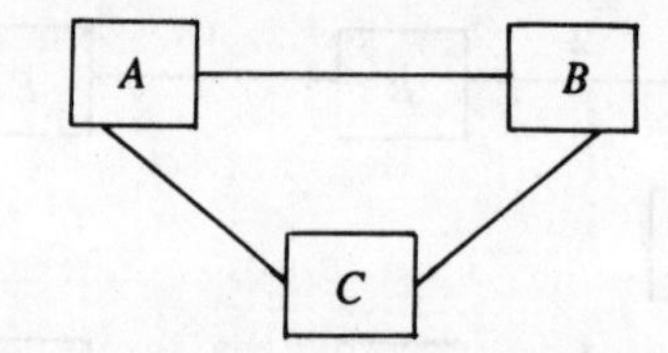

Fig. 12 A three-node subnetwork.

Suppose that each node has finite buffer and each buffer is full with datagrams to be sent to other nodes. In order for node *A* to be able to send a datagram to node *B*, the latter must have a free buffer to accommodate the datagram. To have a free buffer, node *B* needs to send a datagram to node *C*. Since node *C* is also full and also cannot send a datagram to node *A* to create a free buffer, no movement of datagrams within the subnetwork is possible. In other words, there is a deadlock. It is a condition that must never occur.

There are several ways to avoid deadlock. One simple way is to monitor the vacant buffer at each node. Inward flow of the datagrams can be stopped when the buffer nears becoming completely full.

An alternative is to discard some of the datagrams. The free buffers so created allow at least some datagrams to get through so that error recovery procedures may be brought into play by the end systems. The end systems have built in timers and if a data packet is not acknowledged within a specified time, procedure for retransmission of the packet is initiated. Therefore, the discarded datagrams will be taken care of eventually. In fact, all the datagrams which have spent more than their lifetime in the subnetwork can be automatically discarded.

Another approach for avoiding deadlocks is to divide the buffers at each node into several classes. If the maximum number of hops in a subnetwork is *N*, the buffer is divided into *N*+1 classes. A packet is allowed to occupy a buffer numbered *H* after traversing *H* hops. Therefore, a packet always looks for the next higher level of buffer on each hop. Consider a simple situation of two nodes trying to send packets to each other. The packets in the opposite nodes do not contest for each other's buffer, they are looking for a buffer at the next higher level; so deadlocks do not occur.

4.2 Virtual Circuit Routing

The issues that we just discussed concerning deadlocks and discarding datagrams are some of the reasons for adopting virtual circuit routing. In the virtual circuit approach, a logical connection is established through the subnetwork and all the data packets are transported on the same route. Unlike the datagram approach, the nodes do not make the routing decision for each packet. It is made once at the time of establishing the logical connection for all the packets.

Figure 13 shows a simple network with end system *AS* attached to node *A* and end system

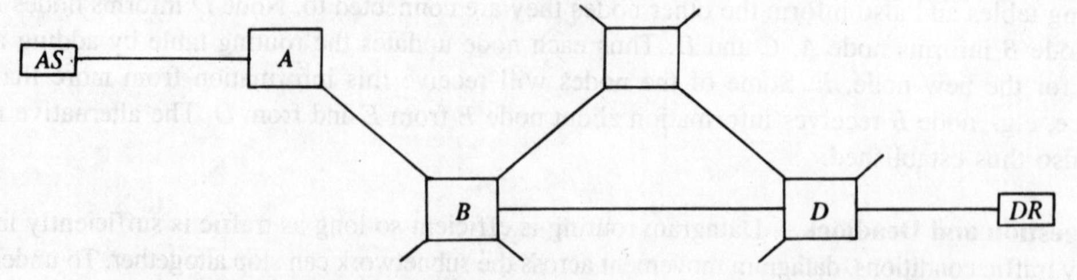

Fig. 13 Example of a subnetwork employing virtual circuit routing.

DR to node *D*. Suppose *AS* wishes to exchange data with *DR*. To establish a connection to *DR*, *AS* sends a CONNECT REQUEST packet to node *A* specifying the destination address *DR*. The CONNECT REQUEST packet also specifies the label number which is later used for identifying the packets meant for a particular destination. Let us say it is N_1. In the X.25 protocol, as we shall see later, N_1 is called the logical channel identifier. Thus the CONNECT REQUEST is essentially:

"Connect *AS* to *DR*. Label number is N_1."

On receipt of the CONNECT REQUEST packet, node *A* takes note of the label N_1. Node *A* also examines the destination address specified in the CONNECT REQUEST packet and works out the next link in the chain leading to the destination from its routing table. For this link to node *B*, node *A* selects another "free" label N_2 which is unique on this link. All the future packets of the connection being established and going between node *A* and node *B* shall bear label N_2. Node *A*, thus, sends a modified CONNECT REQUEST packet to node *B*:

"Connect AS to DR. Label number is N_2."

Note that the label number on the CONNECT REQUEST packet has been changed by *A*. On receipt of this packet, node *B* works out the route and forwards the packet to node *D* using another label N_3. Node *B* also keeps a record relating label N_2 and N_3. Node *D* sends an INCOMING CALL packet to destination *DR*. It does so by using still another label N_4 which is unique on the link between node *D* and destination *DR*. If *DR* decides to accept the call, it returns an ACCEPTANCE packet to node *D*. It uses the label N_4 already being used for this link of the connection. Node *D* sends this ACCEPTANCE packet to node *B* using label N_3. Node *B* forwards this to node *A* using label N_2. Node *A* finally sends a confirmation to *AS* of having established a connection to the destination. The CONFIRMATION packet bears the label N_1.

Thus, in the connection establishment phase, a route to the destination is finalized and a confirmation is received from the destination. The connection is in the form of link tables relating to the labels of data packets. These tables are maintained at each node (Fig. 14). Whenever a packet is received by a node, it looks up in these tables and routes the packet to the link as mentioned

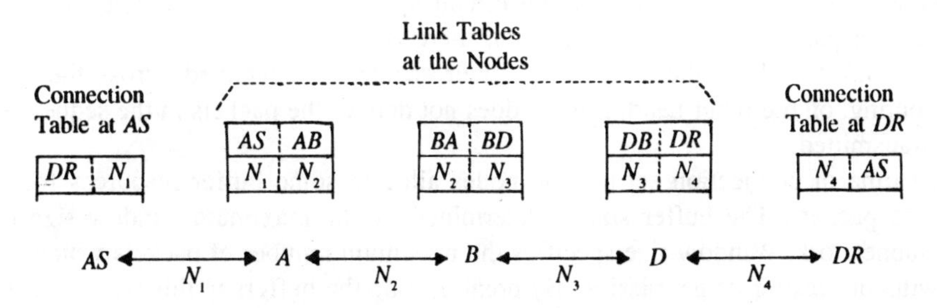

Fig. 14 Link tables maintained at the nodes.

therein. It also gives to the packet a new label which is also indicated in the table. The end systems maintain connection tables which contain the destination addresses and the labels of the connections which are operational.

After the connection CONFIRMATION packet is received, *AS* can send data packets bearing the label N_1. *DR* will use label N_4 on its packets. Destination address is not needed in the data packets.

The motivation behind attaching labels is that several connections can be operated simultaneously on each link between the nodes, and between an end system and a node. This is illustrated in Fig. 15.

Node *P* is operating two connections, one through nodes *Q*, *R* and *S* and the other through

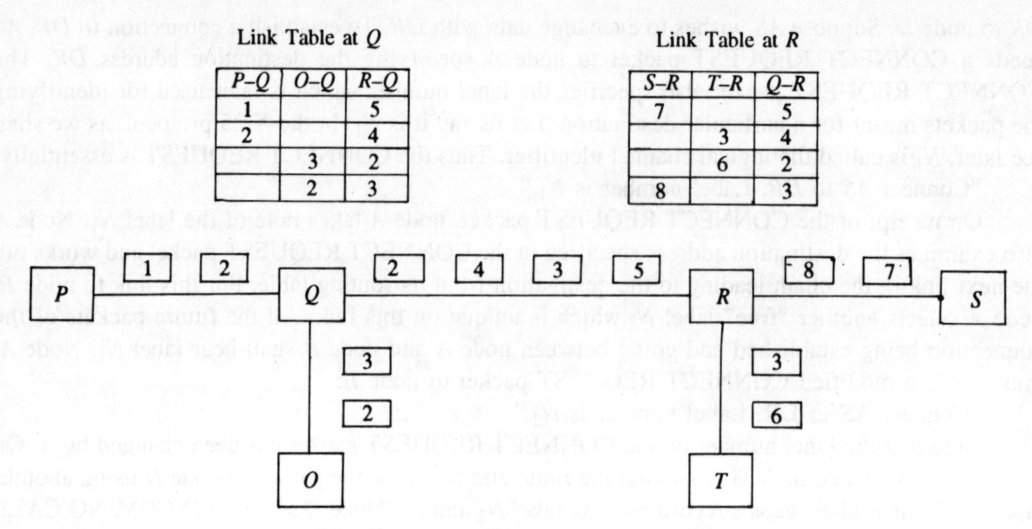

Fig. 15 Multiple virtual connections.

nodes *Q*, *R* and *T*. For the first connection, the packets are labelled "1" on link *P–Q*, "5" on link *Q–R* and "7" on link *R–S*. For the second connection, packets are labelled "2" on link *P–Q*, "4" on link *Q–R* and "3" on link *T–R*. Node *O* is also maintaining two connections, one through *Q*, *R* and *S* and the other through *Q*, *R* and *T*. Note that four connections are simutaneously working through link *Q–R*. Packets are identified as belonging to a particular connection by their labels on each interconnecting link.

Another very important point to be noted is that in a particular connection all packets always take the same route which is decided at the time of establishing the connection. Therefore, delivery delay is more or less constant in the virtual circuit approach. Minor variations could be due to retransmission of packets between two nodes if there is an error.

As the packets follow the same route, their sequence is retained across the subnetwork. Datagram routing, on the other hand, usually does not deliver the packets in the sequence in which they were transmitted.

At the time of connection set up, the nodes allocate some buffer resources for temporary storage of the packets. The buffer size is determined by the maximum window size to be used across the subnetwork. Window size specifies the maximum number of packets which can be sent to a node without seeking its permission. By preallocating the buffers in this manner, it is possible to avoid deadlocks.

4.3 Packet Switching Services

The service offered to the users by the packet switched data subnetworks depends on the routing approach. Datagram service is connectionless mode of service having features of datagram routing shown in Table 3. Sequencing of packets is an enhanced service provided by the subnetwork. Virtual circuit service, on the other hand, is a connection mode service having the features of virtual circuit routing as given in Table 4.

Services are what a subnetwork offers externally. It is quite possible that subnetwork may employ datagram routing but provide virtual circuit service by enhancing the capabilities of the entry and exit nodes. CCITT Recommendation X.25 specifies the protocol for virtual circuit service.

Table 3 Features of Datagram Service

- No call connection or release phases
- Finite and fluctuating delivery delay
- Finite error rate due to lost and duplicate packets
- Disordering of the packets
- Source and destination data rates can differ
- Destination and source addresses specified on each packet
- Non-reliable service as there are no acknowledgements

Table 4 Features of Virtual Circuit Service

- Connection establishment and release phases
- Destination and source addresses specified only once
- Sequenced delivery of the packets
- Source and destination data rates can differ
- Finite and almost constant delivery delay
- Delivery assured using acknowledgements

It does not address the packet routing protocol within the subnetwork. We will examine this recommendation in the next chapter.

5 INTERNETWORKING

We have considered transport of data units between end systems which are connected to the same subnetwork. There are, however, situations when two end systems connected to two distinct but interconnected subnetworks need to exchange data units. The two subnetworks are interconnected using an intermediate system. The intermediate system relays the data units from one subnetwork to the other and sorts out the differences in the services of the subnetworks. We will address the internetworking problem in another chapter but we will lay the foundation in this chapter.

6 PURPOSE OF THE NETWORK LAYER

We have considered so far the operation of the various types of subnetworks primarily from the point of view of the mechanisms employed for transporting data units. To understand their complete functionality we need to examine their layered architecture. To simplify the discussion, we will assume that connection mode service is provided at the Physical and Data Link layer interfaces. Connectionless mode service is considered later in the chapter on Local Area Networks.

Consider the data network, shown in Fig. 1, where two end systems *A* and *B* need to exchange information. These systems are connected through the access and transit nodes of the subnetwork. Let us examine the layered architecture of each section of the end to end path taken by a data packet.

6.1 The End System to Access Node Link

As we have seen earlier, the Physical layer provides the capability to exchange bits on physical transmission media which interconnect the two devices. In this case the physical connection extends from end system A to the access node of the subnetwork (Fig. 16).

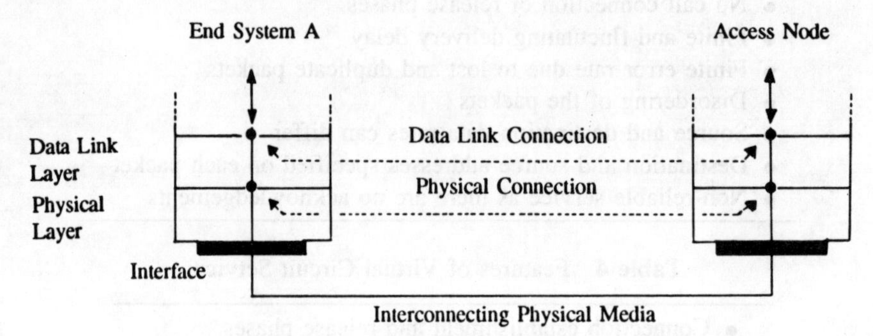

Fig. 16 End system-to-access node Data Link connection.

The Data Link layer of end system A carries out error control to take care of the errors introduced during transmission of the bits over the Physical connection. It interacts with the Data Link layer of the access node for this purpose. Thus, together with the Physical layer, the Data Link layer provides an error free Data Link connection from end system A to the access node of the subnetwork. This Data Link connection is established whenever end system A desires to have a connection to another end system through the subnetwork. One typical example of the protocol used for this purpose is LAP–B, *Link Access Procedure–Balanced.*

6.2 Node to Node Connection

All the pairs of adjacent nodes of a subnetwork have Data Link connections between them so that errors introduced during transmission of data units between two nodes are taken care of. Data units received on one Data Link connection are passed on to the next Data Link connection. Thus, the layered architecture of two interconnected nodes can be drawn as shown in Fig. 17.

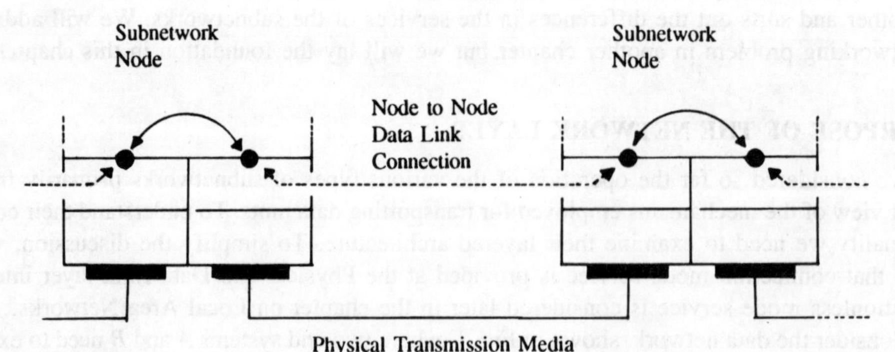

Fig. 17 Node-to-node Data Link connection.

6.3 End System to End System Connection

The end system to end system access across the subnetwork is achieved by routing the data units through series of data link connections (Fig. 18). As each node is connected to several other nodes, there is need to decide the Data Link connection to be chosen at each node for further transport

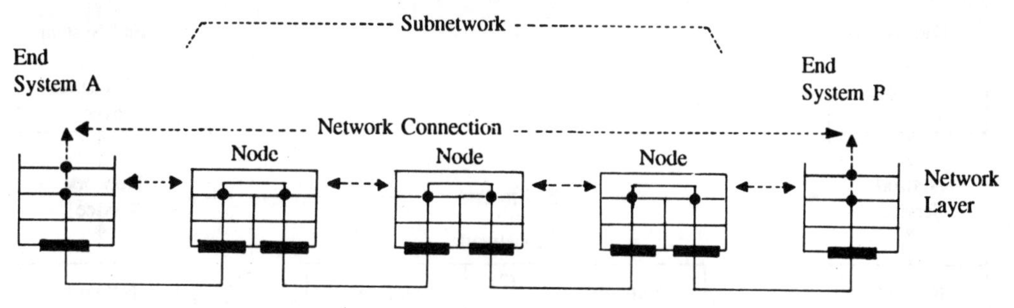

Fig. 18 End system-to-end system Network connection.

of the data units. This routing decision is taken by the Network layer of the nodes. The Network layer of the end systems interacts with the Network layer of the access nodes. For example, this interaction is in the form of sending CONNECT REQUEST, ACCEPTANCE, INCOMING CALL, CONFIRMATION packets in packet switched data subnetwork, providing virtual circuit service.

The route through the subnetwork may be established on per call basis in which case the destination address is given once by the Network layer of the originating end system to the Network layer of the access node during the connection establishment phase. The Network layers of the nodes, then, establish an end-to-end Network connection to transport the data units as shown in Fig. 18. Alternatively, each data unit from the end systems carries the destination address which enables the Network layer of the subnetwork nodes to route the data unit. These two modes of operation are called connection-oriented mode and connection-less mode of data transfer respectively.

Thus, the overall purpose of the Network layer is to provide the capability to exchange data units between any two end systems by routing them through a subnetwork. It must be remembered that the OSI Reference model addresses the architecture of the Network layer of the end systems and the interface of the nodes towards the end system. It does not specify the internal architecture of the subnetwork nodes. Nevertheless, the Network layer of the nodes does serve this basic purpose.

7 NETWORK SERVICE

"Services" are the visible capabilities provided to the next higher layer which is the Transport layer. The service provided by the Network layer to the Transport layer is called Network service (Fig. 19). To appreciate functions of the Network layer, it is necessary to first understand the Network service because functions are the activities performed in a layer in order to provide the service. Moreover, there are several types of subnetworks which have different characteristics but the Network layer is required to provide uniform end-to-end Network service.

The Network service provides for transparent transfer of Network Service Data Units (N-SDU) between two Transport entities which reside in the end systems (Fig. 20). The N-SDUs are received at Network Service Access Point (N-SAP) and are delivered at N-SAP.

Network service can be of two types:

1. Connection-mode Network Service (CONS)
2. Connectionless-mode Network Service (CLNS).

7.1 Connection-Mode Network Service (CONS)

Connection-mode Network service is specified in the ISO 8348 document. The corresponding

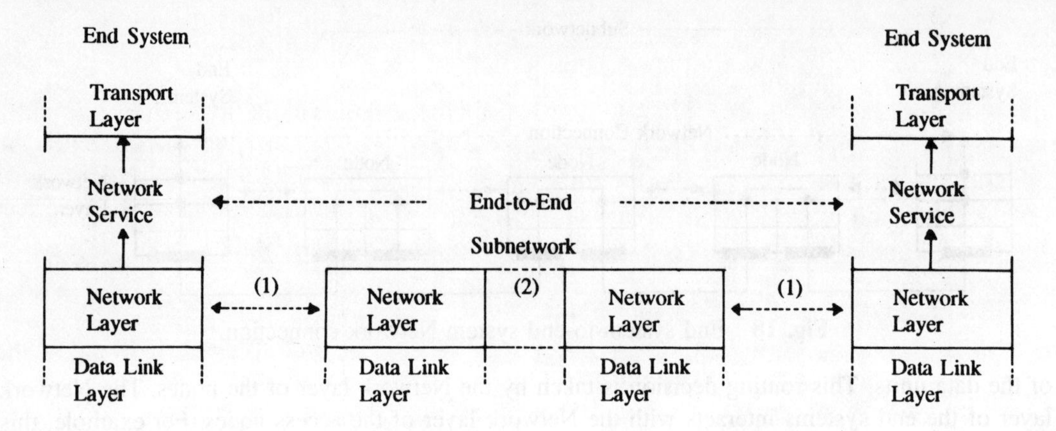

Fig. 19 End-to-end Network service.

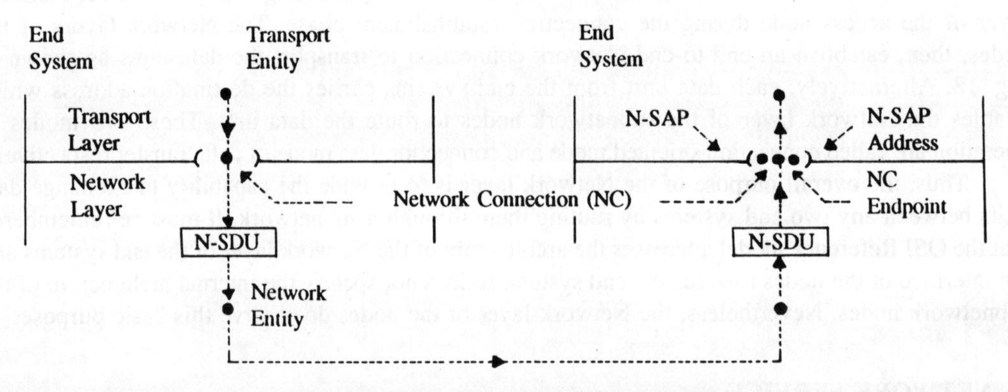

Fig. 20 Delivery of N-SDU at N-ASP.

CCITT recommendation is X.213. In the connection-mode Network service, a Network connection is first established between the communicating Transport entities and subsequently N-SDUs received from the Transport layers are transported on the connection. The N-SDUs are always delivered in the same sequence and an attempt is made to ensure that they are not duplicated or lost. CONS is a "reliable" service in the sense that it has built in error recovery procedures and in case of the Network connection failure, the Transport entities are informed.

7.2 Connectionless-Mode Network Service (CLNS)

Connectionless-mode Network service is specified in ISO 8348 Addendum-1 and ISO 8473 documents. In the connectionless-mode Network service, each N-SDU carries the destination and source addresses and is delivered independently of other N-SDUs. In other words, the Network layer of the nodes makes the routing decisions independently for each N-SDU. As already mentioned, three things can go wrong during operation of CLNS:

1. Some N-SDUs may be lost

2. Some N-SDUs may be delivered out of sequence
3. Some N-SDUs may be duplicated.

The Network layer which provides CLNS cannot report these failures to the Transport entity because the N-SDUs are not sequence-numbered. In other words, there is no guarantee that the N-SDUs will be delivered correctly or delivered at all. Therefore, CLNS cannot be considered reliable. "Reliable" here means that the Transport entities have to make their own efforts to correct the delivery of N-SDUs and they cannot rely on the Network service for this purpose.

7.3 Basic Features of the Network Service

The basic features of the two types of the Network service, are described below. Some of the features are specific to one type of the service.

Independence of Underlying Subnetworks: The Network layer contains functions necessary to mask differences in the characteristics of different subnetworks into a consistent Network service. As a Network service user, the Transport layer is relieved from all concerns regarding how various subnetworks are to be used.

Transparency: Transparency implies that Network service does not restrict the content, format or coding of the user data. The mode of operation of a specific type of subnetwork may place special requirements on formatting the data presented to it. The Network layer takes care of these requirements and does not pass them to the Transport layer.

Network Address: The Transport entities are known to the Network layer by means of their N-SAP addresses which are unique and identify each Transport entity (Fig. 20). Since Transport entities exist in end systems only, an N-SAP address also identifies the end system. Network addresses are known to the Transport entities and the Network entities.

Network Connection: The Network layer establishes, maintains and releases Network connections at the request of Network service users. A Network connection is point-to-point. It is possible to have multiple connections between a pair of N-SAP addresses. Network connection endpoint identifiers specify each individual connection (Fig. 20).

Quality of Service: The Network connection provides Network service of the quality selected by the Transport entity and the quality of service is maintained for the duration of the connection. If the Network layer is unable to maintain the quality, it is obliged to so inform the Transport entities. Quality of service is determined by the following parameters:

- Residual Error Rate (RER)
- Throughput
- Connection set-up delay
- Transit delay
- Reliability of Network connection in terms of mean time to connection loss or reset.

In CLNS, the quality of service parameters includes transit delay, cost and probability of losing, duplicating or damaging the N-SDUs.

The requested quality of service parameters may be expressed as target average values or the worst acceptable values.

Error Notification: Unrecoverable errors detected by the Network layer are reported to the Transport layer. Error notification may or may not lead to release of the Network connection.

Flow Control: The Network service provides for flow control across the layer interface. The Transport entity which is at the receiving end can cause the Network service to stop transferring N-SDUs across the service access point. This flow control condition may or may not be propagated to the transmitting Transport entity according to the specification of the Network service.

Sequencing: When requested by the Transport entity, the Network layer is obliged to provide sequenced delivery of N-SDUs over a Network connection.

Network Connection Release: A Network connection release can be initiated by the Transport entities or by the Network layer entities. When release is requested by the Transport entity the Network service does not guarantee delivery of N-SDUs preceding the release request or still in transit. The Network entity-initiated release may be caused by its inability to maintain the quality of service.

Optional Services. The following optional services are also possible.

Reset: When requested by a Transport entity, RESET request causes the Network layer to discard all N-SDUs still in transit on the Network connection. The Network layer also informs the Transport entity at the other end of the Network connection that a reset has occurred.

Expedited Data Transfer: The expedited data transfer service provides another means of exchanging information which is not subject to the same flow control as other service data units. The Network layer guarantees to deliver an expedited service data unit before other service data units issued subsequently.

Receipt of Confirmation: A Transport entity may confirm receipt of data over a Network connection to the sending-end entity but the service must be agreed by the Transport entities in advance.

7.4 Connection-Mode Network Service Primitives

For the connection-mode Network service, the Transport entities establish a Network connection between them and then transfer the transport protocol data units on the connection. Thus the data transfer operation involves three phases:

1. Connection establishment phase
2. Data transfer phase
3. Connection release phase.

In the connection establishment phase, one service user establishes a connection with another service user, and in the process it may negotiate various characteristics of the connection. As shown in Fig. 21, the initiating Transport entity in end system A forwards N-CONNECT request to the Network entity indicating N-SAP addresses to which the initiating and responding Transport entities are attached. The responding Transport entity is informed about the request by the Network entity with N-CONNECT indication. If the responding Transport entity is agreeable to the connection, it issues N-CONNECT response which results in N-CONNECT confirmation to the initiating Transport entity and thus a connection is established. Note that connection establishment is a confirmed service.

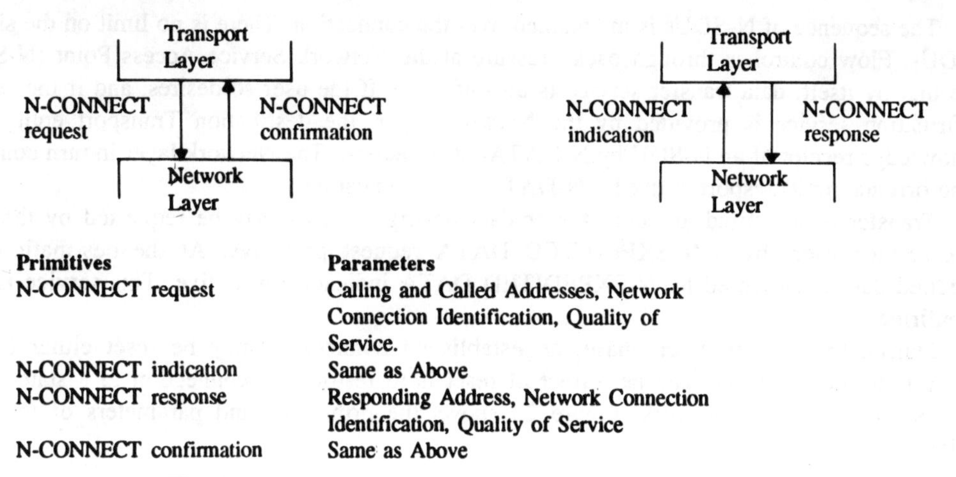

Primitives	Parameters
N-CONNECT request	Calling and Called Addresses, Network Connection Identification, Quality of Service.
N-CONNECT indication	Same as Above
N-CONNECT response	Responding Address, Network Connection Identification, Quality of Service
N-CONNECT confirmation	Same as Above

Fig. 21 Primitives for connection establishment service.

If the responding Transport entity is not in a position to accept the connection, it issues N-DISCONNECT request in response to the N-CONNECT indication. The initiating Transport entity is informed of disconnection by the Network entity through N-DISCONNECT indication.

Once the connection is established, data transfer can take place in both the directions. Figure 22 shows the service primitives and the parameters of this phase.

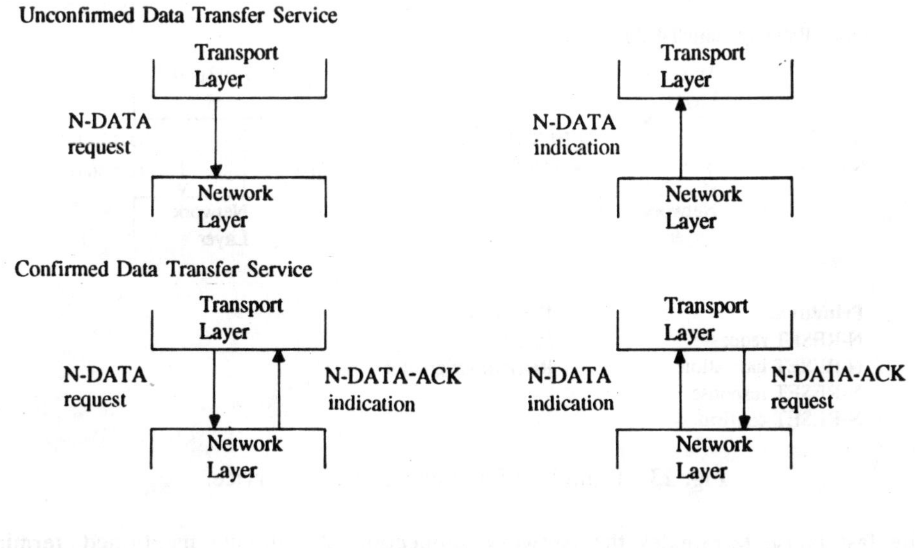

Primitives	Parameters
N-DATA request	User Data, Network Connection Identification
N-DATA indication	Same as Above
N-DATA-ACK request	—
N-DATA-ACK indication	—

Fig. 22 Primitives for data transfer service.

The sequence of N-SDUs is maintained over the connection. There is no limit on the size of N-SDUs. Flow control is through back pressure at the Network Service Access Point (N-SAP). Note that by itself, data transfer service is unconfirmed. If the user so desires, and if the receipt confirmation service is provided by the Network layer, the destination Transport entity may acknowledge receipt of an N-SDU by N-DATA-ACK request. The Network layer in turn confirms to the originating Transport entity by N-DATA-ACK indication.

Transfer of a limited amount of user data on urgent basis may be requested by the Network service users by a N-EXPEDITED DATA request primitive. At the destination, the expedited data is delivered by N-EXPEDITED DATA indication primitive. This service is also unconfirmed.

During the data transfer phase, an established connection may be reset either by the service users or providers. The net effect of reset is to restore the connection to a state where there is no data in the network. Figure 23 shows the primitives and parameters of the reset service.

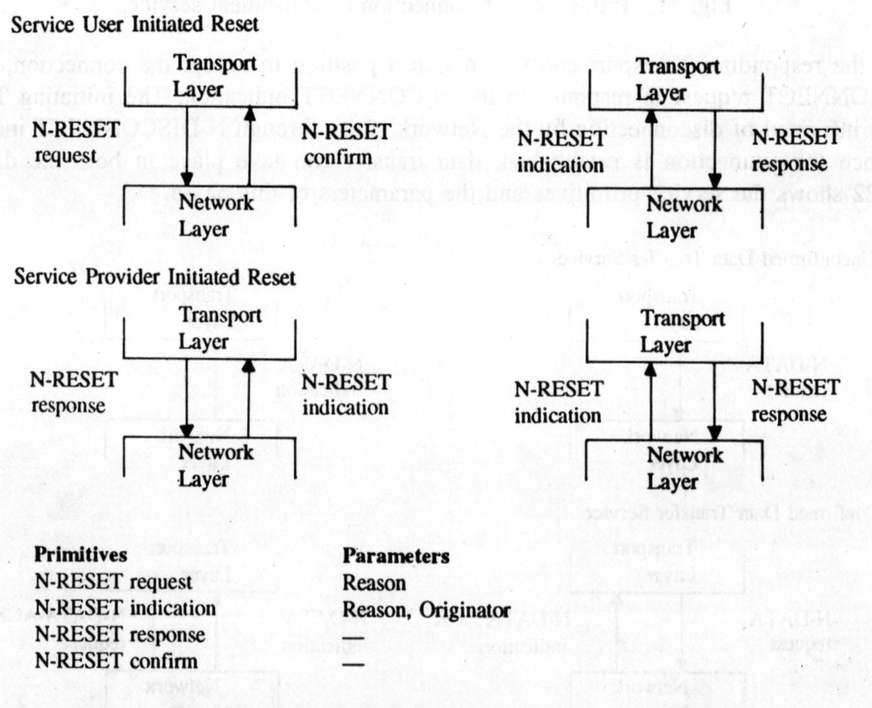

Primitives	**Parameters**
N-RESET request	Reason
N-RESET indication	Reason, Originator
N-RESET response	—
N-RESET confirm	—

Fig. 23 Primitives for connection reset service.

The last phase terminates the Network connection. As already mentioned, termination can be initiated either by the Transport entities or by the Network entities. Figure 24 shows the Network service primitives and the parameters of this phase. Disconnection service is an unconfirmed service.

7.5 Connectionless-Mode Network Service Primitives

For the connectionless-mode Network service, transfer of each N-SDU is a self-contained operation

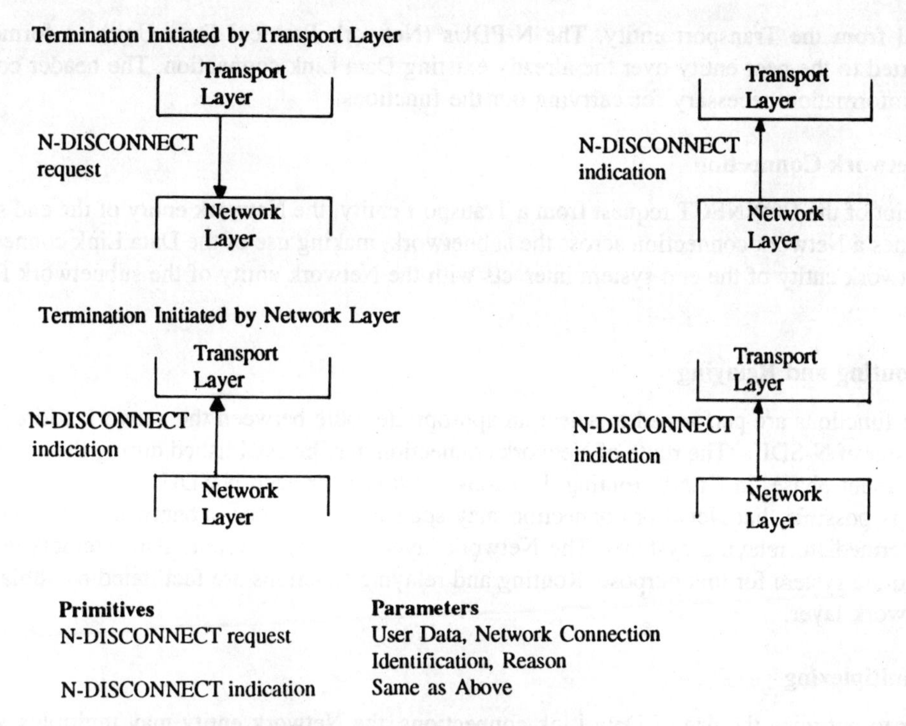

Fig. 24 Primitives for connection release service.

without establishing, maintaining and releasing a connection. Each data unit is completely independent of other units and the sequence of data units is not maintained. Figure 25 indicates the service primitives and associated parameters used for the connectionless-mode Network service.

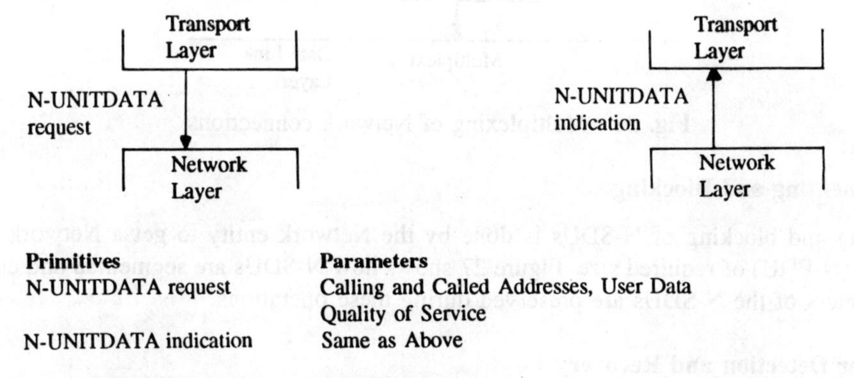

Fig. 25 Primitives for connectionless-mode Network service.

8 FUNCTIONS OF THE NETWORK LAYER

The functions carried out by a layer are different from its services. Functions are those activities which are carried out by a layer in order to provide the services. The Network layer functions are carried out by adding a header, in the form of Protocol Control Information (PCI) to every N-SDU received from the Transport entity. The N-PDUs (Network Protocol Data Unit) so formed are transported to the peer entity over the already existing Data Link connection. The header contains all the information necessary for carrying out the functions.

8.1 Network Connection

On receipt of the CONNECT request from a Transport entity, the Network entity of the end system establishes a Network connection across the subnetwork, making use of the Data Link connections. The Network entity of the end system interacts with the Network entity of the subnetwork for this purpose.

8.2 Routing and Relaying

Routing functions are performed to select an appropriate route between the network addresses for the transfer of N-SDUs. The route of Network connection may be established during the connection establishment phase. In CLNS, routing decisions are taken for each N-SDU.

It is possible that Network connection may span more than one subnetwork interconnected with intermediate relaying systems. The Network layer in an end system also interacts with the intermediate system for this purpose. Routing and relaying functions are facilitated by sublayering the Network layer.

8.3 Multiplexing

In order to optimize the use of Data Link connections, the Network entity may multiplex several Network connections on a Data Link connection (Fig. 26).

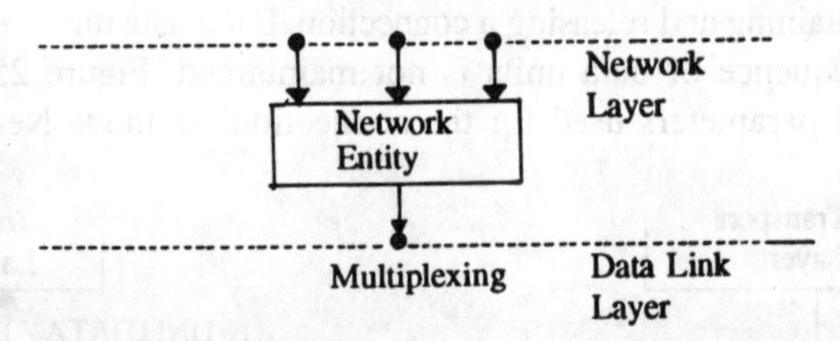

Fig. 26 Multiplexing of Network connections.

8.4 Segmenting and Blocking

Segmenting and blocking of N-SDUs is done by the Network entity to get a Network Protocol Data Unit (N-PDU) of required size. Figure 27 shows how N-SDUs are segmented and combined. The delimiters of the N-SDUs are preserved during these operations.

8.5 Error Detection and Recovery

Error detection functions are used to check that the quality of service provided over the Network connection is maintained. The Data Link layer notifies the residual errors and depending on the quality of the service to be provided, mechanisms for error recovery may be incorporated.

8.6 Other Functions

When requested by the Transport entity, the Network entity carries out sequencing and flow control

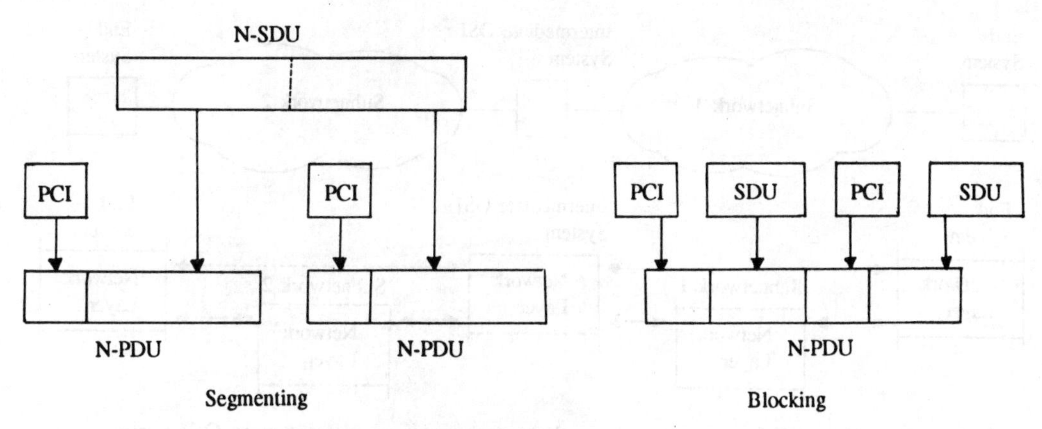

Fig. 27 Segmenting and blocking of N-SDUs.

of the N-SDUs. The Network connection can be reset at the request of the Transport entities. The Network layer also provides for transfer of the expedited N-SDUs to the destination Transport entity.

9 SUBLAYERING OF THE NETWORK LAYER

Compared to the other layers, organization of the OSI Network layer is somewhat complex. There are several reasons for this:

1. Many subnetworks were already operational when the Network service was defined. These subnetworks provided services which were different from one another and from the Network service. One of the functions of the Network layer is to take care of all the differences and provide a uniform Network service.

2. There are several subnetwork access protocols and, therefore, a different set of protocols is required to be implemented in the end systems, depending on the type of subnetwork.

3. Several subnetworks may be interconnected using intermediate systems which transfer N-SDUs from one subnetwork to another (Fig. 28). In such situations, the Network layer of the end systems needs to interact with the intermediate system. We will discuss this interaction in greater detail in the chapter on Internetworking.

Thus, the Network layer interacts with the subnetwork access node as well as with the intermediate system. Because of the multiplicity of the Network layer functions, it becomes necessary to specify the internal architecture of the Network layer so that entities which undertake these functions are identified. For this purpose, the Network is partitioned into sublayers, each sublayer representing a distinct functional entity.

The internal architecture of the Network layer comprises three sublayers (Fig. 29). They are:

1. Subnetwork Independent Convergence (SNIC) functions sublayer.
2. Subnetwork Dependent Convergence (SNDC) functions sublayer.
3. Subnetwork Access (SNAC) functions sublayer.

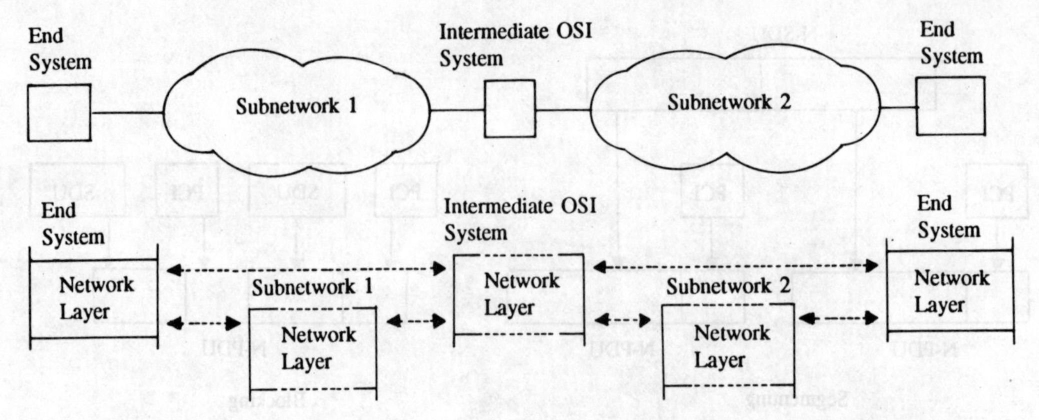

Fig. 28 Network layer protocols in the presence of an intermediate OSI system.

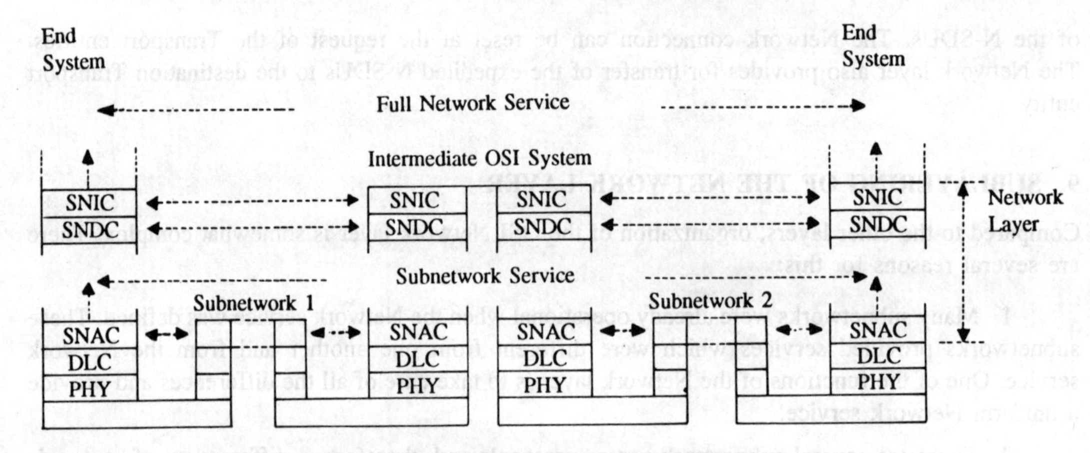

Fig. 29 Sublayering of the Network layer.

The names of sub-layers 1 and 2 contain the word "convergence" which means *to bring together into a common interpretation*. Each of these sub-layers brings the various forms of subnetwork services into a common form.

9.1 Subnetwork Independent Convergence Functions Sublayer (SNIC)

SNIC functions are those functions of the Network layer which can be defined independently of the subnetwork. Some of these functions are:

1. SNIC entity in the end system provides CONS or CLNS as requested by the Transport entity utilizing the well-defined service of the underlying sublayer irrespective of whether the underlying service is connection-mode or connectionless-mode.

2. Relaying function which involves forwarding of received N-SDU by the SNIC entity of the intermediate OSI system from the SNIC entity of one end system to the SNIC entity of another end system.

3. Routing functions which decide over which of the possibly many subnetworks particular

information is to travel. Routing functions are concerned with interpretation of address, quality of service parameters and other parameters to determine the path through the tandem subnetworks.

9.2 Subnetwork Dependent Convergence Functions Sublayer (SNDC)

The SNDC sublayer performs functions required to convert the subnetwork service into a well defined service expected by the SNIC sublayer. SNDC protocols are defined for a specific subnetwork service. Specifically, SNDC protocols may be used to

1. add to, correct or mute functions provided by the subnetwork so that a uniform basic Network service boundary is provided, and

2. relate the services provided by the subnetwork to the provisions of the Network service.

SNDC protocols are above the subnetwork. The SNDC sublayer may be completely absent also.

9.3 Subnetwork Access Control Functions Sublayer (SNAC)

SNAC sublayer performs all the functions and protocol interactions with the corresponding layer of the subnetwork node. The service provided by the subnetwork may or may not coincide with the OSI Network service. Examples of SNAC protocol are X.25 (Packet Switched Data Network) and X.21 (Circuit Switched Data Network).

10 NETWORK LAYER PROTOCOLS

From the internal architecture of the Network layer, it would be obvious that due to the multiplicity of functions and the sublayers carrying out these functions, several protocols are required for the Network layer. The family of protocols for CLNS are ISO 8473, ISO 9542 and ISO 10589. ISO 8473 is also called the Internet protocol.

- ISO 8473 defines protocols for SNIC and SDNC sublayers.
- ISO 9542 defines the protocol for end system and the intermediate system.
- ISO 10589 defines the protocol for interaction of one intermediate system with another.

For CONS, the protocols are ISO 8208 and ISO 8878 and the corresponding CCITT recommendations are X.25 and X.223. X.25 is the protocol for connection-oriented data transfer. X.223 specifies the use of X.25 to provide connection-mode Network service as specified in X.213 and ISO 8348.

11 SUBNETWORK INTERNAL ARCHITECTURE

The OSI reference model does not apply to the internal architecture of the subnetwork but considering the value of layering, the subnetworks do have layered architectures. The protocols and interfaces within the subnetwork are vendor-specific and it is not usually possible to interconnect network nodes supplied by different vendors.

The subnetwork interfaces towards the end systems have a layered architecture consisting of the first three layers of the OSI reference model (Fig. 30). The Network layer of the access node receives the N-PDU and relays the semantics of the PDU across the subnetwork.

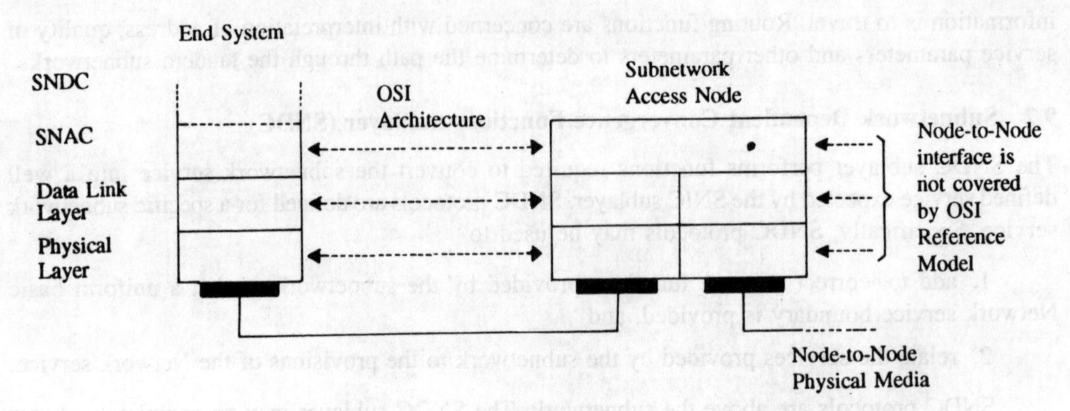

Fig. 30 End system-to-node layered architecture.

The internal architecture of the subnetwork depends on the mechanisms used for transport of the data units. Figure 31 shows the layered architecture of a circuit switched subnetwork.

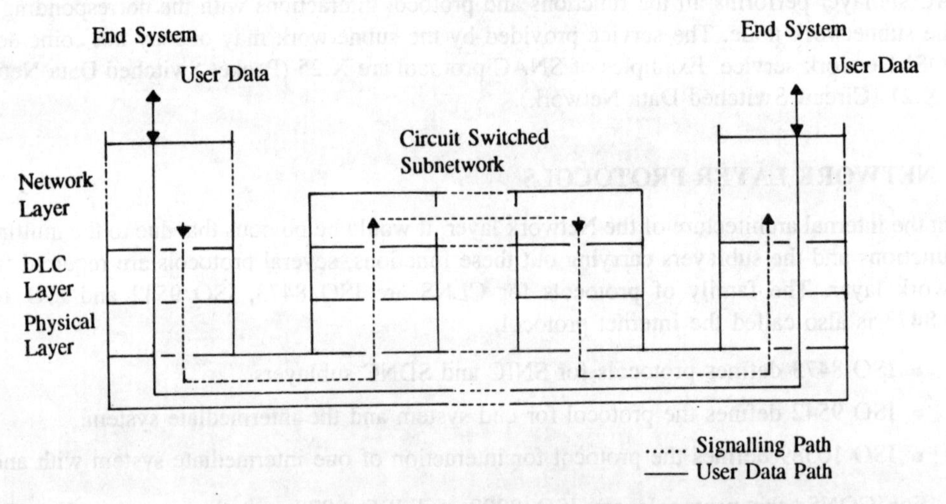

Fig. 31 Layered architecture of circuit switched data subnetwork.

The signals for connection establishment, release and control are generated and used by the Network layer of the end systems and the nodes. The lower two layers provide error-free transport of these signals. The second and third layers of the nodes are not involved in transport of user data units. The nodes provide cross-connections at the Physical layer level or at the physical media level. Therefore there is no node-to-node error or flow control. These functions are carried out end-to-end by the Data Link layers of the end systems.

Figure 32 shows the layered architecture of a packet switched subnetwork which accepts messages, divides them into packets, routes the packets through the subnetwork and reassembles the packets into messages. Partitioning and reassembly of the messages is done in the Network layer of the access and the exit nodes of the subnetwork. There is node-to-node error and flow control provided by the Data Link layer of the nodes.

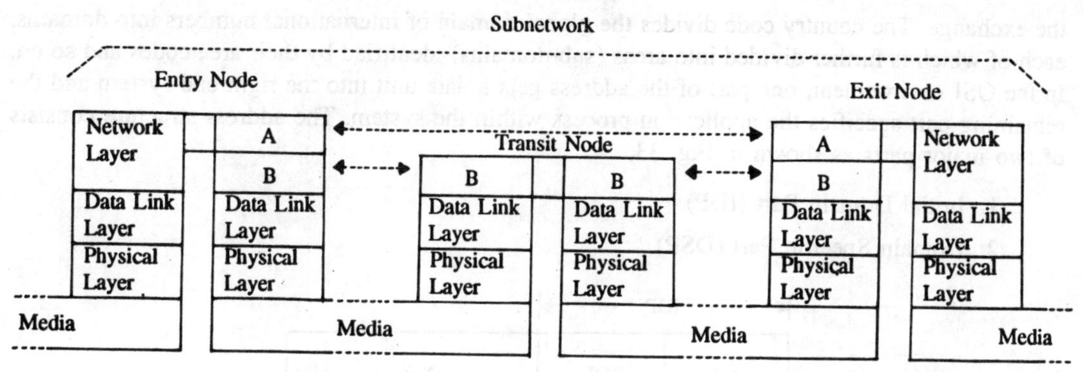

Fig. 32 Layered architecture of packet switched data subnetwork.

Partitioning and reassembly of the messages is not an essential feature of packet switched data networks. The Network layer of the end systems can generate the packets. X.25 recommendation of CCITT specifies the protocol between a packet mode end system and an access node.

12 NAMING AND ADDRESSING

For the Network layer to perform its basic function, specifying the address of the destination is a very important issue. Naming and addressing were not included in the original OSI reference model document but are subjects of a later addendum, ISO 7498-3, *Naming and Addressing*. Names are user-defined text strings and addresses are binary, hex or decimal representations of the actual location of an Application entity in a network. Mapping between the names and the numeric addresses is performed by a directory service which is a constituent of the Application layer. It is analogous to using a telephone directory to find out the telephone number of a person.

The addressing scheme for the OSI model enables locating of the Transport entity which has access to the desired Application entity through the Session and Presentation entities. Access from the Transport entity to the Application entity is on a one-to-one basis as there is no multiplexing above the Transport layer. The Transport entity is located by the N-SAP address to which the Transport entity is attached. It is necessary to understand the distinction between the subnetwork address and the N-SAP address. The subnetwork address refers to the point at which the subnetwork service is offered. It enables the subnetwork to locate the point at which an end system is attached to the subnetwork.

12.1 Hierarchical Addressing Scheme

ISO 8348, *Network Service Definition, Addendum 2: Network Layer Addressing* specifies the structure of the global N-SAP address. The corresponding CCITT recommendation is X.213, *Annex. A.* The address structure is designed to enable definition of global network addresses which identify N-SAPs unambiguously.

The N-SAP addressing scheme is based on the concept of hierarchical addressing domains. This approach is similar to the telephone numbering scheme. A long distance telephone call number consists of the country code, area code, exchange code and finally the subscriber's number within the exchange. The country code divides the global domain of international numbers into domains, each of which is further divided into areas (sub-domains) identified by their area codes and so on. In the OSI environment, one part of the address gets a data unit into the right end system and the remaining part specifies the application process within the system. The address structure consists of two major parts as shown in Fig. 33.

1. Initial Domain Part (IDP)
2. Domain Specific Part (DSP).

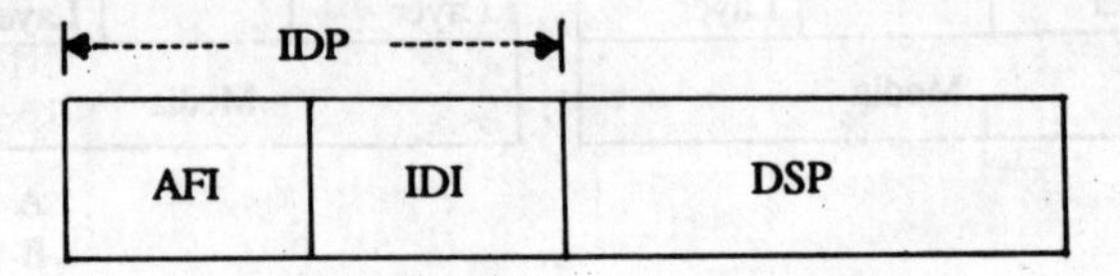

Fig. 33 N-SAP address structure.

IDP is further divided into the following two parts:

1. Authority and Format Identifier (AFI)
2. Initial Domain Identifier (IDI)

Authority and Format Identifier (AFI). AFI is a two-digit decimal number which specifies the format of IDI, the authority (e.g., CCITT) responsible for allocating the values of IDI and the syntax of DSP. Table 5 lists some typical values of AFI.

Table 5 Typical Values of Authority and Format Identifier

AFI	Authority	IDI Format	DSP Syntax	Max. size of DSP
36	CCITT	X.121, 14 digits	Decimal	24 digits
37	CCITT	X.121, 14 digits	Binary	9 octets
38	ISO	ISO DCC, 3 digits	Decimal	35 digits
39	ISO	ISO DCC, 3 digits	Binary	14 octets
40	CCITT	F.69, 8 digits	Decimal	30 digits
41	CCITT	F.69, 8 digits	Binary	12 octets
42	CCITT	E.163, 12 digits	Decimal	26 digits
43	CCITT	E.163, 12 digits	Binary	10 octets
44	CCITT	E.164, 15 digits	Decimal	23 digits
45	CCITT	E.164, 15 digits	Binary	9 octets
46	ISO	ISO 6523, 4 digits	Decimal	34 digits
47	ISO	ISO 6523, 4 digits	Binary	13 octets
48	Local	Null	Decimal	38 digits
49	Local	Null	Binary	15 octets
50	Local	Null	ISO 646 characters	19 characters
51	Local	Null	National characters	7 characters

Currently allocated values of AFI as indicated above refer to the following addressing schemes:

- CCITT X.121—International Data Numbering Plan
- ISO 3166 DCC—This is a subset of the three-digit Data Country Codes of CCITT X.121 for administration by ISO
- CCITT F.69—International Telex Numbering Plan
- CCITT E.163—International Telephone Numbering Plan
- CCITT E.164—International Numbering Plan for ISDN
- ISO 6523 ICD—International Code Designator (ICD) identifies an organizational authority for allocating and assigning values of the DSP. This is an organization-based non-geographical addressing plan. For example, 0006 is assigned to the US Department of Defence.

Initial Domain Identifier (IDI). Initial Domain Identifier (IDI) whose format is indicated by AFI values, specifies the network addressing domain from which the values of DSP are allocated and the authority responsible for allocating these values. It is null (in the case of local formats) or a string of decimal digits as indicated in Table 5. For example, in X.121, IDI consists of a decimal number of upto 14 digits which unambiguously identifies an end system or an internetworking device connected to a switched data subnetwork.

Domain Specific Part (DSP). DSP is the sub-domain address and identifies the N-SAP within the domain defined by the IDI number. Its syntax is defined by the AFI value and can be a binary, decimal or character mode. The maximum size of DSP for various IDI formats is indicated in Table 5. For example, the N-SAP address based on X.121 decimal format for an end system having subnetwork address 310 5 4567890123 and domain specific part 1234567890123456, is as follows.

363105456789012312345678901234567

13 SUMMARY

Switched data subnetworks provide flexible interconnection topology and enable sharing of the subnetwork resources. Switching techniques are primarily of two kinds—circuit switching and store-and-forward switching. In the former case a dedicated path is established between the end systems for transport of data units. In store-and-forward switching, data units which carry the source and destination addresses, are temporarily stored in the subnetwork nodes and then forwarded to the next node.

Message switching is based on the basic philosophy of store-and-forward switching but has significant delivery delay. To reduce the delivery delay, the messages are partitioned into packets. There are two approaches to routing of the packets through the subnetwork—datagram routing and virtual circuit routing. In datagram routing, each packet takes an independent route and as such sequence of the packets may get disturbed. In virtual circuit routing, a virtual path for routing the packets is established and the packets remain in sequence.

The Network layer provides the means to route data units from one end system to another end system which are interconnected through a subnetwork. The Network service provides Transport layer independence from the functions relating to routing, switching and accessing the subnetwork.

Due to the multiplicity of Network layer functions, its internal architecture consisting of three sublayers has been defined. The lowermost sublayer (SNAC) interacts with the access node of the subnetwork. The next sublayer (SNDC) augments the subnetwork service to provide a uniform well defined service to the next sublayer (SNIC) which in turn provides the Network Service.

PROBLEMS

1. Match the features given in the first column to the switching technologies given in the second column.

• Source and destination data rate must be the same.	Circuit switching
• Significant and random delivery delay may occur.	Message switching
• No connection establishment and release phases.	Virtual circuit service
• Source and destination address are specified once and there is minimal delivery delay.	
• Sequenced delivery but source destination data rates can be different.	

2. Draw the routing table for node *B* of the subnetwork illustrated in Fig. 9.

3. Develop the static routing table for node *A* of the following subnetwork. Use the algorithm described in the chapter. Draw the routing diagram also.

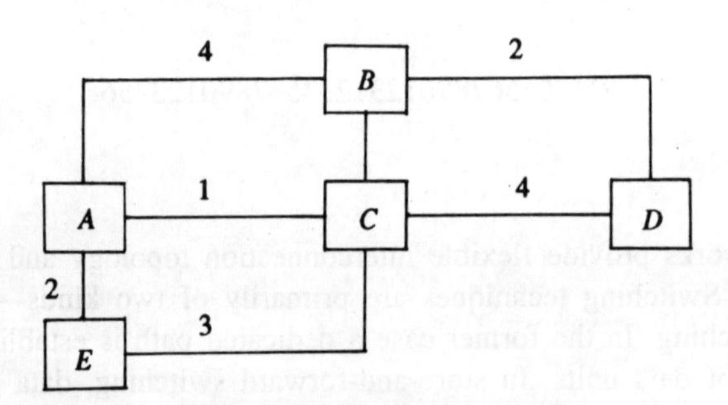

Fig. P.3

4. Repeat the exercise in Problem 3 for node *D*.

5. In Fig. 2, the subnetwork is a packet-switched subnetwork employing virtual circuit routing. End system 4*A* establishes a connection to end system 3*D* through the subnetwork. The connection is established through nodes 4, 1, 2 and 3. Assuming some logical channel identifiers for each intervening link, prepare the link tables for the virtual circuit.

6. In the network shown below, the following link tables are maintained at its nodes. Indicate the routes taken by the packets belonging to the connections established by the end systems *P*, *Q* and *R*.

Link Table at A

A–P	A–B	A–C
3	6	
8		4

Link Table at B

B–Q	B–A	B–C	B–D
	6	2	
2		3	
5			8

Link Table at C

C–A	C–B	C–D
	2	5
4	3	

Link Table at D

D–R	D–B	D–C
1		5
9	8	

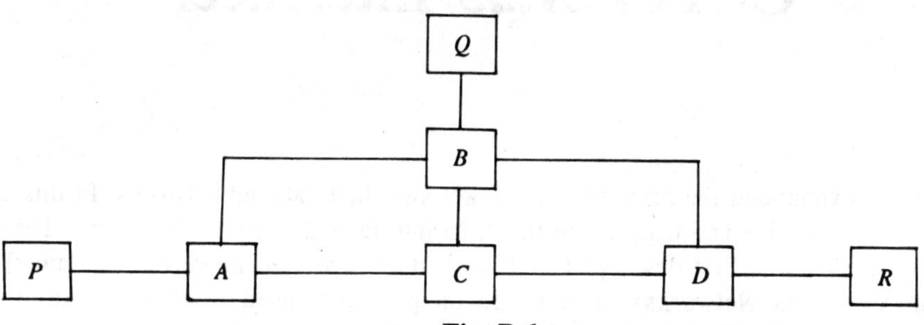

Fig. P.6

CHAPTER 10

CCITT X.25 Interface

In Chapter 9, we expounded the principles of packet switched data subnetworks. In this chapter, we build on this knowledge and learn about the X.25 interface of such a subnetwork. The CCITT X.25 recommendation was first developed in 1976 and later revised in 1978, 1980 and 1984 by Study Group VII (Data Networks). It is based on protocols used in early packet switched networks—ARPANET, DATAPAC, TRANSPAC etc. The corresponding ISO standard is 8208. We will examine the application, services, operation and facilities of X.25 interface. Though the X.25 recommendation covers the first three layers of the OSI Reference Model, we will concentrate on the third layer. The protocols for the first two layers have already been discussed in earlier chapters. After a detailed coverage of X.25, we discuss an intermediary device called Packet Assembler and Disassembler (PAD). It enables interfacing of non-packet mode terminals to the X.25 interface of a subnetwork. We will specifically concentrate on the operation of CCITT X.3/X.28/X.29 PAD which is designed for character mode terminals.

1 TITLE OF X.25 INTERFACE

> *Interface* between *Data Terminal Equipment* (DTE) and *Data Circuit-Terminating Equipment* (DCE) for Terminals *Operating in the Packet Mode* and Connected to Public Data Networks by *Dedicated Circuit*

X.25 defines the interface for exchange of packets between a packet mode end system (DTE—Data Terminal Equipment operated by the user) and a switched data subnetwork node (DCE—Data Circuit Terminating Equipment operated by the data subnetwork service provider). The DTE and the DCE must operate in the packet mode i.e., generate and expect to receive packets. The DTE and the DCE must be connected by a dedicated circuit. Note that X.25 is neither a network architecture nor protocol governing the routing of packets within a packet data subnetwork.

2 LOCATION OF X.25 INTERFACE

The X.25 interface is always contiguous to the DTE having capability to generate and receive data packets. Figure 1 shows various possible configurations of the access to the packet switched subnetwork and location of X.25 interface.

A non-packet mode terminal can also have access to the subnetwork but an intermediary device called a PAD (Packet Assembler and Disassembler) is required. The PAD can be a distinct device as shown in Fig. 1. The PAD as packet mode DTE has an X.25 interface. It can also be built into the access node in which case there is no X.25 interface.

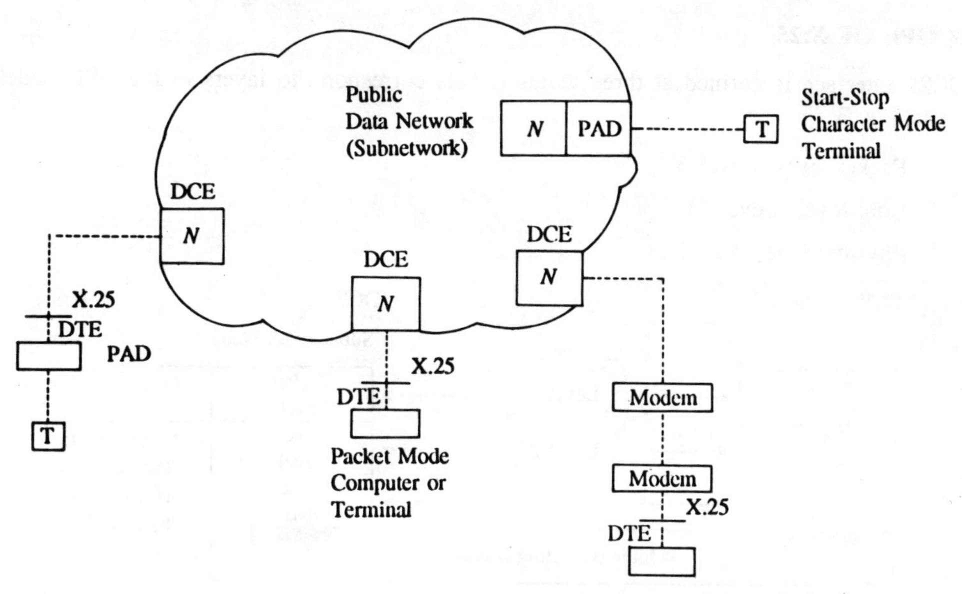

Fig. 1 Location of X.25 interface.

3 X.25 SERVICES

X.25 provides a virtual circuit "connection mode" service by establishing an end-to-end logical communication path through the subnetwork. The logical path can be of two types (Fig. 2):

1. Switched Virtual Circuit (SVC)
2. Permanent Virtual Circuit (PVC).

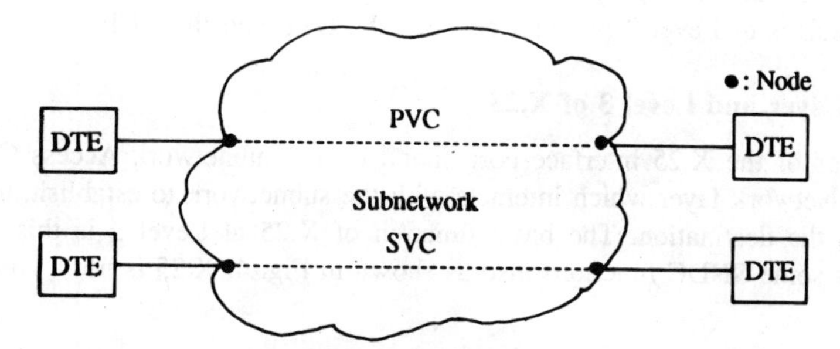

Fig. 2 Switching and permanent virtual circuits.

Switched Virtual Circuit (SVC). A switched virtual circuit is established at the request of a DTE and is cleared at the end of the call. The subnetwork resources are allocated for an SVC for the duration of the call.

Permanent Virtual Circuit (PVC). A permanent virtual circuit is a constant logical connection between two DTEs. It is like a leased circuit connection between two terminals. It is not to be established or cleared and is always in the data transfer phase.

4 SCOPE OF X.25

The X.25 interface is defined at three levels (levels correspond to layers in the OSI model) as shown in Fig. 3:

1. Packet level (Level 3)
2. Link level (Level 2)
3. Physical level (Level 1)

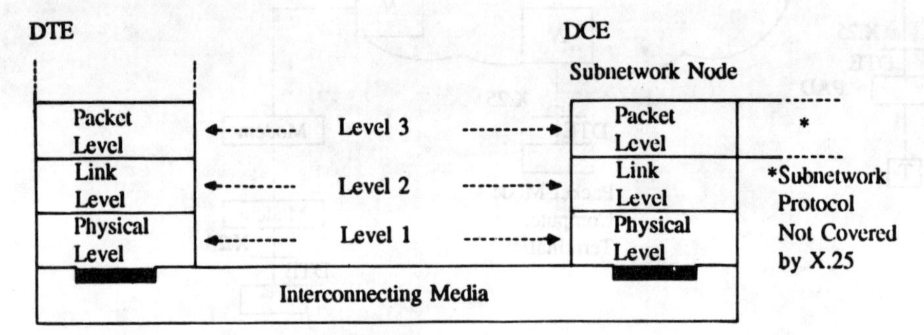

Fig. 3 The three levels of X.25 interface.

At the Physical level, the X.25 makes use of Level 1 of CCITT Recommendations X.21 and X.21*bis* which were defined for circuit switched data networks.

At the link level, X.25 specifies the Link Access Procedure-B (LAP-B) protocol which is a subset of the HDLC protocol.

At the network level, X.25 defines the protocol for access to the packet data subnetwork. It defines format, content and procedures for exchange of control and data transfer packets to establish, utilize and clear the virtual circuits through the subnetwork. This protocol is operated on the point-to-point Data Link connection provided by Level 2 of the DTE and the DCE. In this chapter we will restrict ourselves to Level 3 protocol between the DTE and the DCE.

4.1 Network Layer and Level 3 of X.25

Level 3 functions of the X.25 interface correspond to the Subnetwork Access Control (SNAC) sublayer of the Network layer which interacts with the subnetwork to establish, utilize and clear a connection to the destination. The basic function of X.25 at Level 3 is this function. X.25, however, covers some SNDC functions also as shown in Fig. 4. X.25 is not a complete Network layer protocol.

5 LOGICAL CHANNELS

At Level 3 there is exchange of control information and data in the form of packets. Each packet carries a number which identifies the virtual circuit to which it belongs. Thus, multiple virtual connections can be realized by statistical multiplexing of the packet as shown in Fig. 5.

The DTE A is operating three connections, one each to DTE B, DTE C, and DTE D. Within the subnetwork, these connections are extended on the virtual circuits, either permanent or switched. Packets to and from DTE B, DTE C and DTE D are assigned numbers N_1, N_2 and N_3 respectively

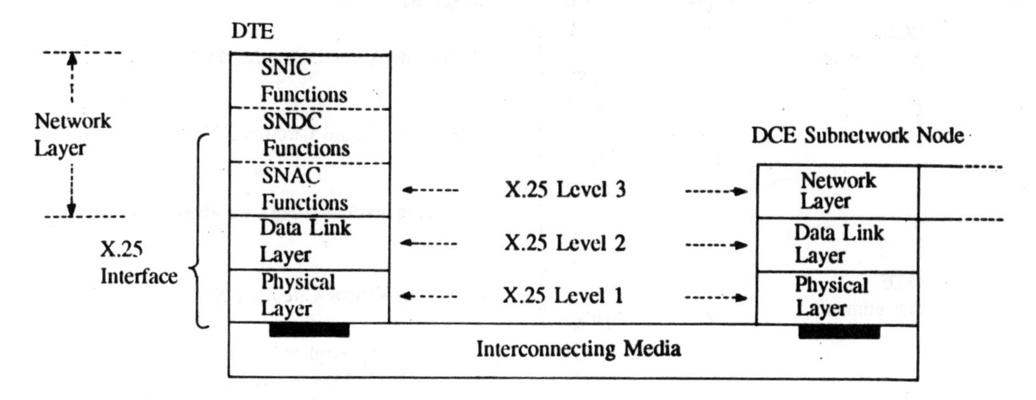

Fig. 4 Network layer functions covered by X.25 interface.

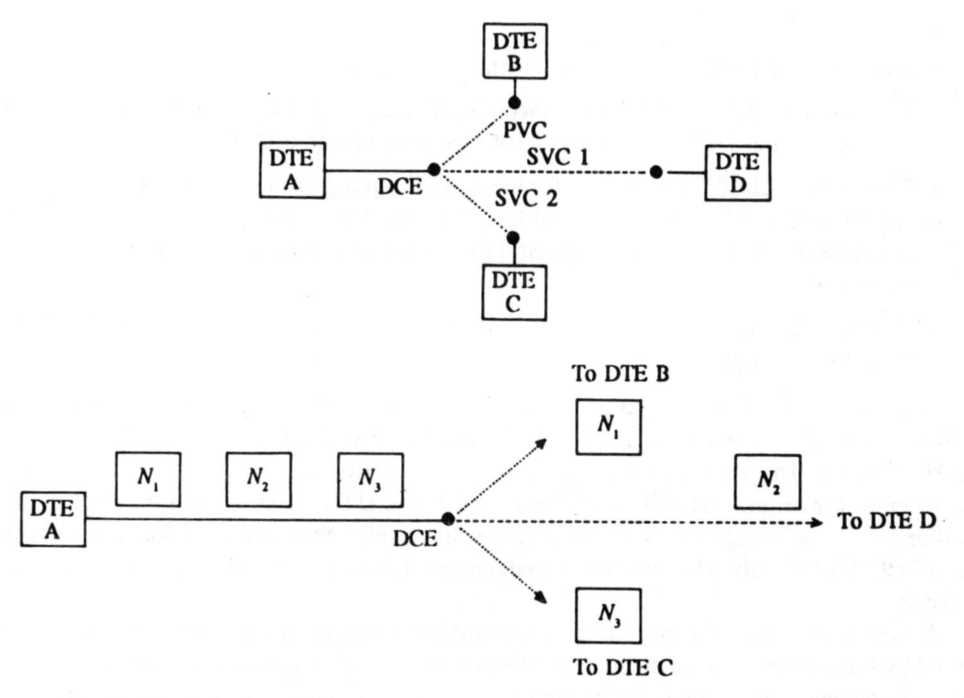

Fig. 5 Statistical multiplexing of packets to realize multiple logical channels.

and are multiplexed statistically between the DCE and the DTE A. Channels so realized are called logical channels and these numbers are called logical channel identifiers.

5.1 Grouping of Logical Channels

Theoretically, 4096 channel assignments (logical channel identifiers) are available, on each X.25 port of a DCE. 4096 channel assignments are divided into several groups as shown in Fig. 6. Grouping is in the form of ranges.

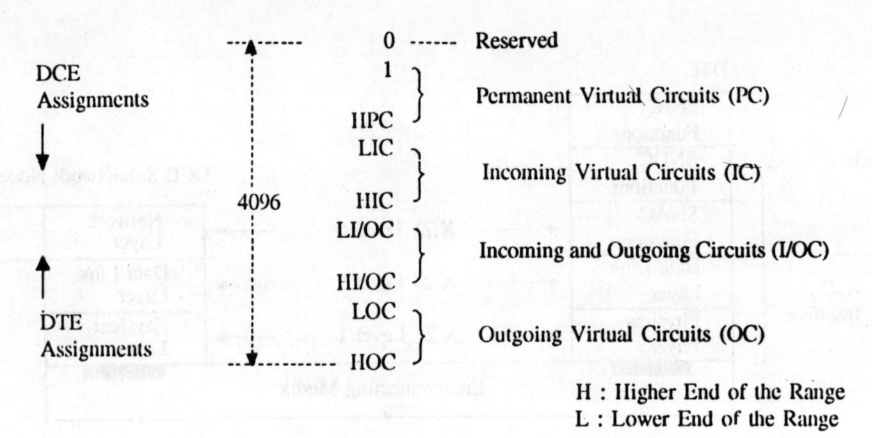

Fig. 6 Grouping of logical channel assignments.

- Channel 0 is reserved.
- First range (1 to HPC) is for permanent virtual circuits.
- Second range (LIC to HIC) is for incoming virtual circuits from DCE. These assignments are used by the DCE when there is an incoming call to a DTE.
- Third range (LI/OC to HI/OC) is for incoming/outgoing virtual circuits. These assignments can be used for incoming calls by the DCE and for outgoing calls by the DTE. These assignments are used after the specific ranges for incoming and outgoing calls have been exhausted.
- Fourth range (LOC to HOC) is for the outgoing virtual circuits and is used only by the DTE for its outgoing calls.

Whenever there is an incoming call, a DCE uses the lowest free logical channel from the LIC-HIC range. For an outgoing call the DTE selects the highest free logical channel from the LOC-HOC range. The DTE and DCE always search for free logical channel assignments from opposite ends to minimize the risk of collision—DCE and DTE selecting the same logical channel identifier for an incoming and an outgoing call respectively. Note that collision can occur in the range LI/OC–HI/OC only when exclusive assignments for incoming and outgoing calls have been exhausted.

A user at the time of subscribing to the service, indicates the number of logical channels required by him and pays for each channel. The number of logical channels dictates to the user how many simultaneous connections can be set up.

6 GENERAL PACKET FORMAT

All interactions between a DTE and a DCE are in the form of exchange of packets which can be broadly categorized as control packets and data transfer packets.

Data transfer packets are used for carrying user data, while control packets are used for control of the virtual circuit and for flow control. A packet consists of a number of contiguous octets (Fig. 7). Bit 8 of each octet is followed by bit 1 of the following octet. Note that the first bit and the first octet of a packet have been shown on the right-hand side.

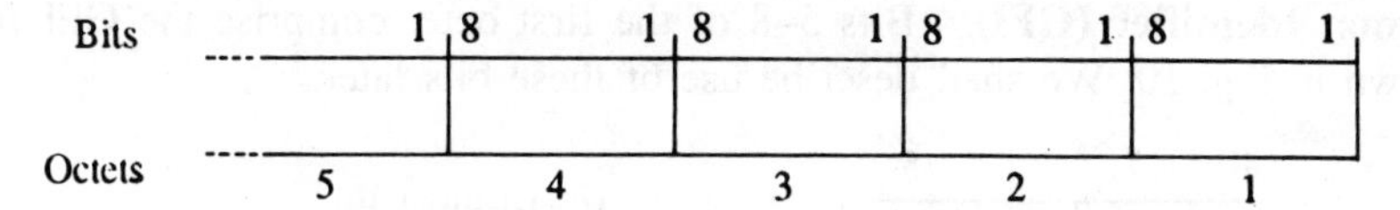

Fig. 7 Numbering of octets and bits of a packet.

Instead of representing a packet as a sequence of octets, a more compact representation of packets is used in X.25 (Fig. 8). The packets are drawn as a stack of octets. The first octet is shown on the top and the first bit is on the right hand side.

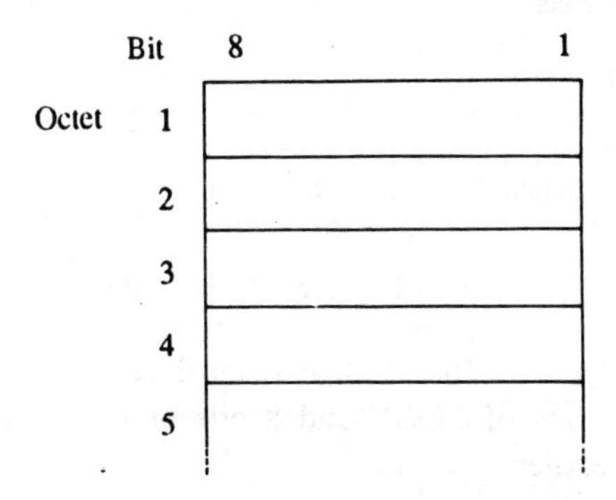

Fig. 8 Packet as a stack of octets.

Figure 9 shows the format of an X.25 packet in general. A packet consists of a number of fields.

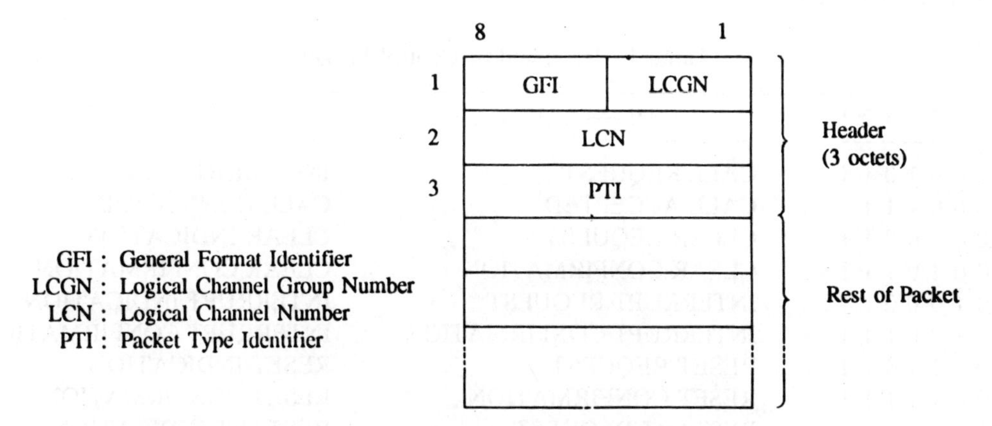

Fig. 9 General format of a packet.

A header is always present in all the packets. It consists of three octets (except in modulo-128 numbering scheme where it has 4 octets). It contains the following fields.

Logical Channel Group Number (LCGN). The logical channels are divided into 16 groups, each group containing 256 logical channels. Bits 1 to 4 of the first octet contain the logical channel group number.

General Format Identifier (GFI). Bits 5–8 of the first octet comprise the GFI field and are coded as shown in Fig. 10. We shall describe use of these bits later.

8	7	6 5
Q	D	SN

Q : Qualifier Bit
D : Delivery Confirmation Bit
SN : Sequence Numbering

General Format Identifier	Octet 1 8 7 6 5
Call Set-up Packets	0 X S N
Data Packets	X X S N
Other Packets	0 0 S N

X : 0 or 1
SN : 0 1 for Modulo-8 Sequence Numbering Scheme
1 0 for Modulo-128 Sequence Numbering Scheme

Fig. 10 Contents of general format identifier field.

Logical Channel Number (LCN). Bits 1–8 of octet 2 comprise the logical channel number associated with the LCGN. 4 bits of LCGN and 8 bits of LCN together constitute the logical channel identifier introduced earlier.

Packet Type Identifier (PTI). The packet type is identified by the coding of the third octet which is called the Packet Type Identifier field. If the first bit of this field is "0", it is a data packet, otherwise, it is one of the control packets. Table 1 gives PTI fields of various control packets used in the X.25 interface.

Table 1 PTI Field of Control Packets

8 7 6 5 4 3 2 1	DTE ⟶ DCE	DCE ⟶ DTE
0 0 0 0 1 0 1 1	CALL REQUEST	INCOMING CALL
0 0 0 0 1 1 1 1	CALL ACCEPTED	CALL CONNECTED
0 0 0 1 0 0 1 1	CLEAR REQUEST	CLEAR INDICATION
0 0 0 1 0 1 1 1	CLEAR CONFIRMATION	CLEAR CONFIRMATION
0 0 1 0 0 0 1 1	INTERRUPT REQUEST	INTERRUPT INDICATION
0 0 1 0 0 1 1 1	INTERRUPT CONFIRMATION	INTERRUPT CONFIRMATION
0 0 0 1 1 0 1 1	RESET REQUEST	RESET INDICATION
0 0 0 1 1 1 1 1	RESET CONFIRMATION	RESET CONFIRMATION
1 1 1 1 1 0 1 1	RESTART REQUEST	RESTART INDICATION
1 1 1 1 1 1 1 1	RESTART CONFIRMATION	RESTART CONFIRMATION

Note that the same PTI is used for a pair of logically related packets e.g., a CALL REQUEST packet from a DTE to a DCE results in an INCOMING CALL packet from the remote DCE to the DTE and these two packets have the same PTI field.

Besides a header, a packet consists of other fields containing addresses, facilities required and user data. While a header is always present in all packets, these fields may or may not be present in a packet depending on its type. We will describe these fields as and when we come across them while discussing operation of the X.25 interface.

However, before we go into the actual operation of the Level 3 X.25 interface, it must be borne in mind that packets, whatever the type, are assembled in the Network layer and handed over as data to the Data Link layer. The Data Link layer having already established the Data Link connection by exchange of mode-setting commands and responses, sends one packet at a time in the information field of an I-frame (Fig. 11). If the I-frame is received without errors at the receiving end, the information field of the frame is handed over to the Network layer. The procedures described below apply to the packets successfully tranferred across the Data Link connection.

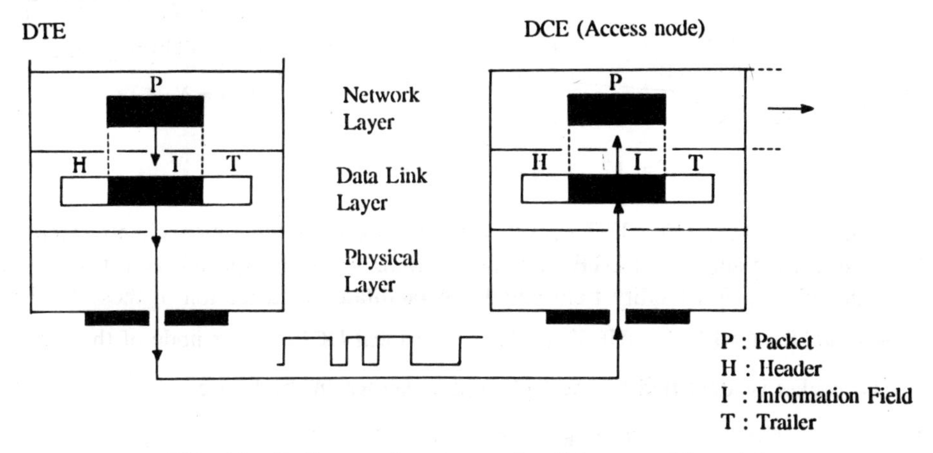

Fig. 11 Delivery of packets using I-frames of Level 2.

7 PROCEDURES FOR SWITCHED VIRTUAL CIRCUITS

Operation of switched virtual connections involves three phases: call establishment, data transfer, and call clearing. These phases involve the following procedures.

7.1 Call Establishment Phase

Overall Basic Scenario. The calling DTE issues a CALL REQUEST packet (Fig. 12) to the local DCE, access node of the subnetwork. The subnetwork establishes a virtual circuit upto the remote DCE which serves the called DTE. The remote DCE issues an INCOMING CALL packet to the called DTE. If the called DTE decides to accept the call, it returns a CALL ACCEPTED packet. The CALL ACCEPTED packet is passed on to the local DCE on the same virtual circuit. The DCE notifies the calling DTE with a CALL CONNECTED packet.

If the called DTE does not wish to receive the call, or the subnetwork cannot establish a virtual circuit to the remote DCE, the call clearing procedure is initiated.

Detailed Scenario

1. The calling DTE
   - selects a logical channel identifier N_1 from the high end of the range, HOC upwards (Fig. 6),

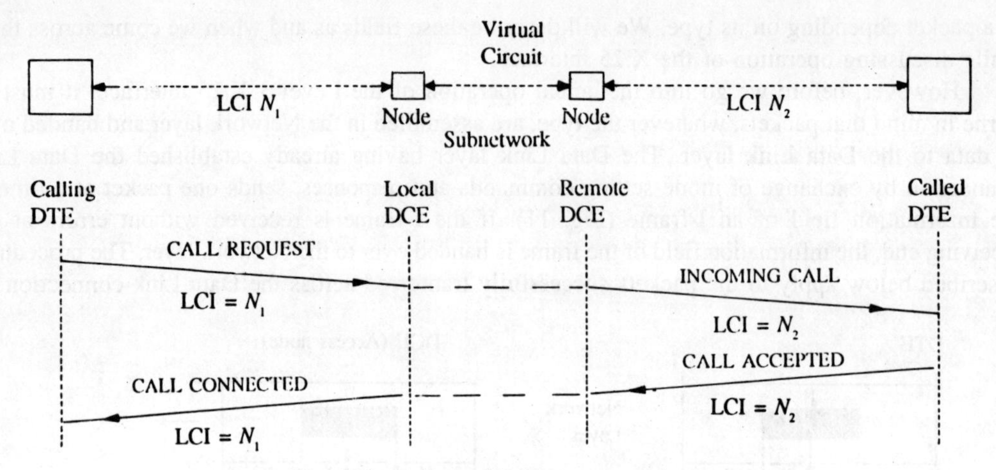

Fig. 12 Call establishment procedure.

- builds a CALL REQUEST packet (Fig. 13) specifying the logical channel identifier N_1, the calling and called DTE addresses in BCD form, the special facilities desired, and the address and facility field lengths to facilitate identification of these fields; and
- sends the CALL REQUEST packet to the local DCE, access node of the subnetwork.

The packet is identified by the PTI field as shown in the figure.

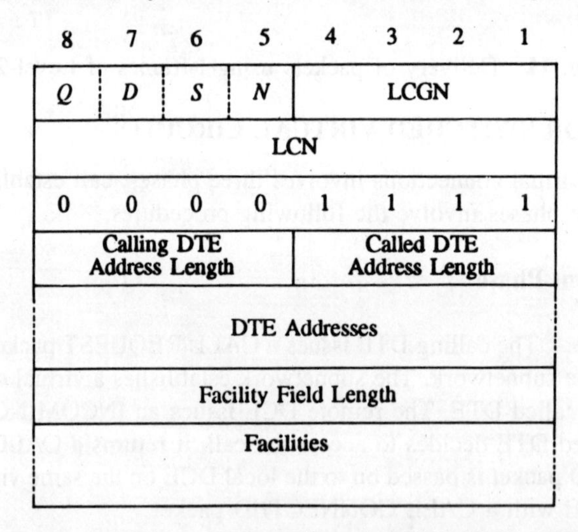

Fig. 13 Format of CALL REQUEST and INCOMING CALL packets.

2. On receipt of the CALL REQUEST packet, the local DCE, the access node of the subnetwork:
   - records the logical channel identifier N_1,
   - checks the validity of the desired facilities,
   - inserts the calling DTE address in the packet if it has not been indicated by the calling DTE,

- sets up a virtual circuit through the subnetwork to the remote DCE serving the called DTE, and
- forwards the CALL REQUEST packet to the remote DCE.

3. On receipt of the CALL REQUEST packet, the remote DCE:
   - checks the validity of the requested facilities,
   - selects a free logical channel identifier N_2 from the lower end of the IC range, LIC downwards (Fig. 6),
   - assembles an INCOMING CALL packet (Fig. 13), and
   - forwards the INCOMING CALL packet to the called DTE.

Note that the two logical channel identifiers, N_1 in CALL REQUEST packet and N_2 in INCOMING CALL packet are independent of each other.

4. On receipt of the INCOMING CALL packet, and if the remote DTE is willing to accept the call, it
   - records the logical channel identifier N_2,
   - assembles a CALL ACCEPTED packet using the logical channel identifier N_2 on which the INCOMING CALL packet was received (Fig. 14),
   - indicates in the packet the facilities which are acceptable, and
   - sends the packet to the DCE.

The DCE forwards this acceptance to the local DCE of the calling DTE using the virtual circuit already established for the CALL REQUEST packet.

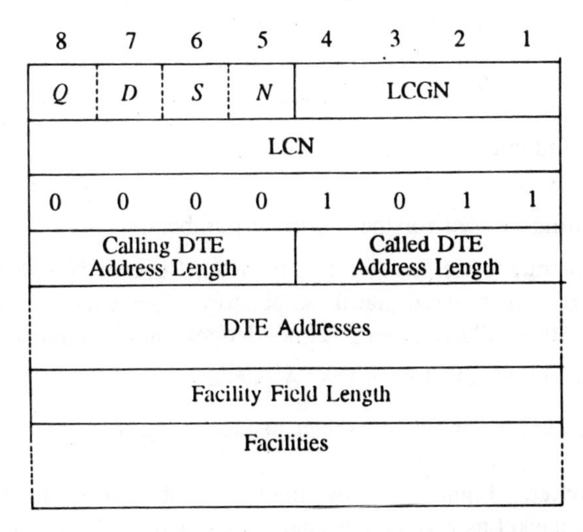

Fig. 14 Format of CALL ACCEPTED and CALL CONNECTED packets.

5. On receipt of the CALL ACCEPTED packet, the local DCE sends a CALL CONNECTED packet (Fig. 14) using the logical channel identifier N_1. It also indicates the accepted facilities in the packet. From the logical channel identifier on this CALL CONNECTED packet, the local DTE will come to know that the connection it requested has been established. It will also note the facilities which have been agreed to.

7.2 Data Transfer Phase

Once the connection is established, the source and destination addresses are no longer required. The logical channel identifiers N_1 and N_2 identify the connection and only they are specified in the subsequent packets.

The data transfer phase involves exchange of DATA packets between the DTE and the DCE. This exchange of packets is on the logical channels already identified in the call establishment phase. The local and remote DCEs exchange DATA packets on the virtual circuit already established (Fig. 15).

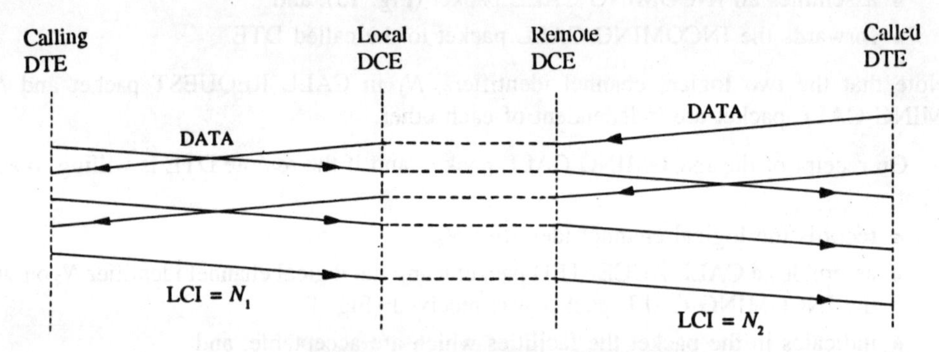

Fig. 15 Data transfer procedure.

Flow Control. A DTE cannot be allowed to transmit an unlimited number of DATA packets into the subnetwork because this may lead to congestion. Therefore, a flow control mechanism is required. The sliding window flow control mechanism which we studied in data link protocols is used here also.

- Each DATA packet and acknowledgement is given a sequence number.
- Widows are maintained at the DTE and the DCE. The window size for each logical channel is usually two.
- Modulo-8 counting is used for the sequence numbers.
- The acknowledgements are Receive Ready (RR), Receive Not Ready (RNR) and Reject (REJ). It must be remembered that these acknowledgements belong to Layer 3 and have nothing to do with similarly named acknowledgements of Layer 2.
- RR can be sent piggybacked on a DATA packet.

The flow control packets consist of 3 octet headers only (Fig. 16).

Format of a DATA Packet. Figure 17 shows the format of a DATA packet. The first bit of the third octet identifies the packet as a DATA packet. It is always "0". $P(S)$ is the sequence number of the DATA packet and $P(R)$ is the sequence number of the piggybacked acknowledgement (RR).

The maximum size of the data field is usually 128 octets, but the users have options for 16, 32, 64, 256, 512, 1024, 2048 and 4096 octets. The maximum size of the data field is negotiated during the call establishment phase by inserting the desired size in the facility field.

Local and Remote Acknowledgements. When a DTE sends a DATA packet, there are two

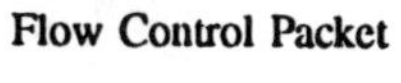

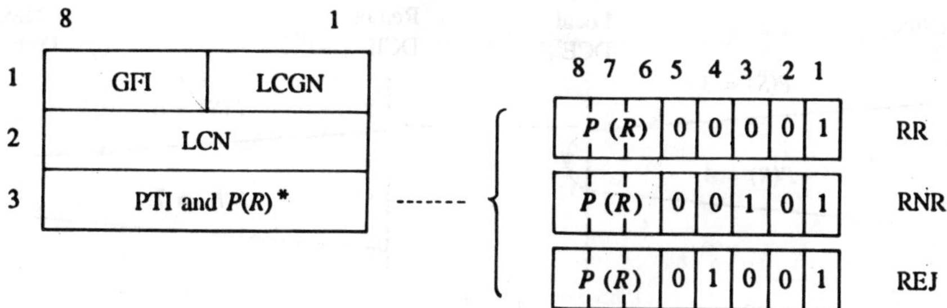

* $P(R)$ is the sequence number associated with the acknowledgement

Fig. 16 Format flow control packets.

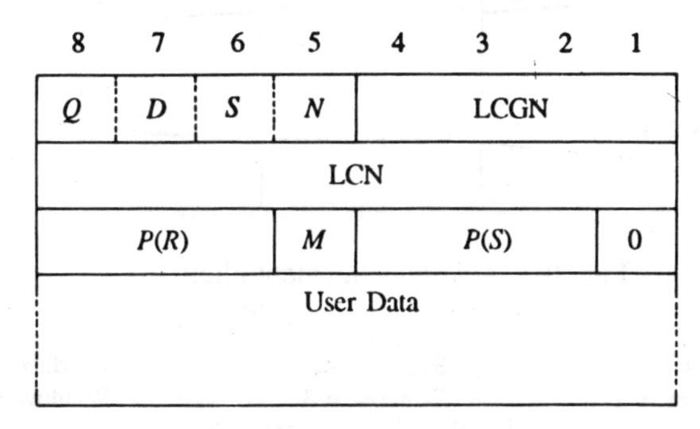

Fig. 17 Format of DATA packet.

alternatives for sending the acknowlegement. It can be either given by the local DCE when it receives the packet, or the local DCE waits for acknowledgement from the called DTE and when received, forwards it to the local DTE. Figure 18 illustrates the two options.

The D (delivery confirmation) bit, which is the 7th bit of octet 1 (Fig. 13), specifies whether the acknowledgement has local or remote significance. When the D bit is set to "1", the acknowledgement for the DATA packets comes from the called DTE. When it is set to "0", the local DCE sends the acknowledgement.

Remote acknowledgement is more useful because it guarantees that the DATA packet has reached the destination although waiting for acknowledgement slows down the data transfer. Figures 19 and 20 show data transfer situations when $D = 0$ and $D = 1$. Note that window sizes for the interface and within the subnetwork are different. At the DTE-DCE interface the window size is 2 and within the subnetwork the window size is 3. For $D = 0$, the local DCE cannot accept more than three packets from the DTE due to window size restriction of the subnetwork. It acknowledges only the first packet with $P(S) = 0$ so that the DTE is permitted to send only two more packets with $P(S) = 1$ and $P(S) = 2$. The local DCE sends the next acknowledgement only when a place for another packet in the subnetwork window is vacated. Effective throughput, in this case, is determined by the subnetwork window size and round trip delay from DTE to DTE.

For $D = 1$, the local DCE can release acknowledgement only after it receives acknowledgement from the remote DCE (Fig. 20). In this case, effective throughput is governed by the size of the smaller of the two windows and the round trip delay from DTE to DTE.

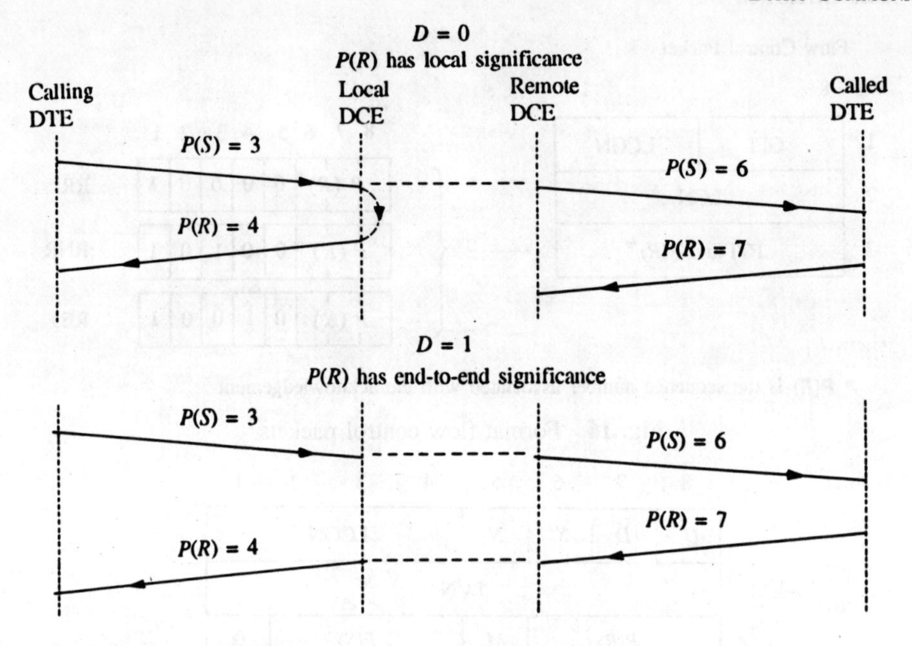

Fig. 18 Acknowledgement alternatives.

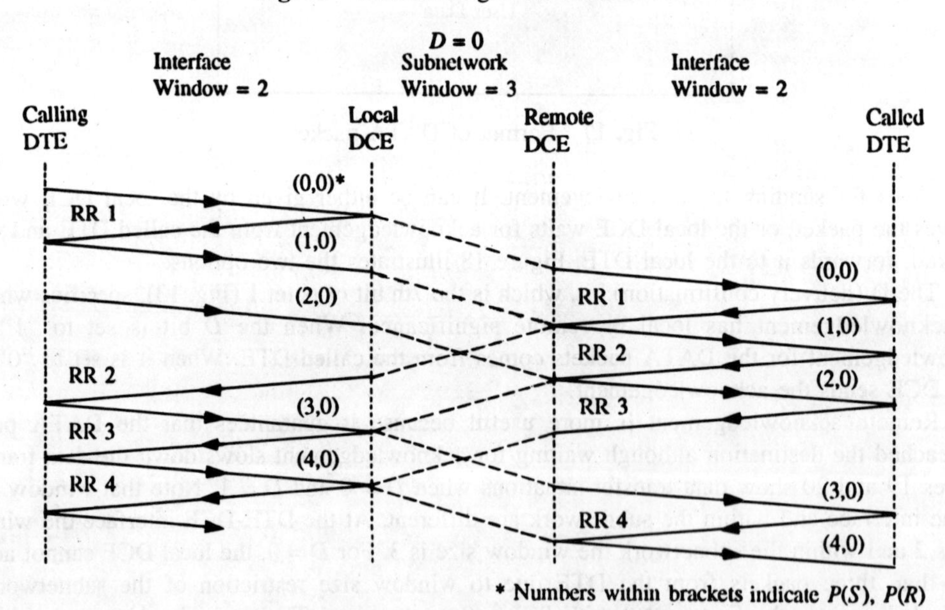

Fig. 19 Effect of window size on throughput in case of local acknowledgement.

INTERRUPT Packet. When a DTE does not receive the acknowledgement and has exhausted its window, there is still a way to getting through to the remote DTE. It can send an INTERRUPT packet which is not controlled by the data packet flow control mechanism and can even overtake other DATA packets still in transit. Only one INTERRUPT packet can be sent at a time and it is acknowledged by an INTERRUPT CONFIRMATION packet (Fig. 21). The INTERRUPT packet can contain 32 octets of user data.

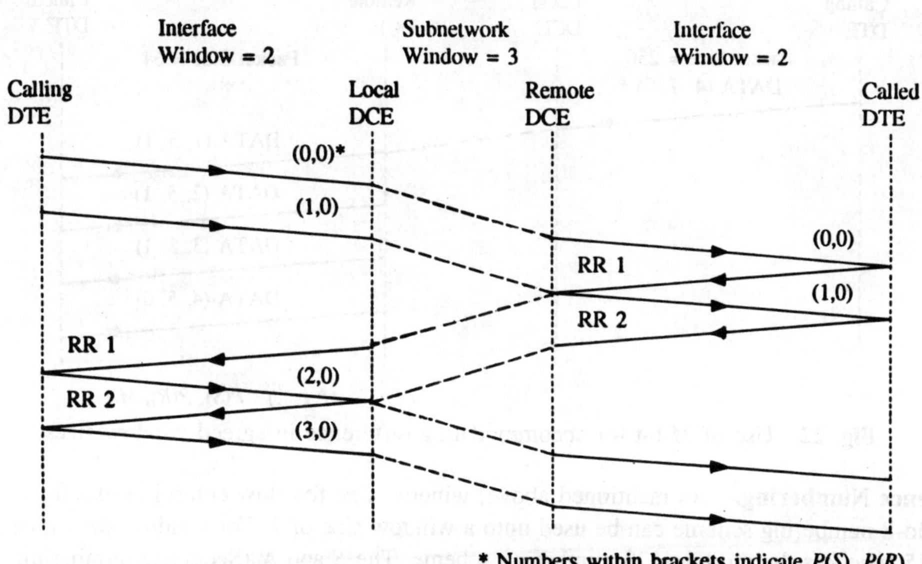

Fig. 20 Effect of window size on throughput in case of remote acknowledgement.

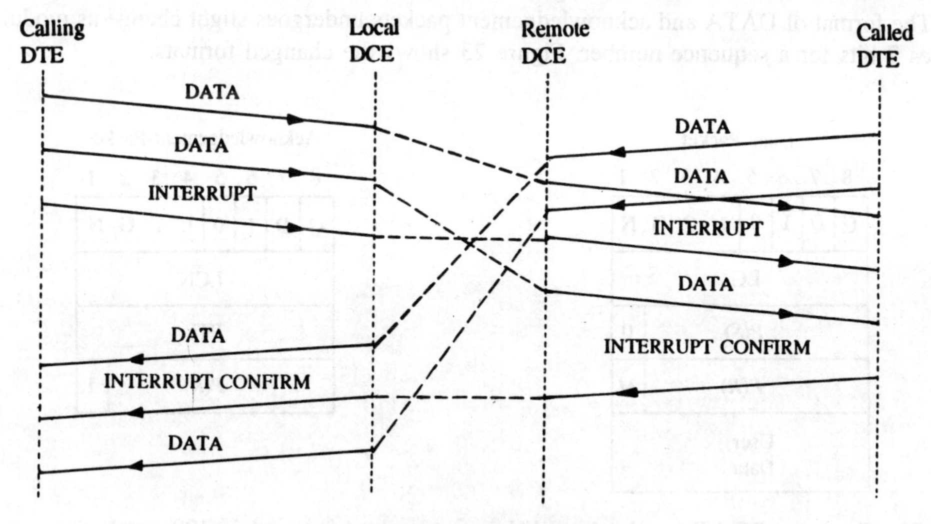

Fig. 21 INTERRUPT packet.

More Data Bit (*M*). The fifth bit of the third octet (Fig. 17) is called the *M* bit or more data bit. When it is "1", it indicates that there is a sequence of more than one packet. To illustrate its use, let us assume that the two interfaces at the originating and destination ends have different agreed sizes of the DATA packets (Fig. 22). The interface at the destination has a smaller packet size. Therefore, the remote DCE splits each received DATA packet into four packets and sets *M* bit to "1" in each of the first three packets indicating "more packets of the sequence to follow". In the last packet it resets the *M* bit to "0".

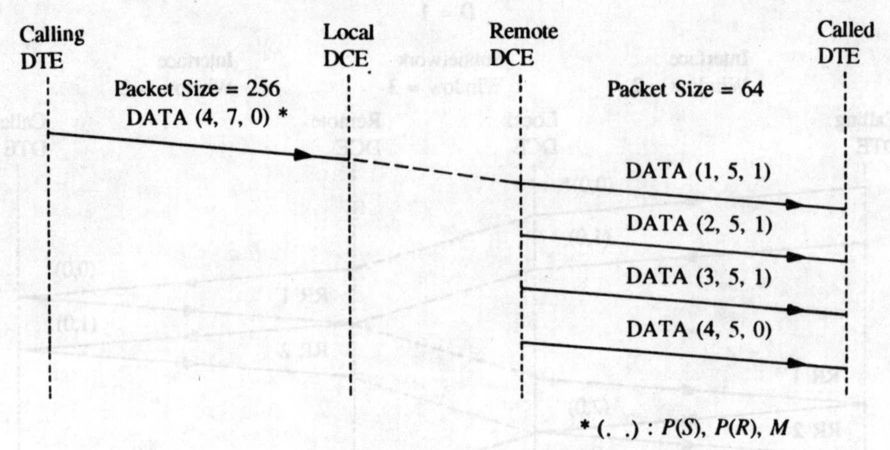

Fig. 22 Use of *M*-bit for accommodating difference in agreed window sizes.

Sequence Numbering. As mentioned above, window size for flow control is usually 2 and the modulo-8 numbering scheme can be used upto a window size of 7. For window sizes greater than 7, X.25 specifies the modulo-128 numbering scheme. The *S* and *N* (Sequence numbering) bits of GFI indicate the numbering scheme being followed (Fig. 8). For modulo-8 numbering, the *SN* bits are "01" and for modulo-128 numbering, the *SN* bits are "10".

The format of DATA and acknowledgement packets undergoes slight change as modulo-128 requires 7 bits for a sequence number. Figure 23 shows the changed formats.

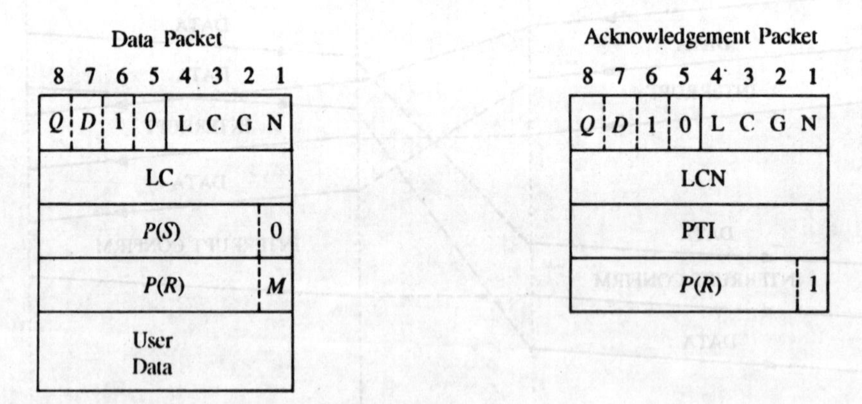

Fig. 23 Formats of DATA and acknowledgement packets for modulo-128 numbering scheme.

7.3 Call Clearing Phase

A virtual circuit can be cleared by either party at any time by sending CLEAR REQUEST packet on the particular logical channel. Figure 24a shows the sequence of packets exchanged during the call clearing phase when a DTE requests for call clearing. Under certain conditions, the subnetwork can also clear the call. Figure 24b shows call clearing by the subnetwork. The clearing process is destructive. Any undelivered DATA packets in the subnetwork are discarded.

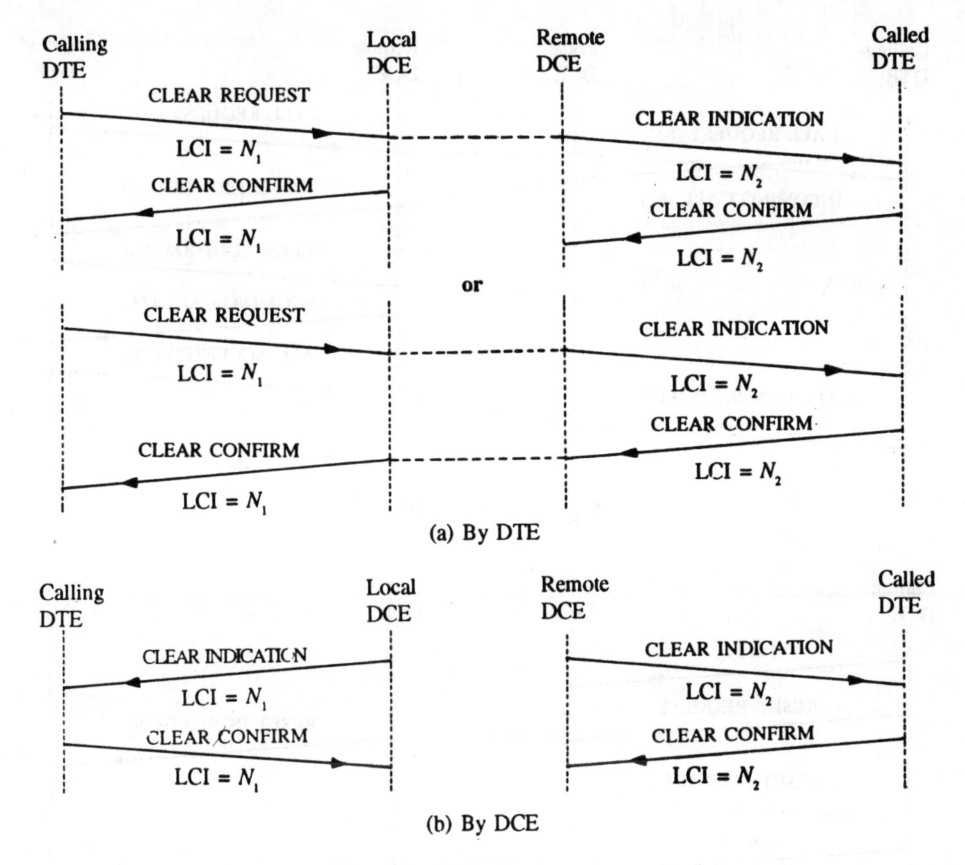

Fig. 24 Call clearing phase.

7.4 Call Collision

Call collision can occur if the same logical channel is simultaneously selected by both the DTE (for an outgoing call) and the DCE (for an incoming call) from the LI/OC-HI/OC range (Fig. 6). If call collision occurs, the outgoing call is given preference over the incoming call. The incoming call is cleared by the subnetwork (Fig. 25).

7.5 Procedures for Permanent Virtual Circuit (PVC)

A PVC is always in data transfer phase. Therefore, it does not have call establishment and call clearing phases. The procedures described for data transfer phase of the SVCs also apply to PVCs.

7.6 Virtual Circuit Reset

The reset procedure is used to re-initialize an SVC or a PVC. The counters $P(S)$ and $P(R)$ are reset. Either user is free to reset the logical channel at any time by transmitting a RESET REQUEST packet (Fig. 26). Reset is a destructive process and any data packets in the subnetwork awaiting delivery are discarded.

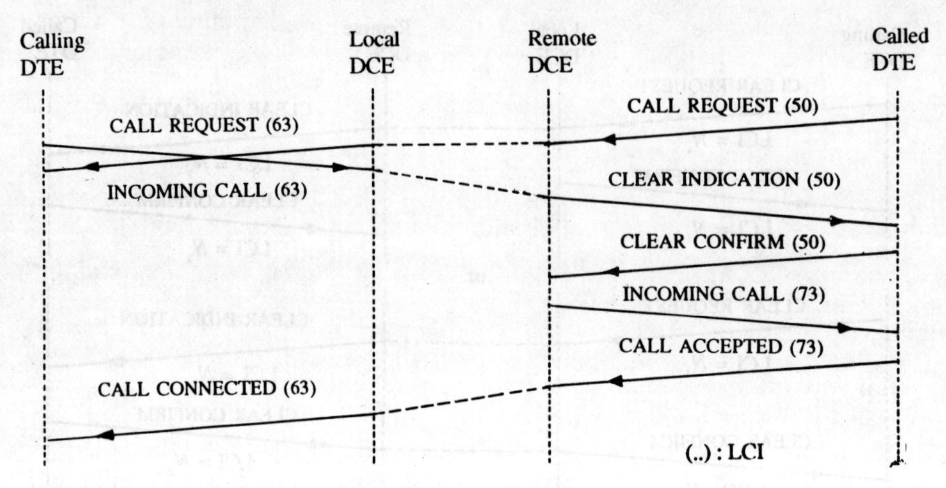

Fig. 25 Call collision.

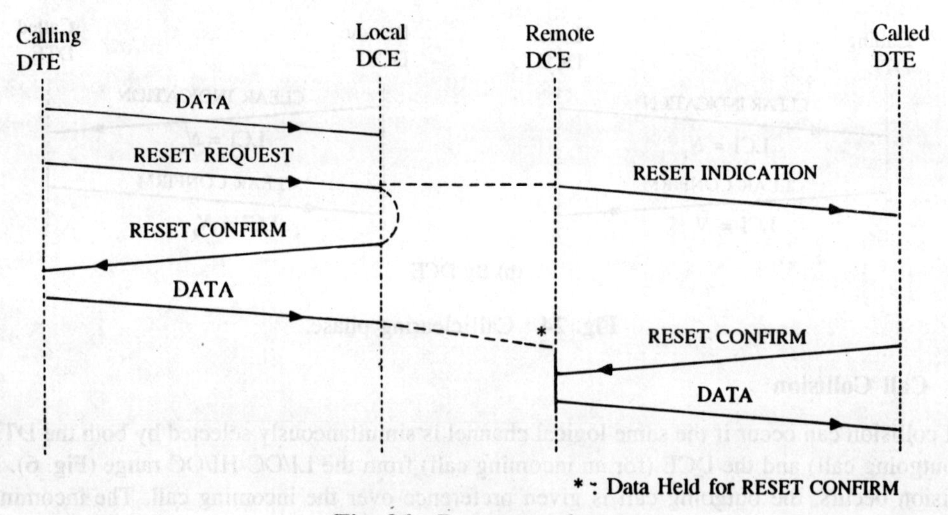

Fig. 26 Reset procedure.

The subnetwork can also reset a connection if it considers a user has infringed the protocol, e.g., transmitting of a wrongly formatted packet may result in reset. The subnetwork resets a connection by transmitting RESET INDICATION packet to both the DTEs simultaneously (Fig. 27).

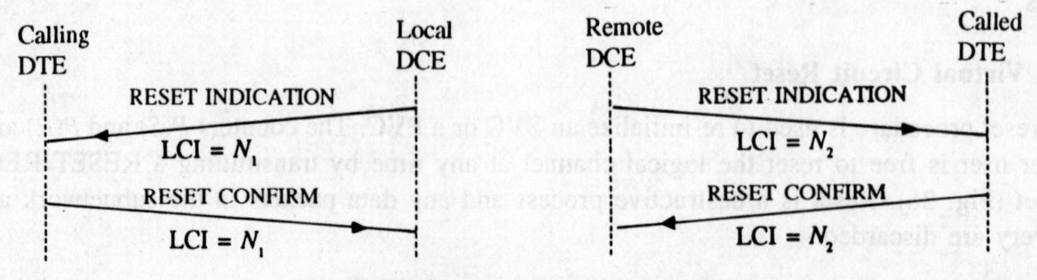

Fig. 27 Reset by the subnetwork.

The RESET REQUEST, RESET INDICATION and RESET CONFIRMATION packets consist of the usual 3 octet header plus 2 additional octets which indicate the reason of reset.

7.7 Restart

The most destructive action a DTE or DCE can take is to transmit a RESTART packet. This packet is always transmitted on logical channel zero which is reserved for this packet. When a RESTART packet is transmitted or received, all the switched virtual circuits operating between a DTE and a DCE are cleared and the permanent virtual circuits are reset (Fig. 28).

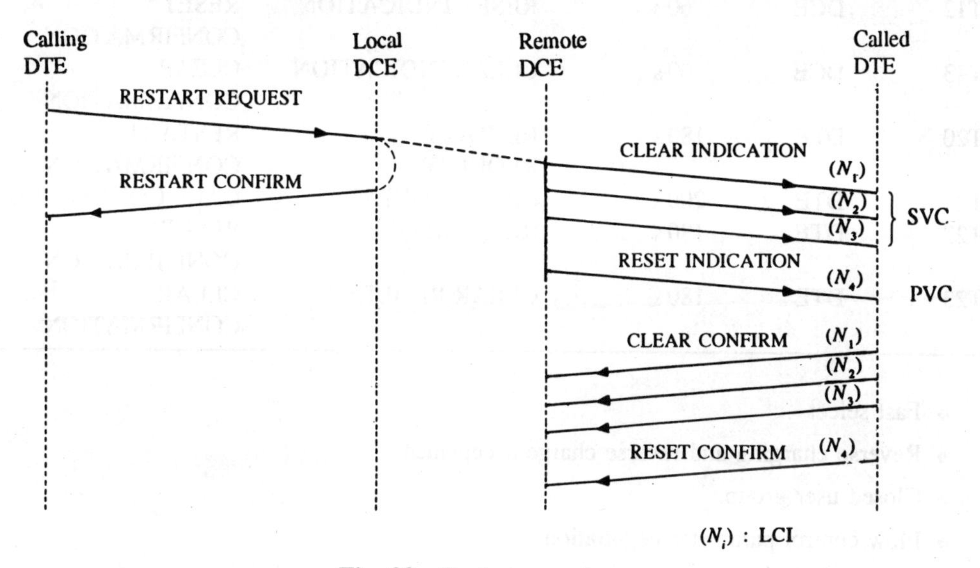

Fig. 28 Restart procedure.

A DTE transmits the RESTART packet when it is switched on. This ensures that all logical channels start in a known state. A DCE may send the RESTART packet in the unlikely event of severe traffic congestion problem in the subnetwork.

7.8 Error Recovery by Timers

A number of timers have been defined at level 3 of X.25 to permit recovery when there is no response. The timers can be grouped into two categories:

1. DCE time-outs
2. DTE time-outs.

DCE time-outs refer to a DCE when it issues a packet and waits for the response. Similarly, DTE time-outs refer to a DTE. Table 2 gives a list of the timers and typical values of the timer parameters. The actual values may be different as set by the subnetwork operator.

8 USER FACILITIES

There are a number of optional facilities to which a user can subscribe. Some of the important facilities are:

Table 2 Error Recovery Timers

Timer	DTE/ DCE	Timer value	Starting instant issue of	Terminating instant receipt of
T10	DCE	60 s	RESTART INDICATION	RESTART CONFIRMATION
T11	DCE	180 s	INCOMING CALL	CALL ACCEPTED or CLEAR REQUEST
T12	DCE	60 s	RESET INDICATION	RESET CONFIRMATION
T13	DCE	60 s	CLEAR INDICATION	CLEAR CONFIRMATION
T20	DTE	180 s	RESTART REQUEST	RESTART CONFIRMATION
T21	DTE	200 s	CALL REQUEST	CALL CONNECTED
T22	DTE	180 s	RESET REQUEST	RESET CONFIRMATION
T23	DTE	180 s	CLEAR REQUEST	CLEAR CONFIRMATION

- Fast select
- Reverse charging and reverse charge acceptance
- Closed user group
- Flow control parameter negotiation.

If a user wants to make use of a facility, he must indicate it at the time of subscription. Parameters of some of the facilities are indicated at the time of subscription while for some they are negotiated for each call during its establishment. The CALL REQUEST, INCOMING CALL, CALL ACCEPTED and CALL CONNECTED packets have a facility field (Figs. 13 and 14). In the facility field of these packets, the desired parameters of the facilities are indicated and accepted.

8.1 Fast Select

To send just one packet of data, a DTE must exchange at least four more packets to establish and later release the logical connection. Subscribers to the fast select facility can avoid this overhead. They can append upto 128 octets of data in the CALL REQUEST/INCOMING CALL packet after the facility field (Fig. 13).

Figure 29 shows the exchange of packets when fast select facility is used. Note that when the remote DCE sends the INCOMING CALL packet with the data field to the called DTE, it is obliged to respond with CLEAR REQUEST packet which can also have upto 128 octets of user data.

A variant of fast select facility allows the called user to respond with a CALL ACCEPTED packet in which case a normal exchange of DATA packets follows (Fig. 30). The call is cleared in normal fashion with the CLEAR REQUEST packet.

The type of fast select facility is selected in the CALL REQUEST packet by indicating the appropriate fast select facility code.

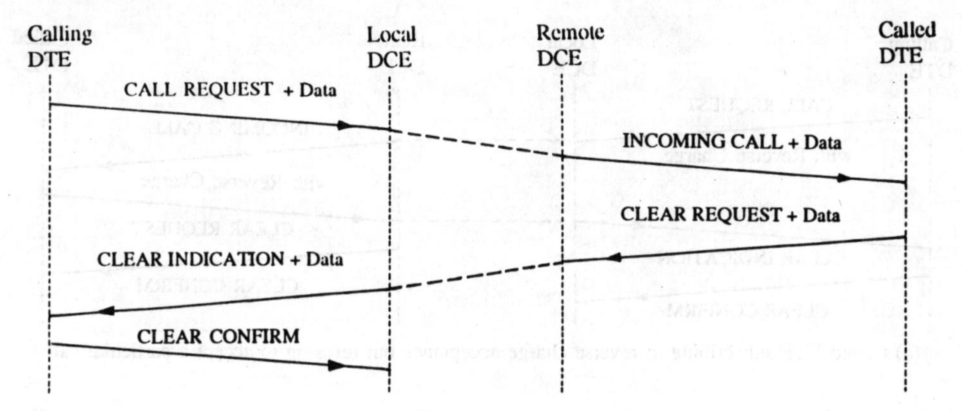

Fig. 29 Fast select procedure.

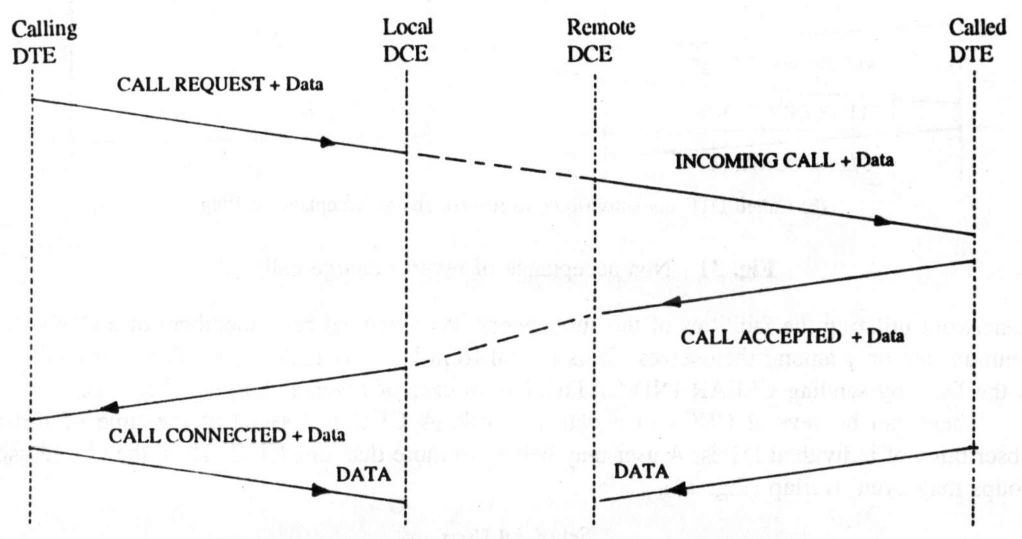

Fig. 30 Fast select procedure with call establishment.

8.2 Reverse Charging and Reverse Charge Acceptance

In reverse charging, the called user is billed for the call. By inserting the appropriate code into the facility field of the CALL REQUEST packet, a calling user who has subscribed to the facility may request for reverse charging. The remote DCE forwards the request to the called DTE in the INCOMING CALL packet if the called DTE has subscribed to reverse charge acceptance facility. If the called DTE is ready to accept the call charges, it returns CALL ACCEPTANCE packet else it returns CLEAR REQUEST packet which results in clearing the virtual circuit (Fig. 31a).

If the called DTE is not a subscriber to the reverse charge acceptance facility, the remote DCE itself clears the call (Fig. 31b).

8.3 Closed User Groups

A group of users may form themselves into a Closed User Group (CUG) creating an psuedo "private"

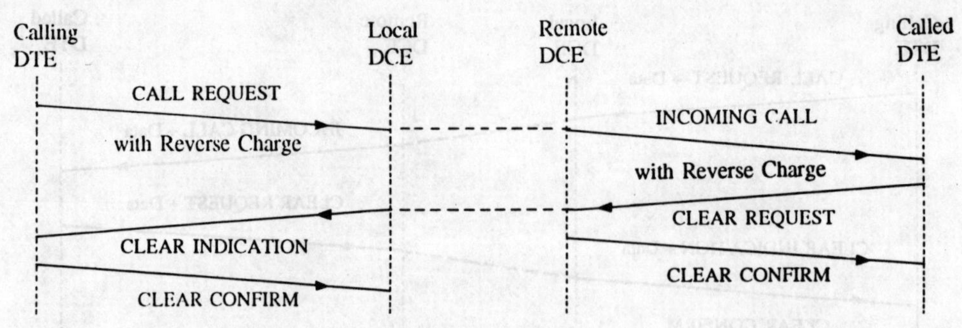

(a) Called DTE subscribing to reverse charge acceptance but refusing to accept a particular call

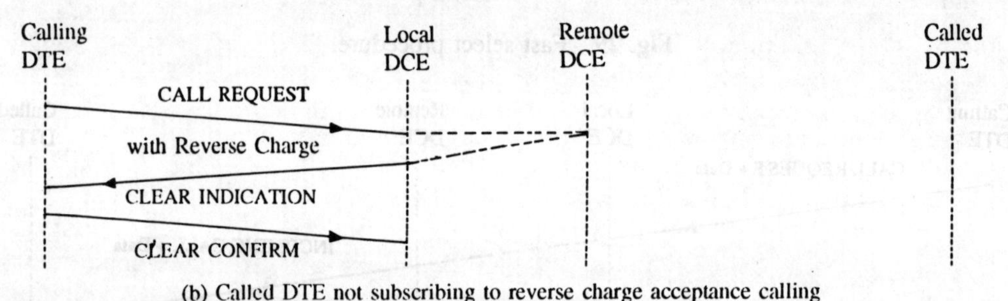

(b) Called DTE not subscribing to reverse charge acceptance calling

Fig. 31 Non-acceptance of reverse charge call.

subnetwork utilizing the facilities of the subnetwork. As a general rule, members of a CUG may communicate only among themselves. Calls to and from DTEs outside the CUG are turned back by the DCE by sending CLEAR INDICATION as in case of reverse charging (Fig. 31b).

There can be several CUGs in a data network. A CUG is formed at the time of initial subscription of individual DTEs. A user may belong to more than one CUG. Thus, the closed user groups may even overlap (Fig. 32).

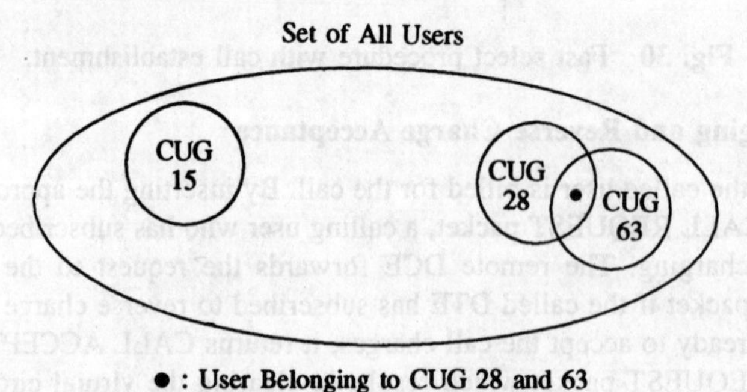

Fig. 32 Closed user groups.

There are three variants of the basic CUG facility (Fig. 33):

1. *DTE with outgoing access*: A DTE with this facility can also originate calls to the DTEs outside the CUG (DTE A).

2. *DTE with incoming access*: A DTE with this facility can also receive calls from the DTEs outside the CUG (DTE B).

3. *DTE with outgoing and incoming access*: A DTE with this facility can also originate and receive calls from the DTEs outside the CUG (DTE C).

Note that it is not necessary that all DTEs within a CUG should have same facilities.

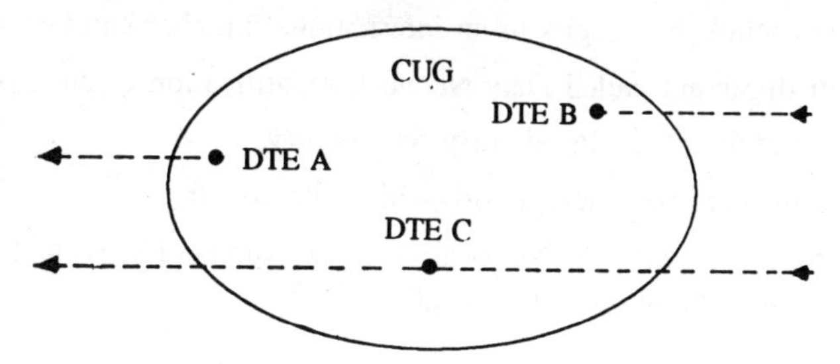

Fig. 33 Additional access facilities for closed user groups.

8.4 Flow Control Parameter Negotiation

The flow control parameters are: (a) Maximum size of the DATA packets, and (b) Window size of the logical channels. The size of data packet refers to the size of data field which can, at the most be 4096 octets long. These parameters can be negotiated between a DTE and a DCE on a per call basis if this facility has been subscribed to. These parameters need not be the same in both directions of transmission and at the ends of the virtual circuit.

Packet and window sizes are negotiated during the call set up phase. To negotiate the packet and window sizes, the calling DTE encodes the values of these parameters in the CALL REQUEST packet, and the local DCE indicates the accepted values in the CALL CONNECTED packet.

When a called DTE has subscribed to this facility, each INCOMING CALL packet indicates the proposed window and packet sizes. No relationship need exist between the parameters requested in the CALL REQUEST packet and those indicated in the INCOMING CALL packet. The called DTE accepts the parameter values in the CALL ACCEPTED packet.

The valid values of parameters in the responses during parameter negotiation are given in Table 3.

Table 3 Valid Ranges of Parameter Values

Requested value (CALL REQUEST and INCOMING CALL packets)	Accepted value (CALL CONNECTED and CALL ACCEPTED packets)
Window size	
$W_R \geq 2$	$2 \leq W_A \leq W_R$
$W_R = 1$	$W_A \leq 2$
Packet size	
$P_R \geq 128$	$128 \leq P_A \leq P_R$
$P_R \leq 128$	$128 \geq P_A \geq P_R$

9 ADDRESSING IN X.25

The address field is present in CALL REQUEST, INCOMING CALL, CALL ACCEPTED and CALL CONNECTED packets only. It contains addresses of the calling and called DTEs. In X.25, addressing is based on CCITT Recommendation X.121 which provides for international addressing of DTEs. X.121 numbering scheme has the following features:

- The maximum number of digits in an international number can be 14 or less.
- The first four digits are called Data Network Identification Code, DNIC (Fig. 34).
- The first three digits of DNIC identify the country.
- The fourth digit identifies the network within the country.
- Subsequent digits are called the National Terminal Number (NTN) and are used to identify the DTE. The maximum allowable length of NTN is 10.

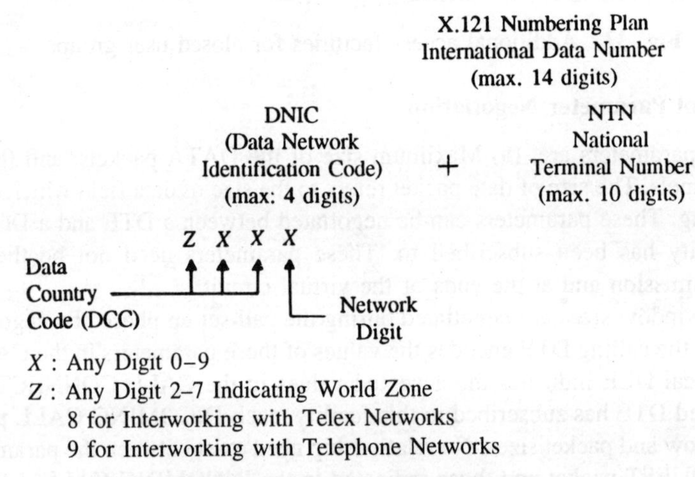

Fig. 34 X.121 numbering plan.

The lengths of the calling and the called addresses are specified in octet 4 of the call establishment packets. For example, if the calling and called address lengths are 9 and 12 digits respectively, the octet 4 will be 1001 1100. Each digit of the address is binary coded and each octet of the address field contains two digits. If the total number of digits is odd, the last octet is padded with zeros. Figure 35 shows an example of the address field used in a CALL REQUEST packet.

The calling DTE may indicate only the called DTE address in the CALL REQUEST packet. It is not essential for the calling DTE to identify itself to the local DCE. On receipt of the CALL REQUEST packet, the local DCE inserts the calling DTE address. Similarly, there is no need for a called DTE to be identified by its DCE, the calling DTE address is sufficient.

10 PACKET ASSEMBLER AND DISASSEMBLER (PAD)

X.25 interface requires the DTE to be a packet mode device, i.e., a device which transmits data and receives data in the form of packets, but in real life situations, we come across many devices which do not have this capability and also cannot be upgraded to have this capability. To extend services

International Data Number
DNIC NTN
4321 9 8 7 6 5 4 3 2 1 : Called DTE Address (13 Digits)

Address Fields

Octet 4	0 0 0 0 1 1 0 1	Called DTE Address Length (13 Digits)
Octet 5	0 1 0 0 0 0 1 1	Digits 4 3
Octet 6	0 0 1 0 0 0 0 1	Digits 2 1
Octet 7	1 0 0 1 1 0 0 0	Digits 9 8
Octet 8	0 1 1 1 0 1 1 0	Digits 7 6
Octet 9	0 1 0 1 0 1 0 0	Digits 5 4
Octet 10	0 0 1 1 0 0 1 0	Digits 3 2
Octet 11	0 0 0 1 0 0 0 0	Digit 1 *

* Octet 11 is padded with zeros to complete the address field

Fig. 35 Example of address field of a CALL REQUEST packet.

of a packet switched data subnetwork to such devices, an additional intermediary device called Packet Assembler and Disassembler, PAD in short, is required (Fig. 36).

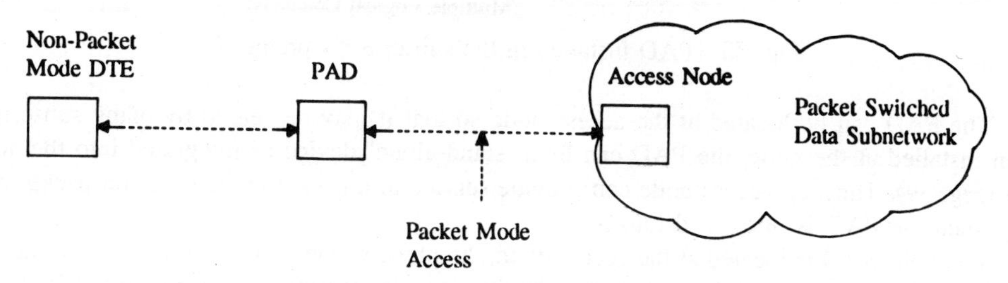

Fig. 36 PAD as an intermediary device.

The interface between the DTE and the PAD is the native-mode protocol of the DTE. The interface between the PAD and access node is the packet mode protocol, e.g., the X.25. PAD, thus, handles the protocol mapping function required between the non-packet mode terminal and packet mode access on the subnetwork side.

Packet mode access allows simultaneous operation of multiple logical connections on one physical connection, and PADs are therefore configured to handle multiple non-packet mode DTEs simultaneously (Fig. 37).

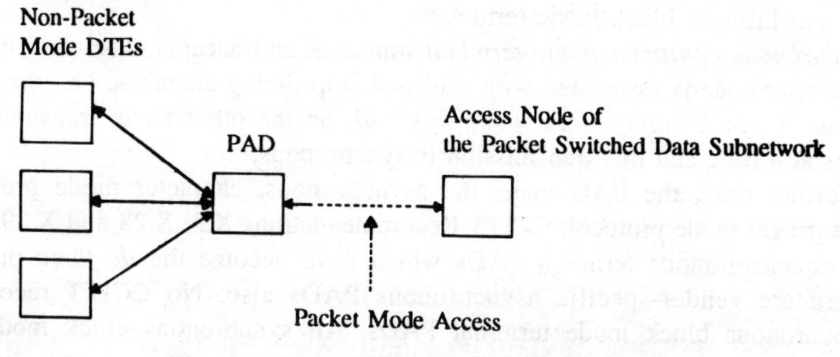

Fig. 37 Non-packet-mode ports of a PAD.

10.1 Location of a PAD

A PAD can be located either at the subscriber's premises or at the access node of the subnetwork. When a PAD is installed at the subscriber's premises (Fig. 38) it

- facilitates connecting multiple non-packet mode DTEs to the PAD; and
- reduces the number of dedicated physical circuits between the subscriber's premises and the access node to single circuit.

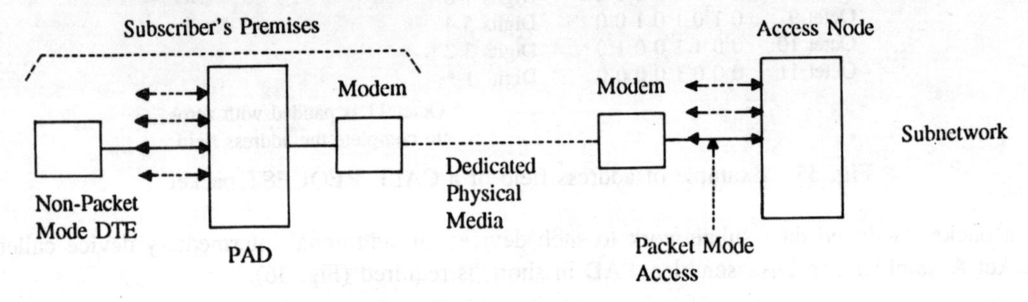

Fig. 38 PAD installed in the subscriber's premises.

The PAD can be located at the access node so that it may be shared by many subscribers. When installed at the node, the PAD can be a 'stand-alone' device or integrated into the access node (Fig. 39). Thus, an access node can provide service to a packet mode or a non-packet mode DTE when the PAD is integrated into it.

When the PAD is located at the access node, the physical medium from the DTE to the PAD is usually a dedicated circuit, but subscribers having occasional traffic can access the PAD located at the node using the Public Switched Telephone Network (PSTN). The DTE, however, cannot have an incoming call because a PAD cannot originate a PSTN call.

The destination terminal for whom the call is meant can be a packet mode terminal, DTE P, or a non-packet mode terminal. In the second case, the destination DTE is connected to the packet switched data subnetwork through a PAD. Thus, there can be PADs at both the ends.

10.2 Types of PADs

A PAD is identified by the DTE protocols it supports. There are asynchronous character mode terminals, and synchronous block mode terminals.

An *asynchronous character mode terminal* transmits and accepts character codes one at a time. Each character code is associated with start and stop timing elements, i.e., the transmission is asynchronous. A *synchronous block mode terminal*, on the other hand, transmits a block of character codes at a time and this transmission is synchronous.

In the former case, the PAD maps the asynchronous, character mode protocol of the terminal to the packet mode protocol. CCITT Recommendations X.3, X.28 and X.29 are for such asynchronous character mode terminal PADs which have become the *de facto* public domain standard. There are vendor-specific asynchronous PADs also. No CCITT recommendation exists for synchronous block mode terminal PADs. All synchronous block mode PADs are proprietary.

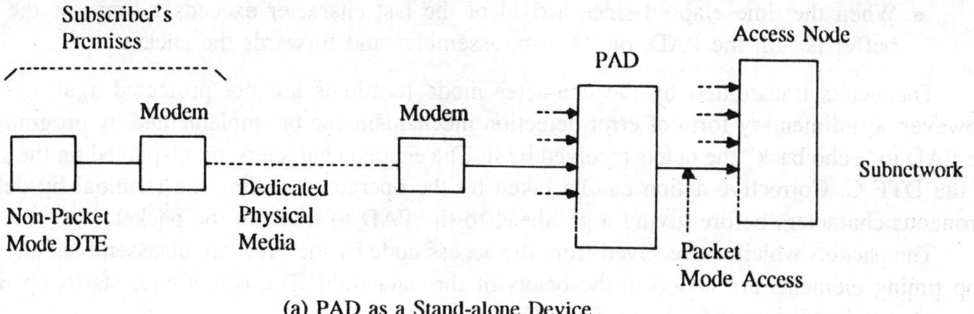

(a) PAD as a Stand-alone Device

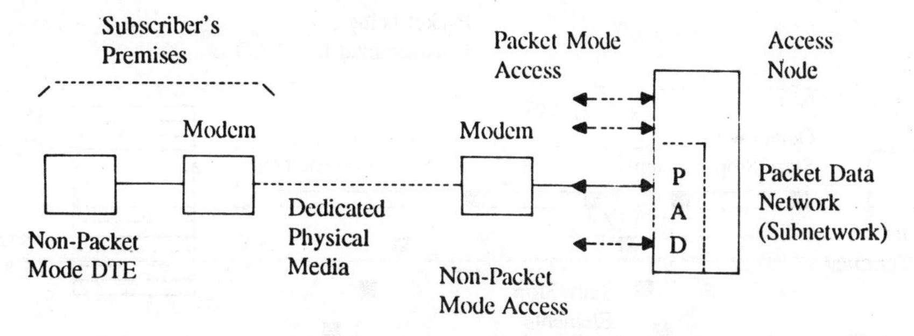

(b) PAD software built into access node

Fig. 39 PAD configuration at access node.

11 ASYNCHRONOUS CHARACTER MODE TERMINAL PAD

An asynchronous character mode terminal (DTE C) transmits and receives data in the form of octets with start/stop timing elements. The PAD assembles the octets received from the DTE C into a packet after removing the start/stop timing elements and adding the header (Fig. 40).

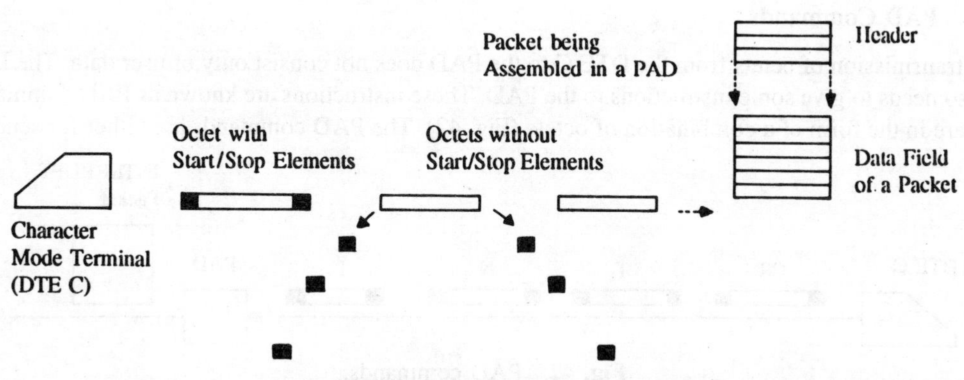

Fig. 40 Packet assembly in a PAD.

The incoming characters are stored in a buffer in the PAD until a decision is made to forward them as a packet.

- The DTE C may ask the PAD to transmit the packet by sending a special character.

- When the time elapsed since arrival of the last character exceeds a limit, or the PAD buffer is full, the PAD, on its own, assembles and forwards the packet.

The octets transmitted by the character mode terminals are not protected against errors. However, a rudimentary form of error detection mechanism can be implemented by programming the PAD to "echo back" the octets received by it. The echoed characters are displayed on the screen of the DTE C. Corrective action can be taken by the operator handling the terminal by deleting erroneous characters before giving a go ahead to the PAD to transmit the packet.

The packets which are received from the access node by the PAD are disassembled and start/stop timing elements are added to the octets of the data field. The octets with start/stop timing elements are handed over to the DTE C (Fig. 41).

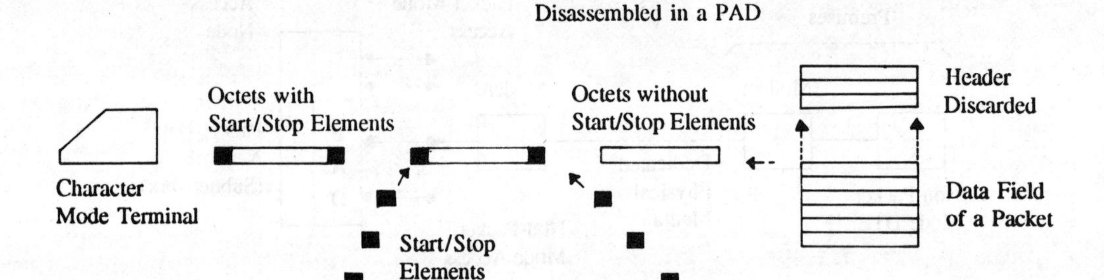

Fig. 41 Packet disassembly in a PAD.

The exceptions to this simple situation are that sometimes additional line feed, carriage return and other "idle" characters are inserted in the data characters stream by the PAD to enable proper formatting of the incoming message on the screen of the DTE C or on the printer attached to it. Obviously, the PAD should know the characteristics of the DTE C in advance. This is done by configuring the PAD. A PAD is configured by setting its parameters.

11.1 PAD Commands

The transmission of octets from the DTE C to the PAD does not consist only of user data. The DTE C also needs to give some instructions to the PAD. These instructions are known as PAD commands and are in the form of a combination of octets (Fig. 42). The PAD commands are either for sending

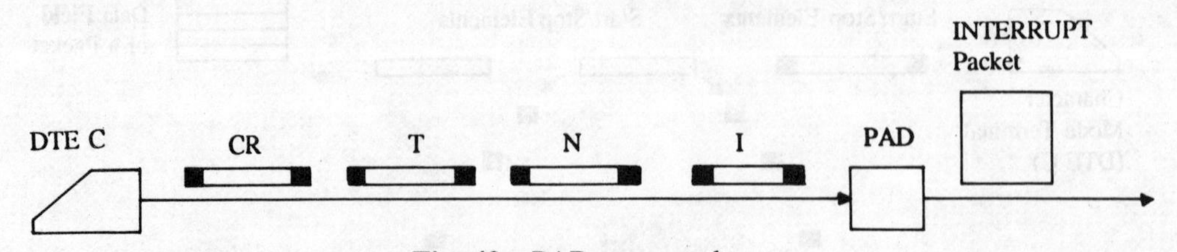

Fig. 42 PAD commands.

some control packets to the access node or for setting/reading the PAD parameters. Table 4 shows some of the important PAD commands and their typical formats and functions. The PAD commands are delimited using Carriage Return (CR).

Note that PAD commands are not further transmitted but the PAD actions them. The user data octets on the other hand, are packetized and transmitted to access node. Therefore, the PAD needs

Table 4 PAD Commands

PAD commands	Function
Read PAD parameters	
PAR?	Request to read back all the PAD parameters
PAR? 8, 9	Request to read back specific PAD parameters
Set and read PAD parameters	
SET?	Request to set and read back the PAD parameters to the initial profile
SET? 8 : 0, 9 : 3	Request to set the specified PAD parameters to the values indicated and read them back
Set PAD parameters	
SET	Request to set the PAD parameters to the initial profile
SET 8 : 0, 9 : 3	Request to set the specified PAD parameters to the values indicated
Set PAD profile	
PROF90	Request to set all the PAD parameters as per the indicated PAD profile
Selection	
N ⟨NUI⟩ R G ⟨CUG⟩ ⟨ADDRESS⟩ D ⟨DATA⟩	Request to establish a connection indicating – Caller's Network User Identifier (NUI) – Reverse charge request (optional) – CUG reference if any – Address of DTE P with sub address if any – Call user data (optional)
Clear request	
CLR	Request to clear the call
ICLR	Request to send invitation to clear message
Status request	
STAT	Request for status information
INTERRUPT request	
INT	Request to send the INTERRUPT packet
RESET request	
RESET	Request to reset the virtual circuit

to distinguish between the octets of PAD commands and of user data during the data transfer phase. To give a PAD command, the DTE C prefixes the 'Data Link Escape' (DLE) character before the command. When the PAD receives this character, it escapes from the 'data transfer' state to the 'waiting for command' state. The delimiter CR at the end of the command returns the PAD to the data transfer state. The PAD can be configured to accept other characters in place of DLE and CR by setting the appropriate PAD parameters.

11.2 PAD Service Signals

Other than disassembled user data octets, the PAD also transmits messages to the DTE C. These messages could be responses to PAD commands or could be indications of the control packets received from the access node, e.g., + COM indicates receipt of CALL CONNECTED packet by the PAD (Fig. 43). These messages to the DTE C are called PAD service signals.

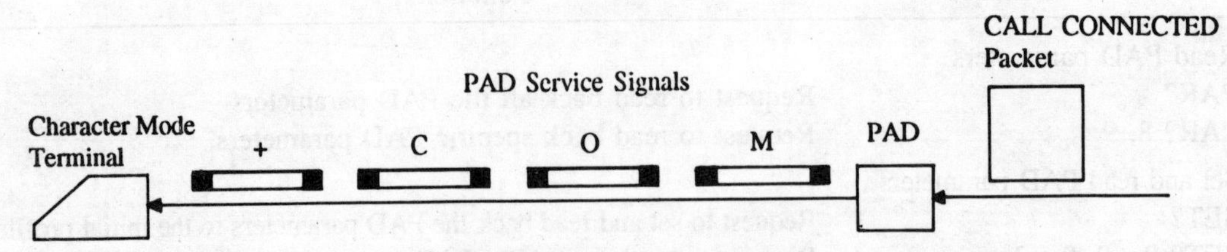

Fig. 43 PAD service signals.

Table 5 shows some of the important PAD service signals, their typical formats and functions.

Table 5 PAD Service Signals

PAD service	Function
Acknowledgement	
FE	PAD acknowledges receipt of the following command signals by sending FE: – Set PAD profile – Set PAD parameters – Selection – RESET request – INTERRUPT request
Parameter values	
PAR 1:1, 8:0	PAD indicates the requested parameter values
Reset	
RESET ⟨CAUSE⟩ ⟨DIAGNOSTIC⟩ ⟨TEXT⟩	PAD indicates that the virtual circuit has been reset by the remote DTE P or by the subnetwork due to local procedure error or congestion
Status	
FREE ⟨TEXT⟩ ENGAGED ⟨TEXT⟩	As response to PAD status command, PAD indicates its free (no call stablished) or engaged (call established) status
CALL CONNECTED and INCOMING CALL	
⟨ADDRESS⟩ + COM FE	PAD Indicates Receipt of CALL CONNECTED packet or of INCOMING CALL packet
Clear confirmation	
CLR CONF ⟨TEXT⟩	PAD confirms clearing of the call
Clear indication	
CLR ⟨CAUSE⟩ ⟨CAUSE CODE⟩ ⟨DIAGNOSTIC⟩ ⟨TEXT⟩	PAD indicates that the call has been cleared due to reasons as suffixed
Prompt	
*	PAD indicates its readiness to accept a PAD command

PAD service signals are delimited using a Format Effector (FE) which is Carriage Return (CR) followed by Line Feed (LF).

Thus, besides exchanging user data octets, the PAD and DTE C exchange PAD commands and PAD service signals. This exchange is governed by the protocol defined by CCITT Recommendation X.28.

11.3 PAD Messages

The remote DTE P also communicates with the local PAD through the subnetwork. This communication takes place on the virtual circuit between the PAD and DTE P using DATA packets. Thus, two types of data are transferred between the PAD and DTE P:

- User data which is data passing transparently between DTE C and DTE P through the PAD.
- PAD messages which are exchanged between PAD and DTE P and are not passed to DTE C. Protocol for this exchange is defined by CCITT Recommendation X.29.

Both the types of data are merely groups of contiguous octets carried in the data field of DATA packets. The packets containing the above two types of data are distinguished by means of the '*Q*' bit—bit 8 of the octet 1 of the DATA packet (Fig. 44).

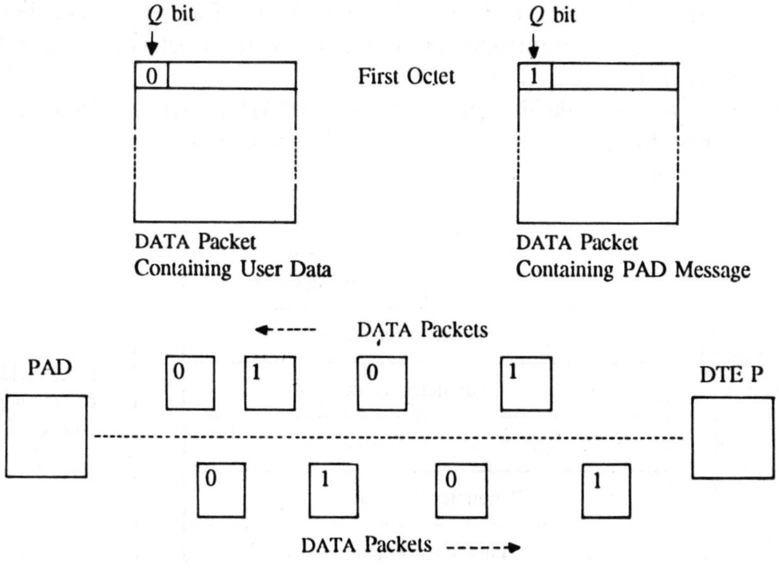

Fig. 44 Qualifier (*Q*) bit.

PAD messages from DTE P are for setting and reading the PAD parameters and for inviting the PAD to clear the call after it has transmitted the user data octets to the DTE C. Messages from the PAD are the responses to the messages from DTE P. Table 6 gives some of the PAD messages and their functions.

Set, read, set and read and parameter indication messages are used by DTE P to set and/or read the PAD parameters. These messages are followed by one or more parameter fields. Each parameter field consists of a parameter reference octet and a parameter value octet. The parameter

Table 6 PAD Messages

PAD message and message code	Function
Set 0010	Request from DTE P for setting values of the specified parameters
Read 0100	Request from DTE P for reading values of the specified parameters
Set and read 0110	Request from DTE P for setting and reading values of the specified parameters
Parameter indication 0000	Response from PAD indicating values of the specified parameters
Invitation to clear 0001	Request for issuing clear request packet after the PAD has transmitted the pending octets to the DTE C

value octet is coded as all binary "0"s in a read message. Parameter indication message is the response to a read message. It contains the parameter reference and its value.

The invitation to clear message from DTE P to the PAD is used to request the PAD to clear the virtual call, after completing transmission of all pending data octets to the DTE C. The virtual circuit can also be cleared by the DTE P by sending a CLEAR REQUEST packet but in this case, the PAD immediately clears the virtual circuit discarding whatever octets may have been pending transmission to the DTE C.

Figure 45 shows format of the data field containing a PAD message. A PAD message consists of a message code possibly followed by a parameter field where the parameter reference and the parameter value are indicated.

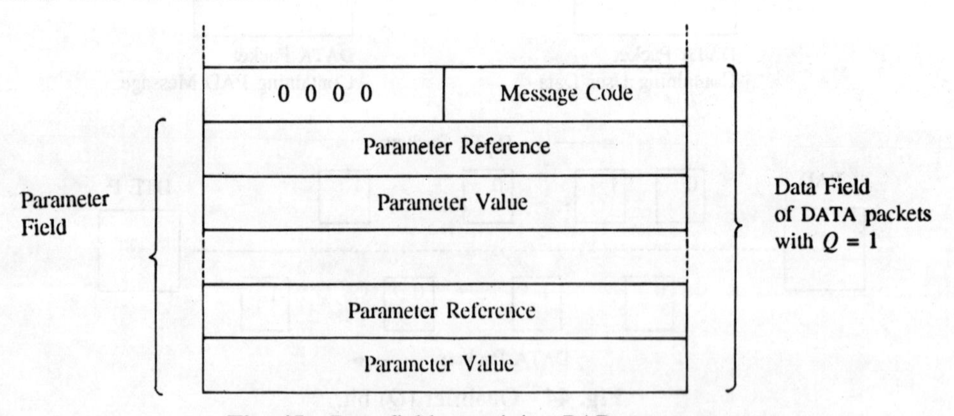

Fig. 45 Data field containing PAD messages.

11.4 CCITT PAD Parameters

The PAD is required to interface with a variety of character mode asynchronous devices—dumb terminal, printer, paper tape reader or a PC. In order to be able to serve these devices the PAD has many built-in functions which can be tailored for specific requirements of a DTE C. Each function is associated with a PAD parameter which can be set. These functions and the corresponding PAD parameters are detailed in CCITT Recommendation X.3 and are summarized below. Some typical values of the parameters and corresponding functional attributes of the PAD are indicated. The reader may go through the X.3 document for all the possible values of PAD parameters.

Parameter 1: PAD recall. This function allows a DTE C to initiate escape from the data transfer state in order to send a PAD command.

Typical values :	0	Escape not allowed
	1	Escape allowed

Parameter 2: Echo. The PAD transmits back the characters received from the DTE C for display on the DTE C screen.

Possible values :	0	No Echo
	1	Echo

Parameter 3: Selection of data forwarding characters. A PAD assembles a packet and transmits it after it receives a data forwarding character from the DTE C. This function allows for defining the data forwarding character.

Typical values :	0	No data forwarding signal
	2	Character CR (Carriage Return)

Parameter 4: Selection of idle timer delay. When no character is received from the DTE C for a certain period of time, the PAD assembles the packet and transmits it. This function allows for selection of this idle time delay.

Possible values :	0–255	Delay in units of 0.3 s.

Parameter 5: Flow control, DTE C. A PAD indicates its willingness/unwillingness to receive characters from the DTE C by transmitting special characters (X-ON, X-OFF).

Possible values:	0	Not operational
	1	Operational using X-ON and X-OFF

Parameter 6: Control of PAD service signals. This function provides the DTE C ability to decide whether or not, and in what format, the PAD service signals are transmitted to itself.

Typical values :	0	No service signal
	1	Transmit service signal

Parameter 7: PAD operation on break signal. When PAD receives a break signal from the DTE C, its operation is selected by this function. On receipt of the break signal, a PAD may send RESET or INTERRUPT packets or escape from the data transfer state.

Typical values:	0	No action
	1	Transmit INTERRUPT packet
	2	Reset
	8	Escape from data transfer state
	16	Discard output to DTE C

Parameter 8: Discard output. This function provides for a PAD to discard the content of user sequences in a packet upon request rather than disassembling and transmitting them to the DTE C. This parameter works in conjunction with parameter 7 when it is set to 16.

Possible values :	0	Normal data delivery
	1	Discard output to DTE C

Parameter 9: Padding after CR. A PAD can automatically insert padding characters in the stream of characters being transmitted to the DTE C after occurrence of a Carriage Return (CR) character for the printing mechanism of the DTE C to function properly.

Possible values :	0	No padding
	1–7	Padding characters after CR

Parameter 10: Line folding. This function of the PAD allows for automatic insertion of line folding effectors after a predetermined maximum number of graphic characters.

Possible values:	0	No line folding
	1–255	Number of characters per line

Parameter 11: DTE C speed. This parameter is read only and is used to indicate the DTE C access speed.

Typical values:	2	300 bps
	3	1200 bps
	12	2400 bps

Parameter 12: Flow control of PAD. This function of the PAD allows the DTE C to flow control the character stream from the PAD by transmitting special characters.

Possible values :	0	No flow control
	1	Flow control using X-ON and X-OFF

Parameter 13: Line feed after CR. In the data transfer phase, the PAD inserts line feed automatically after the CR signal in the character stream being transmitted to the DTE C or being received from it.

Typical values:	0	No line feed insertion
	1	Insert line feed after CR in data stream sent to the DTE C
	2	Insert line feed after CR in the data stream received from the DTE C
	4	Insert line feed after CR echoed back to the DTE C

Parameter 14: Padding after LF. This function provides for insertion of padding characters after the line feed character in the character stream transmitted to the DTE C to enable the printer to perform correctly. This function applies only in the data transfer phase.

Possible values:	0	Option not selected
	1–7	Number of padding characters

Parameter 15: Editing. This function provides for character delete, line delete and line display editing capabilities in the PAD. Parameter 15 activates and deactivates the editing function.

Possible values:	0	No editing in data transfer
	1	Editing in data transfer

Parameter 16: Character delete. This function allows specification of the character to be used for deleting a character.

Possible values:	0–127	A character from IA5 to be used as "character delete" character

Parameter 17: Line delete. This function allows specification of the character to be used for deleting a line.

Possible values:	0–127	A character from IA5 to be used as "line delete" character

Parameter 18: Line display. This function allows specification of the character to be used for "line display".

Possible values:	0–127	A character from IA5 to be used as "line display" character

Parameter 19: Terminal type for editing PAD service signals. When the editing function is used, the action of the PAD reflected in the PAD service signals should depend on the type of terminal. For example, a PAD cannot delete a character on the printer output. It can at the most put a cross. On a display terminal the PAD can actually delete the character.

Typical values:	0	No editing signals
	1	Printing terminal
	2	Display terminal

Parameter 20: Echo mask. When echo is enabled, this function provides for masking a defined set of the characters not to be echoed back.

Typical values:	0	All characters echoed
	1	CR is not echoed
	2	LF is not echoed
	64	Editing characters are not echoed

Parameter 21: Parity treatment. This functions provides for checking of the parity bit in the incoming stream of characters from the DTE C and for generating the parity bits in the characters transmitted to the DTE C.

Possible values:	0	No parity bit checking or generation
	1	Check parity bit
	2	Generate parity bit

Parameter 22: Page wait. To enable page change, the PAD suspends transmission of additional characters to the DTE C after a specified number of line feed characters have been transmitted by the PAD.

Possible values:	0	No page wait
	n	Wait after "*n*" line feeds

As mentioned before, these functions are tailored to the specific requirements of the DTE C by setting associated PAD parameters. As may be noticed, the choices for selection of the parameter values are many and it becomes very difficult for a user to select an appropriate set of parameter values. To help the user in selecting the parameter values, a choice of standard combinations of the parameter values called PAD parameter profiles are offered. The user can select a standard profile which is best suited to his application.

12 LAYERED MODELS

To facilitate better understanding of the PAD protocols, let us look at the layered model of communication through a PAD.

12.1 Layered Model of a DTE C

A DTE C consists of

- a transducer which generates a character code when a key is depressed by the operator,
- a Data Link layer which gives structure to the bits by adding start and stop timing elements to each code, and
- a Physical layer which converts the bits into an electrical signal of suitable polarity and voltage.

See Fig. 46.

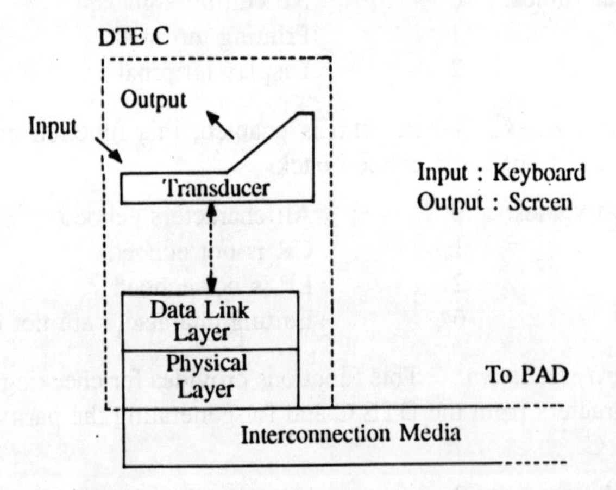

Fig. 46 Layered architecture of DTE C.

These functions are reversed for the signals received on the physical media. The received characters are displayed on the screen for the operator.

12.2 Layered Model of a Stand-Alone PAD

In a stand-alone PAD installed either at the user's premises or at the access node, the interface towards the DTE C comprises Physical and Data Link layers which match with similar layers of the DTE C. The Physical layer transmits and receives the asynchronous octets. The Data Link layer appends start/stop elements to the octets which are to be transmitted to the DTE C. It also removes the start/stop elements from the octets which are received from the Physical layer below.

The interface towards the access node is X.25 with its three levels so that the PAD appears as a packet mode terminal (DTE P) to the access node.

Besides these layers, there is a Terminal Handler (TH) which bridges the two interfaces. It

- receives services of Data Link layer of the PAD interface towards the DTE C, and Network layer of the interface towards the access node;
- accepts commands and user data from the DTE C;
- sends service signals and user data to the DTE C; and
- interacts with the terminal handler of the remote DTE P.

Figure 47 shows a composite layered model of a DTE C, a stand-alone PAD and packet mode access node of the subnetwork. The protocol between the terminal handler and the DTE C is

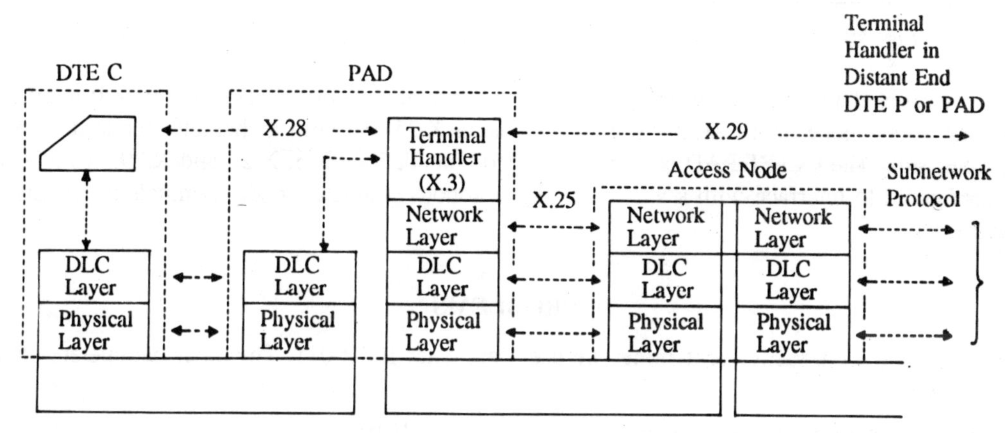

Fig. 47 Layered architecture of DTE C, PAD and X.25 access node.

defined by CCITT Recommendation X.28. The protocol between the terminal handler and the remote DTE P is defined by CCITT Recommendation X.29. The parameters of the terminal handler and its facilities are defined in CCITT Recommendation X.3.

12.3 Layered Model of an Integrated PAD

Note that a stand-alone PAD has the X.25 interface towards the access node. If the PAD is integrated into the access node, there is no need to have the X.25 interface. The terminal handler can directly interact with the proprietary network layer of the access node below it (Fig. 48).

13 SUMMARY

X.25 interface defines the protocol for virtual circuit connection between a packet mode DTE and the DCE which is the access node of the subnetwork. It defines protocols at the physical, data link and packet levels. The packet level protocol corresponds to the Subnetwork Access Control sublayer of the OSI Network layer. The packet level of X.25 provides for sequenced delivery of data packets, flow control of packets using sliding window mechanism, expedited data transfer using interrupt packets and local and remote acknowledgements. It also provides a reasonably wide spectrum of subscribed services.

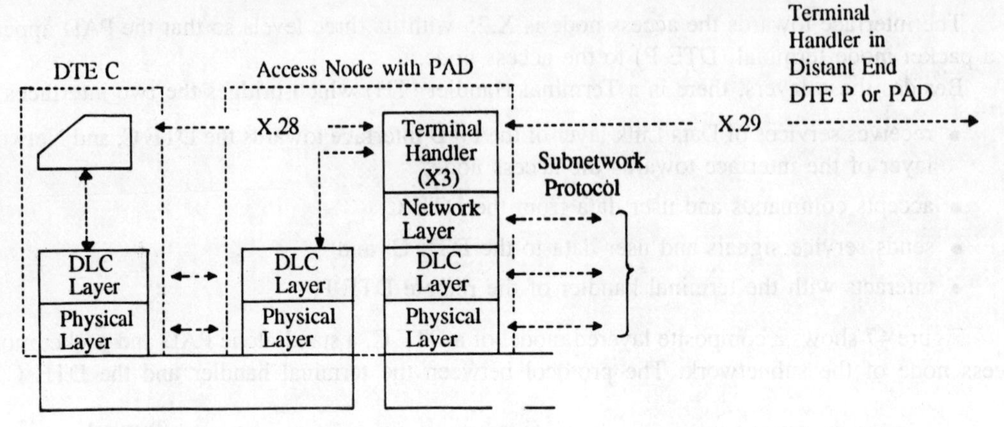

Fig. 48 Layered architecture of DTE C and access node with in-built PAD.

To connect a non-packet mode terminal, an intermediary device PAD is required. It interfaces with the access node of the packet subnetwork using X.25 and with a character mode device on the other side. The CCITT PAD is defined by recommendations X.3, X.28 and X.29. A PAD can be configured to interface with a variety of asynchronous character mode terminals by setting its parameters.

PROBLEMS

1. Draw a DATA packet containing one user data octet 10010011. The other parameters of the packet are:

Logical channel group number	1010
Logical channel number	10111001
Acknowledgement option	Local
Sequence numbering scheme	Modulo 8
Sequence number of the packet	101
Sequence number of the acknowledgement	001
More data bit	0

2. The DATA packet of Problem 1 is transmitted from the DTE P to the access node at 2400 bps. What is the transmission time? Assume single octet control field at level 2.

3. A DTE P having X.25 interface, is operating three virtual connections simultaneously. It sends one data packet on each connection in sequence.

(a) Will the Data Link layer distinguish between the packets belonging to different connections?

(b) If the first frame is given sequence number 0, what will be the sequence numbers of the frames containing DATA packets of other connections?

4. Draw the address fields of a CALL REQUEST packet if the calling Address is 23456789234 and the called Address is 3456678901234.

5. Fill in the packet sequence number, $P(S)$, and acknowledgement number $P(R)$ in the following exchange of DATA packets. Assume $D = 0$.

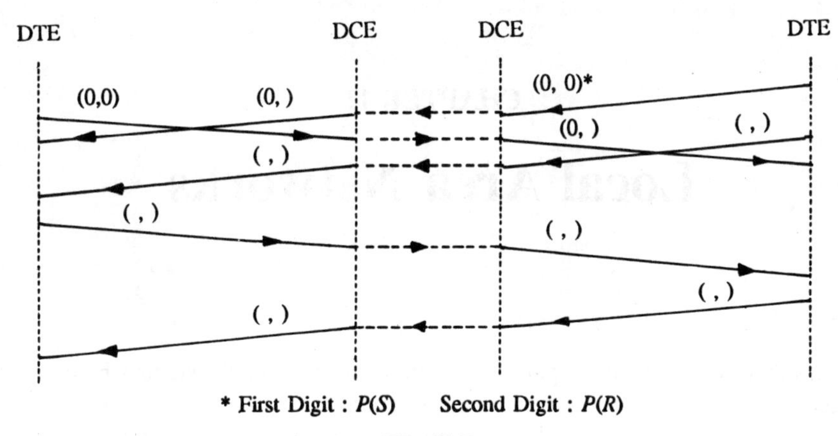

* First Digit : $P(S)$ Second Digit : $P(R)$

Fig. P.5

6. Do Problem 5, with $D = 1$.

7. A DTE needs to send a 1500 bytes long message generated by the user. The maximum size of the data field in the X.25 packet is restricted to 128 bytes. The distant end has a negotiated data field size of 256 bytes with its local DCE. If $D = 0$ and the window size is 3, show the transfer of data and acknowledgement packets assuming that acknowledgements are given after receipt of every third packet. The first packet bears the sequence number "0".

8. Show the data field of a packet containing a PAD message from a DTE P for setting PAD parameter 9 to value 1. How will this packet be distinguished from other DATA packets containing data bytes for DTE C?

CHAPTER 11

Local Area Networks

Switched data subnetworks serve the purpose of interconnecting end systems which may be spread over large geographic area covering even an entire country. They are, therefore, called wide area networks (WANs). There is another class of networks called local area networks (LANs) which interconnect end systems confined to an office premises. Local area networks provide a high speed and high throughput solution for networking such systems.

We begin this chapter with the examination of local networking requirements of an organization and the basic LAN attributes. Bus, ring and star LAN topologies are described next. Then we proceed to examine the layered architecture of the local area networks. Media access control and addressing issues are discussed with the help of layered architecture. After a brief look at IEEE standards for the local area networks, we discuss the services and protocols of the LAN Data Link sublayers. We move next to the transmission media for local area networks and discuss their characteristics and capabilities.

1 NEED FOR LOCAL AREA NETWORKS

Local area network, LAN in short, is a generic term for a network facility spread over a relatively small geographical radius. The LAN concept began with the development of distributed processing in the seventies. With the proliferation of microcomputers to the work sites, the need was felt to interconnect the computers so that data, software and hardware resources within the premises of an organization could be shared.

1.1 LAN Attributes

A LAN consists of a number of computers, graphic stations and user terminal stations interconnected through a cabling system. It has the following characteristic attributes:

- Geographic coverage of local area networks is limited to area less than 5 km.
- The data rates exceed 1 Mbps.
- The physical interconnecting medium is privately owned.
- The physical interconnecting medium is usually shared by the stations connected to the LAN.

1.2 LAN Environment in an Organization

Figure 1 shows the computer networking configuration within an organization. Usually, the various departments of the organization establish their own LANs which interconnect the departmental

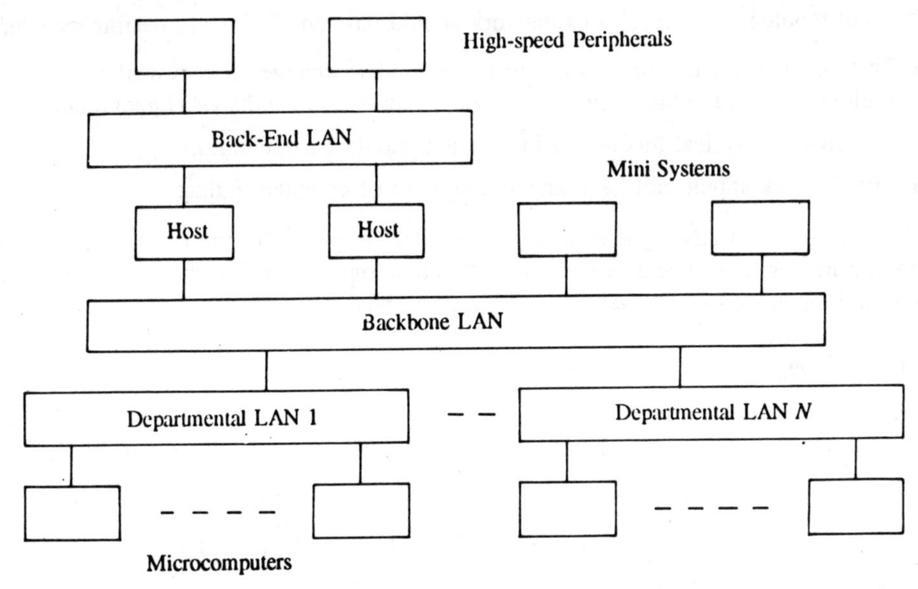

Fig. 1 Computer networking environment within an organization.

microcomputers. Each of these LANs may be self sufficient in its resources and management. However, there is always a need to exchange interdepartmental information and to access one or more mainframe or minisystems at the organizational level. A high-speed backbone LAN serves the purpose of connecting various deparmental LANs, mainframe and minisystems. The mainframe systems may have their own dedicated back-end LAN for sharing data and high-speed peripherals.

Multiple LAN environment in an organization has several advantages, namely,

1. Better data security can be achieved by restricting the interdepartmental LAN access and access to the host.

2. The traffic is dispersed over several LANs which individually meet the intradepartmental communication requirements.

3. Partitioning a big network into several smaller LANs also overcomes the distance and data rate limitations of local area networks.

At the same time, however, interworking of various LANs becomes a major issue. Bridges, routers, and gateways are required to interconnect the LANs. We will address this issue in the chapter on Internetworking.

2 LAN TOPOLOGIES

The physical topology of a local area network refers to the way in which the stations are physically interconnected. In the past, LANs were categorized on the basis of physical topology because it also determined the way in which the LANs operated. But today we have LANs having the same topology but operating in different ways. Therefore, LAN categorization based on physical topology is incomplete. Nevertheless, the attributes of various LAN topologies need to be examined to understand LAN operation.

Physical topology of a local area network should have the following desirable features:

- The topology should be flexible to accommodate changes in physical locations of the stations; increase in the number of stations; and increase in the LAN geographic coverage.
- The cost of physical media and installation should be minimum.
- The network should not have any single point of complete failure.

There are several LAN topologies prevalent in the industry. Bus topology, ring topology and star topology are common. There can be some other topologies as well such as distributed star, tree etc. These are extensions of the basic topologies, i.e., bus, ring and star.

2.1 Bus Topology

In bus topology, a single transmission medium interconnects all the stations which share this medium for transmission of their signals (Fig. 2). The bus operates in broadcast mode, i.e., every

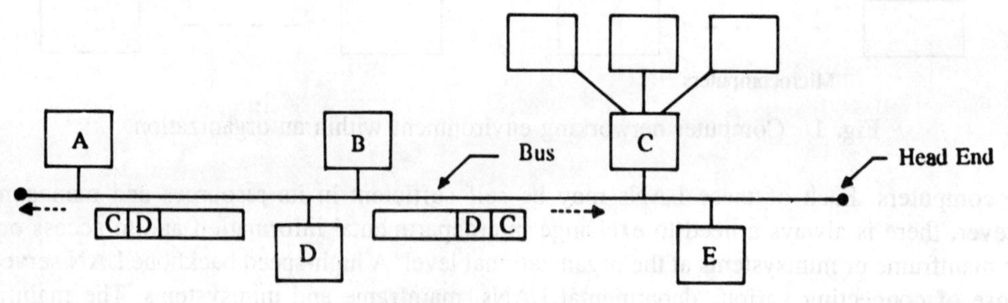

Fig. 2 Bus topology.

station listens to all the transmissions on the bus. Every transmission has source and destination addresses so that stations can pick the data units meant for them and identify their senders.

The bus shown in Fig. 2 is a two way bus, i.e., the signals flow in both directions. The ends of the bus are terminated by an appropriate matching impedance called the head end to avoid signal reflection. As the signal flow is bidirectional, amplifiers cannot be used on the bus to compensate for bus attenuation. If the geographic coverage needs to be expanded, repeaters which interconnect two buses are required (Fig. 3). A repeater is transparent to the rest of the system in the sense that it does not have a buffer and interconnects the two sections to make them into virtually one section.

If signals are amplified along the bus, the bus becomes unidirectional. In this case, two

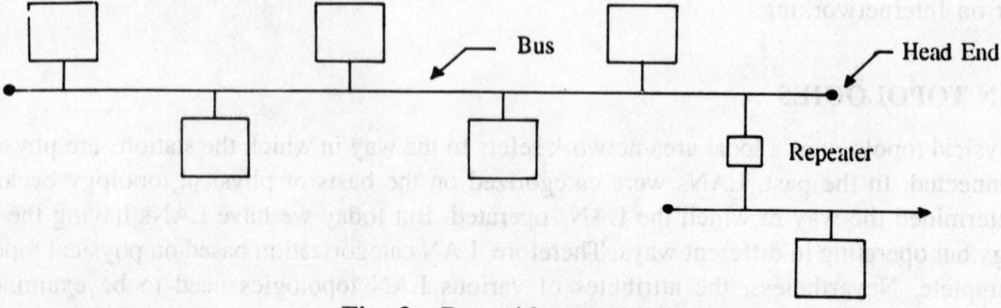

Fig. 3 Bus with a repeater.

separate buses are required—a transmit bus and a receive bus (Fig. 4). Every station injects signals on the transmit bus and listens on the receive bus. The buses are looped at one of the two ends.

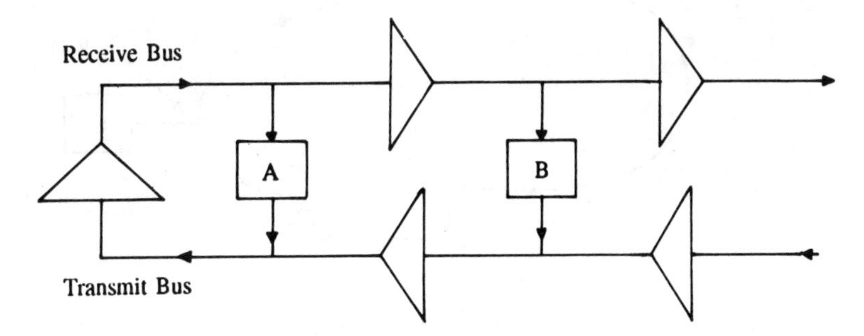

Fig. 4 Dual bus topology with amplifiers.

The transmit and receive buses can also be provided on a single bus by frequency division multiplexing of transmission channels (Fig. 5). The stations send their digital signals by modulating the transmit carrier and receive their digital signals by demodulating the receive carrier. The head end consists of a frequency translator which changes the transmit carrier frequency to the receive carrier frequency. A remodulator is also used in place of a simple frequency translator. The remodulator first demodulates the transmit carrier to get the digital signals and then modulates the receive carrier.

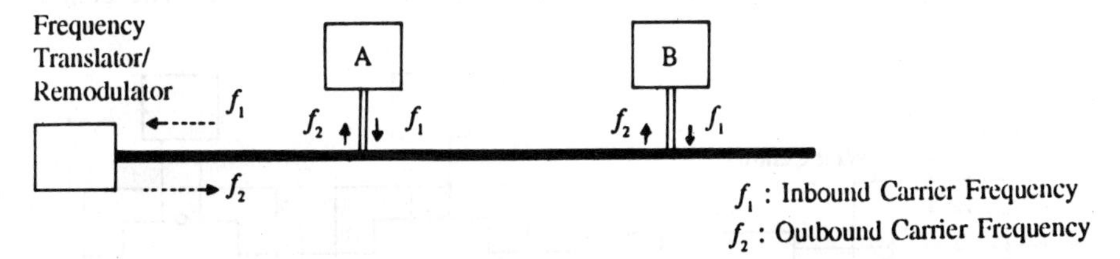

Fig. 5 Dual bus topology using frequency division multiplexing.

Some advantages of bus topology are:

1. Stations are connected to the bus using a passive tap.
2. Least length of physical transmission medium is used.
3. Coverage can be increased by extending the bus through the use of repeaters.
4. New stations are easily added by tapping a working bus.

2.2 Ring Topology

A ring network consists of a number of transmission links joined together in the form of a ring through repeaters called Ring Interface Units, RIU (Fig. 6). The transmission is unidirectional on the ring. Thus, each RIU receives the signals at its input and after regeneration, sends them to the RIU of the next station. Every data unit in the ring contains source and destination addresses. When a circulating data unit passes through an RIU, the station connected to the RIU retains a copy of the data unit if the data unit is meant for it.

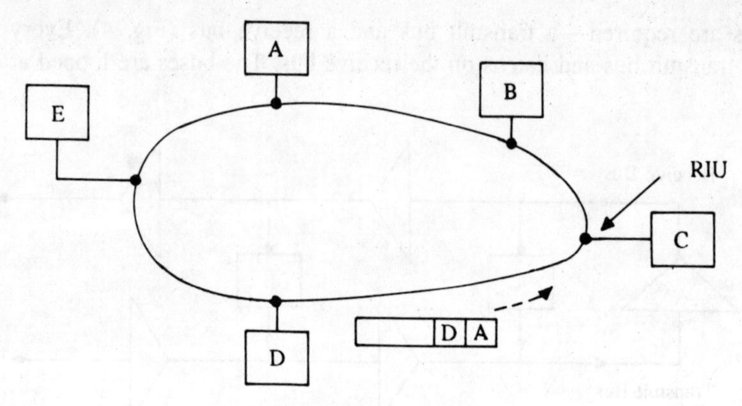

Fig. 6 Ring topology.

Since the ring does not have an end, the data units will circulate continuously unless they are removed from the ring. The responsibility of removing a data unit after it has completed one round is given to the sending station. The destination station is not given this responsibility because it may be out of order or the destination address may be wrong. The possibility of the sending station going out of order after transmitting a data unit cannot be ruled out and, therefore, a monitoring station is also required. It removes the data units going round a second time.

A ring is not as flexible as a bus because to add a station involves breaking the ring and adding an RIU. Wire centres are provided to improve the flexibility of removing or adding a station and to isolate a faulty section (Fig. 7). All the stations are connected to the wire centre.

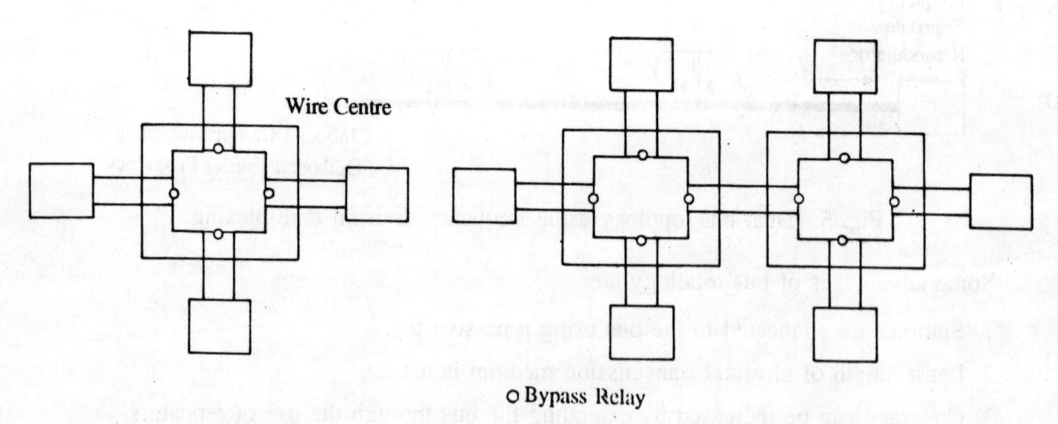

Fig. 7 Wire centres.

If an RIU fails, this can result in total network failure. Therefore, a relay is provided in the wire centre to bypass the failed RIU. Two wire centres can be connected together to increase geographic coverage of the network. As can be seen from the above figure, a ring network does not economize on cables.

2.3 Star Topology

A star network consists of dedicated links from the stations to the central controller (Fig. 8). Each

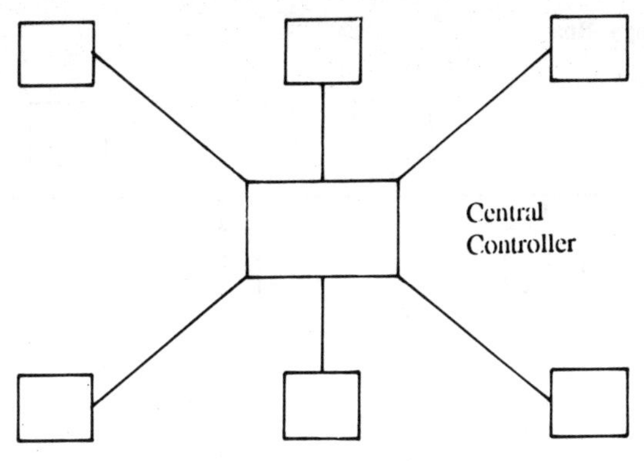

Fig. 8 Star topology

interconnection supports two-way communication. The central controller acts as a switch to route the data units from the source to the destination.

Though star topology is well understood and is based on proven technology used in telecommunications, its use is limited to small LANs only primarily due to the following inherent limitations:

- There is a single point of network failure. If the central controller fails, the full network will be thrown out of service.
- There is no sharing of transmission medium.

2.4 Logical Topology

Logical topology refers to the way the stations are logically interconnected for the purpose of exchanging data units. Physical topology, discussed above, may be different from the logical topology of a network. For example, the LAN in Fig. 9, has bus topology. The bus has a central controller and all the transmissions are to and from it. It receives all the data units and forwards them to their destinations. In other words, the stations have logical connection to the central controller only and, therefore, the logical topology is a star. A logical bus must operate in broadcast mode which this network does not have.

3 MEDIA ACCESS CONTROL AND ROUTING

Unlike switched data subnetworks, the subnetwork of a LAN consists of a physical transmission medium which interconnects various stations of the network. Usually, there are no switching nodes. At most, there can be some repeaters to regenerate the digital signals. The common transmission medium is shared by the stations connected to it. A discipline is followed by the stations so that every station gets fair opportunity to transmit its data and collisions (two stations simultaneously accessing the media) do not take place. Procedures for accessing the medium for signal transmission are called media access control methods.

The overall purpose of a LAN is to provide interconnection just like switched data subnetworks but there is a basic difference in the routing function of the two. In the switched data networks, the data units are routed to the destination end system by the Network layer of the subnetwork nodes. In local area networks, the subnetwork does not have any node. A LAN operates in broadcast

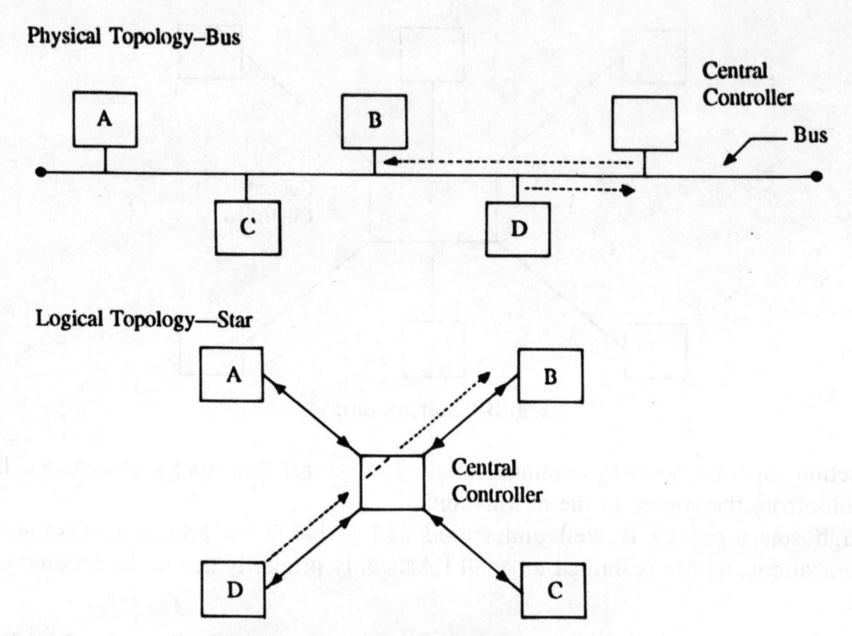

Fig. 9 Physical and logical topologies.

mode. Each data unit carries source and destination addresses. All data units are received by all stations but are accepted by the addressed stations only.

Media access control and addressing functions are implemented in the Data link layer of the stations. These functions are in addition to the error and flow control functions of the Data Link layer.

4 LAYERED ARCHITECTURE OF A LAN

The Data Link layer in the local area networks is divided into two sublayers. These sublayers are called Logical Link Control (LLC) and Media Access Control (MAC). Figure 10 shows the relationship of this division to the OSI reference model.

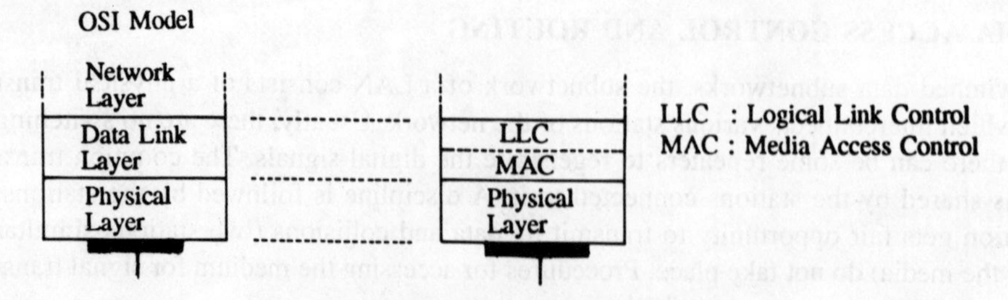

Fig. 10 Layered architecture of a LAN.

The Physical layer transports bits from one station to all the other stations using suitable signal codes. The Data Link layer functions are divided between the two sublayers as follows:

Media Access Control Sublayer
- Control of access to media
- Unique addressing of stations directly connected to the LAN
- Error detection

Logical Link Control Sublayer
- Error recovery
- Flow control
- User addressing

To carry out these functions, each sublayer appends a header to the user data. The MAC sublayer also adds a trailer containing error check bits (Fig. 11).

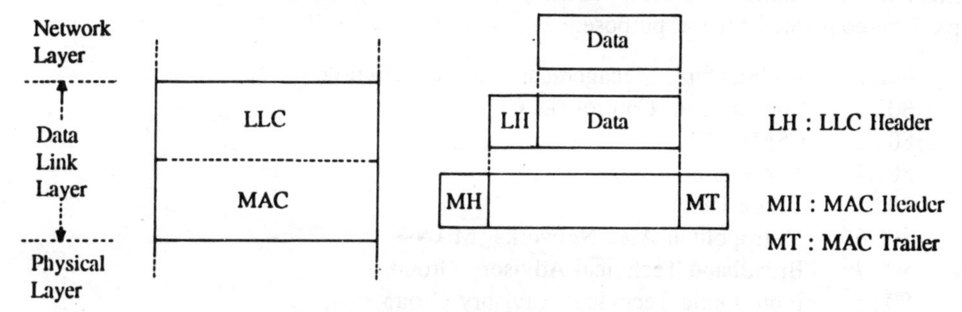

Fig. 11 LLC and MAC headers of a LAN frame.

The user data (*N*-PDU) from the Network layer is passed to the LLC sublayer which adds the protocol control information in the form of a header to it. The LLC protocol data unit so formed is passed to the MAC sublayer which adds a header and a trailer to the LLC PDU to form MAC PDU or the frame. The frame is handed over to the Physical layer for transmission.

The LLC header contains a control field and address fields. The control field is used for error control, flow control and sequencing of LLC service data units. The address fields identify the sending and receiving Network layer entities. Thus, several user entities such as P, Q, R in Fig. 12 can have access to the other user entities S and T on the stations connected to the LAN.

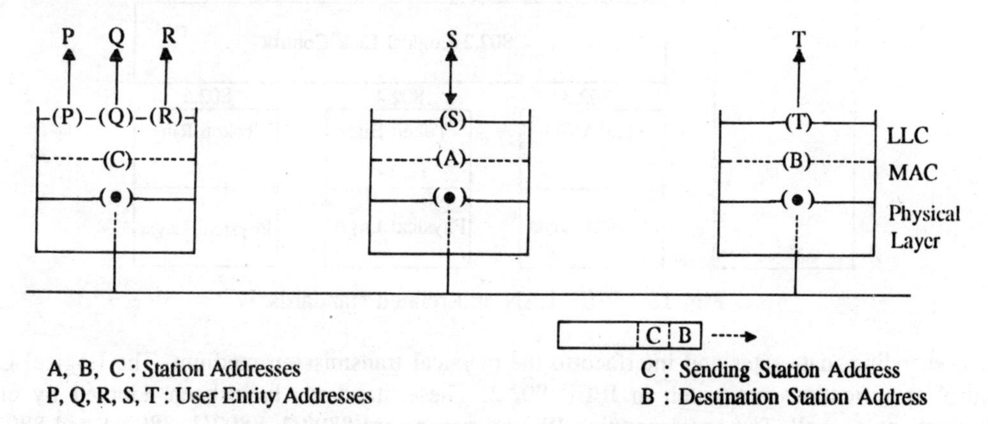

Fig. 12 Addressing at MAC and LLC sublayers.

The MAC header consists of media access control field and station address fields. The address fields identify the sending and receiving stations. Note the difference in addressing at LLC level and at MAC level (Fig. 12).

For error detection, the MAC sublayer appends error check bits as a trailer and carries out error check at the receiving end. If an error is detected, it does not request for retransmission. It just discards the frame and leaves the LLC sublayer to recover from the error. The LLC sublayer will notice a missing frame and send a request for retransmission.

5 IEEE STANDARDS

The layered architecture and other standards of LANs have been developed by the Institution of Electrical and Electronics Engineers (IEEE) under their project 802 set up in 1980. The following groups were constitued for the purpose:

802.1	Architecture, Management and Internetworking
802.2	Logical Link Control (LLC)
802.3	CSMA/CD
802.4	Token Bus
802.5	Token Ring
802.6	Metropolitan Area Networks (MANs)
802.7	Broadband Technical Advisory Group
802.8	Fibre Optic Technical Advisory Group
802.9	Integrated Data and Voice Networks.

For the local area networks, IEEE adopted three mechanisms of media access control, namely, Carrier Sense Multiple Access/Collision Detection (CSMA/CD), Token Bus and Token Ring. The IEEE standards for these media access control schemes and associated Physical layers are IEEE 802.3, IEEE 802.4 and IEEE 802.5 respectively (Fig. 13). The Physical layer specifications include

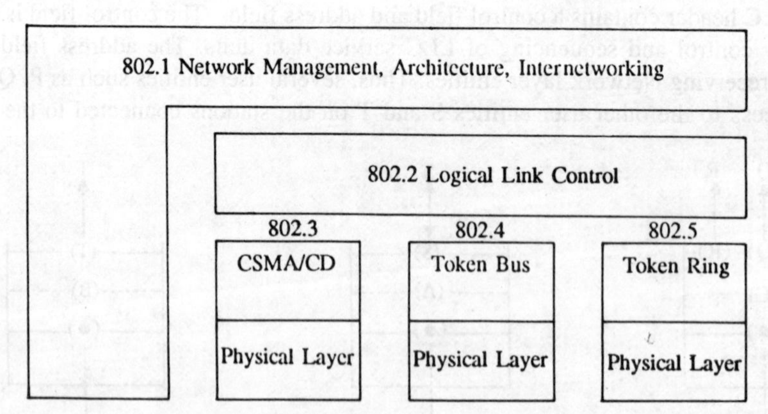

Fig. 13 IEEE LAN and related standards.

signal encoding, data rates and interface to the physical transmission medium. The Logical Link Control specifications are given in IEEE 802.2. These standards have been adopted by other organizations as well. The corresponding ISO references are 8802/2, 8802/3, 8802/4 and 8802/5.

6 LOGICAL LINK CONTROL (LLC) SUBLAYER

As mentioned earlier, the LLC sublayer carries out error control, flow control, sequencing and user addressing functions. It provides service to the Network layer entities and receives services from the MAC sublayer to carry out the assigned functions.

6.1 LLC Service

The LLC sublayer offers the following three types of services to the Network layer entity. These services are made available at the Data Link Layer Service Access Point (DL-SAP).

Type 1—Unacknowledged connectionless-mode service
Type 2—Connection-mode service
Type 3—Acknowledged connectionless-mode service.

A station can provide more than one type of service. Therefore, four classes of service are defined:

Class 1 (LLC1)—Type 1
Class 2 (LLC2)—Types 1 and 2
Class 3 (LLC3)—Types 1 and 3
Class 4 (LLC4)—Types 1–3

Type 1—Unacknowledged Connectionless-Mode Service. It is a non-reliable Data Link service in which there is no guarantee of data delivery. There is no acknowledgement either of delivery or non-delivery of data units. This service is only for sending and receiving user data without bothering about error control, flow control or sequencing at the Data Link layer. These functions become the responsibility of the higher layers. Only two primitives are specified for this service:

- DL-UNITDATA request (source address, destination address, user data, priority)
- DL-UNITDATA indication (source address, destination address, user data, priority).

The request primitive is used at the transmitting end to pass the user data to the LLC sublayer and the indication primitive is used at the receiving end to pass the received user data to the user. The priority parameter is passed down to MAC sublayer which implements the priority mechanisms.

Type 2— Connection-Mode Service. In this service, three phases are involved, viz., establishment of a connection, data transfer and disconnection. Flow control, error control and sequencing are the basic features of this service. The following primitives are used during various phases of connection mode service.

Connection Setup: The following primitives are used for setting up the LLC connection:

- DL-CONNECT request (source address, destination address, priority)
- DL-CONNECT indication (source address, destination address, priority)
- DL-CONNECT confirm (source address, destination address, priority).

Note that priority is decided at the time of setting up of the connection. The data units transferred later during the life of the connection have this priority.

Data Transfer: User data is transferred using the following primitives once the connection is set up:

- DL-DATA request (source address, destination address, data unit)

- DL-DATA indication (source address, destination address, data unit)
- DL-DATA confirm (source address, destination address, data unit, status).

The status parameter indicates whether the data unit was transferred successfully to the LLC entity at the other end or not.

Flow Control: There is no correspondence between the flow control primitives. These primitives are independent and have local significance only. The flow control can be exercised by the LLC entity and the user independently. The flow control is regulated by specifying the amount of data that may be passed.

- DL-FLOW CONTROL request (source address, destination address, amount of data)
- DL-FLOW CONTROL indication (source address, destination address, amount of data).

Reset: Reset causes all undelivered data units to be discarded. The responsibility of recovery is with the user.

- DL-RESET request (source address, destination address, reason)
- DL-RESET indication (source address, destination address, reason)
- DL-RESET confirm (source address, destination address, reason).

Disconnection: The LLC connection is released using the following primitives.

- DL-DISCONNECT request (source address, destination address)
- DL-DISCONNECT indication (source address, destination address, reason)
- DL-DISCONNECT confirm (source address, destination address, status).

Connection establishment, reset, disconnection and data transfer are confirmed services but the confirmation is based on the acknowledgement from the remote LLC entity as there is no response primitive in the LLC service (Fig. 14).

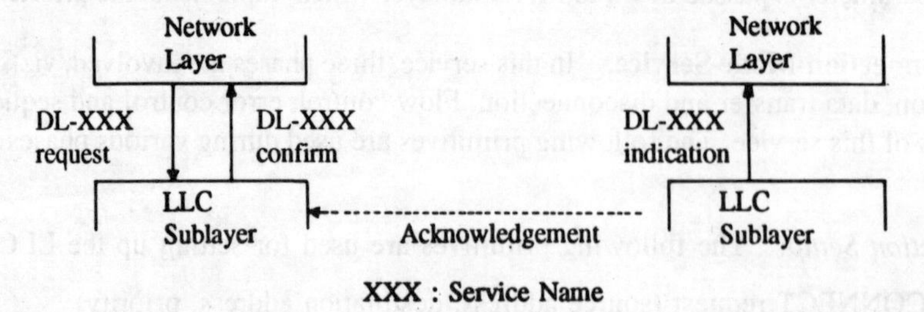

Fig. 14 Service primitives of connection-mode service of LLC sublayer.

Type 3—Acknowledged Connectionless-Mode Service. Acknowledged connectionless-mode service allows an LLC user to request an immediate acknowledgement to a transmission. Data transfer is in connectionless mode and each N-PDU must be acknowledged in the form of status indication across the interface before the next N-PDU is sent. The status indication is based on the acknowledgement received by the local LLC entity from the remote LLC entity as there is no response primitive across the interface. The primitives used for this service are DL-DATA-ACK request, DL-DATA-ACK indication and DL-DATA-ACK-STATUS indication.

Acknowledged connectionless-mode service also provides the poll and response service. A user can request (poll) a PDU from a remote station. The primitives for this service are DL-REPLY request, DL-REPLY indication and DL-REPLY-STATUS indication.

6.2 LLC Protocol

LLC protocol is modelled on the HDLC protocol. Extended asynchronous balanced mode is used in the LLC sublayer. The LLC PDU is shown in Fig. 15. It contains the addresses of the source Data Link layer service access point, the destination Data Link layer service access point, control

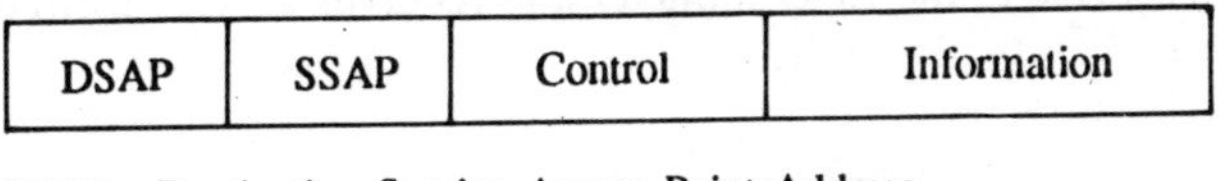

DSAP : Destination Service Access Point Address

SSAP : Source Service Access Point Address

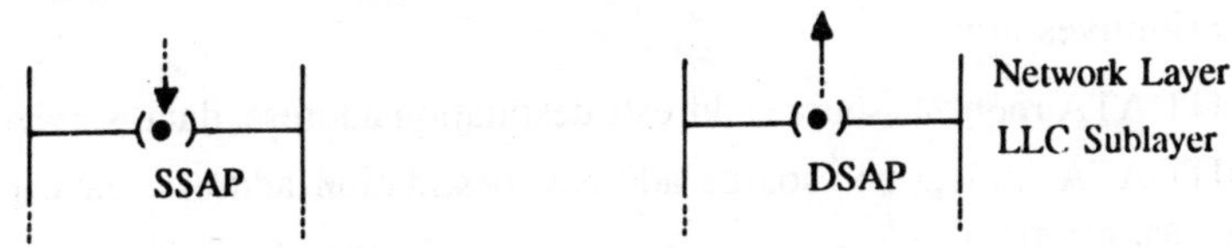

Fig. 15 Format of LLC PDU.

field and information field. Note that the usual frame check sequence field for detection of transmission errors is missing. Error detection function is the responsibility of the MAC sublayer and, therefore, this field is provided in the MAC frame.

The information field contains the N-PDU. The control field is identical to the control field of HDLC frames. It contains one of the commands and responses shown in Table 1.

Table 1 LLC Commands and Responses

Service type	Commands	Responses
1	UI	
	XID	XID
	TEST	TEST
2	I	I
	RR	RR
	RNR	RNR
	REJ	REJ
	SABME	UA
		FRMR
	DISC	UA
		DM
3	AC0	AC0
	AC1	AC1

For the connection-mode service, SABME and DISC commands are used to establish and release the connection. In the connectionless-mode service, UI commands are used for exchanging information. For the acknowledged connectionless-service, two new unnumbered information frames, AC0 and AC1 have been defined. AC stands for Acknowledged Connectionless. The sender alternates the use of AC0 and AC1 command frames, and the receiver responds with the corresponding number.

Asynchronous balanced mode of data transfer was discussed at length in the chapter on HDLC Protocol. The same operation is applicable to the LLC entities.

7 MEDIA ACCESS CONTROL (MAC) SUBLAYER

The media access control, error detection and station addressing are the three basic functions of the MAC sublayer. It provides service to the LLC sublayer and receives service from the Physical layer below it.

7.1 MAC Service

The MAC sublayer provides connectionless-mode service for the transfer of LLC PDUs. The MAC sublayer service primitives are:

- MA-UNITDATA request (source address, destination address, data, service class, priority)
- MA-UNITDATA indication (source address, destination address, data, reception status, service class, priority)
- MA-UNITDATA STATUS confirm (source address, destination address, transmission status, provided service class, provided priority).

The address parameters are the MAC service access point addresses which also identify the stations on the LAN. MA-UNITDATA-STATUS indicates the transmission status of LLC-PDU to the LLC entity (Fig. 16).

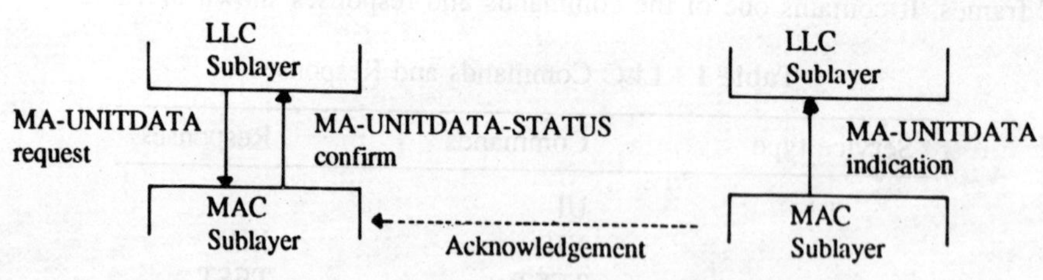

Fig. 16 Service primitives of MAC sublayer.

7.2 MAC Protocol

As mentioned earlier, the MAC frame consists of a header, data field and a trailer. The data field contains LLC PDU. The header specifies source and destination station addresses. It also contains a control field for media access control and a frame delimiter to identify start of the frame. The trailer contains check bits for error detection and an end delimiter to mark end of the frame.

The frame formats, contents of various fields and MAC protocols are different for different media access control mechanisms. We will discuss the media access control mechanisms standardized by IEEE, their frame structure and the Physical layer specifications in the next chapter.

8 TRANSMISSION MEDIA FOR LOCAL AREA NETWORKS

In a local area network, the interconnecting transmission medium can be a twisted pair cable or a coaxial cable or an optical fibre cable. Choice of transmission medium depends on several factors:

Bandwidth. Bandwidth of the transmission medium determines the maximum data rate which can be handled by it. The transmission medium bandwidth should not become a bottleneck in achieving the required data rate. It must be kept in mind that bandwidth of transmission medium is the function of the length of the medium, e.g., it may be possible to achieve very high data rates on a low cost twisted pair but then the maximum length of one transmission segment is limited to not more than a few metres.

Connectivity. Some transmission media are suitable for broadcast mode of operation and point to multipoint links, while others are better suited for point-to-point links. For example, at the present status of technology, optical fibre is suited for point-to-point links only.

Geographic Coverage. In a LAN operating in broadcast mode on a bus, the electrical signals should reach from one end to the other without degradation in quality of the signals below the required limits. Attenuation and group delay characteristics of the medium determine overall distortion in the signals. These chacteristics are a function of distance and, therefore, determine the geographic coverage of the LAN. Propagation delay which is also dependent on the medium characteristics and the length of the medium is an important consideration in some media access control mechanisms. The maximum propagation delay which can be tolerated depends on the minimum size of data frame. Therefore, the geographic coverage gets limited by the minimum size of the frame.

Noise Immunity. Ideally, the transmission medium chosen for LAN should be free from interference from outside sources but, in practice, it is not possible. The degree of immunity to interference varies from medium to medium. Local area networks are susceptible to interference because LAN cabling is usually done in ducts which carry power cables also. The degree of noise immunity required depends on the environment in which a LAN is installed.

Security. In some of the LAN topologies and access methods, it is very easy to tap the LAN and pick up the messages without generating any alarm. Data security considerations are very important in local area networks and, therefore, influence the choice of transmission medium.

Cost. At the present state of technology, the costs of different media are different and are changing continuously. Metallic media are becoming costlier and optical fibre cost is going down. Cost of transmission medium is also related to the cost of the equipment associated with it and its installation. Therefore, an overall view of the cost structure is more important than the cost of transmisssion medium alone.

The following transmission media are used in local area networks:

- Twisted pair cable
- Coaxial cable
- Optical fibre.

8.1 Twisted Pair

By far the most common transmission medium is the twisted pair. It is also the most underestimated medium. Although it is used extensively for voice communication, it can handle frequencies much higher than the voice band.

A twisted pair consists of two insulated wires twisted together in a spiral form (Fig. 17a). Twisting reduces crosstalk and interference problems. More than one pair can be bundled together in the form of a cable (Fig. 17b). A twisted pair can also have a metallic braid as a shield (Fig. 17c) for protection against noise.

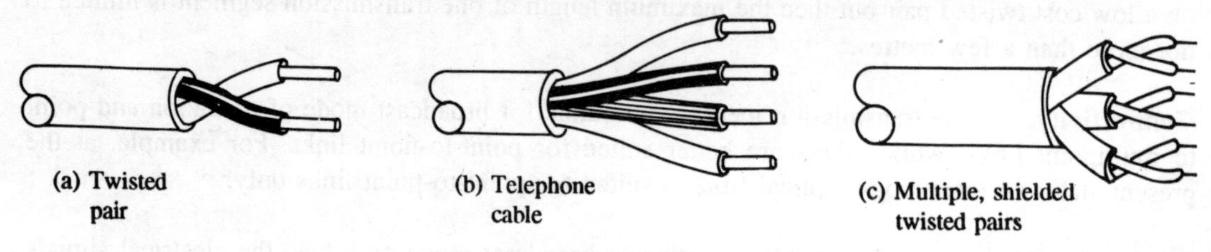

Fig. 17 Twisted pair cables.

Twisted pair cable is used for point-to-point and point-to-multipoint applications. But as a multipoint medium, it supports fewer stations over smaller distances than coaxial cable because of signal impairment. Tapping into a twisted pair cable for additional stations is not easily possible without disturbing other users or without changing transmission characteristics of the medium. Costwise, twisted pair cable is less expensive than coaxial cable or optical fibre.

8.2 Coaxial Cable

Coaxial cable is widely used as transmission medium in local area networks. Coaxial cable has very low loss, high bandwidth and very low susceptibility to external noise and cross-talk.

50 ohm and 75 ohm CATV coaxial cables are popular in local area networks. These impedances refer to characteristic impedance of the cable. The 50 ohm cable is called baseband cable because digital signals are transmitted without any modulation. The outer conductor in 50 ohm cable is metallic braid. Digital signals upto 10 Mbps can be easily transmitted on a baseband cable.

The 75 ohms cable is a broadband cable as it has a large bandwidth. It has a solid outer conductor. Digital signals are transmitted on this cable as modulated carriers. For bidirectional connectivity, the frequency band is divided into inbound and outbound frequency bands. The typical frequencies are 5–174 MHz (inbound carrier) and 232-400 MHz (outbound carrier). The frequency translation takes place at the head end.

Coaxial cables can be used both for point-to-point and point-to-multipoint applications. When used as a bus, 50 ohm cable can support upto 100 devices per segment, the maximum length of a segment being 500 metres. For distances of more than five hundred metres, repeaters are required. A 75 ohm broadband cable can support about 1000 taps over a length of about 4 km.

Geographic coverage of the coaxial cable is better than the twisted pair cable but it depends on end-to-end propagation delay which the network can afford.

Costwise, coaxial cable is more expensive than twisted pair cable but the overall cost of installed cable is marginally different due to significant cable installation costs in either case.

8.3 Optical Fibre

With optical fibres it is possible to realize very high data rates (~ gigabits) over much larger distances compared to about a few kilometres of coaxial cable. Use of optical fibre in local area networks has not been widespread due to its high cost, special installation practices, and non-availability of low loss components. In view of the rapid developments being made in this field, however, optical fibres have a very promising future.

Optical fibre systems has advantage over coaxial and twisted pair cable systems for their following basic characteristics:

- Greater bandwidth
- Smaller size and lighter weigh
- Greater repeater spacing due to low loss
- High immunity to electromagnetic interference
- Secure communication.

Star, ring and bus topologies, are all possible in optical fibre local area networks, but being a new area of development, commercial products are still in the process of standardization for bus topology. Star topology does not justify use of optical fibres because of its low data rate requirements. Ring topology is most suited for optical fibre media and is becoming popular. Optical fibre exhibits less propagation delays, distortion and attenuation and fibre rings can, therefore, be much larger than coaxial rings.

9 TOPOLOGY AND MEDIA PREFERENCES

The choices of transmission medium and topology are not independent of each other. As discussed above, some media are more suitable for point-to-point rather than point-to-multipoint transmission. Other considerations for choice of medium for a topology would include

- underutilization of medium transmission capacities and
- cost considerations.

Table 2 shows preferred combinations in the LAN industry.

Table 2 Preferred Combinations of Media and Topology

Media	Topology		
	Bus	Ring	Star
Twisted pair	×	×	×
Baseband coaxial	×	×	
Broadband coaxial	×		
Optical fibre		×	

Representative features of ring and bus topologies for the various media are given in Tables 3 and 4. These features are being constantly upgraded by the industry with technological developments.

Table 3 Characteristics of Transmission Media for Ring Topology

Medium	Data rate (Mbps)	Repeater spacing (km)	Number of repeaters
Twisted pair (Unshielded)	4	0.1	75
Twisted pair (Shielded)	16	0.3	250
Baseband coaxial cable	16	1.0	250
Optical fibre	100	2.0	1000

Table 4 Characteristics of Transmission Media for Bus Topology

Medium	Data rate (Mbps)	Range (km)	Number of taps
Twisted pair (Unshielded)	1	2	10's
Baseband coaxial cable	10	3	100's
Broadband coaxial cable	20 per chl.	30	1000's

10 SUMMARY

Local area networks provide a high speed and high throughput solution to the networking requirements within a small geographic area. The physical topology of a LANs can take the shape of a bus, a ring or a star.

The physical transmission medium of a LAN is shared by the stations connected on the LAN. Media access control procedures are implemented to ensure that every station gets a fair chance to transmit and collisions do not take place. These procedures are implemented in the MAC sublayer of the Data Link layer. The MAC sublayer also carries out error detection and station addressing functions. The other sublayer of the Data Link layer is the LLC sublayer which carries out error control, flow control, sequencing and user addressing functions. Asynchronous balanced mode of the HDLC protocol is implemented in the LLC sublayer.

IEEE has developed LAN standards which include protocols and services of the LLC and MAC sublayers. These standards also cover the Physical layer specifications.

Transmission media in local area networks can be twisted pair (shielded or unshielded) cable, coaxial cable or optical fibre. Choice of the transmission medium depends on several factors such as geographic coverage, data rate, number of stations, topology etc. At present, coaxial cable is extensively used but optical fibre has very high potential.

PROBLEMS

1. The following terms are associated with one of three basic LAN topologies, bus, ring and star. Indicate the topology against each term.

- Wire centres
- Central controller
- Remodulator
- Bypass relay.

2. Indicate the LAN topology against the following characteristics:

- Unidirectional transmission
- Least length of transmission medium
- Single point of network failure.

3. Consider a 5×8 grid of stations which are connected by a local area network. The stations are 10 metres apart. The interconnecting cables run along the dotted lines as shown. What is the cable length required for bus topology?

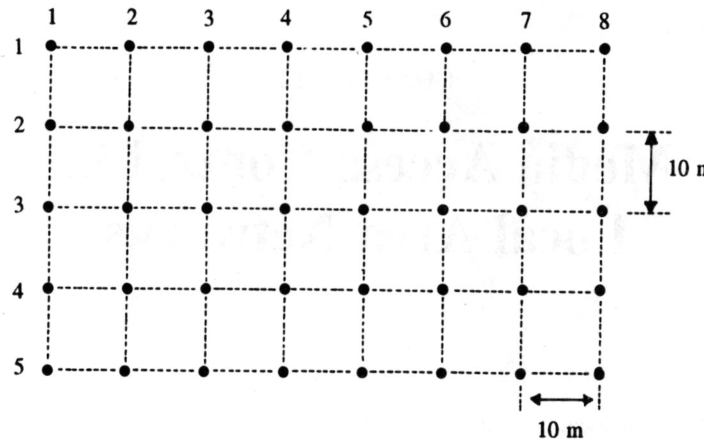

Fig. P.3

4. What will be the length of the cable if ring topology is used in Problem 3?
5. Indicate the transmission medium against the following media characteristics:
   - No electromagnetic interference
   - Least cost
   - Suitable for point-to-point transmission only
   - Low loss, low interference metallic transmission medium.

CHAPTER 12

Media Access Control in Local Area Networks

The physical topology of local area networks can take the shape of a bus or ring or a star but the network attributes in terms of delay, throughput, expandability, etc. are determined not only by the physical topology but also by the media access control methods utilized for sharing the common interconnecting transmission media. In this chapter we examine some of the media access control methods based on bus and ring topologies. Token bus, CSMA/CD and token ring mechanisms are covered in considerable detail. We also examine some less popular media access control method, namely, register insertion and slotted ring mechanisms. In the last we discuss the more recent FDDI-I and FDDI-II optical fibre-based LAN standards.

1 MEDIA ACCESS CONTROL METHODS

There are several methods of media access control in the local area networks. Each of these methods is applicable to a specific LAN topology. We can broadly classify these methods into two categories:

1. Centrally controlled access
2. Distributed access control.

In the first category, access to the media is controlled by a central controller. Polling, demand assigned frequency division or time division multiple access are some such methods. But distributed access control methods are more common in the local area networks. Centrally controlled access methods suffer from a basic limitation in that they have a single point of network failure.

In distributed access control, a discipline is built up among the stations of the LAN so that fair opportunity is given to each station to transmit its data frames. The distributed control methods have an edge over the centrally controlled methods in that there is no single point of network failure.

Distributed access control methods are available for both, bus and ring topologies. For the bus topology, the following two methods dominate the present day market and these have been standardized by IEEE:

- Token passing, IEEE 802.4
- Carrier sense multiple access/collision detection (CSMA/CD), IEEE 802.3.

For ring topology also, there are several media access control methods, as given below, but only token passing has been standardized by IEEE.

- Token passing, IEEE 802.5
- Register insertion
- Empty slot.

The media access control mechanisms standardized by IEEE are implemented in the MAC sublayer of the Data Link layer. The MAC sublayer operates under the LLC sublayer 802.2 which was discussed in the last chapter. IEEE standards for the MAC sublayer also include the physical transmission specifications.

The above media access control methods are applicable to metallic transmission media. For optical fibre-based LANs, there are FDDI-I (Fibre Distributed Data Interface) and the more recent FDDI-II standards which define the media access control and the physical transmission specifications.

2 TOKEN PASSING IN THE BUS TOPOLOGY

In the token passing method, stations connected on a bus are arranged in a logical ring, i.e., the addresses of the stations are assigned a logical sequence with the last number of the sequence followed by the first (Fig. 1). Each station knows the identity of the stations preceding and following it. There is no relation between the physical location of a station on the bus and its sequence number.

Physical Topology

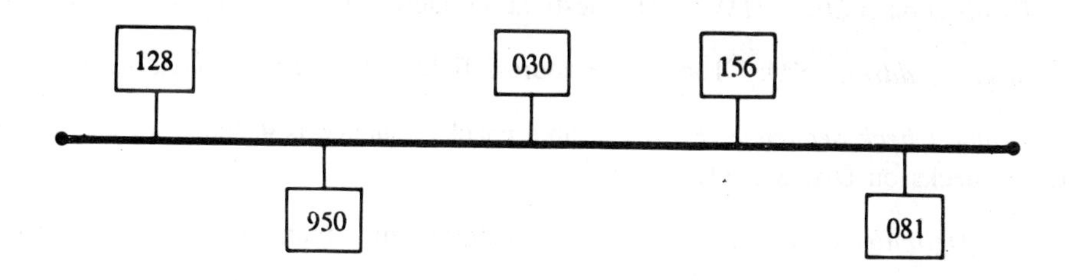

Logical Sequence of Token Passing

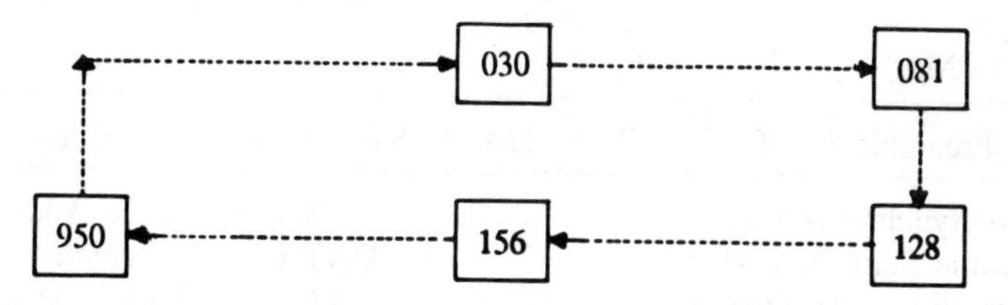

Fig. 1 Token passing sequence in a bus.

2.1 Media Access Control

The basic operation of token passing on the bus is as follows:

- Access to the interconnecting bus is regulated by a control frame known as the token. At any time, only the station which holds the token has the right to transmit its one or more data frames on the bus. Each frame carries source and destination addresses.

- All stations are ready to receive frames at any time except when holding a token.
- The token must be released before time out. The token is released with the address of the next station in the logical sequence and the next station takes it over.
- In one cycle, each station gets an opportunity to transmit. It is possible to give more than one opportunity to a station in one cycle by assigning more than one address to it.
- To maintain continuity of communication, it is necessary for each station to take over the token when its turn comes, even if it does have any data to transmit. It can release the token immediately for the next station.

2.2 Frame Structure

The frame structure as specified by IEEE 802.4 is shown in Fig. 2. It consists of the following fields:

Preamble: The preamble is an at least one octet-long pattern to establish bit synchronization.

Start Delimiter (SD): It is a one octet-long unique bit pattern which marks the start of the frame.

Frame Control (FC): The frame control field indicates the type of the frame—data frame or control frame. The token is one of the control frames. Other examples of control frames are Claim Token, Set Successor and Solicit Successor frame. This field is one octet-long.

Destination Address (DA): The destination address field is 2 or 6 octets long.

Source Address (SA): The source address field is also 2 or 6 octets long.

Frame Check Sequence (FCS): Frame check sequence is 4 octets long and contains CRC code. It checks on DA, SA, FC and data fields.

End Delimiter (ED): It is a unique bit pattern which marks the end of the frame. It is one octet-long.

The total length of the frame from FC to FCS fields is a maximum of 8191 octets.

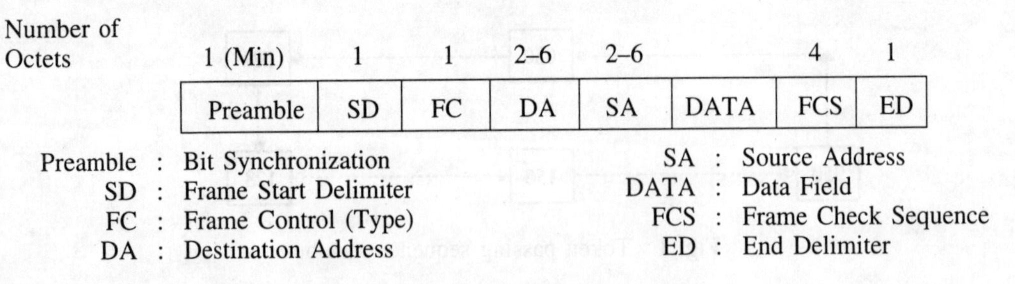

Fig. 2 Format of IEEE 802.4 frame.

2.3 Token Management

Control and management of the token is distributed among the active stations. Each station can initiate and respond to the control frames such as a Claim Token frame, Solicit Successor frame, Set Successor frame and Who Follows frame. These frames are used for initialization of the bus and for adding/removing a station.

The bus is initialized by the Claim Token frames which are sent by one or more stations. The station with the largest address claims the token by sending the longest Claim Token frame. By detecting silence after its transmission of the frame, it assumes takeover of the frame and sends a Solicit Successor frame. It waits for response for a defined period, and decides the next station from the responses received. It then passes the token to the next station. The process is repeated until all the stations have been included in the token passing sequence.

To add new stations in the sequence after initialization, each station periodically transmits a Solicit Successor frame inviting new stations with addresses between itself and the next station already in the sequence. If there is a response, it passes the token to the new station.

After passing the token to the next station, the sending station waits for a valid frame from the next station. If the next station does not release a valid frame within a time-out, the sending station repeats the token. If the second attempt also fails, it is assumed that the next station has failed and a Who Follows frame with the address of the failed station is transmitted. The station next to the failed station responds with Set Successor frame with its address. The token is then passed to the indicated address, thus bypassing the failed station.

2.4 Priority Scheme

An optional priority scheme is specified in the IEEE 802.4 standard. There are four classes of frame priorities—6, 4, 2, 0. Whenever a station receives a token, it can always send priority 6 frames until expiry of its Token Holding Time, THT. After sending these frames, it is allowed to transmit the next lower priority frames depending on the status of another timer called the Token Rotation Timer (TRT) which is the time taken for the token to complete one full rotation. If the TRT plus the time taken to transmit priority 6 frames is below a defined threshold, priority 4 frames can also be transmitted by a station until the threshold. Similar thresholds are defined for priority 2 and 0 frames. Anotehr timer, TRT M, is defined for maintenance purpose. If this timer is not expired, a station sends Solicit Successor frame to include new stations in the token passing sequence. Figure 3 illustrates the normal operation of the bus when a station receives the token.

2.5 Physical Specifications

Token passing LANs operate at data rates 1, 5 or 10 Mbps using analogue signalling over 75 ohms coaxial cable. Two types of transmission systems are used:

1. Carrierband (single channel).
2. Broadband (multiple channel).

Both the systems use modulation techniques to reduce the effect of noise in the factory environment. The carrierband system is based on single channel bidirectional transmission. For 1 Mbps data rate, phase coherent FSK modulation is used. 1 Mbps bus is implemented using flexible semi-rigid coaxial cable. A more expensive version of the bus operates at 5 Mbps or 10 Mbps data rate using phase coherent FSK signalling. The Manufacturing Automation Protocol (MAP) of General Motors is the implementation of IEEE 802.4 token passing bus at 5 and 10 Mbps.

Broadband system employs multiple channel unidirectional transmission using combination of phase and amplitude modulation. It uses conventional CATV components and a remodulator at the head end. Separate carriers are used for the transmit and receive directions. There can be several transmit and receive carriers and each carrier provides 5 or 10 Mbps data rate. Broadband LAN can cover a span of several kilometres.

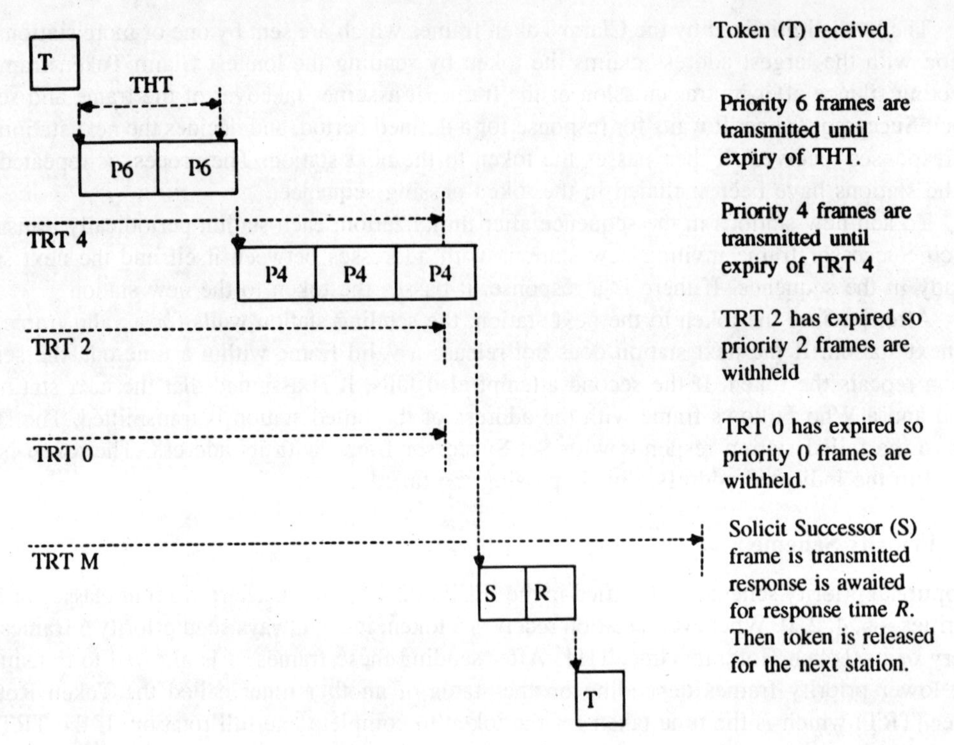

Fig. 3 Priority scheme.

3 CONTENTION ACCESS

In contention access methods there is no scheduled time or sequence for stations to transmit on the medium. They compete for the use of the medium. It is, therefore, quite likely that more than one station will transmit simultaneously and the data frames will "collide". There are several ways of reducing the likelihood of these collisions. Carrier Sense Multiple Access/Collision Detection (CSMA/CD) is one of the ways used in local area networks. To understand this method, we must start from the first contention access method—Pure ALOHA.

3.1 Pure ALOHA

This access mechanism was originally used in ALOHA packet radio network which provided a single radio channel for access by a number of stations to a central computer (Fig. 4). The basic scheme is as under:

- A station can transmit whenever it wants. There is no preassigned time or sequence.
- If a station starts to transmit when another transmission is already in progress, collisions will occur but there will be some instances when transmissions will reach the destination without any collision.
- A mechanism to detect collision is established (e.g., acknowledgement). Collision is assumed to have occurred and the message is retransmitted if the acknowledgement is not received within a specified time (twice the propagation time plus processing time).

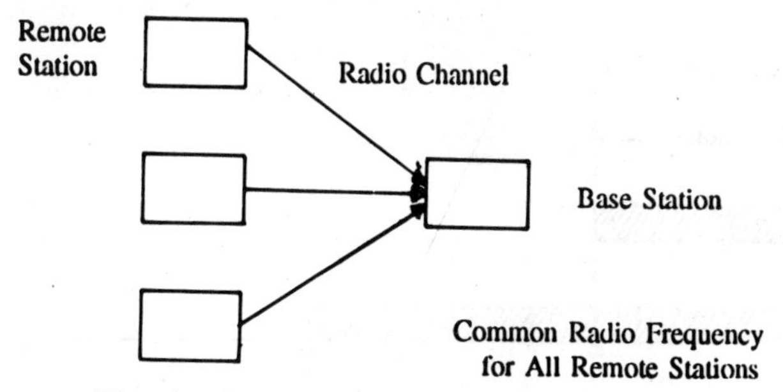

Fig. 4 Configuration of ALOHA network.

The maximum throughput of pure ALOHA access method is somewhat modest, mere 0.18. Appendix A gives the detailed analysis of throughput calculations. The reason for its poor efficiency is large wasted time when a collision occurs (Fig. 5). Note that even though there may have been only a few bits of overlap, the whole of two frames time is wasted in one collision.

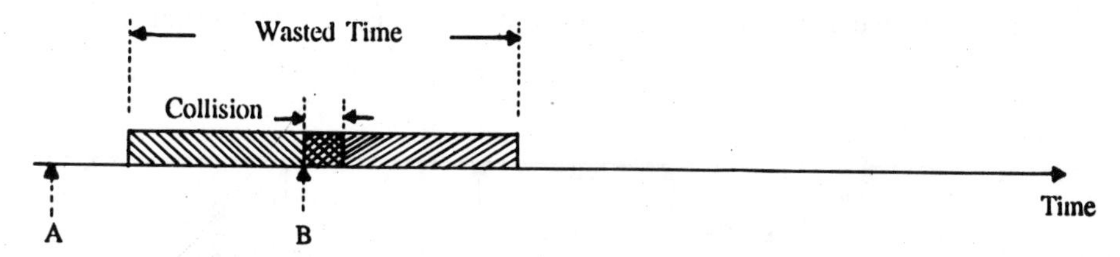

Fig. 5 Time wasted due to a collision in pure ALOHA.

Provided that there is relatively little traffic on the network and the frame size is fairly small, this technique is reliable.

3.2 Slotted ALOHA

Wasted time due to collisions can be reduced if all the transmissions are synchronized. The channel time is divided into time slots and the stations are allowed to transmit at specific instants of time so that all transmissions arrive aligned with the time slot boundaries (Fig. 6). Collisions will still occur but wasted time is reduced to one time slot. Throughput increases to 37 per cent, twice that of pure ALOHA. The throughput analysis is given in Appendix 1.

4 CARRIER SENSE MULTIPLE ACCESS (CSMA)

In the ALOHA channel discussed above, the possibility of collision can be reduced if some discipline is built into the totally random access mechanism. If a station senses the carrier on the medium before starting its own transmission, a collision can be avoided. CSMA, as the name suggests, is based on this principle. We are using the term 'carrier' despite the fact that most of the baseband local area networks do not use a carrier to transmit data.

Consider a situation where frame transmission time is much more than propagation time. If a station has a frame to send and the transmission medium is free, it starts to transmit. It is soon heard by other stations which defer their transmission on sensing the carrier. Contention for the

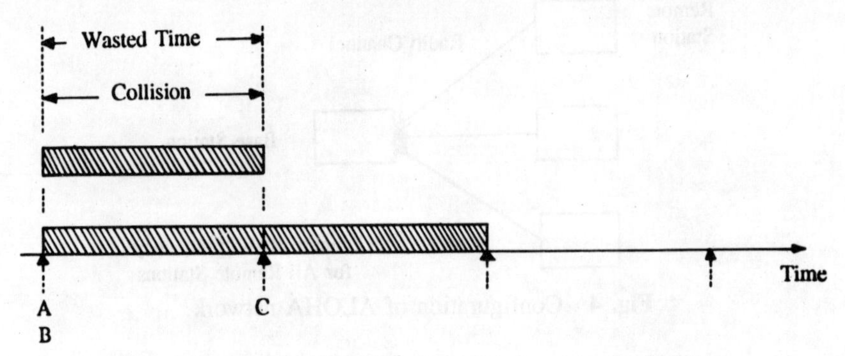

Fig. 6 Time wasted due to collision in slotted ALOHA.

channel can take place only during the first few bits of the frame when the first bit is still in transit. Once the first bit is heard by every station, there cannot be a collision.

In CSMA, an algorithm is needed to specify when a station can transmit once the channel is found busy. There are several ways in which the waiting frames can be transmitted (Fig. 7).

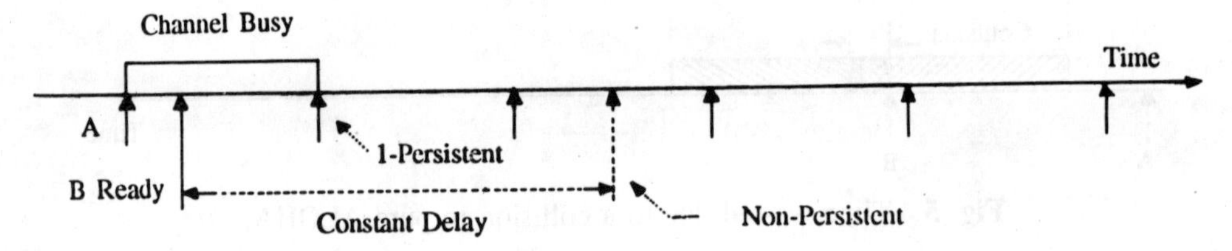

Fig. 7 CSMA persistence schemes.

4.1 Non-Persistent CSMA

In this scheme, when a station having a frame to send finds that the channel is busy, it backs off for a fixed interval of time. It then checks the channel again and if the channel is free, it transmits. The back-off delay is determined by the transmission time of a frame, propagation time and other system parameters. There is likelihood of some wasted idle time when the channel is not in use by any station.

4.2 1-Persistent CSMA

In this scheme, stations wishing to transmit monitor the channel continuously until the channel is idle and then transmit immediately. The problem with this strategy is that if two stations are waiting to transmit, then they will always collide and require retransmission.

4.3 *p*-Persistent CSMA

To reduce the probability of collision in 1-persistent CSMA, not all the waiting stations are allowed to transmit immediately after the channel is idle. A waiting station transmits with probability "p" if the channel is idle. For example, if $p = 1/6$ and if 6 stations are waiting, on average only one station will transmit and the rest will wait. It is equivalent to throwing a dice, and if a station gets six, it transmits. If two stations get six, then both will transmit and collision will take place. Likelihood of such occurrences can be reduced by reducing the transmission probability.

Optimized *p*-persistent CSMA can give throughput of 0.8-0.9 while 1-persistent CSMA achieves throughput of 0.53. Figure 8 shows the graph of throughput (*S*) versus the offered traffic (*G*) for various contention schemes.

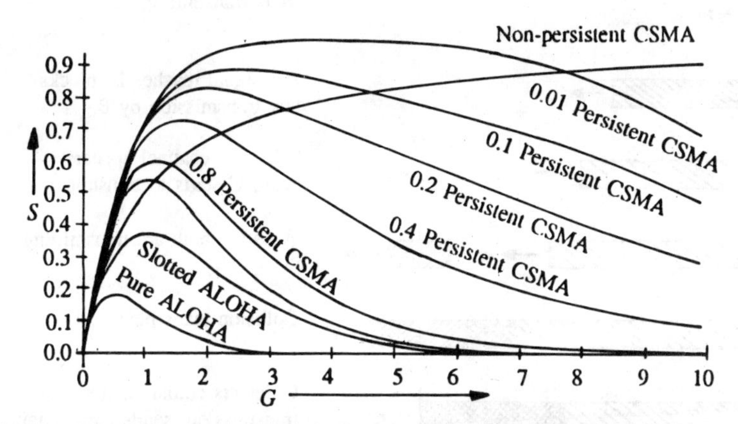

Fig. 8 Throughput as a function of the offered traffic and persistence.

5 CSMA/CD

One of the most commonly used multiple access techniques in the present day local area networks is CSMA/CD, where CD stands for Collision Detection. CSMA/CD specifications were developed jointly by DEC, Intel and Xerox. They called this network *Ethernet*. These specifications were later adopted by IEEE as their standard IEEE 802.3.

5.1 Media Access Control

In the CSMA technique discussed above, a station continues transmission of a frame until the end of the frame even if a collision occurs. This results in unnecessary wastage of channel time. If the station receives other transmissions while it is transmitting, a collision can be detected as soon as it occurs and transmission of the wasted frame can be abandoned. This scheme is known as CSMA/CD and is illustrated in Fig. 9.

When a collision is detected, the transmitting stations release a jam signal to alert the other stations to the collision. For the technique to work properly, the stations should not attempt to transmit again immediately after a collision has occurred. Otherwise, the same frames will collide again. Usually the stations are given a random back off delay for retry. If collision repeats, back off delay is progressively increased. So the network adapts itself to the traffic. In Ethernet, the random back off delay for retry after collision is doubled on each retry upto 10 retries. By careful design, it is possible to achieve efficiencies of more than 90 per cent using CSMA/CD.

5.2 Maximum Cable Segment

It is necessary that a station be transmitting if it is to detect a collision and append a jam signal to the aborted frame. This condition puts a limit on the minimum size of the frame and maximum end-to-end cable segment. If in Fig. 9, stations A and C are at the extreme ends of a cable segment,

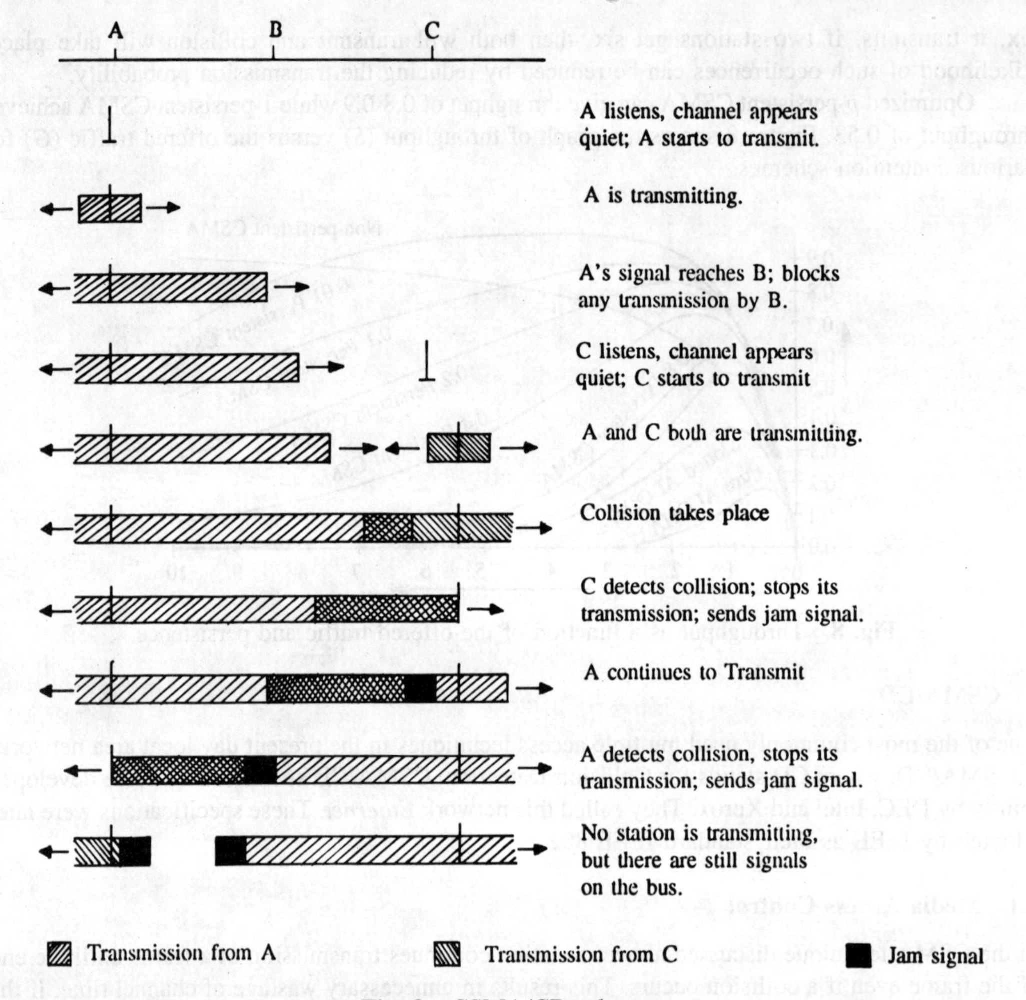

Fig. 9 CSMA/CD scheme.

the first indication of occurrence of the collision at C reaches A after a delay equal to twice the propagation time. Therefore, the frame transmission time should be at least equal to twice the end-to-end propagation time. Thus, depending on the bit rate, transmission characteristics of the cable and the frame size, maximum end-to-end cable segment length can be calculated.

In Ethernet, end-to-end cable segment is 2500 metres for a frame of 64 bytes transmitted at 10 Mbps over a standard 50 ohms cable.

5.3 Frame Format

Frame format as per IEEE 802.3 standard is shown in Fig. 10. It consists of the following fields:

Preamble: The preamble is a seven octets long pattern to establish bit synchronization.

Start Frame Delimiter (SFD): It is a one octet-long unique bit pattern which marks the start of the frame.

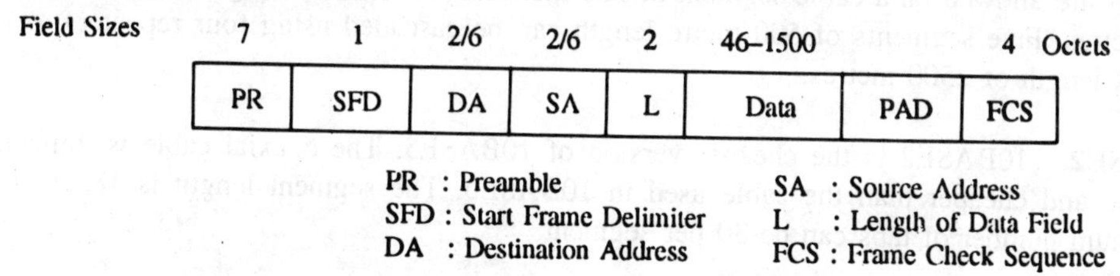

Fig. 10 Format of IEEE 802.3 CSMA/CD frame.

Destination Address (DA): The destination address field is 2 or 6 octets long.

Source Address (SA): The source address field is also 2 or 6 octets long.

Length (L): This field is two octets long and indicates the number of octets in the data field.

Data Field: It can have 46 to 1500 octets if the address field option is 6 octets. If data octets are less than 46, the PAD field makes up the difference. This ensures minimum size of the frame.

Frame Check Sequence (FCS): The frame check sequence is 4 octets long and contains the CRC code. It checks on DA, SA, length, data and PAD fields.

The frame size is variable consisting of between 64 and 1518 octets. The data field can be from 46 to 1500 bytes. If the number of data bytes is less than 46, PAD field is added to make up a frame of the minimum size. The Ethernet frame format is somewhat different from IEEE 802.3. Therefore, Ethernet and IEEE CSMA/CD stations cannot be mixed on the same network. Ethernet frame structure is shown in Fig. 11. The Type field identifies the higher level protocol type associated with the frame. It determines how the data field is to be interpreted.

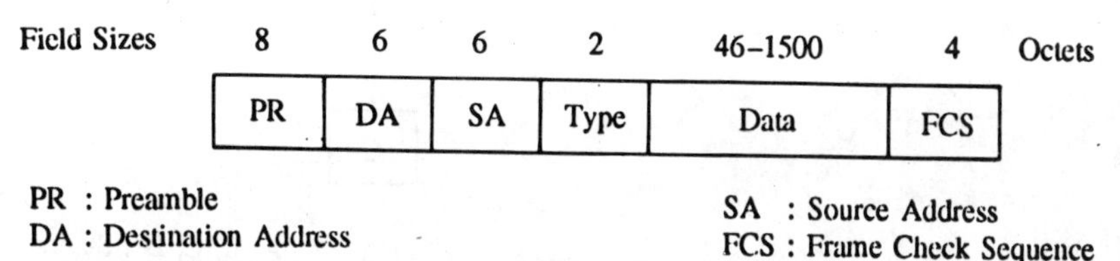

Fig. 11 Format of ethernet CSMA/CD frame.

5.4 Physical Specifications

CSMA/CD offers four options in terms of bit rate, signalling method and maximum electrical cable segment length. The options are 10BASE5, 10BASE2, 10BROAD36 and 1BASE5. The numeric field in the beginning indicates the bit rate in Mbps, baseband or broadband signalling is indicated next and finally the electrical cable segment length in x100 metres.

Manchester signal code is used at the baseband level for transmission. In broadband transmission, differential phase shift keying is used to convert the Manchester encoded signal into analogue form.

10BASE5. Fifty ohms coaxial cable is used in the 10BASE5 standard. A maximum of 100 stations are allowed on a cable segment of 500 metres. Minimum spacing between the stations is 2.5 metres. Five segments of 500 metre length can be cascaded using four repeaters to give an overall length of 2500 metres.

10BASE2. 10BASE2 is the cheaper version of 10BASE5. The coaxial cable is thinner, more flexible and cheaper than the cable used in 10BASE5. The segment length is 185 metres and maximum number of taps can be 30 per segment.

10BROAD36. 10BROAD36 uses 75 ohms CATV cable with a remodulator as the head end. The maximum cable span is 3750 metres in two segments of 1875 metres from the head end. Other services such as TV and voice can also be integrated on the same cable using frequency division multiplexing.

1BASE5. 1BASE5 is commonly referred to as StarLAN and provides 1 Mbps data rate over the unshielded twisted pair. The physical structure is a distributed star but logically it operates as a CSMA/CD bus (Fig. 12). The stations are connected to hubs which act as repeaters and detect collisions. The hubs transmit 'collision present' signals to all the stations when a collision is detected. There can be 8 to 16 stations connected to one hub. The hubs themselves are connected in the form of a tree having upto 5 levels.

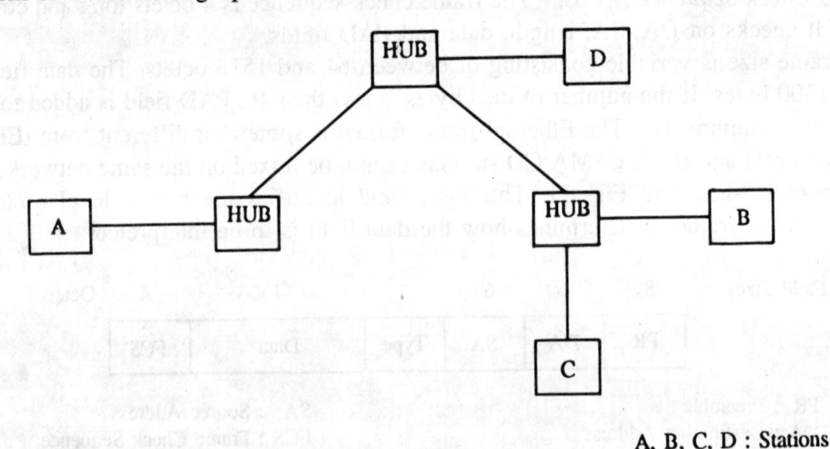

Fig. 12 Typical configuration of StarLAN.

6 TOKEN PASSING ON A RING

A token passing ring consists of a number of point-to-point links interconnecting adjacent stations (Fig. 13). Each station is connected to the ring through a Ring Interface Unit (RIU). Each RIU regenerates the data frames it receives and sends them onto the next link after a delay of at least one bit. RIU also makes available the incoming frame to the local station which accepts the frame if it is addressed to it.

When a station transmits, it breaks the ring and inserts its own frame with destination and source addresses. When the frame eventually returns to the originating station after completing the round, the station removes the frame and closes the ring.

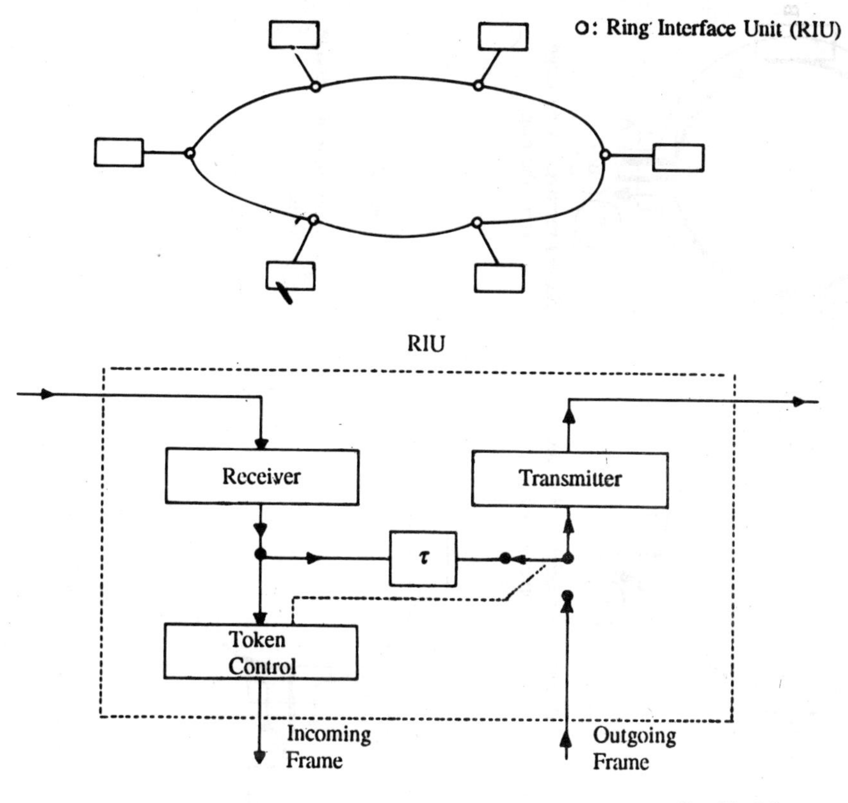

Fig. 13 Configuration of the ring interface unit.

6.1 Media Access Control

Access to the medium is controlled by use of a token as in the token bus. The token is passed from station to station around the ring. The sequence of token passing is determined by the physical locations of the stations on the ring.

Figure 14 shows the sequence of events when station A sends a frame to station C and then the token is picked up by station D. Note that size of the ring in terms of bits is less than the size of the frame and therefore, the frame comes back to the originating station after traversing the ring even before it has been completely transmitted.

While in possession of the token, a station may transmit one or more data frames but must eventually release the token by transmitting the token frame before expiry of Token Holding Timer (THT). The recommended THT is 10 ms which limits the maximum number of bytes to 5000 at 4 Mbps.

6.2 Frame Format

Formats of the token frame and the data frame as standardized in IEEE 802.5 are shown in Fig. 15. The following fields comprise these frames.

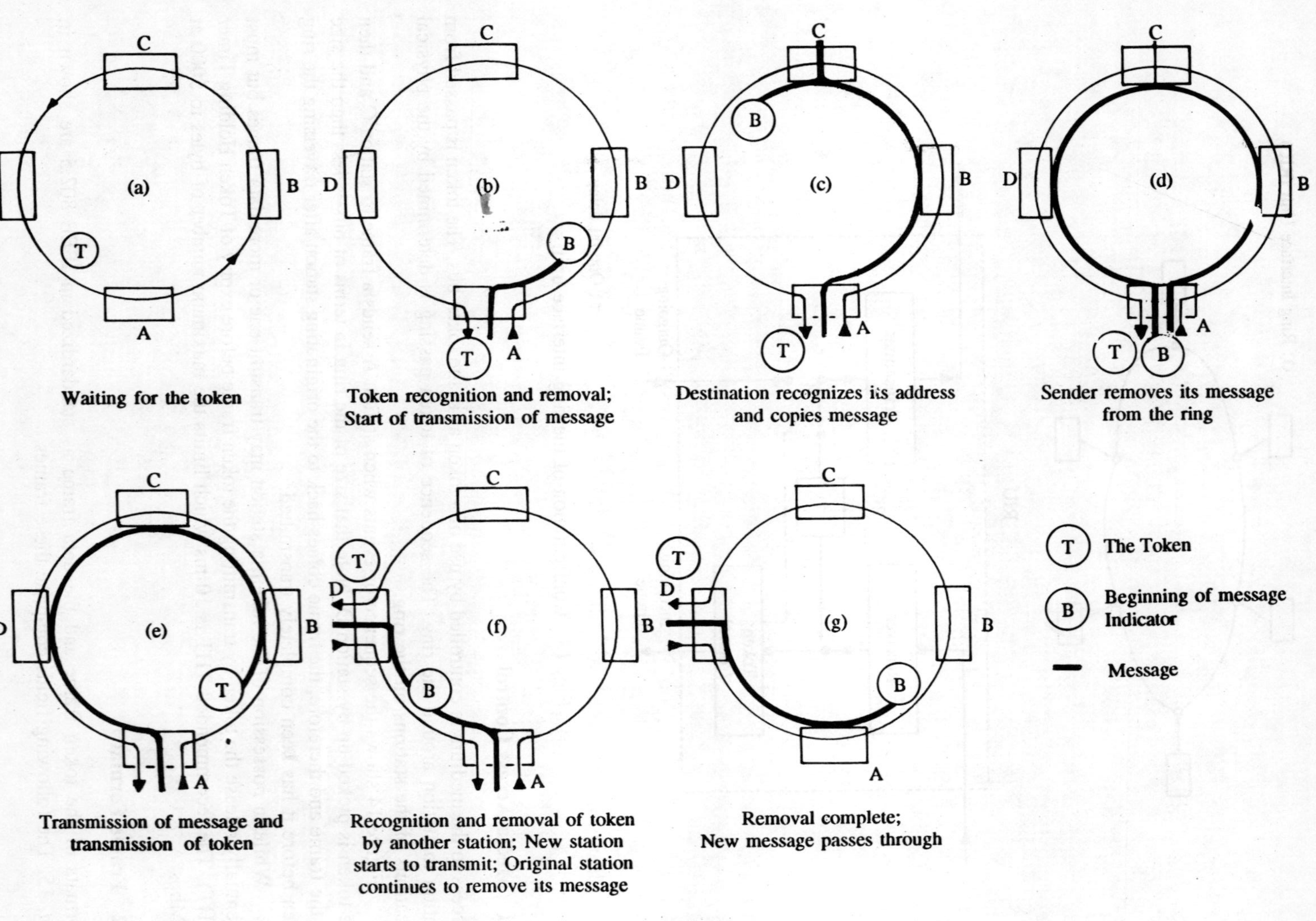

Fig. 14 Token passing in a ring.

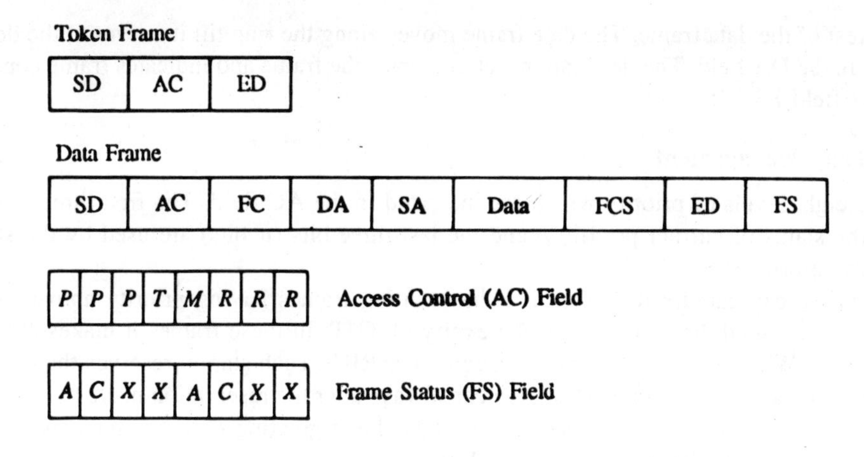

SD : Start Delimiter
AC : Access Control
ED : End Delimiter
FC : Frame Control
DA : Destination Address
SA : Source Address
FCS : Frame Check Sequence
FS : Frame Status

Fig. 15 Formats of IEEE 802.5 frames.

Start Delimiter (SD): It is a one octet long unique bit pattern which marks the start of the data or token frame.

Access Control (AC): It is a one octet long field containing priority bits (*P*), token bit (*T*), monitoring bit (*M*), and reservation bits (*R*).

Frame Control: It is a one octet-long field and indicates the type of frame, data frame or control frame. It also distinguishes different types of control frames.

Destination Address (DA): The destination address field is 2 or 6 octets long.

Source Address (SA): The source address field is also 2 or 6 octets long.

Data Field: It can have 0 or more octets. There is no maximum size but the frame transmission time is limited by the token holding timer.

Frame Check Sequence (FCS): The frame check sequence is 4 octets long and contains the CRC code. It checks on DA, SA, FC and data fields.

End Delimiter (ED): It is one octet-long and contains a unique bit pattern marking the end of a token or data frame.

Frame Status: This field is one octet-long and contains two address recognized bits (*A*), two frame copied bits (*C*) and reserved bits (*X*).

The fourth bit of the AC field is called the token bit. It distinguishes a token frame from the data or control frame. It is "0" in a token frame and "1" in other frames. A station which has a frame to send waits for the the token bit. If the token bit is "0", it seizes the token by disconnecting the ring at the RIU (Fig. 11). The one bit delay ensures that the token bit is not passed to the output before change over of the relay contact. The station inserts "1" in place of "0" and then continues with the rest of the data frame. The data frame moves along the ring till it arrives at the destination indicated in the DA field. The destination station copies the frame and indicates frame copied status in the last field FS.

6.3 Priority Management

There are eight levels of priority which are indicated in the AC field. The first three bits (*P* bits) indicate the status of current priority *p* and the last three bits (*R* bits) are used by the stations to reserve the priority *r*.

A station can capture the token if it has waiting frames having priority above the current priority *p*. It can send these frames till the expiry of THT. In these frames it makes *R* bits of the AC field zero. When these frames pass through other RIUs, each station reserves the next priority *r* in the AC fields of these frames. A station can always alter the reservation made by the preceding stations but only upward revision can be done. Thus, the originating station will receive the highest priority request *r* when its frame comes back to it.

When time comes for releasing the token, the station fixes the new priority in the token equal to the highest of the following and releases the token.

- Current priority *p*
- Reservation priority *r*
- Priority *s* of the waiting frames in the station.

But if a station raises the priority level of the token, it is responsible for subsequently downgrading it to the previous value. If it is not done, the priority level of the token will rise eventually to the highest level and will remain there. To avoid this, when a station releases a token with raised priority level, it puts the old value into its memory stack and waits for the token to pass by. This station is called the stacking station and it alone can reduce the priority level of the token to the old value. When the station sees the token with the same raised priority level passing by, it changes the priority bits of the token and restores the priority level.

6.4 Ring Management

One of the stations on the ring acts as an active monitor station. Its job is to identify and rectify various error conditions which include:

- Removing persistently circulating frames
- Detecting lost tokens
- Monitoring the priority mechanism.

Persistently circulating frames are detected by the monitor bit (*M*) of the AC field. This bit is set to "1" in a data frame by the frame-originating station. When the data frame passes by the active monitor station, it changes this bit to "0". If this frame is not removed by the originating station, the second round of the frame is immediately detected by the monitor station because the *M* bit is zero.

The active monitor station is also the master source of the clock integrated in the Manchester signal code of the data signal. It periodically sends Active Monitor Present frames to indicate its presence to the other stations. Any station can become an active monitor station. Other stations are in the standby monitor state, ready to take over the active monitor status if the current active monitor fails.

6.5 Physical Specifications

The data signal is transmitted as a differential Manchester encoded baseband signal. The type of the physical transmission medium is not specified in the IEEE standard 802.5. In practice, a shielded twisted pair cable is used. The data rate is from 1 to 4 Mbps.

7 OTHER RING ACCESS METHODS

Over the years several other mechanisms have been proposed for controlling access to the ring. Most of these are not based on any international standard. Register insertion and slotted ring are the two methods of some significance and these are discussed below.

7.1 Register Insertion Ring

In register insertion technique, a shift register is placed in parallel to the ring at each station. The shift register can be switched in/out of the ring as shown in Fig. 16. Normally, the register is switched out of the ring. When a station has a frame to send, it places the frame in the register and waits for a gap in the data stream being transmitted on the ring. When it finds a gap, it inserts the shift register into the ring. The frame stored in the register is sent bit by bit on the ring at the outgoing end of the shift register. The incoming data is read into the shift register from the other end. The register remains in the ring till the frame comes back. As soon as the returned frame gets stored in the shift register, the register is switched out of the ring so that the frame is removed from the ring.

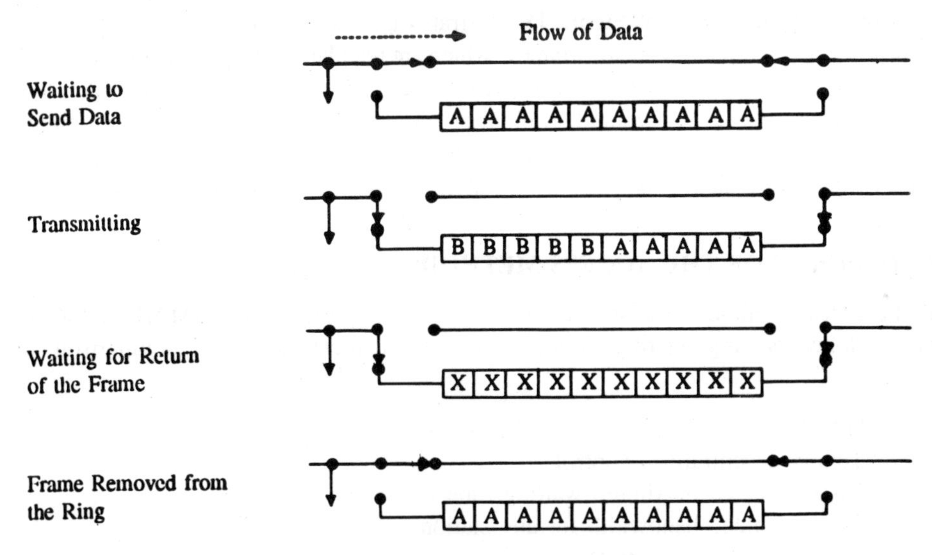

Fig. 16 Register insertion ring.

The figure above shows the configuration in simplified form. In real implementation, additional registers are required. Also note that in effect, the length of the ring increases every time a station inserts its register into it. Thus, the ring is able to accommodate more bits. The time to transport a frame from source to destination increases as more and more registers are inserted into the ring.

7.2 Slotted Ring

One particular version of the slotted ring access mechanism is also known as the Cambridge Ring. In this method, one or more empty slots circulate continuously round the ring. The format of the slot is shown in Fig. 17. If a station has data to send, it waits for the empty slot passed to it. The empty status of the slot is indicated by the status (ST) bit. The station changes the status bit indicating that the slot is being used and inserts source and destination addresses and data into the slot.

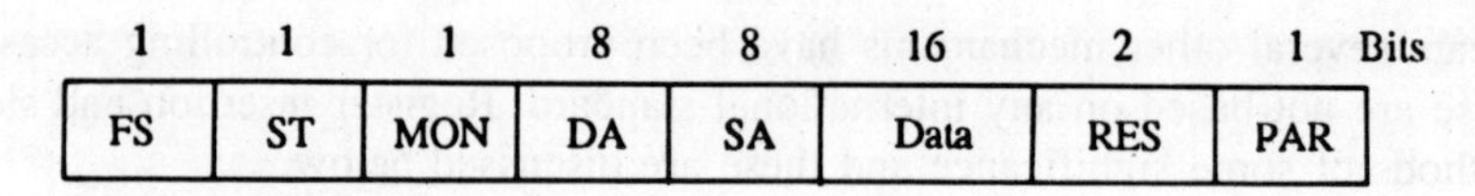

FS : Frame Start
ST : Frame Status (in use/not in use)
MON : Monitor Bit
DA : Destination Address
SA : Source Address
RES : Response Bits
PAR : Parity

Fig. 17 Format of the slotted ring frame.

The frame is passed from station to station until it reaches the destination. The destination station reads the frame and switches another marker (response bits) in the frame indicating that the frame has been read. The frame returns to the source which resets the status bit in the frame and hands it over to the next station. Figure 18 illustrates the mechanism.

It is necessary to monitor the "lost frames" which are not claimed by any station and continuously circulate in the ring. A frame with "in use" status needs to be reset after it has taken one round of the ring. Usually, these rings include a monitor station. This station generates an empty slot in the first place and then monitors the slot continuously. Use of the monitor bit (M) has been explained earlier. Here also, it is used in the same manner by the monitoring station.

8 COMPARISON OF THE ACCESS METHODS

Out of the access methods discussed above, the most important are CSMA/CD, token passing on bus and token passing on ring. Some comments on their qualitative comparison are given below:

- CSMA/CD
  - — Totally decentralized control
  - — No guaranteed maximum waiting time for access
  - — No guaranteed bandwidth for any station
  - — Short delay for light traffic.
- Token passing on bus
  - — Guaranteed maximum waiting time for access
  - — Easy expandability
  - — Moderate overhead of token
  - — High reliability
  - — Potential for high utilization of media.

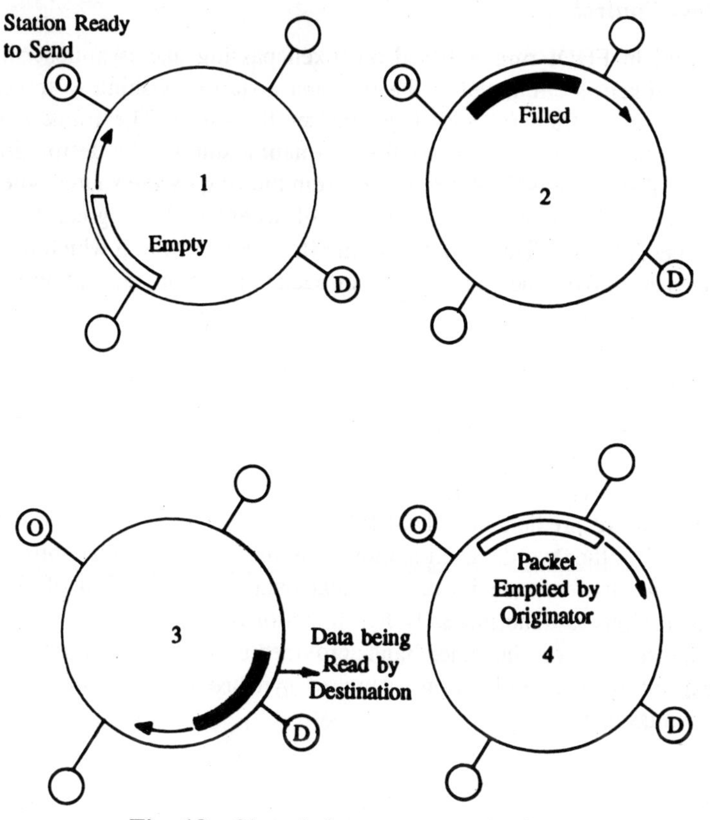

Fig. 18 Slotted ring access mechanism.

- Token passing on ring
  — Guaranteed maximum waiting time for access
  — Low overhead of token passing
  — Maximum utilization of media bandwidth
  — Easy expandability
  — High reliability.

9 FIBRE DISTRIBUTED DATA INTERFACE (FDDI)

Fibre Distributed Data Interface is the LAN standard based on optical fibre as the transmission medium. This standard is developed by ANSI and is given in X3.139-1987. It operates under IEEE 802.2 LLC sublayer allowing it to be integrated easily with other LAN standards.

FDDI was originally conceived as a back-end network interconnecting several hosts and high-speed peripherals. It can also be used as a backbone network interconnecting several front-end LANs. It has ring topology and 1000 stations can be connected on the ring.

Optical fibre as transmission media in a LAN enjoys several advantages over copper media. Its potential bandwidth is immense. It is thinner, lighter in weight and immune to electromagnetic interference.

9.1 Media Access Control

Media access control in FDDI ring is based on token passing and is almost similar to the one discussed earlier. A token is circulated in the ring and a station wishing to transmit captures the token. It transmits its data frame(s) and then releases the token. The frame carries source and destination addresses so that when it passes the destination station, the destination station retains a copy of the frame. The frame is finally removed from the ring by the source when it comes back. Because of the high data rate and the large number of accesses, the ring can have several frames circulating simultaneously in it. This is not so in the IEEE 802.5 ring in which the ring size is kept less than the frame size so that when a frame comes back to a station, the station is still transmitting the frame.

9.2 Types of Services

FDDI LAN provides for two types of services, namely, synchronous service, and asynchronous service.

Synchronous service is for real time applications in which bandwidth and response time are critical parameters and are predictable. FDDI LAN provides guaranteed bandwidth and response time to each station. In other words, each station is assured of token availability and a minimum token holding time for transmitting its time-critical data. The allocation of ring bandwidth for synchronous transmissions is done mutually by all stations.

Asynchronous service, on the other hand, provides dynamic bandwidth and is suitable for bursty traffic and interactive applications. Unused synchronous bandwidth is transferred for asynchronous transmissions.

9.3 Traffic Control

FDDI controls the traffic in such a manner that the token is rotated round the ring once in the Target Token Rotation Time (TTRT). In other words, total time for transmitting synchronous data and some asynchronous data by all the stations is less than TTRT. All the stations maintain two timers—Token Rotation Timer (TRT) and Token Holding Timer (THT). If the token arrives earlier than TTRT at a station, the leftover time is transferred to the THT. First, the synchronous data is transmitted upto the allotted time, and then asynchronous data is transmitted till expiry of the THT. The token is released immediately thereafter for the next station.

If the token arrives late, only the synchronous data is transmitted for the allotted time and the token is released thereafter. The maximum token rotation time can be 2 × TTRT beyond which a corrective action is necessitated. The stations send claim token frames indicating the required TTRT values. The station with the lowest TTRT wins and claims the token.

9.4 Priority Management

As mentioned above, synchronous data gets the top priority. For the asynchronous data, an optional priority scheme with eight levels of priority can be implemented. The basic scheme is the same as in the IEEE 802.5 token ring.

9.5 Ring Management

The ring management function includes ring initialization, ring monitoring, and token monitoring functions as explained in the IEEE token ring media access control. These functions are also implemented in the FDDI ring. The ring initialization function also involves allocation of synchronous bandwidth to each station.

9.6 Frame Format

The frame format is shown in Fig. 19. Each field is expressed in terms of four-bit long symbols. The following fields comprise the frame:

Preamble (PA): It is for bit synchronization and consists of 16 idle symbols.

Start Delimiter (SD): It is a flag indicating start of the frame and is two symbols long (11000 10001).

Frame Control (FC): It is a two symbols long field. It contains the following bits:

- a class bit (*C*) indicating whether the data is synchronous or asynchronous;
- an address length bit (*L*) indicating the size of address fields—16 or 48 bits;
- frame format bits (*FF*) and control bits (*NNNN*) indicating the type of frame—control or data frame, type of the control frame and control function.

Destination Address (DA): It is 4 or 12 symbols long.

Source Address (SA): It is 4 or 12 symbols long.

Information: It contains 0 or more data symbol pairs.

Frame Check Sequence (FCS): It is eight symbols long and checks on address fields, frame control field and information field.

End Delimiter (ED): It marks end of the frame and is one or two symbols long depending on whether the frame is a token frame or a data frame.

Frame Status (FS): It consists of three or more symbols and includes error detected, address recognized, and frame copied indicators.

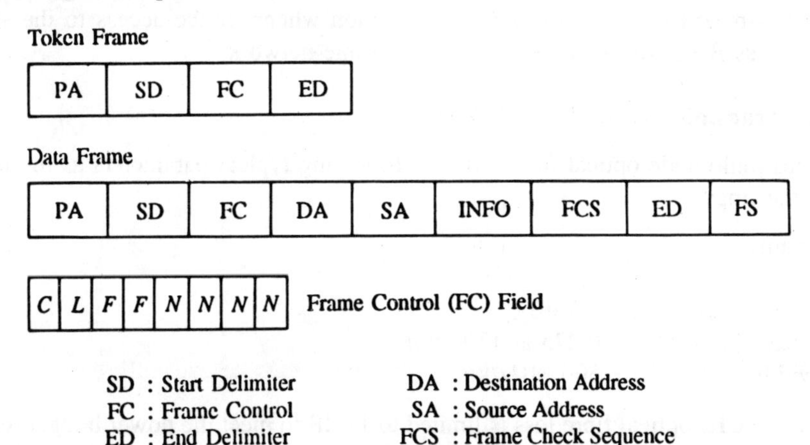

Fig. 19 Format of FDDI frame.

9.7 Physical Topology

The FDDI LAN is based on the dual ring concept to increase reliability (Fig. 20). The rings consist of point-to-point links, each less than two kilometres between neighbouring stations. The primary ring is used for all data transmissions. The secondary ring is used as a back-up in the event of a link or station failure. This reconfiguration of the ring occurs automatically and is an integral part of FDDI specifications. Optical bypass switches across each Channel Interface Unit (CIU) are provided as additional protection. When a station is down or is likely to be inactive for extended period, it is bypassed.

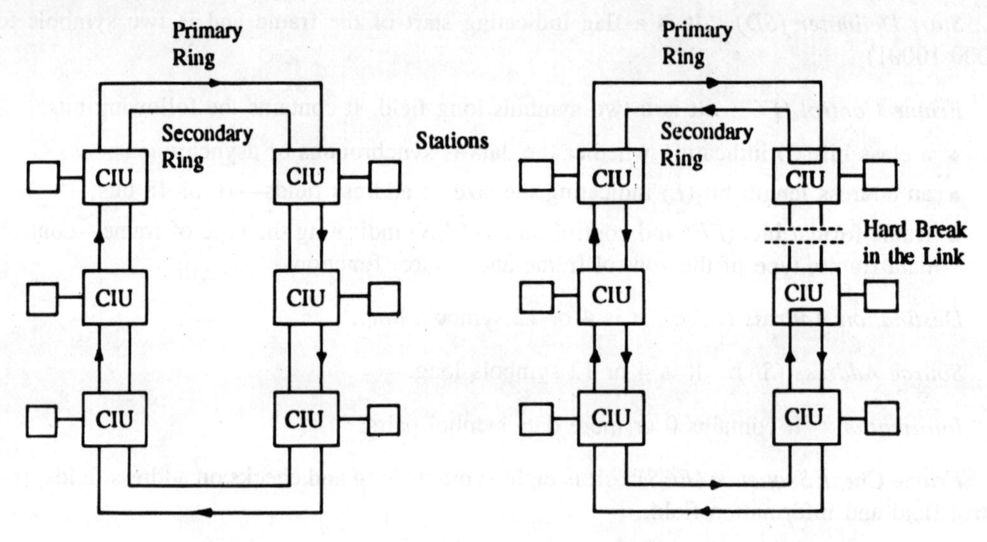

Fig. 20 Ring reconfiguration in the event of a link failure.

There are two classes of stations, Class A and Class B. Class A stations are connected using the dual ring concept illustrated above. Class B stations, on the other hand, do not have dual ring facility and are connected to the primary ring through a wiring concentrator (Fig. 21). The wiring concentrator has means of bypassing a Class B station whenever the access to the station fails. Class A and Class B stations can be mixed in the same network.

9.8 Optical Transmission Media and Devices

FDDI specifies multimode optical fibre with the following typical parameters as the transmission medium for the ring:

Core diameter	:	62.5 micron
Cladding diameter	:	125 micron
Loss	:	2.5 dB
Numerical aperture	:	0.275 at 1300 nm
Bandwidth	:	500 MHz/km

The CIU to CIU optical fibre loss is limited to 11 dB to meet the power budget requirements. The maximum length of an optical section is 2 km and the maximum total length of the ring including the secondary path can be upto 200 km.

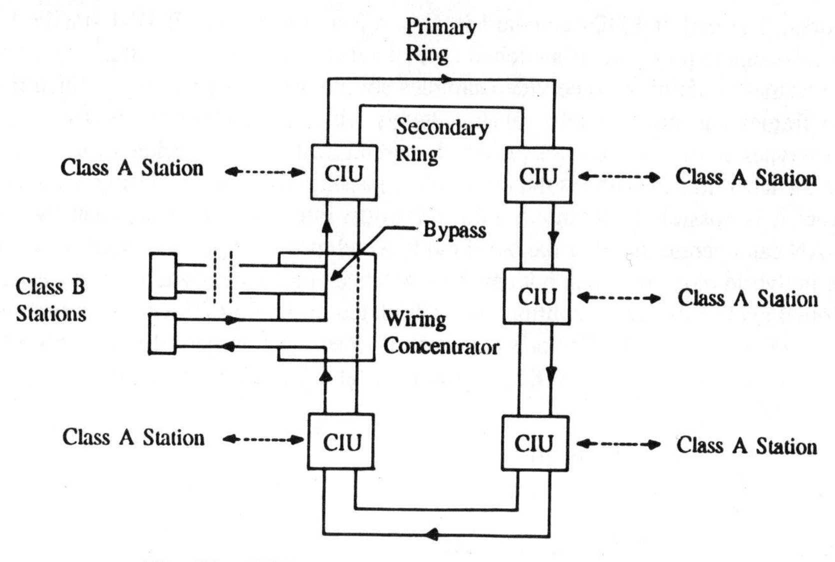

Fig. 21 Wiring concentrator for class B stations.

Low cost LED is specified as the transmission source. The operating wavelength of the optical signals is 1330 nm. The spectral width of the source is typically 140 nm.

9.9 Electrical Interface

The FDDI ring operates at the data bit rate of 100 Mbps. The electrical signal is encoded using 4B5B NRZ-M code to ensure clock recovery (Table 1). This scheme increases the baud rate to 125 Mbaud.

Table 1 4B5B Encoding Scheme of FDDI Ring

Symbol	5 Bit code
0000	11110
0001	01001
0010	10100
0011	10101
0100	01010
0101	01011
0110	01110
0111	01111
1000	10010
1001	10011
1010	10110
1011	10111

Symbol	5 Bit code
1100	11010
1101	11011
1110	11100
1111	11101
Start delimiter	11000 10001
End delimiter	01101
Non-data R	00111
Non-data S	11001
Idle	11111
Quiet	00000
Halt	00100

10 FDDI-II

FDDI-II is a new development which provides for transmission of data, voice and video over local area networks. The earlier FDDI standard is now referred to as the FDDI-I standard. Voice and video transmissions require circuit-switched type of service so that frame jitter[†] is within stipulated limits. The frames containing voice/video samples should arrive at precisely defined time instants. Usual data frames can have widely varying delays without affecting the performance. Thus, in FDDI-II, two types of transmissions are provided—circuit-switched and packet-switched transmissions. The packet-switched transmission is the usual token-passing data transfer. Using the circuit-switched transmission, it is possible to maintain a constant data rate connection between two stations. An FDDI-II LAN can operate in either the basic mode which provides packet-switched service just like FDDI-I, or in hybrid mode in which both packet-switched and circuit-switched services are available.

A slotted cycle structure is continuously circulated in the ring. The rotation time for the cycle structure is 125 microseconds. For one octet of information, this period represents 64 Kbps data rate. One voice channel requires 64 Kbps basic data rate. The cycle structure is shown in Fig. 22. It comprises the following:

1. Preamble (P), 5 symbols (20 bits)
2. Cycle Header (CH), 12 octets
3. Packet Data Group (PDG), 12 octets
4. 96 Cyclic Groups (CG).

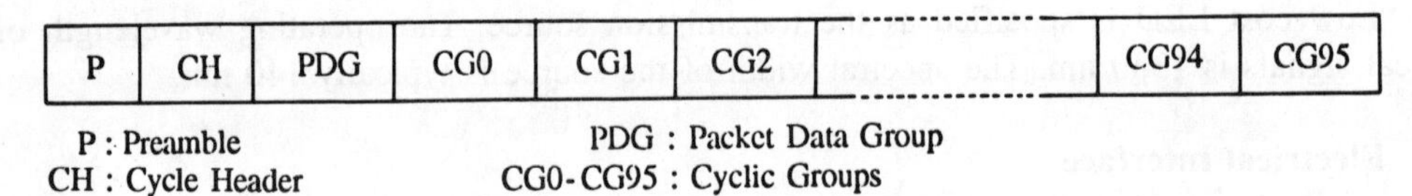

Fig. 22 FDDI-II cycle structure.

Each Cyclic Group contains 16 one-octet slots. Each slot represents one Wide Band Channel (WBC). Thus, there are 16 wide band channels and each cycle structure contains 96 bytes of each of the 16 WBCs. The stations use one of the WBCs for circuit switching applications. The WBCs are allocated to the stations on their request by a cycle master station. Whenever there are spare WBCs, they are used for packet-switching applications. The cycle header contains a Cycle Programming Template which shows the allocation of WBCs for circuit-switching and packet-switching. Since the rotation time of the cycle structure is 125 microsecond, 96 octets of one WBC will be delivered in 125 μs. Thus, bandwidth of each WBC is 6.144 Mbps. The PDG contains the minimum provision of the packet-switched data transfer applications. The format of the cycle header is shown in Fig. 23. It consists of the following fields:

Start Delimiter (SD): It indicates the beginning of a cycle and contains a unique pair of non-data symbols.

Synchronization Control Field (C1): It shows the synchronization state of the ring. It is set to non-data symbol *S* by the cycle master station when the synchronization is achieved. When a station notices that it is not receiving the cycle in 125 μs, it resets the field using the non-data symbol *R*.

[†]It refers to early or late arrival of frames.

Cycle Header

SD	C1	C2	CS	P0	P1	P2	...	P15	IMC

SD : Start Delimiter
C1 : Synchronization Control
C2 : Sequence Control
CS : Cycle Sequence
P0–P15 : Programming Template
IMC : Isochronous Maintenance Channel

Programming Template

R	*R*	*S*	*R*	*S*	*S*	*S*	*R*	*R*	*S*	*S*	*R*	*S*	*R*	*S*	*S*

Sample Cyclic Group

P	P	C	P	C	C	C	P	P	C	C	P	C	P	C	C

16 Slots

P : Packet Switched Data Channel
C : Circuit Switched Channel

Fig. 23 Format of the cycle header.

Sequence Control Field (C2): It indicates status of the cycle sequence field. Non-data symbol *S* shows a valid cycle sequence number and a non-data *R* an invalid sequence number.

Cycle Sequence (CS): During normal operation, the CS field contains a cycle sequence number which is incremented by the cycle master for each new cycle. The number varies from 64 to 255 and then restarts from 64. During the initialization phase, the CS field contains the rank of a monitor station. The station with the highest rank is given the cycle master status. The ranks vary from 0 to 63.

Programming Template Field (P0-P15): It consists of 16 symbols, one for each WBC. An *R* value indicates that the channel is part of the packet data channel. An *S* value shows that the channel is for circuit switched application. All the stations read the template, but only the cycle master can modify it.

Isochronous Maintenance Channel (IMC): It contains isochronous traffic for maintenance purpose.

The FDDI-II ring is operated at 100 Mbps. This bandwidth is divided into 16 WBCs and one packet-switching channel. The bandwidth allocation for circuit-switching applications and packet-switching applications is done by the cycle master station. The WBC capacity can be subdivided at bit level also. Each bit of a WBC represents a data channel of 8 Kbps.

The FDDI-II expands the power of the FDDI-I by supporting circuit-switched applications like voice and video on a local area network. It can be easily integrated with ISDN (Integrated Services Digital Network).

11 SUMMARY

Media access control methods based on distributed control do not have a single point of failure and are preferred. IEEE has standardized three methods of media access control, token passing on the bus, token passing on the ring, and CSMA/CD. All the three methods work under the LLC sublayer as per IEEE 802.2. In token passing methods, a token is circulated among the stations. The station holding the token has the right to transmit data frames. CSMA/CD is based on contention access. A station checks if the transmission medium is free and if the medium is free, it transmits its frames. If collision is detected during transmission, it aborts the frame and then tries again.

CSMA/CD has totally decentralized control but does not assure guaranteed bandwidth and maximum waiting time. Token passing on the bus and the ring assure maximum waiting period and very high utilization of the network.

Optical fibre as transmission medium offers significant bandwidth advantage. Optical fibre-based LAN technology is becoming popular and has a very high potential of growth. At present, LANs based on optical fibre use token passing on the ring media access control. FDDI-I, Fibre Distributed Data Interface is an approved ANSI standard and provides a bit rate of 100 Mbps. It provides synchronous and asynchronous services and is suited for real time applications. FDDI-II is a new development and is still under consideration for standardization. It offers the bandwidth for voice and video transmission in addition to the usual packet transmission bandwidth as in the other local area networks.

APPENDIX

Throughput of Pure ALOHA Channel

Throughput S is defined as average successful traffic transmitted between stations per unit time. The unit of time is slot-time which is the time required to transmit a frame. All the frames are assumed to have the same size. Clearly, we can successfully send at the most one frame per slot, and so the maximum value of S is 1. When collisions occur, some of the frames are lost and therefore, part of the available channel time is wasted. The resulting value of the parameter S is always less than 1.

Let us denote the offered traffic in the network by G. G is the average number of frames per slot time which are presented to the network for transmission by the stations. G can have any value from 0 to infinity. Subjectively, we would expect that when G is low, i.e., less than 1, there should be few collisions, and $S \approx G$. When G is high ($G \geq 1$), there will be many collisions, lot of wasted time and, therefore, S will be less than 1.

Let us assume a large number of stations N which transmit a frame with probability s per slot time. Obviously,

$$G = Ns$$

Channel time is wasted whenever there is overlap of two frames (Fig. A1). Overlap of any amount is always fatal and results in wastage of channel time. For successful transmission of

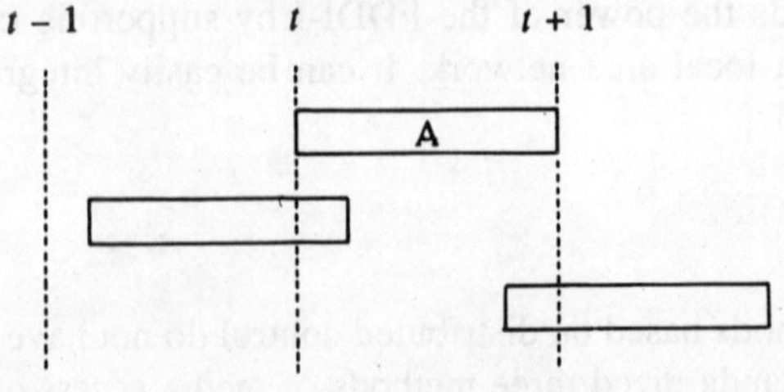

Fig. A1 Frames overlap.

frame A, it is necessary that there should not be any other frame after time $t - 1$ and upto time $t + 1$.

If we assume that the stations become ready to transmit independently of each other, the probability of transmission of a frame in a slot time by a station is s, and the probability of no transmission by the remaining $N - 1$ stations is $(1 - s)^{N-1}$. But the $N - 1$ stations should not transmit during the previous slot-time also. Therefore, the probability of no transmission by the $N - 1$ stations during two consecutive slot times is $(1 - s)^{2(N-1)}$. Thus the overall probability of single successful transmission is $s(1 - s)^{2(N-1)}$. Since there are N stations, the throughput of frames is given by

$$S = Ns\,(1 - s)^{2(N-1)}$$

Since N is large and s is small, the above expression for throughput can be written as

$$S = G\,e^{-2G}, \quad G = Ns$$

The plot of S with respect to G is shown in Fig. A2.

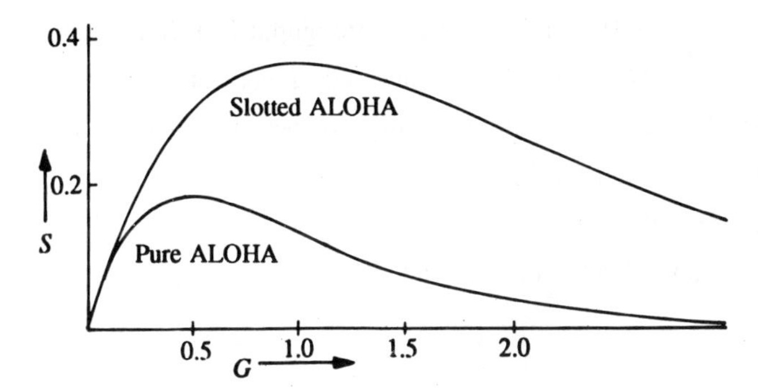

Fig. A2 Throughput vs. offered traffic.

The maximum throughput occurs at $G = 0.5$ and is equal to 0.184. This means the maximum throughput occurs when the stations on average generate 500 frames in 1000 slot times and out of these 500 frames, 184 frames would be successfully transmitted.

Throughput for Slotted ALOHA

In slotted ALOHA, the frames arrive in a synchronized fashion (Fig. 6). Note that collision can take place if and only if there are more than one transmission in a slot time. The probability of single transmission in a slot time is $s(1 - s)^{N-1}$ which gives the throughput S as given below for large N and small s.

$$S = Ns\,(1 - s)^{N-1} = Ge^{-G}$$

Figure A2 depicts the plot of S versus G. In this case, the maximum occurs at $G = 1$ and the maximum value of S is 0.368.

PROBLEMS

1. Consider a hypothetical token-passing bus LAN in which the data frames have a size of 1000 octets. The control frames have a size of 15 octets. The response window is 50 octets. The timers count down in terms of octets to arrive at the threshold at "0". The THT is 2000 octets. When the token is received by a station, the status of its TRTs is as follows:

TRT-4	3000 octets
TRT-2	3500 octets
TRT-0	4000 octets
TRT-M	4100 octets.

The station has three frames of each priority level for transmission. Indicate the frames transmitted by the station before it passes the token to the next station.

2. (a) Calculate the throughput S for a pure ALOHA network if the offered traffic G is 0.75.

(b) What is the average number of frames offered per second and the average number of frames successfully transmitted per second if the bit rate is 9600 bps and the frame size is 960 bits?

3. Show that for pure ALOHA, the maximum throughput is $1/2e$ and it occurs at $G = 0.5$.

4. Show that for slotted ALOHA, the maximum throughput is $1/e$ and it occurs at $G = 1.0$.

5. A slotted ALOHA channel has, on an average 10 per cent of the slots idle.

(a) What is the the offered traffic?

(b) What is the throughput?

(c) Is the channel underloaded or overloaded?

CHAPTER 13

Internetworking

So far we have studied the various types of local area networks and switched data subnetworks. These subnetworks more often need to be interconnected allowing users to communicate not just within one subnetwork but across several subnetworks. The extended network so formed is often called an Internetwork. Mere electrical connection of two subnetworks is not sufficient because the subnetworks differ in so many ways. Their technologies, topologies, protocols and their services all are different. Therefore, some special intermediary devices are required. These devices sort out the differences and enable the users on one subnetwork to communicate with users on another subnetwork.

We begin this chapter with identification of the internetworking problem. The simplest internetwork device, the bridge is discussed first. We examine its layered architecture, operation, and routing alternatives. Then we move on to another internetwork device called the router. The discussion on the router is divided into two broad sections, one devoted to providing connectionless-mode Network service and the other devoted to providing connection-mode Network service. We discuss some very important protocols and their applications in internetworking and we also examine how the internetwork can utilize the services of the X.25 packet-switched data subnetwork.

1 INTERNETWORKING DEVICES

Figure 1 shows three subnetworks that are interconnected. The devices which interconnect the subnetworks are often called bridges, routers or gateways. The network of the subnetworks so formed is called internetwork.

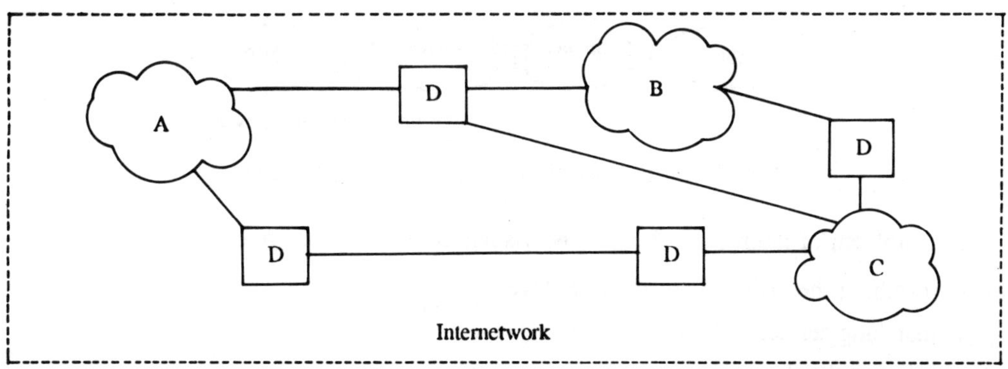

Fig. 1 Interconnected subnetworks.

Figure 2 shows some possible internetworking configurations. In Fig. 2a two similar or dissimilar LANs which are adjacent to each other are interconnected. One would think that it should be simpler to form a bigger LAN but it may not be always possible to do so because of the

- constraint of the maximum number of stations which can be connected on a single LAN,
- constraint of the maximum physical size of a LAN,
- security reasons, and
- investments already made.

Figure 2b shows an internetwork of LANs which are geographically separated. They may be located in different corners of a city or in different cities. In this case the two LANs require a connection which can be a leased telephone circuit or a data circuit through the packet switched data subnetwork (Fig. 2c). Figure 2d shows the case of interconnecting a host and a LAN under similar circumstances. If the two LANs in Fig. 2b are located in areas served by different subnetwork operators, interconnection may also involve connecting the two packet-switched data subnetworks as shown in Fig. 2e.

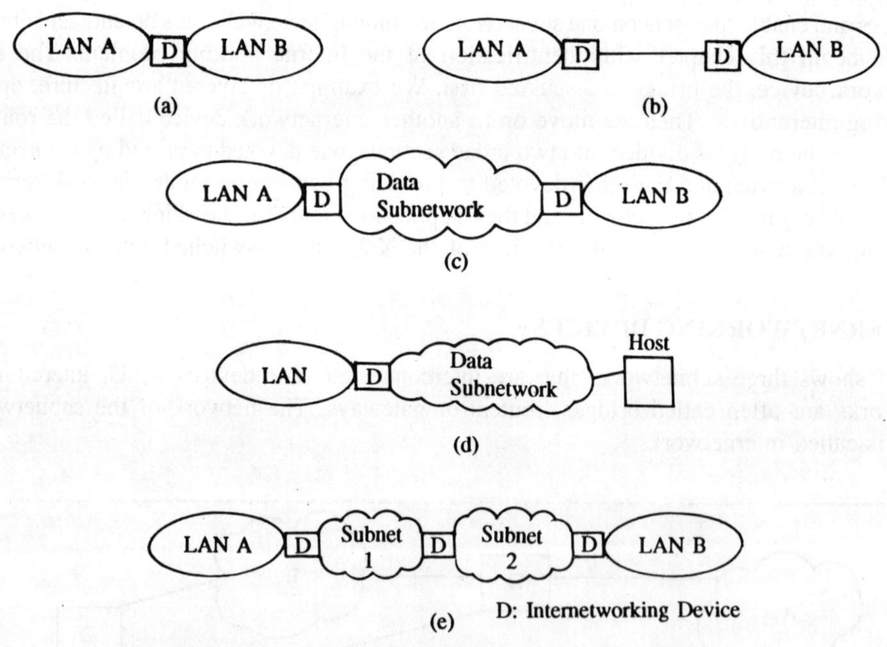

Fig. 2 Some possible internetwork configurations.

The problem of interconnection can be visualized as the problem of

- matching the protocols of the peer layers,
- matching services offered by different layers,
- relaying the data units from one subnetwork to another, and
- routing the data units from one subnetwork to another.

These functions are carried out by bridges, routers and gateways. All these devices have two or more physical ports and offer matching sets of protocols (Fig. 3).

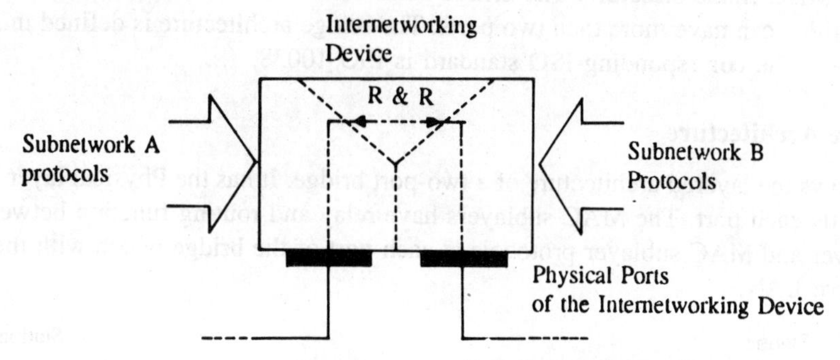

Fig. 3 Internetworking device with matching set of protocols.

It may be mentioned that a repeater is not an internetworking device because it operates within a subnetwork. The primary function of a repeater is to regenerate data signals when two segments of the same LAN are joined together to extend the LAN coverage. Figure 4 shows the layered architecture of a repeater. Note that it has only the Physical layer to regenerate the signals.

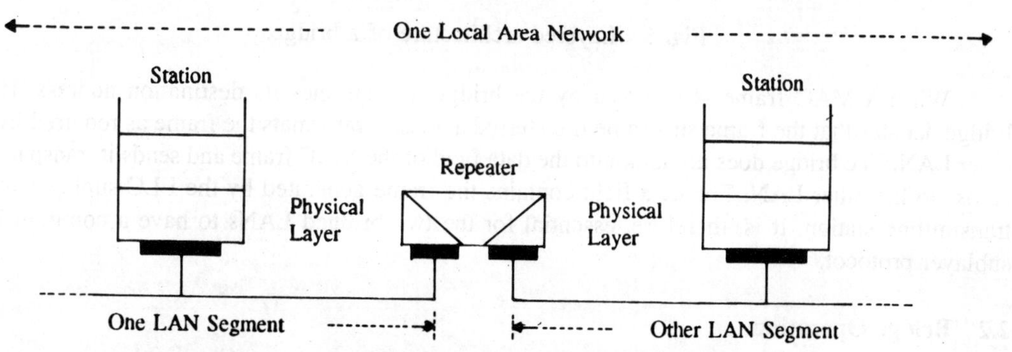

Fig. 4 Repeater as signal regenerating device.

The different internetworking devices mentioned above are not interchangeable, although they serve the same purpose. Their capabilities are different. A bridge can interconnect two networks at the level of the MAC sublayer. The router carries out internetworking at the Network layer. A gateway interconnects two network architectures at the highest layer. We will first take up the simplest internetworking device, the bridge.

2 BRIDGE

The bridge is a device that is attached to two or more local area networks to create an *extended* LAN. It takes MAC frames from one LAN and sends them across to the other LAN. A frame is transferred across the bridge only if its destination is in the other LAN. Note the difference between a bridge and a repeater. A repeater does not look at the address field and regenerates all the frames.

The interconnected LANs need not be of the same type e.g., a CSMA/CD LAN can be interconnected using a bridge to an FDDI LAN. As we saw in the last chapter, each type of LAN has its own MAC frame structure. The bridge takes care of these differences by reformatting the frames. A bridge can have more than two ports. The bridge architecture is defined in IEEE 802.1d, *MAC Bridges.* The corresponding ISO standard is ISO 10038.

2.1 Bridge Architecture

Figure 5 shows the layered architecture of a two-port bridge. It has the Physical layer and the MAC sublayer at its each port. The MAC sublayers have relay and routing function between them. The Physical layer and MAC sublayer protocols at each port of the bridge match with the protocols of the respective LAN.

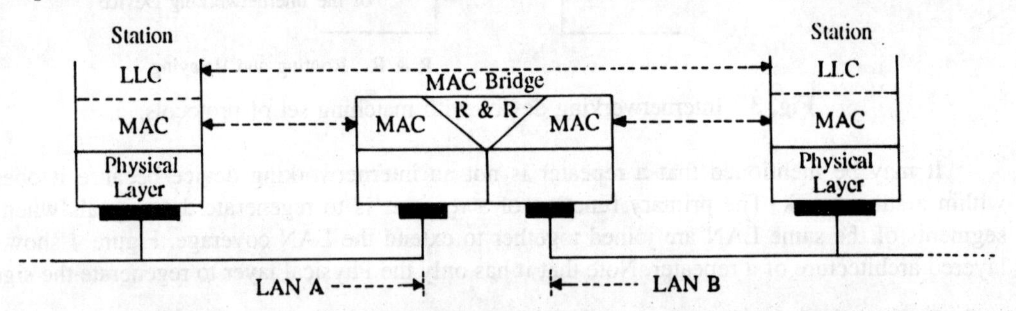

Fig. 5 Layered architecture of a bridge.

When a MAC frame is received by the bridge, it examines its destination address. If the bridge decides that the frame should be transferred across, it reformats the frame as required by the other LAN. The bridge does not look into the data field of the MAC frame and sends it transparently across to the other LAN. The data field contains the frame generated by the LLC sublayer of the transmitting station. It is, therefore, essential for the two bridged LANs to have a common LLC sublayer protocol.

2.2 Bridge Operation

The operation of LAN bridges is usually transparent to the stations, i.e., the stations on the extended LAN communicate with each other as if the bridge is not present. A bridge performs three basic functions:

- Frame filtering and forwarding
- Learning the addresses of the stations
- Routing.

Frame Filtering and Forwarding. As LANs operate in broadcast mode, all frames, irrespective of their destination, appear at the port of the bridge connected to the LAN. The bridge maintains a data base of the MAC addresses of all stations on the extended LAN. The data base also contains information on the physical port at which a particular address is available. When the bridge receives a frame at any of its ports, it takes one of the following actions:

1. If the destination address on the frame is a global address, it sends the frame on all its physical ports except the one from which it received the frame.

2. If the destination address is available on the same port through which it received the frame, the bridge discards the frame.

3. If the bridge determines that the destination address is available on a different physical port than the one through which the frame was received, it forwards the frame on to that port.

4. If the bridge does not find the address in its data base, it forwards the frame over all its physical ports except the one from which it received the frame.

Thus, the bridge ensures that the frames are seen by their destination stations wherever they may be located in the extended LAN.

Learning Addresses. When a bridge first comes up, its data base is empty. It builds up the data base by examining the source address field of the MAC frames. Whenever the bridge finds that the source address is not in the data base, it updates the data base along with the port at which the frame was received. Thus, after it is switched on, a bridge builds its data base by self-learning when the stations exchange the frames.

All the entries in the data base have a lifetime and are deleted from the data base if they are not updated. If a station is moved from one LAN to another, its old entry is deleted and the new entry is rebuilt automatically. Similarly, if a station is down for some time, its entry will be deleted automatically. Entries of the active stations are continuously updated by the frames released by them.

The above scenario where a bridge makes frame forwarding decisions based simply on the destination address and the data base is applicable to two LANs interconnected by a single bridge. When there are multiple LANs interconnected by multiple bridges, the duplicate routes between a pair of stations create some problems in the self address learning mechanism.

Consider that there are two bridges B1 and B2 interconnecting two LANs—A and B (Fig. 6). When station X on LAN A transmits a frame addressed to station Y on LAN B, both the bridges will note that station X is available on LAN A and will forward the frame to LAN B. These frames will be received by station Y and by the bridges as well. When the frame released by bridge B2 in LAN B is received by bridge B1, bridge B1 will immediately update its data base thinking

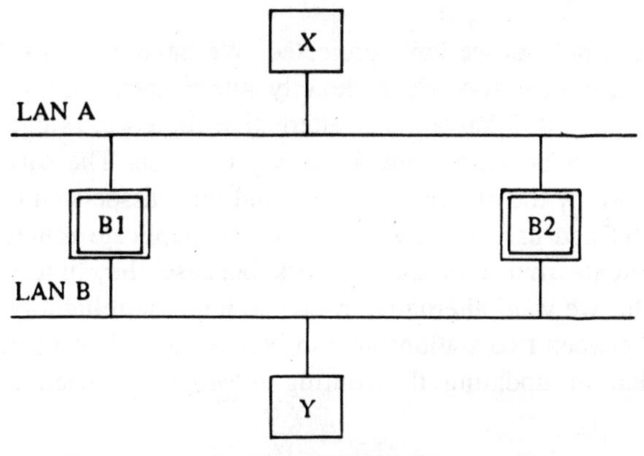

Fig. 6 Duplicate routes through bridges.

that station X is available on LAN B. Bridge B2 also does the same when it receives the frame released by B1 from LAN B. Although station X is on LAN A, the data bases of the bridges do not indicate so. To overcome this problem, it is necessary to eliminate the duplicate paths between any two stations. We will see in the next section how the routing methods achieve this.

Routing. When there are multiple LANs and multiple interconnecting bridges, the bridges need to be equipped with routing capability. Consider the configuration shown in Fig. 7. Suppose

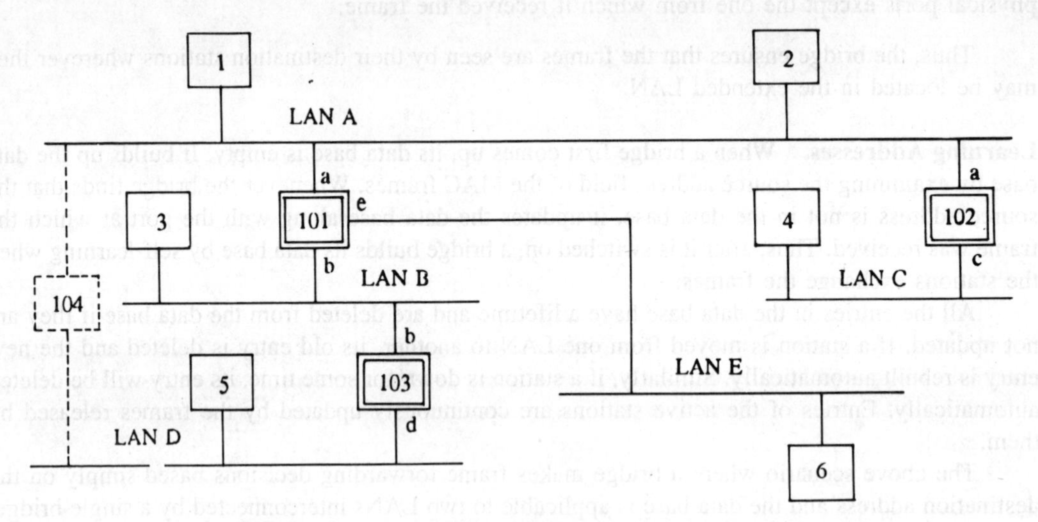

Fig. 7 Routing in multiple interconnected LANs.

station 1 on LAN A sends a frame for station 5 which is on LAN D. The frame will be read by bridges 101 and 102. For each bridge, the addressed station is not on the interconnected LAN. Therefore, both these bridges must make a decision whether or not to relay the frame to the next LAN. Bridge 102 should refrain from relaying the frame. Bridge 101 has two alternatives, LAN B or LAN E, for forwarding the frame. It must make the routing decision in favour of LAN B because it is connected to LAN D. Thus, we see that a bridge must be equipped with the routing information.

Routing is not as simple as we have projected. We have not considered alternative routes. Suppose there is another bridge 104 which directly interconnects LAN A to LAN D. Now the frame from station 1 to station 5 has another alternative route via bridge 104. Perhaps the route through bridge 104 is better because it involves only one hop. The earlier route had two hops. Thus, the bridges must know the alternative routes and their associated costs in terms of number of hops or any other defined parameter. Alternative and duplicate routes must be distinguished. We do not want duplicate routes in the network because they interfere in the self address learning mechanism. But we want alternative routes to increase reliability. Therefore, only one of the alternative routes between two stations should be used at a time to transfer a frame. Another routing problem is that of updating the routing information when changes in the network configuration occur.

3 ROUTING STRATEGIES

A variety of routing strategies have been proposed and implemented in recent years. The following three are the most important.

- Fixed routing
- Spanning tree routing
- Source routing

We examine these strategies in turn.

3.1 Fixed Routing

In fixed routing, each bridge maintains a routing table which indicates the destinations and the ports to be chosen for each destination. Out of the alternative routes, the one with the minimum number of hops is indicated in the table. The routes are fixed and the tables are updated as and when there are changes in the topology. Table 1 shows an example of the routing table maintained in bridge 101 of Fig. 7. We have assumed that bridge 104 is also present.

Table 1 Routing Table Maintained at Bridge 101

Frames arriving at port a		Frames arriving at port b		Frames arriving at port e	
Destination LAN	Port	Destination LAN	Port	Destination LAN	Port
B	b	A	a	A	a
C		C	a	C	a
D		D		B	b
E	e	E	e	D	b

The fixed routing strategy is widely used due to its simplicity. But it has very limited capabilities to handle configuration changes which may due to station failures, subnetwork failures and new additions to the subnetworks.

3.2 Spanning Tree Routing

A spanning tree is a graph structure that includes all the bridges and stations on the extended LAN but in which there is never more than one active path connecting any two stations. In Fig. 7, if either bridge 103 or bridge 104 is removed, the resulting structure of the extended LAN will be a spanning tree. The frames can be easily routed through the spanning tree without any worry about the duplicate paths. What we now need is an algorithm by which the bridges will be able to derive the spanning tree automatically. The algorithm must be dynamic so that topological changes are automatically discovered and taken care of. Before we go into the algorithm, let us understand some of the definitions.

Root Bridge: Each bridge has a unique identifier. The bridge with the lowest identifier is called the root bridge.

Root Path Cost: Each port of a bridge has an associated cost parameter which is the cost of transmitting a frame through the particular port. When a frame traverses a path through several bridges, the path cost is the sum of all the intervening port cost parameters. Note that the cost is added only for transmission of a frame through a port and not for its reception at a port. Root path cost is the minimum path cost from a bridge to the root bridge.

Root Port: Each bridge determines its port through which, if a frame is transmitted, it will reach the root bridge incurring the root path cost. This port of the bridge is called the root port.

Designated Bridge and Designated Port: If a LAN has several bridges connecting it to the root bridge, one of the bridges is called the designated bridge and all the frames from the LAN are transmitted through the designated bridge. The corresponding port of the bridge is called the designated port. The designated bridge is chosen based on minimum path cost and in case more than one bridge have the same path costs, the bridge with lower bridge identifier is chosen as the designated bridge.

Constructing the Spanning Tree: Constructing the spanning tree involves the following steps:

1. Identification of the root bridge
2. Determination of the root ports
3. Identification of the designated bridges and ports.

To carry out these steps, bridges exchange Bridge Protocol Data Units (BPDUs). A BPDU transmitted by a bridge contains the following information:

1. Bridge identifier
2. Bridge identifier of the root as per this bridge
3. Root path cost of the bridge.

Let us consider the configuration shown in Fig. 8. Each bridge has a bridge identifier and each port has a port cost as shown in the figure.

To begin, each bridge asserts that it is a root bridge by sending a BPDU. On any given LAN, only one claimant will have the lowest bridge identifier and will maintain this belief. The others will accept this fact by comparing the bridge identifiers in the BPDUs they receive. Thus, bridge 101 will be identified as the root by bridges 102 and 103. They will identify the root port as well. Bridge 104, however, will consider bridge 102 as the root.

In the next step, bridge 102 and 103 release BPDUs indicating the root bridge identifier and the path cost to the root. These BPDUs are released only through ports other than the root port. Thus, the BPDU released by bridge 102 in LAN C will indicate root bridge identifier as 101 and the root path cost as 5. Similarly, the BPDU released by bridge 103 in LAN C and LAN D will indicate the root bridge identifier as 101 and the root path cost as 10. When bridge 104 receives these BPDUs, it will realize that the root identifier is 101 and the root is accessible through bridge 102 at a lower path cost of 5.

Therefore, bridge 102 is the designated bridge for LAN C and the port of bridge 102 connected to LAN C is chosen as the designated port for transmission of the frames to the root. The port of bridge 103 connected to LAN C is put in the blocked state.

Bridge 104 will further propagate this information to other bridges connected to LAN E. It

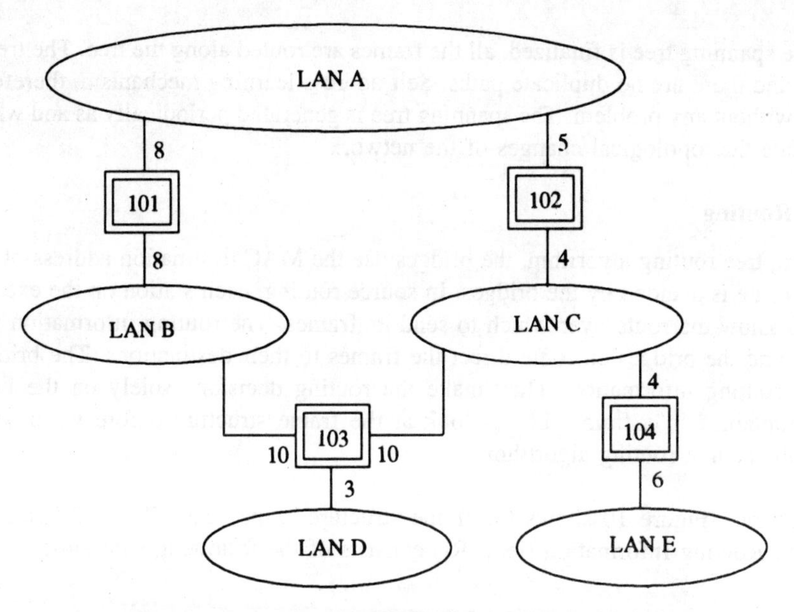

Fig. 8 Example of multiple interconnected LANs.

will indicate the root identifier as 101 and the root path cost as 9. This process continues and finally we have the spanning tree with no loops (Fig. 9).

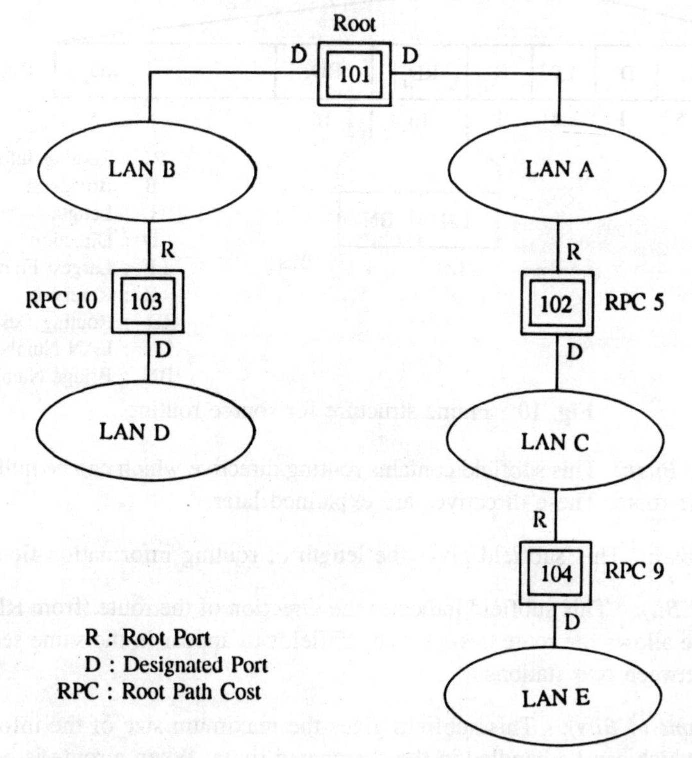

Fig. 9 Spanning tree for the configuration shown in Fig. 8.

Once the spanning tree is finalized, all the frames are routed along the tree. The tree connects all the LANs and there are no duplicate paths. Self address learning mechanism, therefore, can be implemented without any problem. The spanning tree is generated periodically as and when needed to accommodate the topological changes of the network.

3.3 Source Routing

In the spanning tree routing algorithm, the bridges use the MAC destination address of a frame to direct it. The route is decided by the bridges. In source routing, each station on the extended LAN is expected to know the route over which to send its frames. The routing information is included in the frames and the bridges en route direct the frames to their destinations. The bridges do not maintain any routing information. They make the routing decisions solely on the basis of the information contained in a frame. Let us look at the frame structure before we proceed to the operation of the source routing algorithm.

Frame Structure. Figure 10 shows the frame structure as per the IEEE 802.5 source routing document. The Routing Information field (RI) consists of the following subfields:

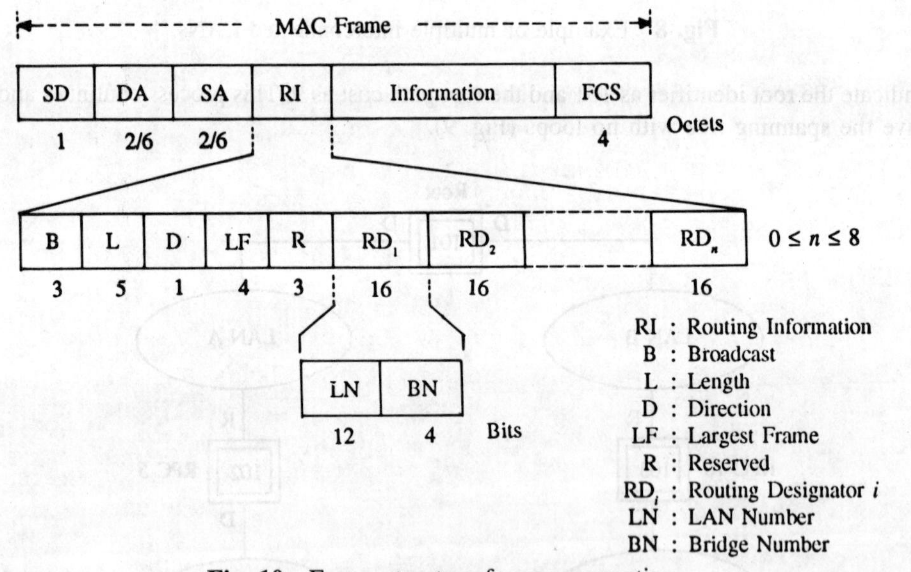

Fig. 10 Frame structure for source routing.

Broadcast (3 Bits): This subfield contains routing directive which can be null, non-broadcast, all-routes or single route. These directives are explained later.

Length (5 Bits): This subfield gives the length of routing information field in octets.

Direction (1 Bit): This subfield indicates the direction of the route (from RD_1 to RD_n or the other way). Its use allows the route designation subfields to appear in the same sequence in the to and fro frames between two stations.

Largest Frame (4 Bits): This subfield gives the maximum size of the information field of the MAC frame which can be handled in the designated route. When a route is being discovered, the bridges en route indicate the largest size of the information field which can be sent on the route. Each bridge updates this field if the current value in the subfield exceeds what the bridge or the adjoining LAN can handle.

Reserved (3 Bits): This subfield is reserved.

Route Designator Subfield, RD_i (16 Bits): It contains a 12-bit LAN number and a 4-bit bridge number. The route is from RD_1 to RD_n or the other way depending on the direction bit.

Routing Directives. Each frame contains one of the following four routing directives. These directives are coded in the broadcast subfield (B) of the MAC frame.

(i) *Null*: This directive indicates to the bridges that no routing is required. It is used in the frames when the destination and the source stations are available on the same LAN.

(ii) *Non-Broadcast*: This directive instructs the bridge to follow the route given in the route designator (RD_i) subfield.

(iii) *All-Routes-Broadcast*: This directive to the bridges instructs them to forward the frame to all their ports except the one from which the frame was received. In this case the destination station will receive multiple copies of the frame from all possible routes.

(iv) *Single-Route-Broadcast*: This directive results in appearance of the frame only once in each LAN. Thus, the destination station will get only one copy of the frame. This is achieved by sending the frame through a spanning tree.

Route Discovery. Before a station may transmit a frame, it is necessary to discover the route to the destination station. There are two alternatives:

1. The source station sends an all-routes-broadcast directive in its bid to determine the route to a particular destination station. Each bridge en route adds the Route Designator (RD_i) sub-field in which it indicates its own number and the LAN number in which the frame is released by the bridge. It also updates the Largest Frame (LF) subfield if required. To prevent multiple visits to the bridge due to looping, each bridge also checks if its route designator is already present in the RI field. If so, it does not forward the frame. Eventually, multiple copies of this frame reach the destination station and each copy indicates a different route. The destination station then sends a response frame using non-broadcast directive on each route. When these frames are received, the source station will have a number of alternative routes which it can use to send the information frames.

2. In the second alternative, the source station sends single-route-broadcast directive in its bid to determine the route to a destination station. This frame is released only once in each LAN. When the destination station receives the frame, it responds with an all-routes-broadcast directive. The source station will eventually receive multiple copies of the response through different routes which are indicated in the RI field of these frames. Thus, the source station will have a number of alternative routes.

Once the routes are discovered, the source station selects one of the alternative routes, indicates it in the RI field and sends the frame. The choice of the selected route can be made either randomly or based on the number of hops. To account for the possibility of topological changes in the network, route discovery procedures are repeated after fixed time intervals.

3.4 Spanning Tree vs. Source Routing

1. The spanning tree approach does not alter the MAC frame format while source routing alters the format.

2. The spanning tree approach limits the use of redundant bridges to a standby role. Only the designated bridges forward the traffic. Thus, there is no load sharing. In source routing, a station can dynamically change the route if there is congestion in a particular route. Thus, all the bridges effectively participate in handling traffic.

3. In source routing, an optimal route is discovered and all the frames take the optimal route unless there is congestion. In the spanning tree approach, the objective is to find a route through the root. It may not be the optimal route between two stations.

4. In source routing, the frame size is enlarged, which effectively increases the traffic. Also, the route discovery procedure is very resource-intensive. It can be shown that the number of frames transmitted for route discovery is of the order of B^L, where B is the average number of bridges in each LAN and L is the number of LANs.

4 REMOTE BRIDGES

We have so far considered local bridges which interconnect contiguous LANs. Only one bridge is required between two LANs. Consider another situation where two LANs are located at some distance. To interconnect these LANs one bridge may not serve the purpose because of the distance limitation of the LANs or the non-availability of a high-speed transmission medium. One alternative is take a full duplex leased connection from the telephone network operator and connect the LANs using two bridges, one at each end of the leased connection. These bridges are called remote bridges (Fig. 11).

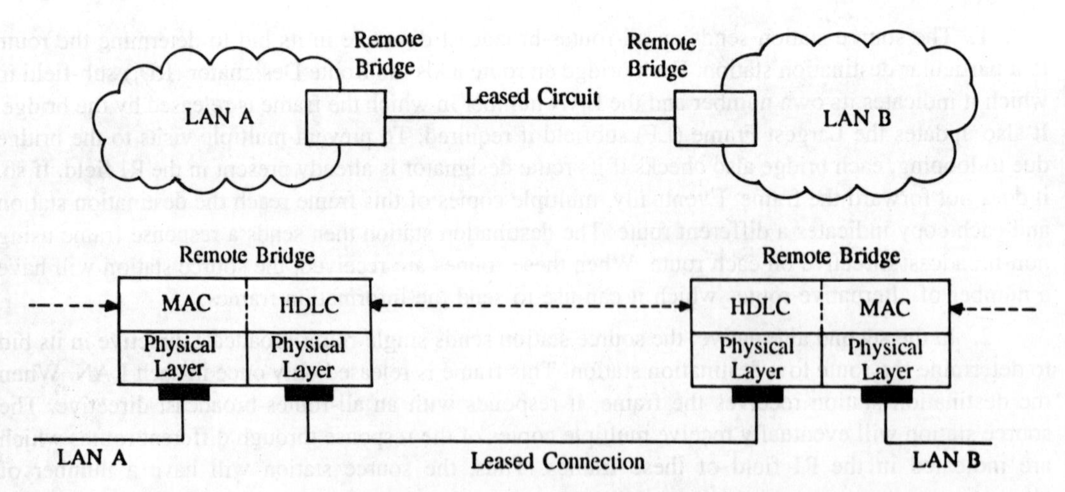

Fig. 11 Layered architecture of internetworking using remote bridges.

At one of the ports of a remote bridge, HDLC protocol is implemented. The other port has the MAC protocol of the local area network. The bridges establish a Data Link connection between them through the leased circuit and then carry out the bridge operation. HDLC protocol takes care of the transmission errors of the leased connection. Note that the HDLC protocol implements the

MAC frame transport service. The MAC frame is encapsulated in the information field of an HDLC frame using a header and trailer at the transmitting end. At the other end, the MAC frame is delivered after stripping off the HDLC header and trailer.

5 ROUTERS

A bridge, as we saw above, can interconnect LANs having different media access control sublayers but with a common LLC protocol. If there are differences in the protocols at the Data Link and Networks layers of two subnetworks, another internetworking device called a router is required to interconnect them. It operates at the Network layer and accommodates all the differences of the subnetworks upto this layer to provide a uniform Network service to the Transport layer entities in the end systems.

Since the router operates at the Network layer, its configuration is determined by

- The type of the Network service, connectionless-mode Network service (CLNS) or connection-mode Network service (CONS) to be provided to the Transport entities; and
- The type of underlying subnetwork service.

The subnetwork service could be a Logical Link Control (LLC) service, Data Link service provided by HDLC protocol or connection oriented virtual circuit service of the X.25 subnetworks. Internetworking the subnetworks, with different underlying services and providing a uniform Network service, call for matching sets of protocols to be implemented in the Network layer of the end systems and in the routers (Fig. 12). The Network entities in the end systems provide the required kind of Network service and interface with the router. The Network layer of the router contains the

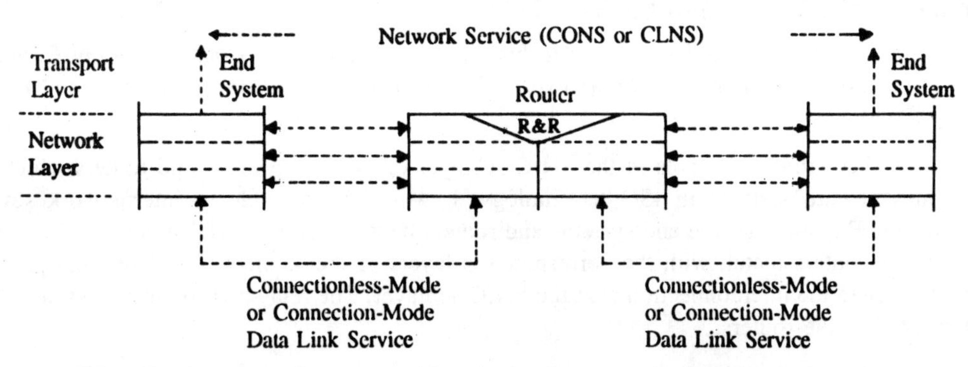

Fig. 12 Router as internetworking device between dissimilar subnetworks.

protocols of the subnetworks being interconnected and also a routing and relaying function which transfers the N-PDUs from one subnetwork to the other. Recall that the Network layer contains three sublayers to provide its diverse functionality. We need to select a suitable stack of the Network layer protocols to meet the above requirements. There is ongoing controversy about the merits and demerits of CONS and CLNS. Nevertheless, it is a fact that most of the LAN installations provide connectionless-mode Network service. ISO had originally formulated CONS but later came out with the standards for CLNS as well. It appears that CLNS will continue to gain ground and push CONS into the corner in the future. We will therefore concentrate more on the ISO Internet Protocol which provides CLNS than on other Network layer protocols which provide CONS.

6 ISO 8473 INTERNET PROTOCOL

The ISO Internet protocol is described in ISO 8473, *Protocol for Providing the Connectionless-mode Network Service.* ISO 8473 specifies the protocols for SNIC sublayer which resides in the end systems and provide CLNS to the Transport entities in the end systems. A part of ISO 8473 concerns subnetwork dependent convergence functions sublayer (SNDC) also. We will discuss it later.

The Internet Protocol operates using the connectionless-mode service provided by the next lower layer which may be SNDC sublayer as per ISO 8473 or LLC sublayer of the Data Link control layer (Fig. 13).

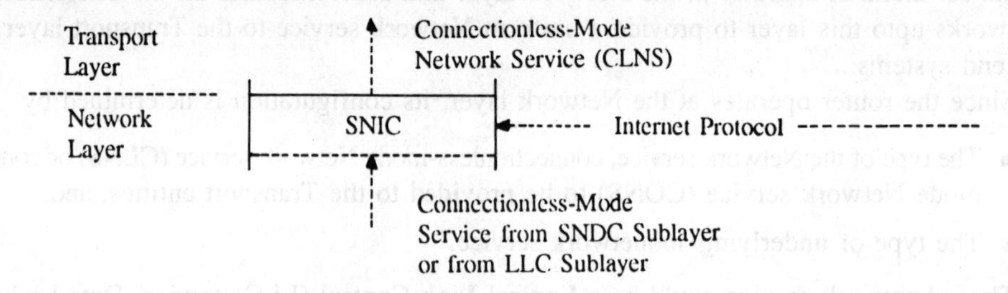

Fig. 13 Internet protocol (ISO 8473) for connectionless-mode Network service.

ISO 8473 is designed to operate between end systems connected by arbitrary number of subnetworks of various kinds. There are two more related protocols:

1. ISO 9574, *End System to Intermediate System Routing Exchange Protocol for Providing the Connectionless-mode Network Service.*

2. ISO 10589, *Intermediate System to Intermediate System Intra-Domain Routing Exchange Protocol for Use in Conjunction with the Protocol for Providing the Connectionless-mode Network Service (ISO 8473).*

Using these three protocols in the SNIC sublayer of the end systems and routers which are called intermediate systems in ISO terminology, it would be possible to internetwork several subnetworks (Fig. 14). All the end systems and routers have matching SNIC sublayers. The lower layers of the routers match with the corresponding layers of the subnetworks. Note that just like the bridges, here the interconnection is at the SNIC sublayer. The relay and routing function is also implemented in the routers.

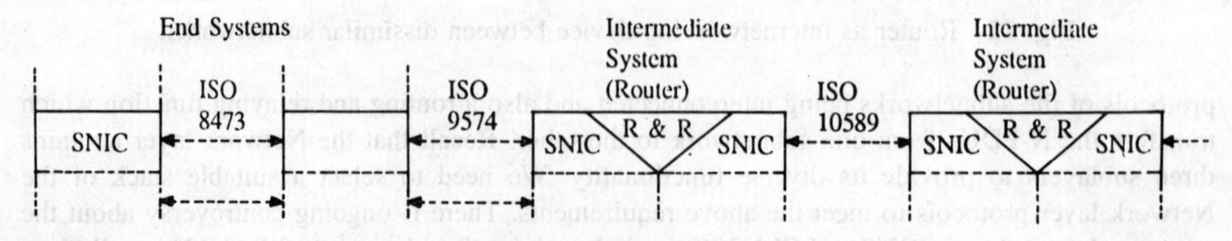

Fig. 14 ISO protocols for internetworking.

6.1 Service Primitives

For the connectionless-mode Network service there are only two primitives N-UNITDATA request and N-UNITDATA indication for transfer of data units across the interface of the Transport layer and the Network layer (Fig. 15). The associated parameters with these primitives are source address, destination address, quality of service and the user data.

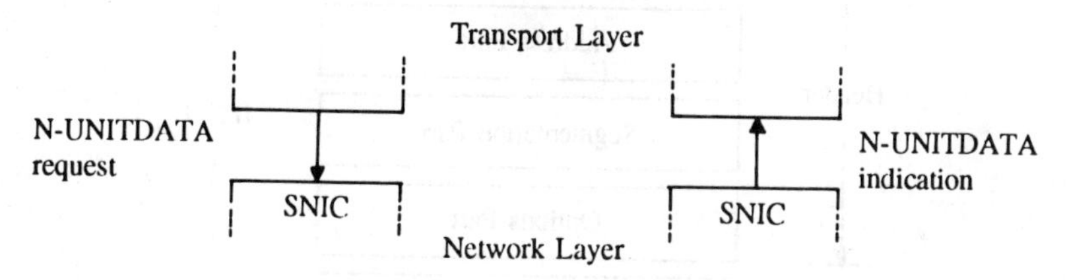

Parameters : Source Address, Destination Address, Quality of Service, User Data

Fig. 15 Service primitives for connectionless-mode Network service.

6.2 Formation of IPDU

The SNIC sublayer in an end system adds a header (PCI) to the N-SDU to form an Internet Protocol Data Unit (IPDU) as shown in Fig. 16. If the IPDU size exceeds the maximum size of the data unit which can be handled by the underlying service, the N-SDU is segmented and two or more IPDUs are formed. Segmentation can be performed at the router as well if the incoming IPDU from a subnetwork exceeds the size which can be handled by the service provided by the other subnetwork.

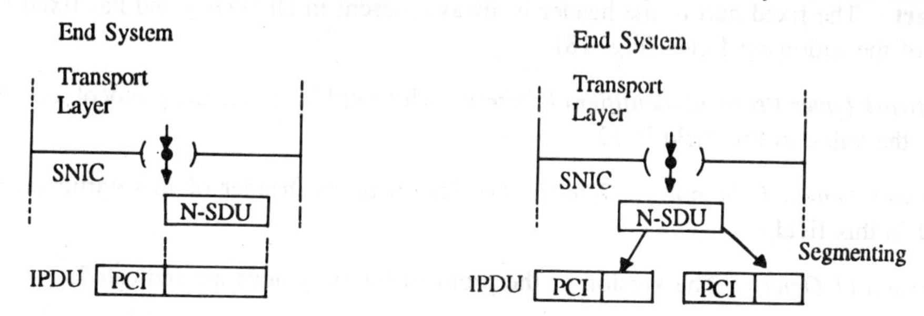

Fig. 16 Formation of Internet Protocol Data Unit (IPDU).

6.3 Types of IPDU

The ISO 8473 Internet protocol defines two types of IPDUs to control its operation.

Data IPDU. It carries N-SDUs between the two users of the Network service in the end systems. These users are the respective Transport entities which generate the N-SDUs.

Error Report IPDU. Error report IPDU is returned to the originating end system when a data IPDU is discarded at any intermediate or end system. There can be several reasons for the discarding of an IPDU. We will come across them as we go through this protocol. The error report IPDU is generated only if the originating end system has requested for the error report in its data IPDU should this IPDU be discarded.

6.4 Format of ISO 8473 IPDU

The format of IPDU can be divided into five parts. The first four parts constitute the header and the fifth part is the N-SDU received from the Transport layer (Fig. 17). The four parts of the header are:

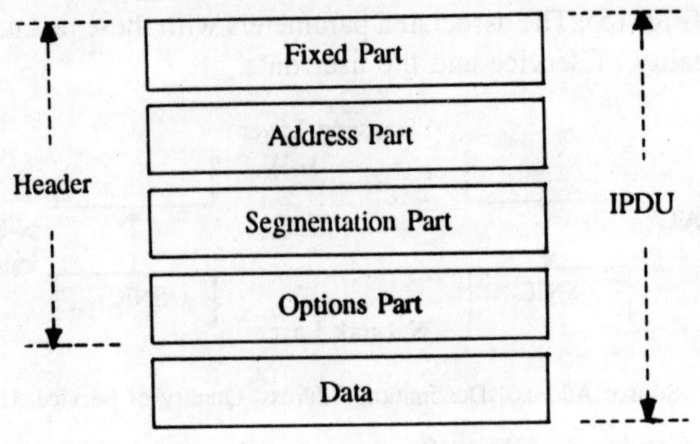

Fig. 17 Structure of IPDU.

1. Fixed part
2. Address part
3. Segmentation part
4. Options part.

Fixed Part. The fixed part of the header is always present in all IPDUs and has fixed length. It consists of the following fields (Fig. 18):

Network Layer Protocol Identifier (1 Octet): This field identifies the protocol. For ISO 8473 protocol, the value in this field is 12.

Header Length Indicator (1 Octet): The length of the header of this particular IPDU is specified in this field.

Version (1 Octet): The version or the protocol ID extensions are indicated.

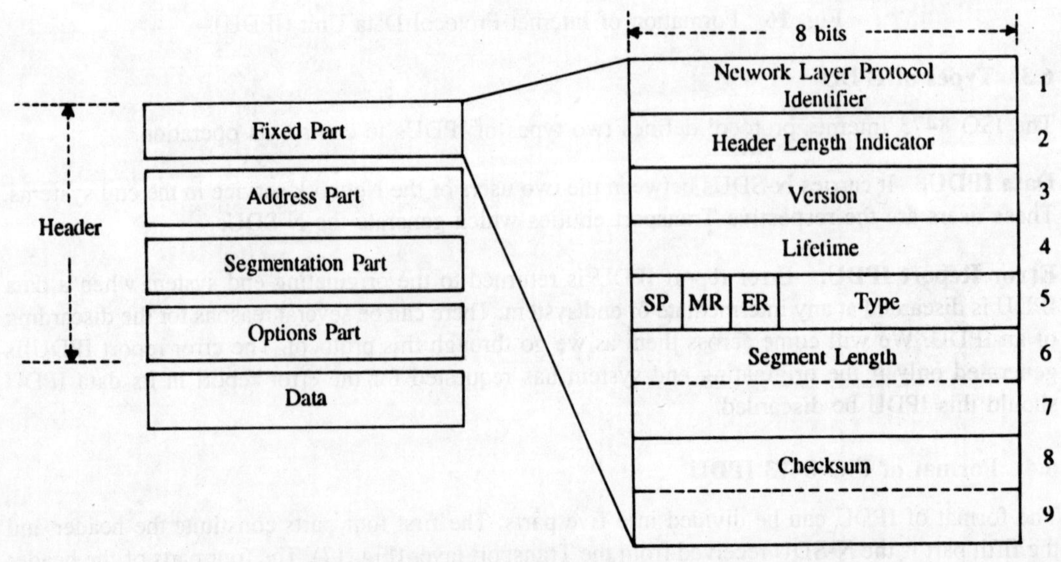

Fig. 18 Format of the fixed part of IPDU.

Lifetime (1 Octet): The remaining lifetime of the IPDU in units of 1/2 seconds is indicated in this field.

Segmentation Permitted, SP (1 Bit): If segmentation of the IPDU is permitted it contains the value "1".

More Segments, MR (1 Bit): This field contains "1" if more segments of a segmented IPDU follow this one. The last segment contains "0" in this field.

Error Report, ER (1 Bit): To request the return of the error report IPDU should this packet be discarded, "1" is inserted in this field.

Type (5 Bits): This field indicates the type of IPDU. It contains value "28" if it is a data IPDU and value "1" if it is an error report IPDU.

Segment Length (2 Octets): The length of the IPDU including the header is indicated in the field.

Checksum (2 Octets): It contains a two-octet checksum of the header portion only for error detection.

Address Part. The address part contains source and destination addresses and is variable in size. The address fields are identified by the address length fields provided in the address part. The format of address part is shown in Fig. 19 and its various fields are now enumerated:

Destination Address Length (1 Octet): The length of the destination address field is specified in this field.

Destination Address (≤ 20 Octets): This field contains the destination address to which this IPDU is being sent. The maximum length of the address field is 20 octets.

Source Address Length (1 Octet): The length of the source address field is specified in this field.

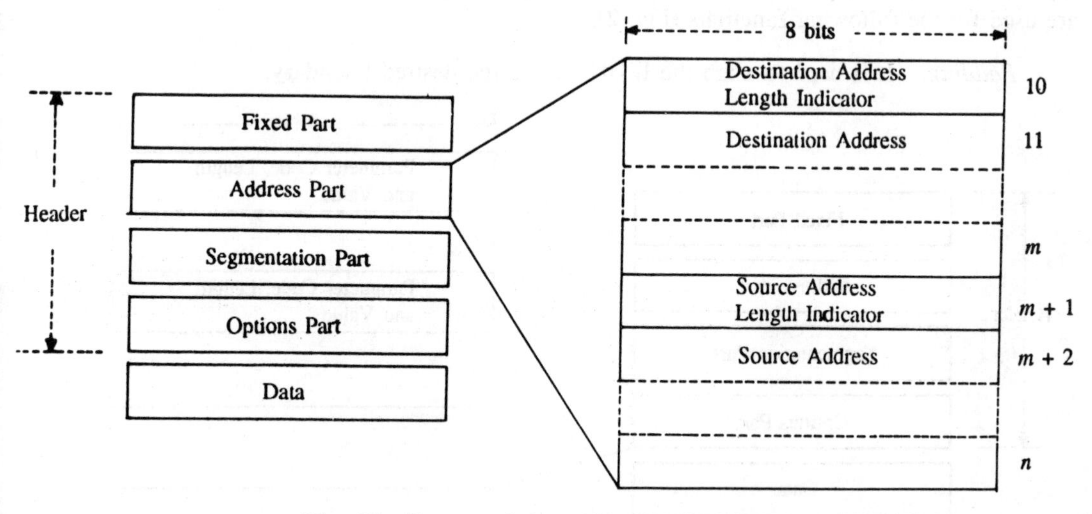

Fig. 19 Format of the address part of IPDU.

Source Address (≤ 20 Octets): This contains the address of the source which originated this IPDU. The maximum length of the address field is 20 octets.

Segmentation Part. If the SP flag is set to "1", a six-octet segment part is included in the header. It contains the following fields (Fig. 20):

Data Unit Identifier (2 Octets): It is a unique identifier generated by the source. It identifies the unsegmented IPDU and all its segments.

Segment Offset (2 Octets): The number in this field gives the relative position of this segment within the unsegmented IPDU.

Total Length (2 Octets): This field contains the total length of the entire IPDU before segmentation.

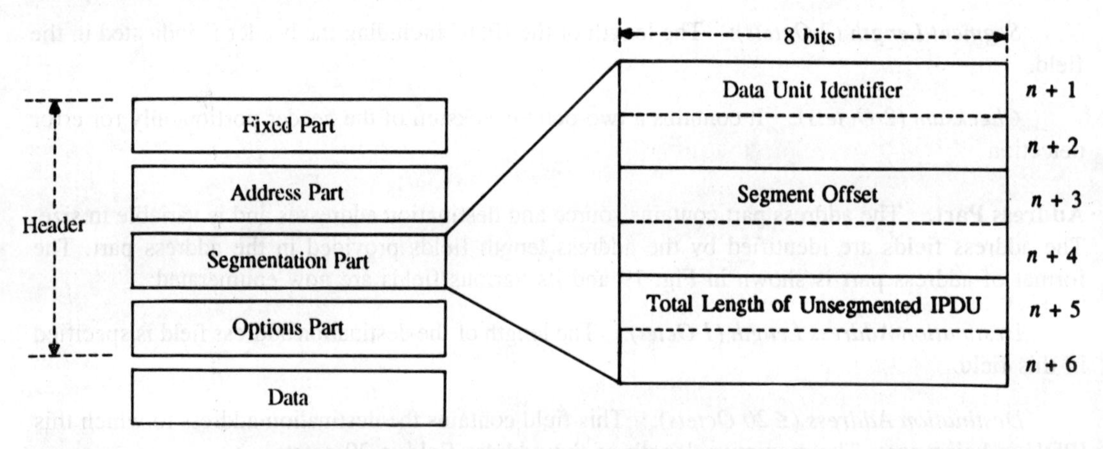

Fig. 20 Format of the segmentation part of IPDU.

Options Part. The options part is of variable length. It contains several groups of fields which are used for the following functions (Fig. 21):

Padding: It is used to align the IPDU size to the desired boundary.

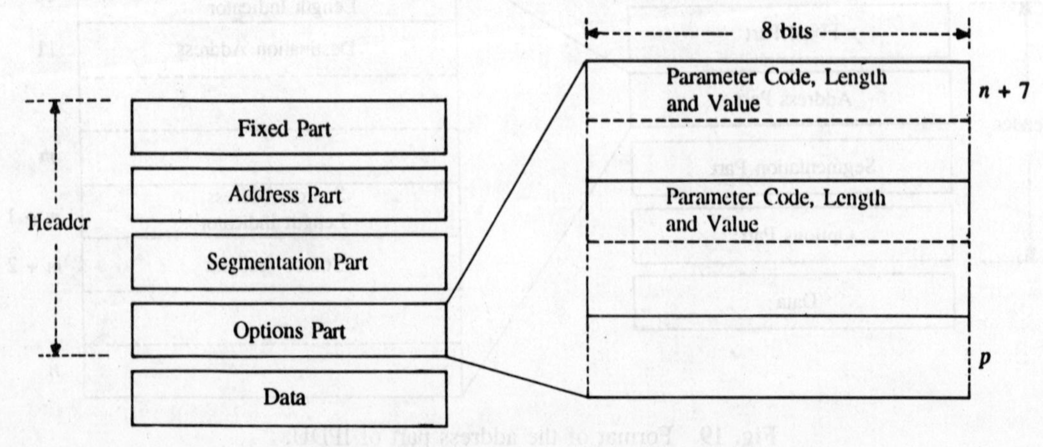

Fig. 21 Format of the options part of IPDU.

Security: This field contains security information.

Source Routing: When source routing is used, it gives the route of the IPDU.

Route Recording: In this field, the route of the IPDU is recorded as it travels through the network.

Quality of Service: This field contains values which describe the quality of service parameters.

Priority: This field contains the relative priority of the IPDU.

Reason for Discard: It is used in the error report IPDUs to indicate the reason a packet was discarded.

Data Part. It contains the user data (N-SDU) received from the Transport layer. There may be a full N-SDU or a segmented N-SDU in this field.

6.5 ISO 8473 Internet Protocol Functions

Table 2 lists the protocol functions. Full descriptions of these functions would fill a large book and therefore only some of the important functions are described here.

Table 2 ISO 8473 Protocol Functions

Function	Description
IPDU composition	Constructing the IPDU by adding protocol control information (PCI).
IPDU decomposition	Removing PCI from the IPDU.
Header format analysis	Checking the format of the header.
IPDU lifetime control	Enforcing the IPDU lifetime control.
Route IPDU	Determining the next node/end system to which the IPDU must be forwarded.
Forward IPDU	Forwarding the IPDU based on the routing decision.
Segmentation	Breaking a bigger N-SDU into smaller data units and forwarding as several IPDUs.
Reassembly	Reassembling the original N-SDU.
Discard IPDU	Discarding an IPDU which cannot be processed because of errors, protocol violation or lack of resources.
Error reporting	When discard IPDU function is used, this function allows return of error report IPDU.
IPDU header error detection	Performing a checksum calculation on the header to detect errors.
Padding function	Padding of bytes in the header so that the data part may be aligned to a convenient boundary.
Security	Enabling protection of data but its implemetation is not specified.
Source routing	Allowing the source end system to specify the route.
Record routing	Allowing recording of the actual route taken by the IPDU header.
Quality of service maintenance	Providing indicators for maintenance of the quality of service.
Priority	Allowing a packet with higher priority to be processed ahead of other packets.
Congestion notification	Setting an indicator in the header to indicate that congestion was experienced in the network.

Routing Function. This function determines the path over which an IPDU flows from the source Network entity to the destination Network entity. It interprets the destination address and the route from the PCI of the received IPDU and then forwards the IPDU. If the destination address corresponds to the local N-SAP, the user data is assumed to have arrived at destination. The end system can also influence routing by indicating the source route in the IPDU.

IPDU Lifetime Control Function. This function places a limit on the amount of time an IPDU can remain in the network and thus ensures that the IPDUs do not circulate endlessly in the network. It is important to the Transport layer that an N-SDU is either delivered in a time-bound manner or discarded.

The lifetime control is exercised using the lifetime field in the header of an IPDU. The source Network entity in the end system encodes the initial value of lifetime in the field. It is decremented at each intervening Network entity which processes the IPDU. If the remaining value of the lifetime becomes zero at any stage before the IPDU is delivered to the destination, the IPDU is discarded. The decrease in the lifetime value by a Network entity is at least one unit. If the IPDU spends more than 500 ms during its processing and forwarding, the decrease in lifetime is at the rate of one unit per 500 ms interval.

Segmentation and Reassembly. Segmentation is performed when the size of IPDU is greater than the maximum size that can be handled by the next lower layer. Segmentation consists of composing two or more IPDUs containing segments of the user data. The IPDUs containing segments of user data are identified as belonging to the same data unit by the data unit identifier field, source, and destination addresses. The location of a user data segment in the unsegmented user data is indicated by the segment offset field (Fig. 22). Total length of the unsegmented IPDU is indicated in the total length field in the header. The first and the intermediate segments are marked by the more segment (MR) flag in the header. It is set to "1". In the IPDU containing the last segment, the flag is set to "0".

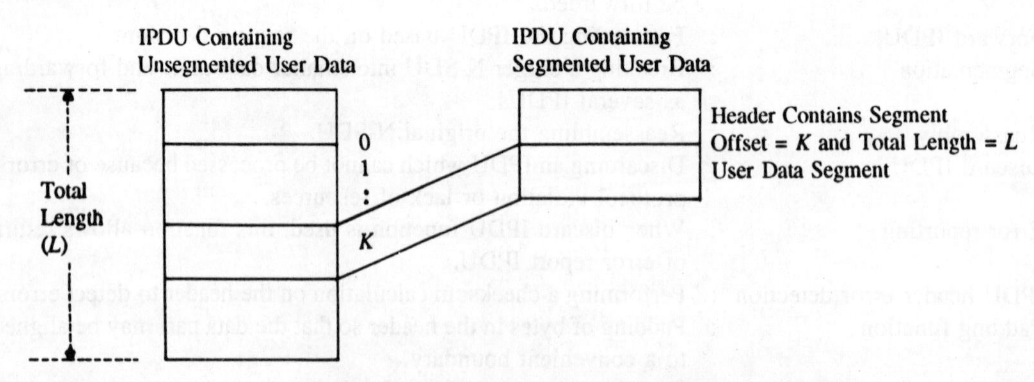

Fig. 22 Segmentation of user data.

Segmentation may be performed by the source Network entity or at the intermediate Network entities. IPDU segments may be further segmented if required at the subsequent stages. If segmentation is not allowed, the segmentation permitted (SP) flag in the header is set to "0". If an IPDU is received, which cannot be segmented (as indicated by the SP field) and the IPDU cannot be forwarded because its length exceeds the maximum subnetwork service data unit size, it is discarded.

Reassembly function consists of reconstructing the unsegmented user data. It is usually performed at the destination. If performed at an intermediate stage, unsegmented IPDU is reconstructed.

Discard IPDU Function. There are several reasons for discarding an IPDU. The following list is not exhaustive but enumerates some important reasons:

- The lifetime of the IPDU has expired.
- The header cannot be analyzed due to errors, wrong formatting or any other reason.
- The IPDU cannot be segmented and cannot be forwarded due to size limitations.
- An IPDU may be discarded to relieve traffic congestion.

6.6 ISO 8473 Subnetwork Dependent Convergence Functions

As mentioned above, a part of ISO 8473 deals with subnetwork dependent convergence functions as well. These functions are primarily concerned with augmenting the underlying service to provide the service expected by the SNIC sublayer.

SNDC Sublayer Service Definition. The interface between SNIC and SNDC sublayers is not formally standardized in ISO 8473 but it appears in the document only for description purposes. The service primitives and parameters are given in Fig. 23. Being a connectionless-mode service, there are only two primitives, SN-UNITDATA request and SN-UNITDATA indication.

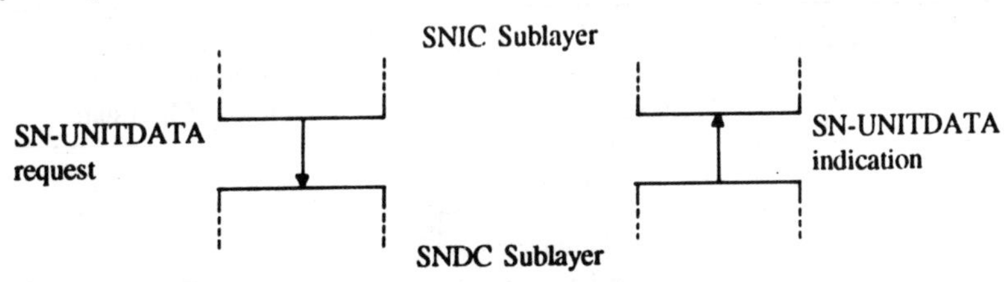

Parameters: Source Address, Destination Address, Quality-of-Service, User Data

Fig. 23 Service primitives between SNIC and SNDC sublayers.

SNDC Sublayer Functions. ISO 8473 defines SNDC functions for three types of subnetworks:

(i) Point-to-Point Subnetworks: These subnetworks are implemented on point-to-point links using HDLC protocol. SNDC sublayer directly interfaces with the Data Link layer to use its service.

(ii) Broadcast Subnetworks: This category of subnetworks comprises LANs which provide Logical Link Control (LLC) service. LLC service primarily being a connectionless-mode service, SNDC function simply maps to the LLC service primitives.

(iii) X.25 Subnetwork: X.25 is a connection oriented data transfer interface. SNDC function for X.25 subnetworks involves using established connections, establishing new connections, data transfer and releasing the connections.

These sets of functions enable the SNDC sublayer to interface and utilize a variety of underlying services to provide a common connectionless-mode service to the SNIC sublayer. Thus, the

ISO 8473 SNDC sublayer above the X.25 SNAC sublayer will allow interfacing the end system or router to the X.25 subnetwork. The service finally provided to the transport layer entity is connectionless-mode Network service.

7 INTERNET PROTOCOL (IP) OF US DEPARTMENT OF DEFENCE

The current internetworking protocol, which is the most popular in the LANs, is the Internet Protocol (IP) of the US Department of Defence (DoD). It is specified in the MIL-STD-1777 document. It is used along with the Transmission Control Protocol (TCP) which roughly corresponds to the Transport, Session and the Presentation layers of the OSI model. TCP/IP protocols are not as per the OSI architecture. With the development of ISO-OSI standards which are rich in functions and more extensive, it is likely that US governmental organizations including the Department of Defence will also move to OSI.

IP is very similar to ISO 8473 in operation. It provides connectionless-mode service to the TCP layer. Its basic purpose is to route datagrams (In IP terminology IPDUs are called datagrams). Other important functions are addressing, life time control, fragmentation (corresponding to segmentation in ISO 8473) and reassembly. Figure 24 shows the structure of the IP datagram header. It is at least 20 bytes long and consists of the following fields:

4	4	8	16	Bits
Version	IHL	Type of Service	Total Length	
Identification			Flags	Fragment Offset
TTL		Protocol	Header Checksum	
Source Address				
Destination Address				
Options				Padding

Fig. 24 Format of IP datagram header.

Version (4 Bits): It indicates the version of IP being used.

Internet Header Length, IHL (4 Bits): The header length is recorded as minimum five 32-bit words. Thus, the minimum length of the header is 20 octets.

Type of Service (8 Bits): This denotes priority, delay, throughput and reliability parameters.

Total Length (16 Bits): The total length of the datagram including the header and data field is indicated in this field.

Identification (16 Bits): It is a unique identification to associate the fragments.

Flags (3 Bits): It is a three-bit field containing more bit and don't fragment flags. The third bit is undefined.

Fragment Offset (13 Bits): The offset of user data fragment relative to the beginning of the unfragmented data is indicated in this field.

Time to Live, TTL (8 Bits): The maximum time a datagram is allowed to remain in transit is specified in the TTL field.

Protocol (8 Bits): This field indicates the next higher layer protocol which is to receive the datagram.

Header Checksum (16 Bits): It is 16 bit 1's complement of the checksum of all 16-bit words in the header.

Source and Destination Addresses (32 Bits each): The source and the destination addresses are given in these fields. Out of 32 bits, 8 bits are used for subnetwork address and the following 24 bits for the local address.

Options (Variable): 'This field specifies the options chosen by the source.

Padding (Variable): Padding ensures that the header ends on a 32-bit boundary.

To sum up, ISO 8473 and IP of DoD are very similar in function and design. They both provide connectionless-mode service but are not the same and, therefore, cannot interwork.

8 ROUTERS FOR PROVISIONING OF CONNECTION-MODE NETWORK SERVICE

Connection-mode Network service (CONS) is specified in ISO 8348 and CCITT X.213 documents. The most common protocol for connection-oriented data transfer is X.25 which we have already covered in considerable detail in another chapter. As X.25 was framed before formulation of the Network service defined in X.213 and ISO 8348, it does not meet all the Network service requirements. To augment the service provided by X.25, another protocol above X.25 was framed. ISO 8878 and CCITT X.223 documents specify this protocol, *Use of X.25 to Provide the OSI Connection-mode Network Service*. ISO 8878 can be viewed as a sublayer running on top of X.25 that defines how the CONS can be provided using the underlying X.25 procedures. This protocol also contains SNDC functions to remove the differences in the various versions of X.25. Table 3 lists the mappings of connection establishment phase CONS primitives to the X.25 packet level protocol as per ISO 8878. Thus if required, CONS can be provided to the Transport entities using X.25 and X.223 protocols.

Table 3 Mapping of CONS Primitives for Connection Establishment Phase

Connection-mode service primitives	X.25 Protocol
N-CONNECT request	CALL REQUEST
N-CONNECT indication	INCOMING CALL
N-CONNECT response	CALL ACCEPTED
N-CONNECT confirmation	CALL CONNECTED

8.1 Use of X.25 Protocol over a LAN

The X.25 packet level protocol (PLP) usually needs the service of LAP-B protocol at the Data Link layer. Local area networks more than often use LLC sublayer protocol over the MAC sublayer. ISO 8881 protocol has been defined for use of the LLC service by the X.25 PLP. This protocol is entitled *Use of the X.25 Packet Level Protocol in Local Area Networks*.

The LLC sublayer provides three types of service to the Network layer. Types 1 and 3 LLC service define unacknowledged and acknowledged connectionless-mode service respectively. Type 2 service is connection-mode service. X.25 PLP is used mainly over Type 2 LLC operation. Its use over Type 1 LLC operation is also possible except that the Data Link layer may lose a packet once in a while.

Recall that X.25 is inherently asymmetric, i.e., the communicating Network entities are designated as DTE and DCE. The subnetwork entity is given the DCE status. Such is not the case with local area networks where all stations are of the same type. ISO 8208 protocol, *X.25 Packet Level Protocol for Data Terminal Equipment,* permits the use of X.25 PLP between two DTEs. With this modification, the X.25 PLP can be made to operate in a local area network as if each pair of stations is directly connected in DTE-DTE mode (Fig. 25).

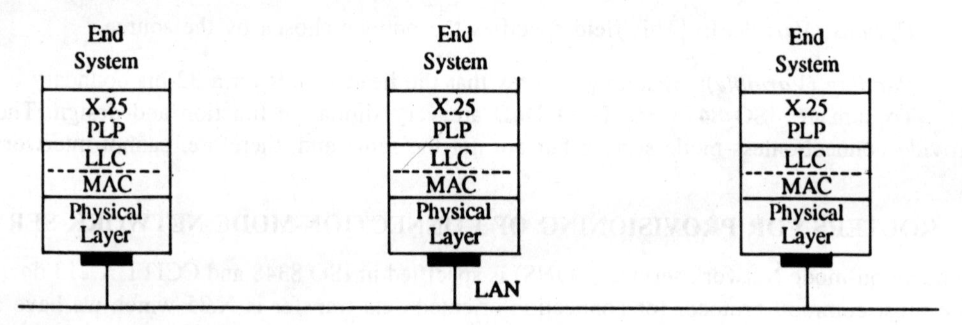

Fig. 25 Use of X.25 protocol in LANs.

Summarizing the above discusssion, we conclude:

1. It is possible to provide CONS using X.25 PLP with ISO 8878.
2. X.25 PLP can utilize LLC service.
3. Asymmetric design of X.25 protocol can be made symmetric using ISO 8208 for use in LANs.

8.2 Internetworking for CONS

With the above background, we can now discuss the internetworking configurations of the networks when the service provided to the Transport entity is CONS. We will consider internetworking in the following configurations:

- LAN to LAN internetworking
- LAN to LAN internetworking using X.25 subnetwork
- LAN to Host internetwoking using X.25 subnetwork.

Figure 26 shows internetworking of two different LANs using a router. X.25 PLP is implemented in all the stations of both the LANs and in the router. Data Link layer service is, of course, Type 2 LLC. The lower two layers of the router take care of the differences in LLC and MAC sublayer protocols of the two LANs.

Figure 27 shows LAN to LAN internetworking through an X.25 subnetwork. A router is required at each LAN to interwork with the X.25 subnetwork. The routers have full X.25 implementation towards the packet subnetwork and X.25 PLP towards the LANs.

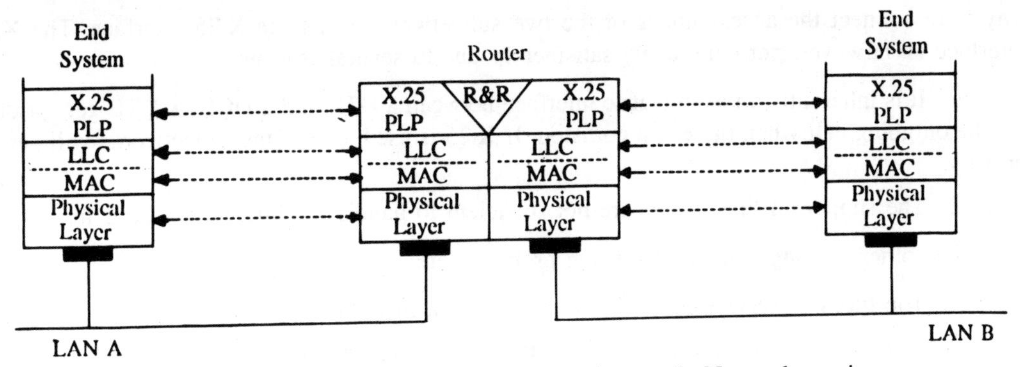

Fig. 26 Internetworking router for connection-mode Network service.

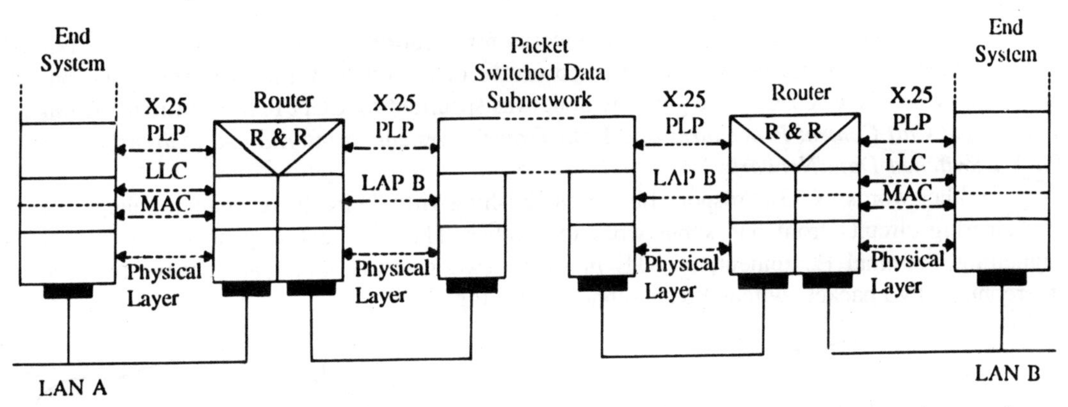

Fig. 27 Layered architecture of LAN to LAN internetworking through X.25 subnetwork.

In Fig. 28 a remote host is shown connected to a LAN through an X.25 subnetwork. Only one router is required at the LAN in this case. The host has full X.25 implementation in its first three layers and, therefore, it can directly interface with the X.25 subnetwork.

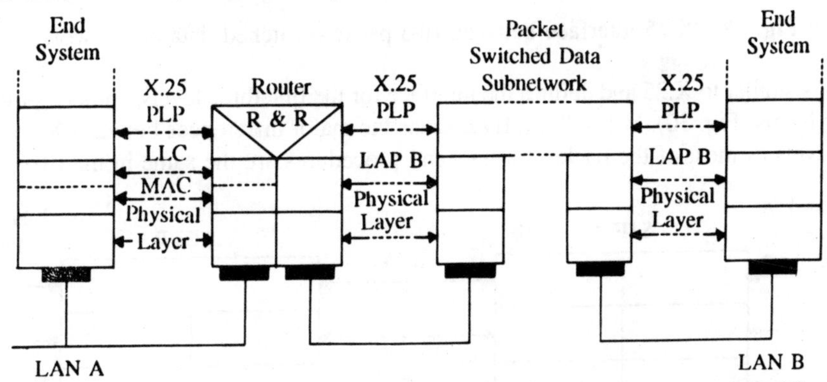

Fig. 28 Layered architecture of LAN to remote host interconnection through X.25 subnetwork.

9 INTERNETWORKING OF X.25 SUBNETWORKS

When two packet-switched subnetworks are to be interconnected to exchange traffic, one possible way is to connect the access nodes of the two subnetworks using the X.25 interface. The X.25 interface is, however, not completely satisfactory due to several reasons:

1. It is inherently an asymmetric interface between a DTE and a DCE. A DTE gets priority for the outgoing call when there is a collision. If X.25 is used, one of the subnetworks will get the priority.

2. The supported line speeds are not sufficient to handle the internetwork traffic.

The other envisaged internetworking problems are:

1. The types of service offered may be different in terms of facilities offered.
2. The grades of service in terms of throughput class may be different.
3. The accounting and billing problems need to be sorted out.

It is also desirable to isolate the subnetworks for security and billing purposes. The failure of one subnetwork should not affect the performance of the other. With these considerations in mind, a new interface X.75 for the gateway node was specified by CCITT. It is entitled *"Terminal and Transit Call Control Procedures and Data Transfer System on International Circuits between Packet Switched Data Networks"*.

In other words, X.75 is responsible for establishing virtual circuits, exchanging data packets and clearing circuits from one subnetwork to another. The X.75 gateway nodes are known as Signalling Terminal Equipment, or STE in short. Two gateway nodes are always required to interconnect two packet-switched data subnetworks (Fig. 29).

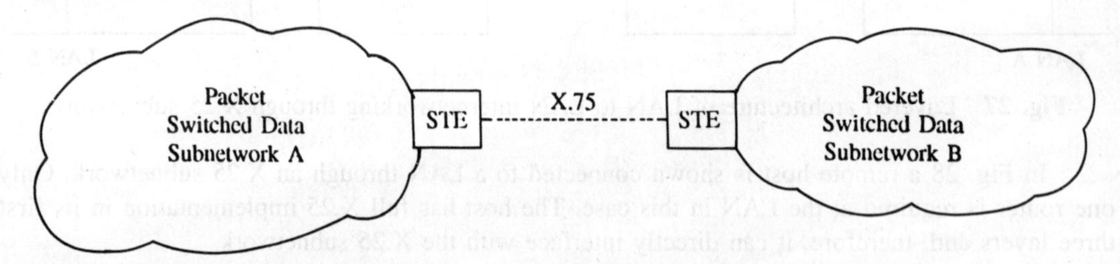

Fig. 29 X.75 interface between two packet-switched data subnetworks.

X.75 is similar to X.25 and defines the interface for the first three levels, namely, physical, frame and packet levels (Fig. 30). We will briefly discuss the major differences between X.75 and X.25 at the three levels as most of the packet formats and procedures are the same in the two protocols.

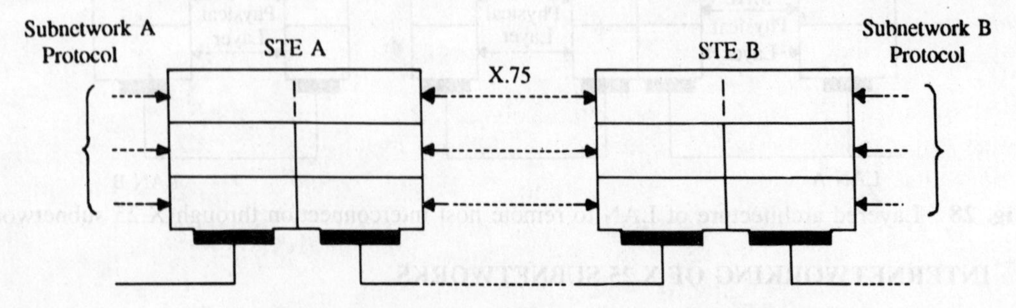

Fig. 30 Layered architecture of X.75 interface.

9.1 Packet Level of X.75

The packet formats and their exchange procedures of X.75 are very similar to those of X.25. The call set-up packets contain an additional utility field (Fig. 31). Some of the important parameters of the utility field are:

Transit Network Identifier: Each transit subnetwork appends its TNIC (equivalent to DNIC) in the CALL REQUEST Packet.

Call Identifier: It is used to uniquely identify a call by all the networks involved for accounting and auditing purposes.

Throughput Class, Window Size and Packet Size: Negotiation of these parameters is identical to X.25.

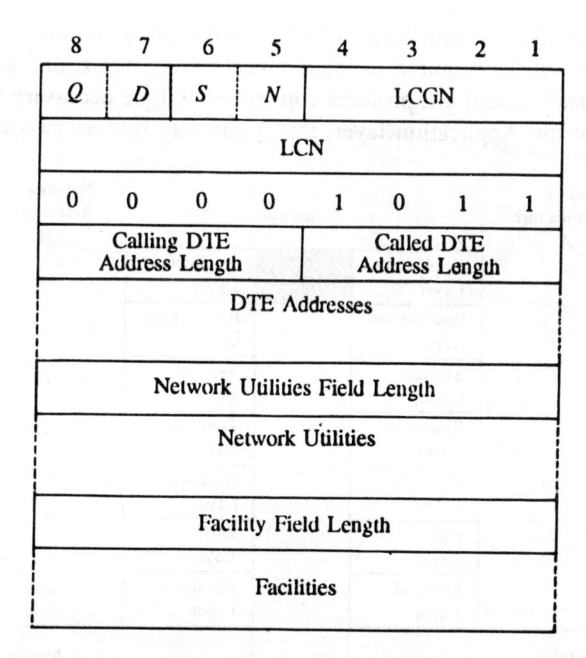

Fig. 31 Format of X.75 CALL REQUEST packet.

The maximum size of the network utility field is 63 octets. The utility field is different from the facility field in the sense that the facility field contains the facility requested by the DTEs (end systems) while the utility field contains the parameters which are negotiated between the two STEs and are locally applicable at the X.75 interface.

Either of the two STEs can establish a call by sending a CALL REQUEST packet. The call is established in the same manner as in X.25. In the event of call collision, i.e. both STEs using the same logical channel identifier, both the calls are cleared. It is different from X.25 where only the incoming call is cleared. To avoid call collisions, the STEs select logical channel identifiers starting from the opposite ends of the ranges. Which STE will use the lowest number and which the highest is mutually agreed upon beforehand.

9.2 Frame Level

At the frame level, X.75 uses the LAP-B protocol. Multilink procedure, discussed in the chapter on HDLC protocol, can be used to enhance reliability.

9.3 Physical Level

The characteristics of the Physical layer interface are in accordance with the CCITT Recommendation G.703 for 64 kbps. Alternatively, V series of CCITT recommendations supporting data rates of 48 kbps or any other data rate are used. To support multilink procedure, multiple physical connections are provided between the two STEs.

10 GATEWAYS

A gateway is a general term describing a device which is used to bridge different *network architectures*. For example, a gateway will be required to interconnect two systems, one based on OSI and the other on SNA. In most such situations, protocol conversion will be necessary at layer 4 and above. More than often, it is at the Application layer, that a gateway bridges two systems (Fig. 32).

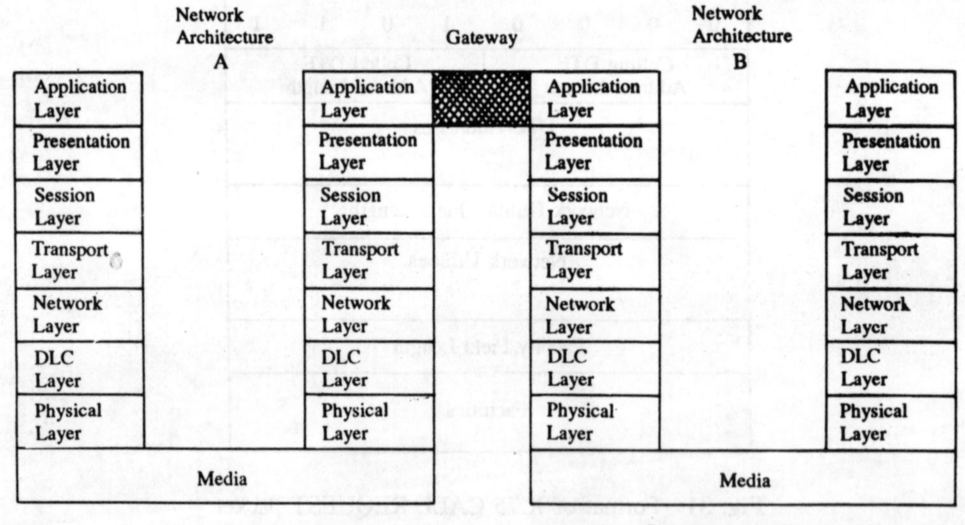

Fig. 32 Gateway architecture.

The gateway incorporates a protocol conversion function at the Application layer. For example, to transfer a file from a system employing the OSI architecture to another using a proprietary architecture, the gateway will accept the file using FTAM (File Transfer, Access and Management) protocol and then transfer it to the other system using the proprietary protocol. Such mapping of protocols may not be, always and in all applications, possible.

Let us consider that network architecture A uses a file transfer protocol FA in which the file is transferred record by record and at the end, end-of-file marker is transmitted. When the receiver responds positively, the file transfer is complete, meaning thereby that the file resides in the new system and is available for use. The other network architecture B uses a file transfer protocol FB in which first, the size of the file is indicated and then numbered records of the file are sent. The receiver checks whether all the records have arrived safely and then acknowledges. Suppose a file is to be transferred from a system using FA to another system using FB. When the gateway is used to bridge these protocols, its protocol conversion function finds that these protocols are incompatible. For FB, the gateway must have the full size of the file right in the beginning. But FA does not provide any such information in advance. Therefore, the gateway must first receive and store the file and count the records. Then it can commence transfer of the file to the other system. The gateway cannot confirm to the file-originating system till this transfer is complete. In this process, the gateway does not remain transparent. The end systems can feel its presence in all transactions.

Thus, the use of a gateway to bridge applications running in two different network architectures has its limitations. The service provided by the gateway for a given application is restricted on the common semantics of the two protocols.

11 SUMMARY

There is always need to interconnect subnetworks having different protocols and services to form an internetwork. Bridges, routers and gateways are the intermediary devices which facilitate internetworking. Bridges are the simplest and interconnect two subnetworks at the media access control sublayer. The subnetworks interconnected by bridges must have same LLC sublayer and the higher layers. If there are differences in the LLC and the Network layer protocols, a router must be used. A router interconnects two subnetworks at the Network layer level. Type of router and its configuration are determined by the Network service desired and the subnetwork service available to the router. ISO, CCITT and other organizations have specified the Network layer protocols for internetworking using routers. ISO 8473 is one such very important internetworking protocol. Gateways are used to bridge network architectures at the Application layer level. They are complex, application-dependent and usually not transparent to the users.

PROBLEMS

1. Draw the spanning tree for the network shown below. Identify the root and designated ports (Fig. P.1).

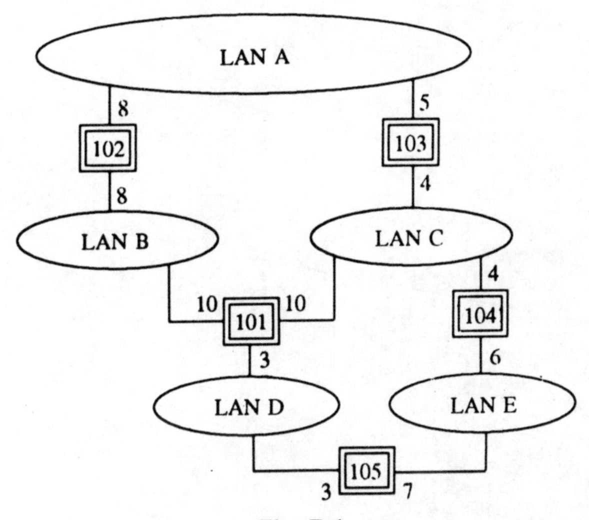

Fig. P.1

2. Station X sends a route discovery frame with all-router-broadcast directive for station Y. Indicate the route designator subfields in the multiple copies of the frames received by Y. Which is the minimum hops route?

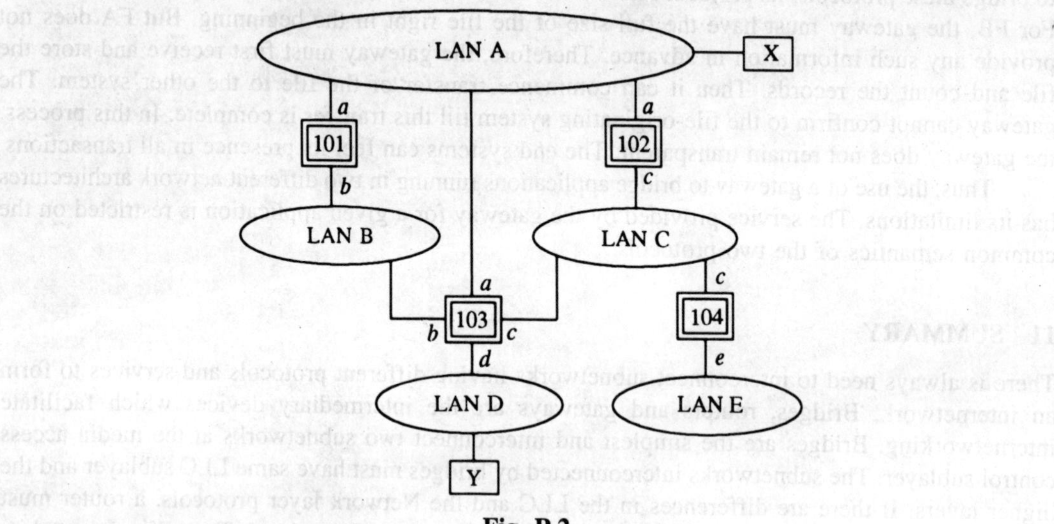

Fig. P.2

CHAPTER 14

The Transport and Upper OSI Layers

Upto the Network layer, the OSI Reference Model is primarily concerned with delivery of data bits or bytes across one or more subnetworks to the destination end system. The issues which we further need to address for meaningful communication include

- matching quality of bit transport service requirements and the offered Network service,
- dialogue discipline, synchronization,
- data representation, and
- remote LOGIN, authentication, etc.

These issues concern the applications processes which reside in the end systems. The Transport, Session, Presentation and Application layers of the OSI reference model implement the protocols for these purposes. In this chapter, we examine these layers. The Transport layer is covered in detail as it is the last layer which is concerned with providing the required quality of end-to-end transport service to the upper three layers. As regards the upper three layers, just an overview of each layer is presented.

1 THE TRANSPORT LAYER

The Transport layer is situated between the Network layer and the Session layer of an end system. While the Network layer and the other lower layers reside in all the end systems and intermediate systems, the Transport layer is implemented only in the end systems where the applications are implemented (Fig. 1). The interactions between peer Transport layers are end-to-end and these are made possible by the data transfer service provided by the Network layer.

As with the other OSI layers, the Transport layer is defined by the service it provides to the users which are the Session entities and by the Transport layer protocol which is used by the Transport entities to provide the service.

1.1 Purpose of the Transport Layer

The Transport layer provides transparent, reliable and cost effective transfer of data between the user entities in the Session layers. Transparency implies that the Transport entities do not place any constraints on the content of the data units received from the user entities. Reliable, in this context, implies that the data units are received as they are sent by the user entities. If any type of error occurs during the transfer, an indication is given to the receiver. Cost effectiveness implies that the underlying Network service is utilized in the most optimum way to provide the quality of service requested by the user entities.

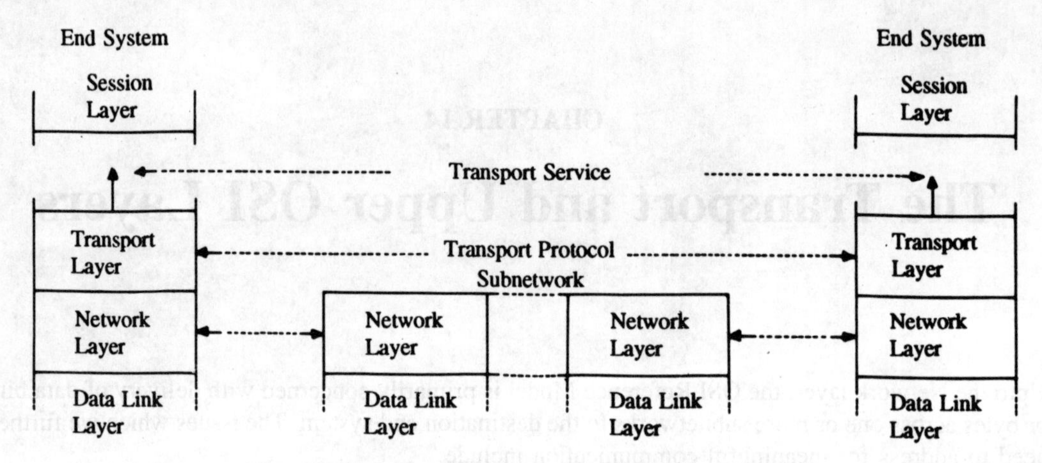

Fig. 1 End-to-end Transport service and protocol.

The Transport layer relieves the Session entities from any concern with the way in which the end-to-end data transfer is to be achieved. It lets the user entity define the quality of service to be provided and then achieves this quality by the optimum utilization of the underlying Network service.

1.2 Operation of the Transport Layer

In a typical use of the Transport service, the Session entity in an end system accesses the service through the Transport-Service-Access-Point (TSAP) and passes a Transport-Service-Data-Unit (TSDU) to the Transport layer entity for transfer to the peer Session entity in the destination end system (Fig. 2). The Transport entity adds protocol control information (PCI) to the TSDU to create a Transport Protocol Data Unit (TPDU). The TPDU is sent to the Transport entity in the destination end system using the underlying Network service which can be connection-mode Network service (CONS) or connectionless-mode Network service (CLNS). The PCI is removed from the

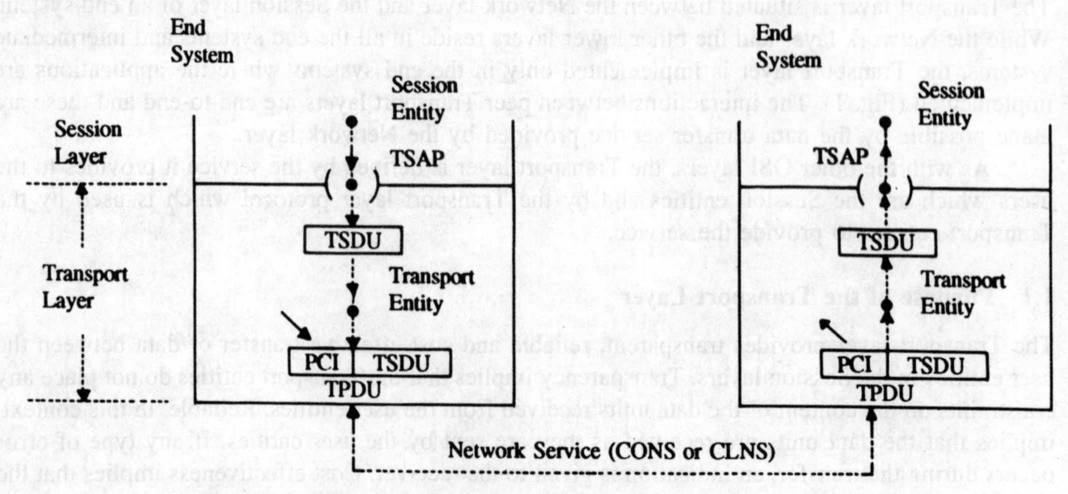

Fig. 2 Formation of Transport Protocol Data Units (TPDUs).

TPDU by the Transport entity of the destination end system and the TSDU is delivered to Session entity.

1.3 Transport Service (TS) Definition

As with the other layers of the OSI model, the ISO standards define both a connectionless-mode and a connection-mode Transport Service (TS). The Transport layer may offer either or both types of services but the greatest number of the present day applications are based on connection-mode Transport service.

Connection-Mode Transport Service. The connection-mode Transport service definition is described in ISO 8072, *Transport Service Definition*. The CCITT recommendation on the connection-mode Transport service is X.214.

In the connection-mode service, a Transport connection is established between the user entities on their request. Transfer of the TSDUs takes place on this connection. The connection can be released either by the users or by the service provider.

The following capabilities are offered to the connection-mode Transport service users.

(i) Transport Connections Establishment and Release

- The TS user can establish and release a Transport connection with another TS user. More than one connection may exist between the same pair of TS users. These connections are distinguished by using endpoint identifiers.
- The TS users can request and negotiate a certain quality of service which may be specified in terms of the following parameters:
  - — Throughput
  - — Transit delay
  - — Residual error rate
  - — Connection establishment delay
  - — Connection termination delay
  - — Failure probability of connection establishment, data transfer and disconnection
  - — Transport connection protection.
- 32 octets of user data can be exchanged during the connection establishment phase.
- Either TS user may unconditionally and unilaterally release the connection. As a consequence, any data still in transit may be destroyed. The connection may be released by the TS provider also.
- 64 octets of user data can be sent along with the DISCONNECT primitives.

The TS primitives and parameters for establishment and release of Transport connection are shown in Fig. 3. Note that connection establishment is a confirmed service but disconnection is an unconfirmed service or is TS provider-initiated.

In the T-CONNECT request, quality of service parameters and expedited data transfer option parameter are indicated. Expedited data transfer option permits TS users to request for sending a TSDU ahead of other TSDUs already in the queue.

In the T-CONNECT response primitive, the accepted quality of service parameters are indicated. The accepted quality of service cannot be better than the requested quality of service. The expedited data option can be either selected or not selected but it cannot be requested in the response primitive.

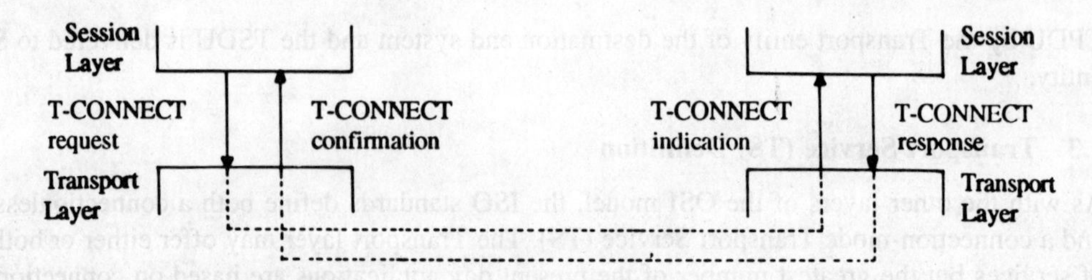

Primitive	Parameters
T-CONNECT request	Calling and called addresses, expedited data option, quality of service, user data
T-CONNECT indication	Same as above
T-CONNECT response	Responding address, expedited data option, quality of service, user data
T-CONNECT confirmation	Same as above

Disconnection Initiated by the TS User

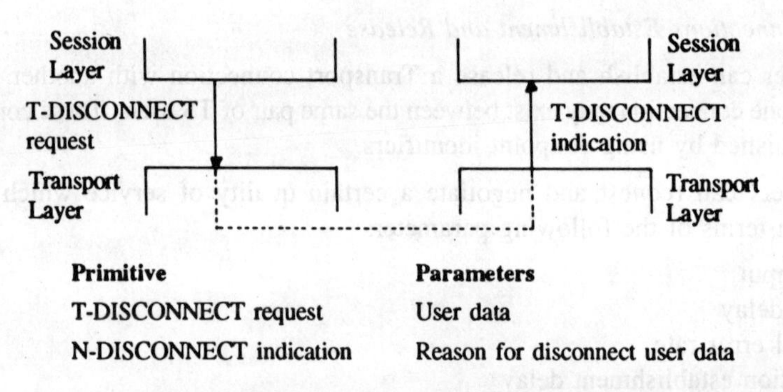

Primitive	Parameters
T-DISCONNECT request	User data
N-DISCONNECT indication	Reason for disconnect user data

Disconnection Initiated by the TS Provider

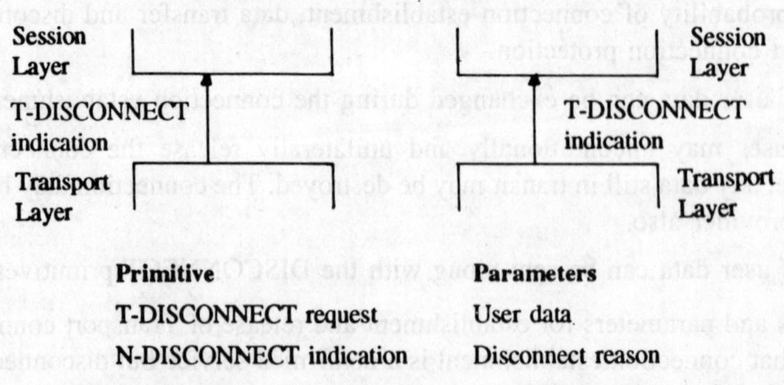

Primitive	Parameters
T-DISCONNECT request	User data
N-DISCONNECT indication	Disconnect reason

Fig. 3 Service primitives for connection establishment and release.

(ii) Normal Data Transfer

- The TS users can request for transfer of user data having any integral number of octets. The transfer is transparent in that the boundaries of the user data are preserved during the transfer and there are no constraints on the content and number of octets.
- The mode of data transfer can be two-way simultaneous.

Figure 4 shows the primitives and parameters of normal data transfer service. Note that the service is unconfirmed but reliable. Unsuccessful transfer is indicated by connection release.

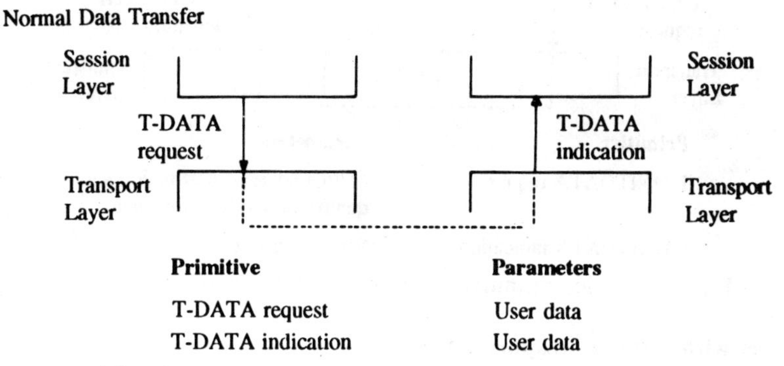

Primitive	Parameters
T-DATA request	User data
T-DATA indication	User data

Fig. 4 Service primitives for normal data transfer.

(iii) Expedited Data Transfer

- The TS user can request transfer of a limited amount of user data in the form of Expedited TSDUs which have a flow control distinct from the usual TSDUs. This service is user- and provider-optional, i.e., it is available if the provider agrees to provide and the users agrees to use it.
- The number of user data octets is restricted to 16.
- Expedited TSDU is guaranteed to be delivered before any subsequent TSDU.
- This service is also not a confirmed service but is a reliable service.

The sequence of expedited data primitives is similar to the normal data transfer. The parameter is user data restricted to 16 octets as mentioned above. The service primitives are T-EXPEDITED-DATA request and T-EXPEDITED-DATA indication.

Connectionless-mode Transport Service. OSI connectionless-mode Transport service is defined in ISO 8072, *Transport Service Definition*. Because of the connectionless operation, there is no connection establishment phase. The users may transparently transfer TSDUs of restricted length. Each TSDU is sent independently of other TSDUs. The service users may request the desired quality of service associated with each TSDU. The service does not provide reliable delivery, sequencing or flow control. This is consistent with the connectionless-mode of operation.

The only primitives for the connectionless-mode transport are T-UNITDATA request and T-UNITDATA indication (Fig. 5). There are four parameters associated with these primitives:

1. Source address
2. Destination address
3. Quality of service
4. User data.

The quality of service parameters include transit delay and residual error rate. These parameters cannot be negotiated. The users have *apriori* knowledge of the capabilities of the Transport service.

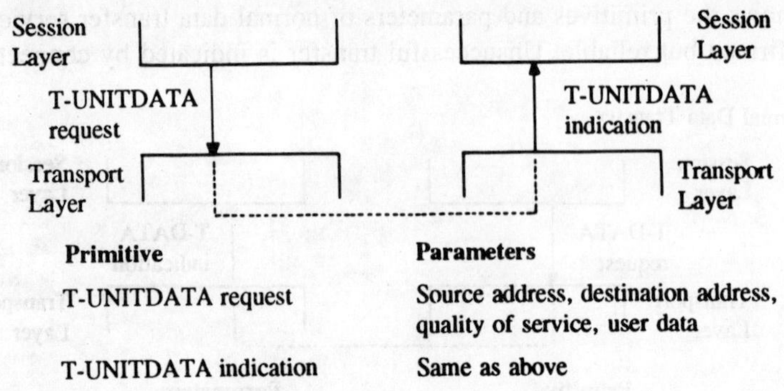

Fig. 5 Service primitives for connectionless-mode data transfer.

1.4 Functions within the Transport Layer

The Transport layer may carry out the following functions in order to provide the Transport service:

- Mapping Transport address onto the Network address
- Assignment of Network connection
- Multiplexing of Transport connections
- Splitting of a Transport connection
- Establishment and release of the Transport connections
- Transfer of normal and expedited TSDUs by attaching a header (PCI) to them
- End-to-end sequence control, error recovery and monitoring the quality of service
- Segmenting and concatenation of the TPDUs
- End-to-end flow control over individual connections.

It is not necessary that all these functions are implemented always in the Transport layer. Some of these functions are essential and always implemented while the use of some is dependent on the need and the type of the Network service available. For example, if the Network service provides acceptable error rate of the TPDUs, the Transport layer would not require the error recovery function. We shall now examine some of these basic functions carried out within the Transport layer. The discussion has been restricted to the functions relating to the connection-oriented data transfer using CONS so that too many alternatives may not hinder understanding of the concepts.

Address Mapping. When a Session entity requests to send a TSDU to another Session entity identified by its TSAP address, the Transport entity determines the Network address (NSAP) of the Transport entity which serves the correspondent Session entity. Because the transfer is end-to-end, no intermediate Transport entity is involved.

Assignment of Network Connection. Transfer of the TPDUs is facilitated on a Network connection. Therefore, the Transport entity assigns a Network connection for carrying the TPDUs. It is necessary that the assigned Network connection must have been established by the Transport entity. If the Transport protocol allows recovery from the network disconnection, a Transport entity may reassign the transfer of TPDUs to a different Network connection whenever a disconnection occurs.

Multiplexing and Splitting. For optimizing the use of the Network connection, a Transport entity may assign several end-to-end Transport connections to one Network connection (Fig. 6a). Multiplexing is resorted to when the underlying Network service offers quality of service (e.g., throughput) in excess of what is required for a Transport connection. TSDUs belonging to different Transport connections are sent on one Network connection. To enable identification of the TSDUs belonging to the different Transport connections by the receiving Transport entity, the sending Transport entity attaches a Transport Connection End Point (TCEP) identifier to each TSDU. TCEP identifier is unique for each connection.

Splitting is done when the quality of service (e.g., throughput) offered by the Network service is less than the required quality of service or when greater resilience is required against Network connection failures. When splitting is permitted, a Transport connection may be assigned to a number of Network connections (Fig. 6b). In other words, TPDUs belonging to one Transport connections may be sent over different Network connections. Splitting will result in reordering of the TSDUs and, therefore, resequencing function is also required.

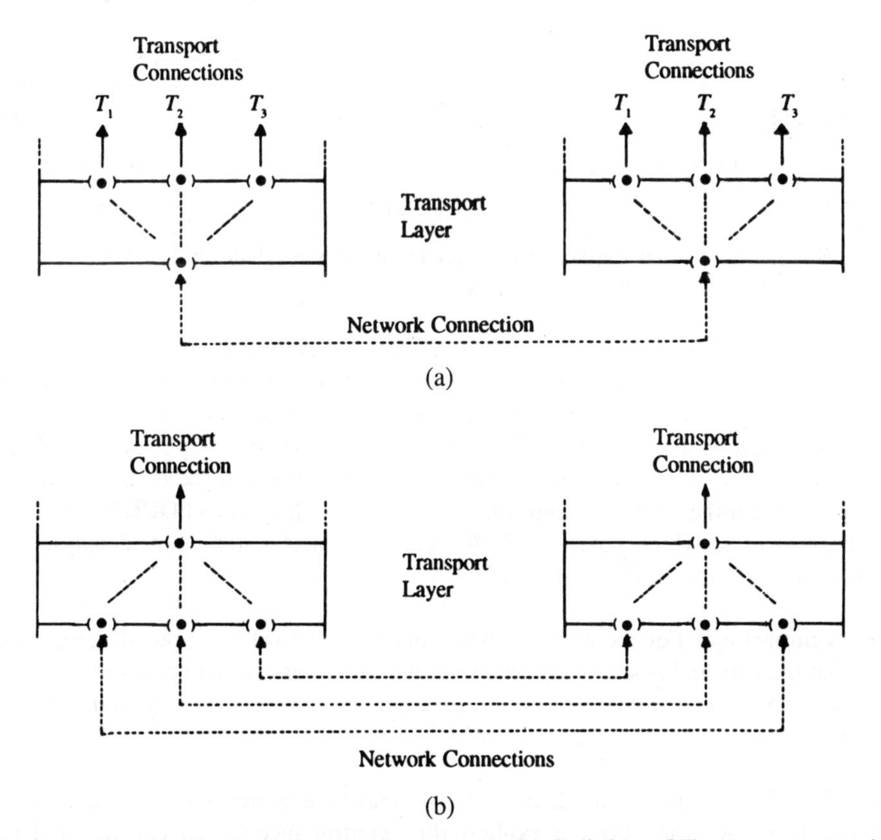

Fig. 6 (a) Multiplexing of Transport connections; (b) Splitting of Transport connections.

Segmenting and Concatenation. The segmenting function enables the Transport entity to divide a TSDU into several TPDUs each with a separate header containing a PCI. These TPDUs are sent on the same Network connection and the TSDU is reassembled by the receiving Transport entity (Fig. 7a). The last segment of a TSDU so transported carries an indication End of TSDU (EOT)

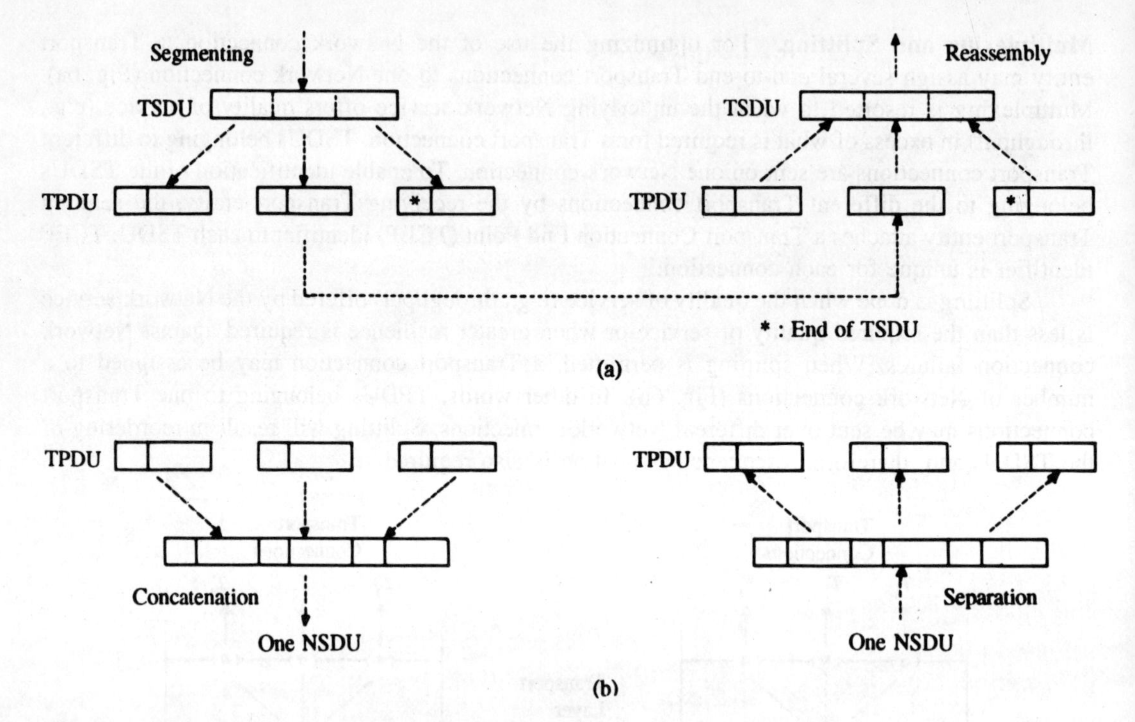

Fig. 7 (a) Segmentation and reassembly of Transport service data units; (b) Concatenation and separation of Transport protocol data units.

for the receiving Transport entity so that it may start the reassembly process. Segmenting is done when the Network service cannot support the size of TPDU containing an unsegmented TSDU.

The concatenation function enables mapping of several TPDUs onto a single NSDU (Fig. 7b). The TPDUs may belong to one or several Transport connections but they must be travelling in the same direction. At the receiving end, the Transport entity separates the concatenated TPDUs. It should be possible for the entity to identify the boundaries of different TPDUs. Therefore, there are some restrictions as to which types of TPDUs can be concatenated. Concatenation is done for efficient utilization of the Network service.

Sequence Numbering. For the error recovery and flow control functions described below it is necessary that the data and acknowledgement TPDUs are given a sequence number. The sequence number is seven bits long in normal mode of operation. It can be 31 bits long in the extended mode of operation.

Flow Control. Flow control at the Transport layer may be experienced due to back pressure from the Network layer. Alternatively, an explicit flow control mechanism can be implemented. A modified form of sliding window protocol is used. The window size is variable and is controlled by the receiver. The receiver sends a credit allocation to the sender. The credit allocation indicates how many TDPUs the receiver is ready to receive. We will examine this mechanism in detail later.

Error Recovery. There can be several ways in which errors can be introduced some of these are:

1. Protocol errors
2. Signalled failures of Network connection
3. TPDU errors.

Protocol errors refer to errors in the format or procedures of the TPDU exchange. Signalled failures refer to the reset or release of the Network connection due to some error at layer 3. Such failures are reported to the Transport layer which needs to take necessary action for recovery. The TPDU errors can be in the form of reordering of their sequence, loss of TPDUs, duplication of TPDUs or content errors in TPDUs.

When a protocol error occurs, the Transport protocol incorporates a mechanism for release of the connection. For protection against signalled failures, the Transport protocol provides for reassignment of the Network connection and for resynchronization, i.e., the Transport entities exchange acknowledgements to indicate the status of the received TPDUs.

If the sequence of the data TPDUs has been disturbed by the Network service, the Transport entities resequence the TPDUs before TSDUs are handed over to the user-entity. The duplicate TPDUs (indicated by their sequence numbers) are discarded. For lost TPDUs, an acknowledgement mechanism with timers is implemented and lost TPDUs are retransmitted after time out. For detecting content errors, error detection bytes are incorporated in the TPDUs. TPDUs received with content errors are discarded and are considered as lost.

2 NETWORK CONNECTION TYPES

The Transport layer provides a great variety in the Transport service by allowing the users to specify the quality of service. It has the responsibility of using the underlying Network service in the most optimum way and of enhancing the service to meet the user requirements. To bridge this gap, the Transport layer implements a number of functions which include flow control, error detection and recovery, multiplexing, segmentation etc. It must be, however, understood that the range of functions required to be implemented in the Transport layer depends on the underlying Network service. For example, if the Network service provides acceptable residual error rate in the content of the TPDUs, there is no need to implement the error detection function. Therefore, several classes of Transport protocols, each designed for a specific quality of Network service, have been defined. Before we discuss the Transport protocol classes, let us examine the categorization of the connection-mode Network service.

The quality of connection-mode Network service is characterized in terms of the residual error rate and the frequency of signalled failures. The residual error rate is estimated by the probability that the user data is transferred with error, or lost, or duplicated or delivered out of sequence. Frequency of signalled failures is the frequency of the Network connection release or reset. Signalled means that these failures are reported by the Network layer to the Transport layer whenever such incidences occur. Based on this characterization, a Network connection can be categorized as one of the following types:

- Type A: Network connection with acceptable residual error rate and acceptable frequency of signalled failures.
- Type B: Network connection with acceptable residual error rate but unacceptable frequency of signalled failures.
- Type C: Network connection with unacceptable residual error rate.

From the viewpoint of the Transport layer, a high residual error rate in the Network connection would require a complex Transport protocol to detect and correct the errors. A Network connection with a high rate of signalled failures will require a Transport protocol which can recover from these failures and maintain continuity of the Transport connection. In Type C Network connection, the rate of signalled failures is immaterial because the residual error rate itself is not within the acceptable limits.

If the underlying Network service is CLNS, design of the Transport protocol is as complex as the one that uses a Type C Network connection because they both exhibit similar error characteristics. CLNS is not reliable and does not provide sequenced delivery of the data units.

3 TRANSPORT PROTOCOLS

The following Transport protocols have been defined to cover the full spectrum of the Transport service and the Network service. The CCITT and ISO documents for these protocols are indicated in parentheses.

1. Transport protocols for providing connection-mode TS using CONS. Depending on the type of Network connection, the following classes of the protocols have been defined (CCITT X.224, ISO 8073).
    - Class 0 : Simple class
    - Class 1 : Basic error recovery class
    - Class 2 : Multiplexing class
    - Class 3 : Error recovery and multiplexing class
    - Class 4 : Error detection and recovery class.
2. Transport protocol for providing connection-mode TS using CLNS (ISO 8073 DAD 2).
3. Transport protocol for providing connectionless-mode TS using CONS or CLNS (ISO 8602).

Majority of the applications today are based on connection-mode Transport service. In the following sections, we will describe the features of the five classes of protocols defined by CCITT X.224 and ISO 8072. Use of a particular class of protocol is negotiable between the Transport entities during the connection establishment phase. The class which can be utilized to provide a given quality of TS is dictated by the type of Network connection.

3.1 Class 0—Simple Class

Class 0 is the simplest form of Transport protocol. It assumes that most of the required functions for supplying a reliable Transport connection are carried out by the Network layer. It requires connection-mode Network service (CONS) and Type A Network connections. As the residual error rate and the signalled failures in the Network connection are within the acceptable limits, the Class 0 Transport protocol does not include the error recovery function. Some of its important features are:

- It includes functions for connection establishment, data transfer with segmentation and reassembly.
- It does not include functions for multiplexing, disconnection, flow control, error recovery and expedited data transfer.

- Life time of the Network connection determines the life time of the Transport connection.
- It does not provide enhancement of the quality of the Network service.
- The maximum size of TPDU is 128 octets including the header.

3.2 Class 1—Basic Error Recovery Class

Class 1 Transport protocol is intended for use over CONS and Type B Network connections in which the frequency of signalled failures is unacceptably high. Therefore, this protocol provides for recovery from signalled Network connection failures. There is no need for improving the residual error rate as it is within the acceptable limits. Thus, it improves upon the Class 0 protocol by providing the following additional features:

- It provides for Transport connections with error recovery from the signalled failures by reassigning the connection to another Network connection and resynchronizing.
- TPDUs are numbered and acknowledgement mechanism is established. One variant, known as confirmation of receipt of the acknowledgement procedure uses N-DATA-ACKNOWLEDGEMENT service, if available.
- It provides for expedited data transfer.
- The Transport connection can be disconnected independently of the Network connection.
- There is no explicit flow control, but the Network layer flow control is visible over the Transport connection.

3.3 Class 2—Multiplexing Class

Class 2 is also an enhancement of Class 0 and permits multiple Transport connections to be established over one Network connection. It is designed for operation on Type A Network connections and, therefore, it does not provide for error detection and recovery. If the Network connection is reset or disconnected, the Transport connection is terminated without any end-to-end exchange, but the probability of such eventuality is within the acceptable limit as the Network connection is of Type A.

For multiplexing several Transport connections, the following additional functions must be implemented and are part of the Class 2 protocol:

- Explicit disconnection of individual Transport connections
- Explicit flow control
- Expedited data transfer
- TPDU numbering and acknowledgement for flow control.

Flow control is mandatory since TPDUs are pipelined and high throughput of one connection may affect the throughput of other multiplexed connections. If response time is not critical and traffic is low, use of the flow control function can be negotiated by the Transport entities.

3.4 Class 3—Error Recovery and Multiplexing Class

It is intended for use over CONS and Type B Network connections. It combines functionality of Class 1 and Class 2 protocols. It provides the multiplexing function and error recovery from the signalled failures.

3.5 Class 4—Error Detection and Recovery

It is intended for use over Type C Network connections having an unacceptable residual error rate. In other words, the TPDUs will be lost, duplicated and reordered. There will be signalled Network connection failures as well. Therefore, this protocol extends the Class 3 protocol by implementing additional mechanisms for error detection, retransmission of the TPDUs and resequencing of TPDUs containing data. This is the only protocol which can also be used to provide a connectionless-mode Transport service using CLNS.

4 TRANSPORT PROTOCOL DATA UNIT (TPDU)

The OSI Transport protocol defines 10 different types of TPDUs. Different subsets of the 10 TPDUs are required to support the five classes of the Transport protocol. Figure 8 shows the general format of an OSI TPDU. It can be divided into four parts. The first part (A) is the header length indicator field. It is of one octet and is followed by a fixed part (B). The fixed part contains several fields of fixed lengths. These fields are different for different types of TPDUs. The third part (C) is the variable part and contains some optional fields. These three parts constitute the header. Following the header is the fourth part (D) which contains the user data (TSDU).

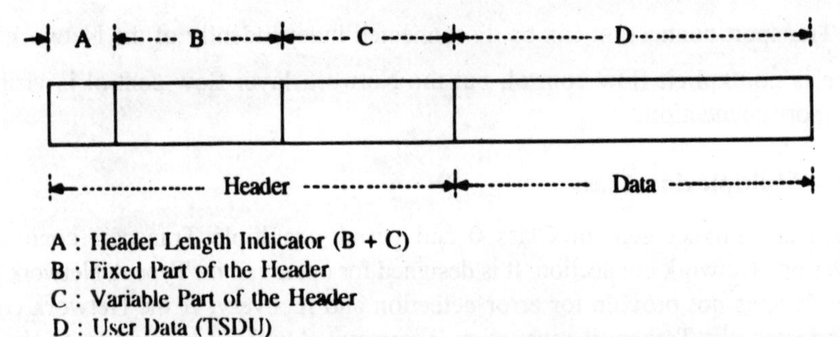

Fig. 8 General format of TPDU.

The first octet of the fixed part contains the 4-bit or 8-bit code which identifies the type of the TPDU. The codes are shown in Table 1. We will discuss the format of the TPDUs while discussing the protocol operation. Not all of the types of TPDUs are always used. The collection of TPDUs required to implement each class of Transport protocols is also indicated in the table.

5 TRANSPORT PROTOCOL MECHANISMS

Having described the general format of TPDUs, let us now examine the Transport protocol mechanisms. As and when we come across specific types of TPDUs, we will discuss their specific formats as well.

The connection oriented data transfer involves the following three phases:

1. Connection establishment
2. Data transfer
3. Connection release.

Table 1 TPDU Type Identifier Codes

Type of TPDU	Code								Applicable in protocol class				
	8	7	6	5	4	3	2	1	0	1	2	3	4
Connection Request (CR)	1	1	1	0	×	×	×	×	*	*	*	*	*
Connection Confirm (CC)	1	1	0	1	×	×	×	×	*	*	*	*	*
Disconnect Request (DR)	1	0	0	0	0	0	0	0	*	*	*	*	*
Disconnect Confirm (DC)	1	1	0	0	0	0	0	0		*	*	*	*
Data (DT)	1	1	1	1	0	0	0	0	*	*	*	*	*
Expedited Data (ED)	0	0	0	1	0	0	0	0		*	NF	*	*
Data Acknowledgement (AK)	0	1	1	0	×	×	×	×		NRC	NF	*	*
Expedited Data Acknowledgement (EA)	0	0	1	0	0	0	0	0		*	NF	*	*
Reject (RJ)	0	1	0	1	×	×	×	×		*		*	
TPDU Error (ER)	0	1	1	1	0	0	0	0	*	*	*	*	*

××××: Credit
NF: Not available when non-explicit flow control is selected.
NRC: Not available when receipt confirmation is selected.

5.1 Connection Establishment Phase

Upon request of the Transport service user, the Transport layer carries out the following functions:

1. It maps the Transport address to a Network address and obtains a Nework connection, taking into account the cost and quality of service requested.

2. It decides whether multiplexing or splitting is needed to optimize the use of the Network connections.

3. It chooses a source reference number to identify the Transport connection. All the data (DT) TPDUs which are later received on this connection will bear this reference number.

4. It sends a connection request (CR) TPDU to the destination Transport entity. The CR TPDU indicates

- the reference number to identify the Transport connection;
- the preferred and alternative transport protocol classes;
- options such as explicit flow control and normal or extended formats;
- the calling and called Transport Service Access Point Identifiers (TSAPs);
- the proposed values of maximum TPDU size and the other operational parameters;
- the proposed quality of service parameters; and
- the user data, if any.

5. On receipt of the CR TPDU, the responding Transport entity sends a connection confirm (CC) TPDU if it accepts the connection. It can refuse the connection by responding with a disconnect request (DR) TPDU.

6. In the CC TPDU, the protocol class, options and other operational and quality of service parameters are accepted.

7. The responding Transport entity also indicates its destination reference number in the CC TPDU. All the TPDUs belonging to this connection and meant for the responding entity should bear this number.

8. The Transport connection is established when the initiating transport entity receives the CC TPDU.

CR TPDU Format. Figure 9 shows the format of the CR TPDU. It consists of fixed and variable parts followed by a user data field. The CC TPDU is similar except that it contains "1101" in the TPDU type field and selected class and options in the respective fields.

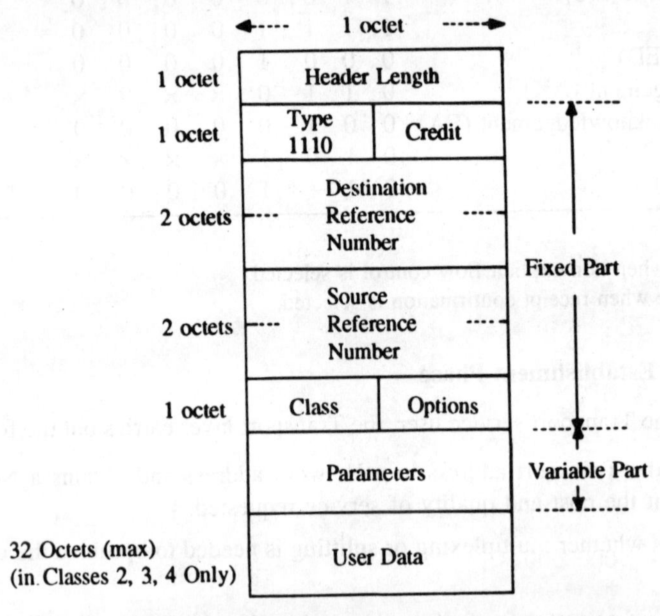

Fig. 9 Format of a connection request TPDU.

- The CR TDPU is identified by the type field code.
- The credit field indicates the initial credit allocation. It is set to zero if the preferred class is 0 or 1. Credit field is explained in Section 5.2.
- The source reference number identifies the TPDUs belonging to the particular Transport connection.
- The destination reference number is yet to be allotted by the destination and, therefore, this field is set to zero for the time being.
- The class field contains the binary code of the class number and indicates the preferred class of protocol. The second and subsequent choices are indicated in the variable part of the TPDU.
- Bits 3 and 4 of the options field are always zero. Bit 2 is set to "1" for using the extended formats. It is set to "0" for normal formats. The use of explicit flow control in Class 2 is indicated by "0" in bit 1. This bit is set to "1" for no explicit flow control. Bits 1 to 4 are always set to zero in Class 0 and have no meaning.

- The variable part conveys the optional parameter values. Each option is indicated by a parameter code field, the parameter length indication field and the parameter value (Fig. 10). Examples of the parameters are:
  - — Calling TSAP (Parameter code 11000001)
  - — Called TSAP (Parameter code 11000010)
  - — TPDU size (Parameter code 11000000)
  - — Maximum acknowledgement time (Parameter code 10000101)
  - — Checksum (Parameter code 11000110).

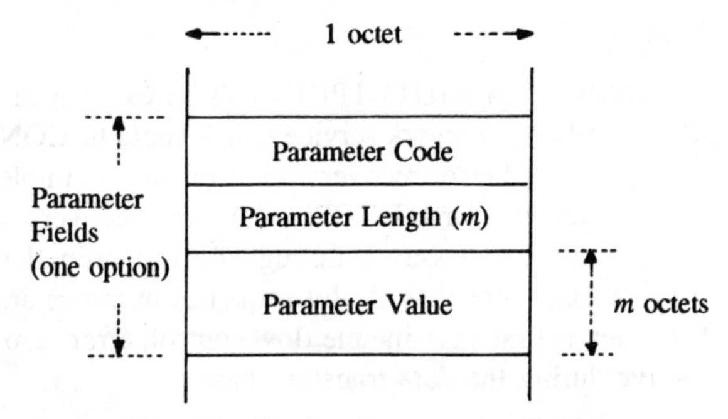

Fig. 10 Optional parameter fields.

- The proposed TPDU size can be from 128 to 8192 octets and is coded as below. The default value is 128 octets.

00001101	8192	(Not allowed in Classes 0 and 1)
00001100	4096	(Not allowed in Classes 0 and 1)
00001011	2048	
00001010	1024	
00001001	512	
00001000	256	
00000111	128	

- The maximum acknowledgement time is expressed in milliseconds and is coded as a binary number of two octets.
- Checksum on CR TPDU is mandatory in Class 4 protocol.
- No user data field is permitted in CR TPDU of Class 0 operation. In other classes, it is optional and may contain a maximum of 32 octets of data.

EXAMPLE 1

Show the variable part of CR TPDU for the following optional parameter values:

TPDU size	1024 octets
Acknowledgement time	2048 ms
Checksum	11001100 00110011

TPDU Size	Parameter Code	11000000
	Parameter Length	00000001
	Parameter Value	00001010

Acknowledgement time	Parameter code	10000101
	Parameter length	00000010
	Parameter value	00001000
		00000000
Checksum	Parameter code	11000110
	Parameter length	00000010
	Parameter value	11001100
		00110011

5.2 Data Transfer Phase

Once the connection is established, data (DT) TPDUs can be exchanged by the communicating Transport entities using the underlying Network service which could be CONS or CLNS. Depending on protocol class, the flow control and error recovery mechanisms are implemented. It is necessary to have a sequence numbering scheme for DT TDPUs for error recovery and flow control. Thus, sequence numbering is applicable to Classes 1 through 4. The normal sequence numbering is modulo-128 (7 bits). In the extended formats, 31-bit sequence numbers are used. Before we look at the formats of the TPDUs, let us first examine the flow control, error recovery and other protocol mechanisms which are active during the data transfer phase.

Flow Control. Explicit flow control procedure is applicable to protocol Classes 2–4. In the Class 2 protocol, the use of flow control procedure is negotiated. In protocol Classes 0 and 1, the flow control of the Network layer is experienced by the DT TPDUs.

Unlike the sliding window flow control mechanism of the Data Link layer, where the receiver may express its inability to accept more data units by sending RNR, the flow control mechanism of the Transport layer protocol controls the size of the sliding window to achieve the same. In this mechanism, the receiver specifies the size of window in addition to its lower edge. Flow control is exercised by using either the acknowledgement (AK) TDPU or reject (RJ) TDPU. These TDPUs carry the sequence number of the next wanted DT TDPU and a credit indication. The credit specifies how many DT TPDUs the receiver is ready to accept starting from the specified sequence number. When connection is being established, the initial credit allocations are made in the CR and CC TPDUs. The 4-bit credit field is used in the normal mode of operation and in the extended mode of operation the 16-bit credit field is used. Figure 11a provides an example of the flow control mechanism. The window status at the transmitting end is shown within parentheses. It indicates the lowest and the highest sequence numbers of the TDPUs within the window.

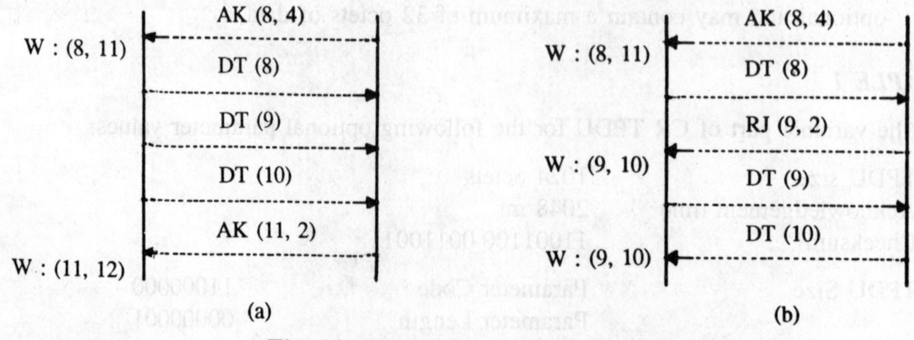

Fig. 11 Flow control mechanisms.

In protocol Class 2, once the credit is granted, the highest sequence numbered TPDU cannot be removed from the window by reducing the credit in subsequent acknowledgement but it is possible to do so in protocol Classes 3 and 4. As one implication of withdrawing credit, a Transport entity may receive a DT TPDU which is beyond the window. In protocol Class 3, a Transport entity may reduce the upper edge of the window and thereby enforce removal of a DT TPDU from the window by sending a reject (RJ) TPDU (Fig. 11b).

In protoçol Class 4, a transport entity may move the upper edge backward and forward by sending AK TPDU. But there is one difficulty. If two AK TPDUs are sent by a Transport entity, the receiving Transport entity must process them in the order they were sent but the underlying Network service being Type C, it is possible that the sequence of AK TPDUs is disturbed by the Network service. If this happens, the final status of the window will not be as the other Transport entity believes it to be. This problem is resolved by giving subsequence numbers to the acknowledgements. The subsequence number is indicated in the variable part of the header of the AK TPDU.

Protocol Errors. A protocol error occurs whenever a Transport entity receives a TPDU which cannot be interpretted or which is invalid. Some examples of protocol errors are:

- If the received TPDU is greater in size than the maximum selected size, the TPDU is considered invalid.
- Reception of an out of sequence DT TPDU in protocol Class 2 is considered to be a protocol error as the Network service is assumed to be error-free.
- Loss of a DT TPDU in the protocol Classes 1 and 3 is considered to be a protocol error because the underlying Network service is Type B. In protocol Class 4, DT TPDU loss is not considered as protocol error as recovery procedures are available to handle such events.

There could be many other causes of protocol errors. Whenever a protocol error is detected, the Transport entity may release the Transport connection or send an error (ER) TPDU and wait for the other entity to release the connection.

Error Recovery from Signalled Failures. As in flow control, TPDU error detection and recovery procedures are different in the different classes of Transport protocols. Class 0 and Class 2 Transport protocols do not provide for any error recovery because they are designed to operate with Type A Network connections which have acceptable residual error rate and acceptable signalled failures. As such, whenever the Network connection resets or disconnects, the Transport connection is released.

Class 1 and Class 3 protocols operate over Type B Network connections which have an acceptable residual error rate but unacceptable signalled failures of the Network connection. The Network entities indicate failures by giving RESET indication or DISCONNECT indication to the Transport entities. When DISCONNECT indication is received, the Transport entities need to reassign the Transport connection to another Network connection and then a resynchronization procedure is required to recover from the failure. If there is RESET indication from the Network entities, only resynchronization procedure is required because the Network connection is not released. The resynchronization procedure involves exchange of RJ TPDUs which indicate the sequence numbers of the next wanted DT TPDUs and the credits. Timers are provided to monitor the reassignment and resynchronization functions and if the timers expire, the Transport connection is considered released.

DT TPDU Errors. As mentioned earlier, the sequence numbers on the DT TPDUs enable recovery from the sequence errors and TPDU duplication. For the content errors, checksum is provided in the DT and AK TPDUs in the Class 4 protocol. The use of checksum for error detection is negotiable during connection establishment. A two-octet checksum using modulo-255 addition is generated over all the octets of the TPDU and is placed in the parameter field. The checksum calculation procedure was explained in Chapter 4. When a receiving Transport entity detects an error in the TPDU, it discards the TPDU. Since the destination reference number itself may be wrong, it is not possible to find the connection to which the TPDU belongs. When no acknowledgement is received from the destination, the source Transport entity retransmits the TPDU after expiry of the retransmission timer. A lost DT TPDU is handled in the same way.

Inactivity Control. A Transport entity maintains an inactivity timer which tracks the time which has elapsed since a TPDU was last received. When the timer runs out, the Transport entity initiates the connection release procedure. To prevent expiration of the remote Transport entity's inactivity timer, when no data is being sent, the local Transport entity must send AK TPDU at regular intervals.

Expedited Data Transfer. Expedited data transfer is applicable to Classes 1 through 4 protocols. Expedited data is the form of a special data TPDU which is not limited by the flow control. It can contain 1 to 16 octets of user data. Only one expedited data (ED) TPDU can be outstanding at any given time. After sending an ED TPDU, an expedited data acknowledgement (EA) TPDU must be received before next expedited or normal data TPDU is transmitted.

TPDU Formats. Figure 12 shows the formats of the DT and AK TPDUs. Extended mode of sequence numbering has been assumed, i.e., the sequence numbers are 31 bits long and the credit field is 8 bits. In the normal format, the sequence number including EOT is one octet and the credit field is 4 bits. The credit field is located in the first 4 bits of the second octet (just before the type field).

- The type field identifies the TPDU type.
- The destination reference number identifies the Transport connection to the remote Transport entity.
- 31-bit sequence number is used for recovery and flow control.
- End of TSDU (EOT) bit indicates the last fragment of the segmented TSDU. It is kept as "1" in the last fragment.
- The parameters field in the DT TPDU contains the checksum in Class 4 protocol.
- The parameters field in the AK TPDU contains a subsequence number, which is 16 bits long, and other parameters.
- The credit field in AK TPDU gives current allocation of the credit.

For the formats of the other TPDUs, the reader may refer to the CCITT X.224 recommendation. It may be mentioned, however, that the formats are similar to what we have seen above except that each type of TPDU is identified by its respective type code as indicated in Table 1.

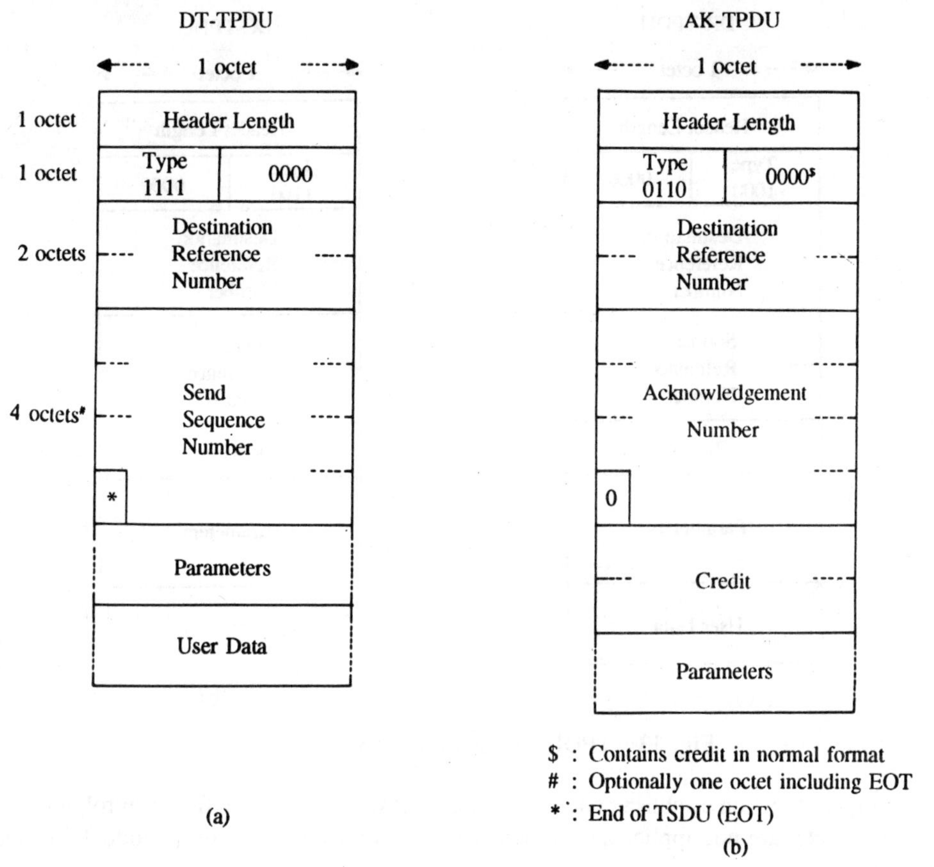

Fig. 12 TPDU formats for data transfer.

5.3 Connection Release Phase

The release procedure is used by the Transport entity to terminate the Transport connection. In Class 0 protocol, the connection is released by disconnecting the Network connection. When *N*-DISCONNECT indication is received, the Transport entity assumes release of the Transport connection as well.

In Class 1 to 4 protocols, the Transport connection can be explicitly released by sending a disconnect request (DR) TPDU. The disconnection is confirmed by sending a disconnect confirm (DC) TPDU. The formats of these TPDUs are shown in Fig. 13. The reason field contains the code indicating reason of disconnection, e.g., protocol error. The parameters field of the DR TPDU may contain a checksum and a user defined parameter. Up to 64 octets of the user data can be sent in the DR TPDU. The parameter field of DC TPDU contains checksum in Class 4 protocol.

6 CONNECTIONLESS DATA TRANSFER PROTOCOL

Connectionless-mode Transport service aims at moving a TSDU from one user entity to the other independent of other TSDUs. The protocol mechanisms for providing this service are extremely simple because the Transport entities do not have to ensure reliable delivery of the TSDUs.

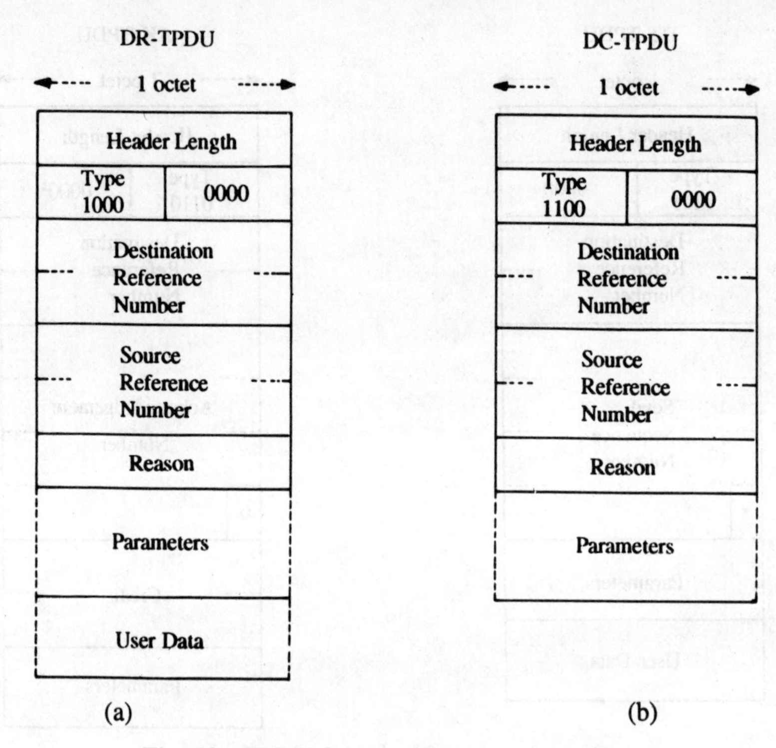

Fig. 13 TPDU formats for connection release.

Consequently, connection establishment, connection release, error control, flow control, multiplexing, error recovery etc. are not applicable. Transport protocol for connectionless-mode TS is specified in ISO 8602.

Only one type of TPDU is defined for connectionless data transfer. This TPDU primarily contains the user data, the source and destination addresses and optionally, a checksum parameter. If the receiving Transport entity detects an error, it discards the TPDU. No positive or negative acknowledgements are sent by the receiving Transport entity.

Figure 14 shows the general format of a data TPDU. Besides the header length field in the beginning and user data at the end, the TPDU contains a series of type-length-value parameter fields which contain source address, destination address and the optional checksum. TPDU code field identifies the TPDU as being connectionless. There is no way of segmenting a TSDU in the connectionless data transfer.

Connectionless Transport protocol is not as common as connection-oriented protocol today but it is becoming increasingly popular to support transaction-oriented services in which one cannot always tolerate the overhead of opening a connection. The reliability or the sequenced delivery assurances of the connection-oriented protocol are either not needed or are provided by the higher level protocols.

7 OSI UPPER LAYER ARCHITECTURE

The first four layers of the OSI Reference Model handle end-to-end transfer of streams of octets through the subnetwork(s). End system addressing, interaction with the subnetwork, optimal utilization

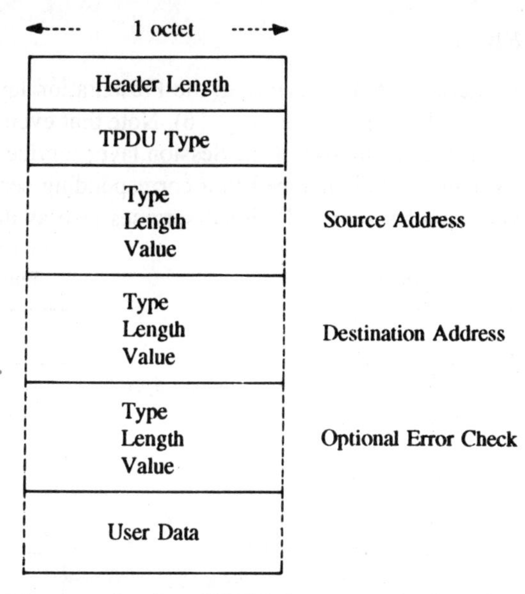

Fig. 14 General format of a data TPDU for connectionless data transfer protocol.

of the Network connections, and other associated functions are taken care of by these layers. The upper three layers of the OSI model are concerned with the application programmes themselves. They enable meaningful communication between the communicating entities in the end systems (Fig. 15).

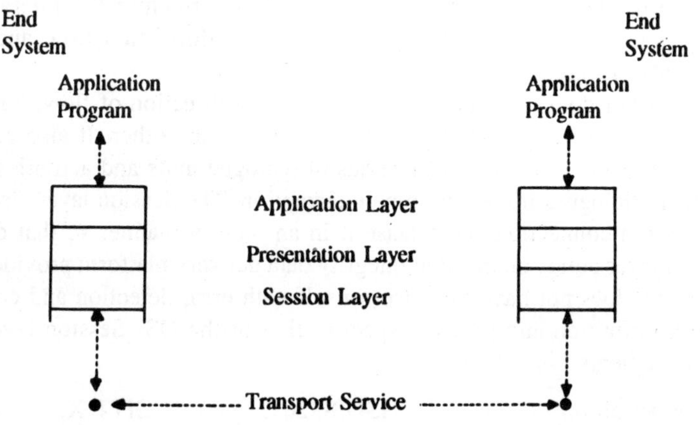

Fig. 15 Architecture of the upper OSI layers.

The Application layer which is on the top of the layered architecture provides LOGIN, password checking, file transfer and other functions to directly support the distributed applications. It incorporates protocols to exchange the semantics of the information.The Presentation layer which is the next layer, is concerned with data the representation function. The Session layer is responsible for organization of the dialogue. All these functions are built on the end-to-end reliable data transport service provided by the first four layers of the OSI Reference Model. In the following sections, we will examine the general functionality of the upper three layers of the OSI model without going into detailed descriptions of the many protocols and services of these layers. An overview of each of these layers should be adequate in the present context.

8 THE SESSION LAYER

The Session layer provides a defined set of services to the Presentation layer using the underlying Transport service and the Session layer protocols (Fig. 16). Note that even though the Presentation layer is above the Session layer, the actual user of the Session layer service is the Application layer. Each service the Session layer provides is mapped to a corresponding service of the Presentation layer. The Presentation layer provides some additional services also as discussed in Section 9.

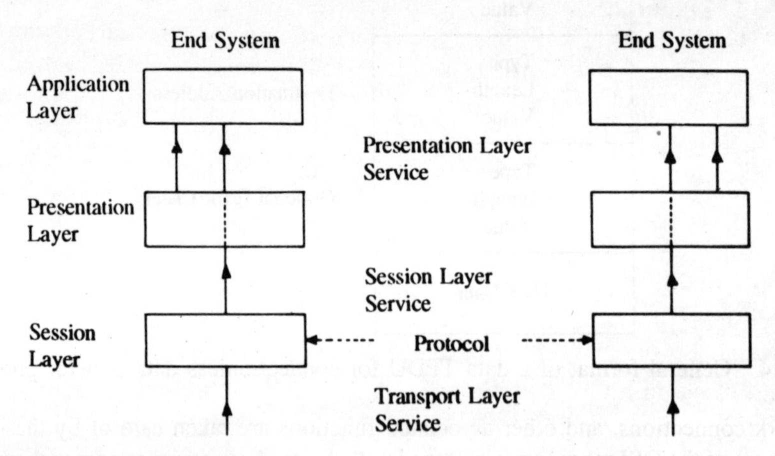

Fig. 16 Mapping of Session service to corresponding Presentation service.

The purpose of the Session layer is to control and structure the interaction between the application programmes. It controls the direction of the information flow and synchronizes the interaction elements or dialogues.

The Session layer does not take decisions regarding direction of flow, but enables the users of the Session service to transfer control from one user to the other. It also enables the users to structure their interaction in the form of a series of dialogue units and activities so that users may resynchronize their dialogues if there is some interruption. The Session layer also enables the users to establish a Session connection and release it in an orderly manner so that data are not lost.

The Session layer builds on the high integrity data delivery platform provided by the Transport layer and, therefore, does not have to concern itself with error detection and correction functions.

The service definition and protocol specification of the OSI Session layer are given in the following two documents:

1. Service definition	ISO 8326	CCITT X.215
2. Protocol specification	ISO 8327	CCITT X.225

8.1 Functionality of the Session Layer

The Session layer provides a spectrum of services to the users. These services are provided by a set of functional units which constitute the Session layer. In fact, the Session layer can be viewed as a general purpose tool kit and the users are allowed to select and use the tools needed by them. The selection of functional units is negotiated at the time of establishment of the Session connection.

Before we look at these functional units, let us examine some of the important functions carried out by the Session layer to provide organized interactions between the user entities. These functions relate to data transfer, orderly release, synchronization and the activity management.

Normal Data Transfer. Data transfer between the user entities may either be Two Way Alternate (TWA) or Two Way Simultaneous (TWS). In TWA, at any time only one user has exclusive rights to initiate transfer of the data. In TWS, simultaneous data transfer is possible in both directions. The data transfer mode is negotiated by the users at the time of establishing the Session connection.

In TWA mode, the Session layer makes available a token for direction control to the users. The user having the token can issue an S-DATA request primitive. When the Session connection is established, the token may be with the initiator (connection-requesting user) or with the responder (connection-accepting user) as indicated in the S-CONNECT primitives. Later, the token is exchanged by the users using the following primitives:

- S-TOKEN-PLEASE : Request for the token.
- S-TOKEN-GIVE : Transfer of the token.

In addition to the data token, there are other tokens as well. The S-CONTROL-GIVE primitive is used by a user to transfer all its tokens to the other user entity. Figure 17 shows an example of TWA data transfer between the users. We have shown the primitives issued by the Session service users instead of the usual protocol data units. This is done at the Session layer because its functionality is controlled by the users.

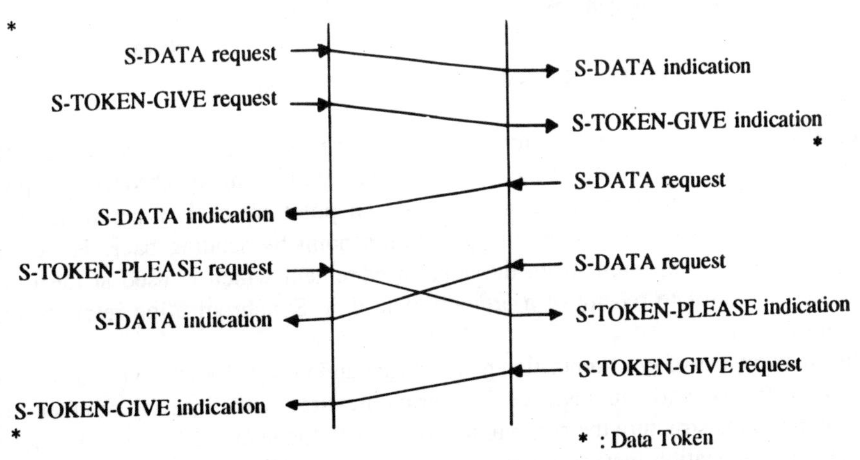

Fig. 17 Token control in TWA mode of data transfer.

Session Release. For releasing the session connection, there are four variations:

1. User abort
2. Provider abort
3. Orderly release
4. Negotiated release.

In the first two cases a Session connection may be aborted unilaterally by any one of the users or providers. The other entities are at the best informed of the connection release. As a consequence, data in transit may be lost.

The orderly release of the connection involves a two-way interaction between the users and, therefore, ensures that the data in transit are delivered before the connection is released. Orderly release has two important characteristics:

- either user may initiate the orderly release of the connection, and
- the other user has no option but to accept the release.

Negotiated release is controlled by a release token. Only the user having the token can initiate a request for release of the Session connection. The other user can either accept the release or refuse it if it has reason to do so (Fig. 18). The ownership of the release token can be requested by an S-TOKEN-PLEASE primitive and can be transferred by an S-TOKEN-GIVE primitive as in the case of data token. Note that each token primitive has a token type parameter to identify the token being requested or given.

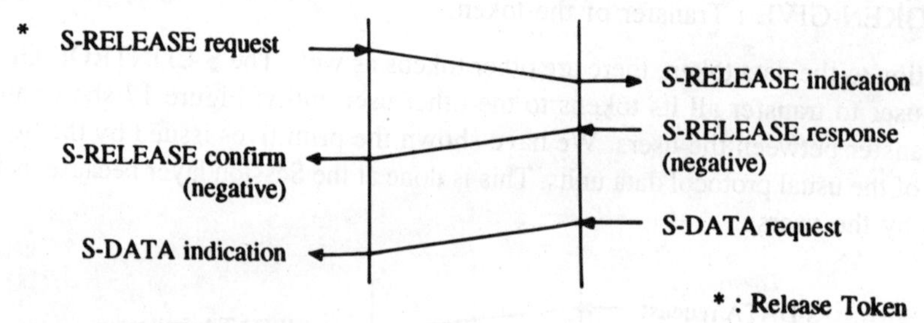

Fig. 18 Refusal of Session connection release.

Synchronization. Synchronization refers to assuring the same state of interaction between the Session service users at any point of time. It is achieved by inserting synchronization points in the interaction. The sending user inserts a synchronization point which is a serial number and the receiving user confirms the receipt of synchronization point by sending back the serial number. Synchronization point is different from the acknowledgement which is used at the lower layers. Acknowledgement refers to receipt of a protocol data unit. Synchronization refers to receipt of a synchronization point.

There are two types of synchronization points: major and minor. Both the types of synchronization points allow users to mark an instance in communication with a serial number. The major synchronization points structure the communication into dialogue units. The minor synchronization points mark communication instances within a dialogue unit (Fig. 19). The semantics associated with the major or minor synchronization points is user-determined.

When a major synchronization point is inserted at the request of one of the users, no further protocol data units may be sent until a confirmation of the synchronization has been obtained. A minor synchronization point, on the other hand may not be confirmed immediately. The confirmation implies to the sender that the data up to that point are safely stored or processed.

Both major and minor synchronization points use a common serial number. The default value at the start of session is 1 and the maximum count is upto 999998 (999999 is reserved). The type and serial number of the synchronization point are indicated in the request primitive by the user. The users and the Session layer maintain a parameter $V(M)$ which is the serial number of the next synchronization point to be used. When the connection is established, the initial value of the parameter is indicated to the users by the Session layer. This parameter is incremented by one whenever a synchronization point request is made. Figure 20 shows an example of insertion of major and minor synchronization points.

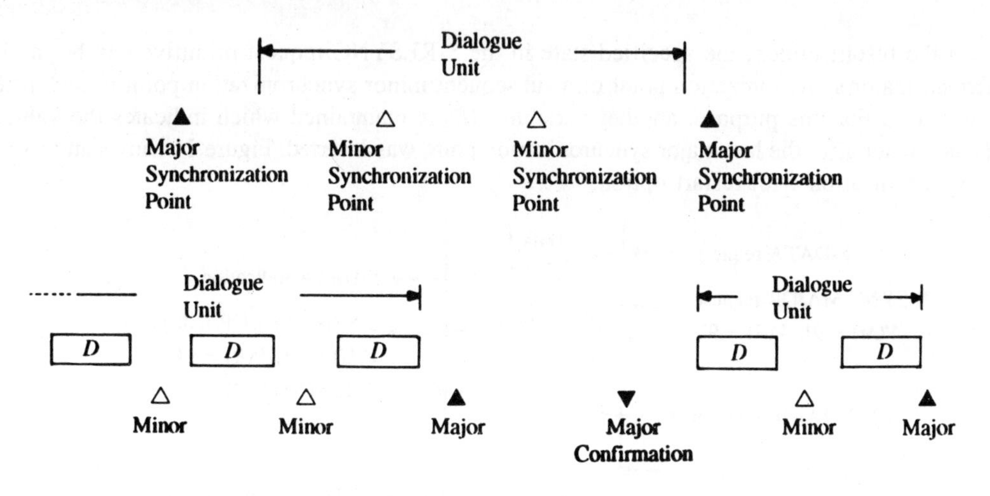

Fig. 19 Major and minor synchronization points.

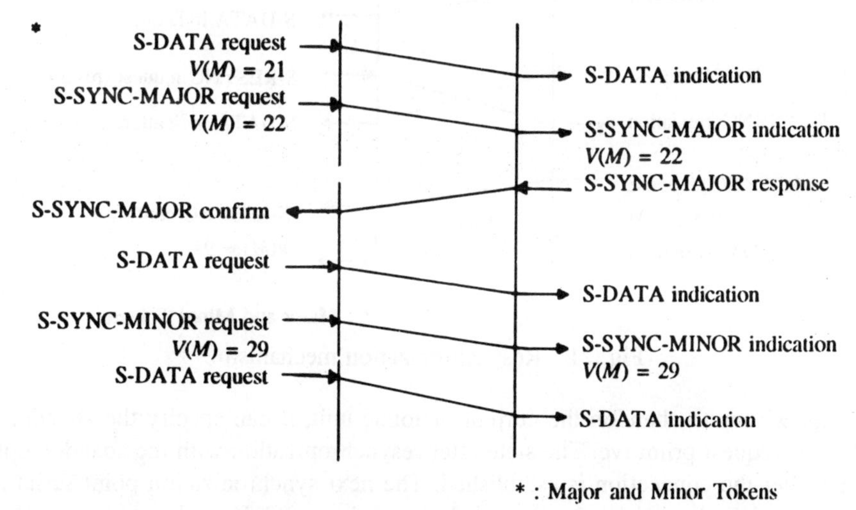

Fig. 20 Insertion of major and minor synchronization points.

The use of major and minor synchronization points is controlled by their respective tokens. A synchronization point can be inserted by a user if it owns the respective token. The synchronization tokens are exchanged as mentioned earlier by issuing S-TOKEN-PLEASE and S-TOKEN-GIVE primitives.

Resynchronization. Due to some reason, it may be necessary that the state of dialogue be restored to some previously defined state. This action is called *resynchronization*, or *backward synchronization*. The desired state is indicated by the users in the S-RESYNC request primitive. There are three options of resynchronization:

1. Restart
2. Abandon
3. Set.

In the restart option, the specified state in the S-RESYNC request primitive can be the last confirmed major synchronization point or a subsequent minor synchronization point in the current dialogue unit. For this purpose, another variable $V(R)$ is maintained which indicates the value of $V(M)$ parameter after the last major synchronization point was inserted. Figure 21 shows an example of resynchronization with restart option.

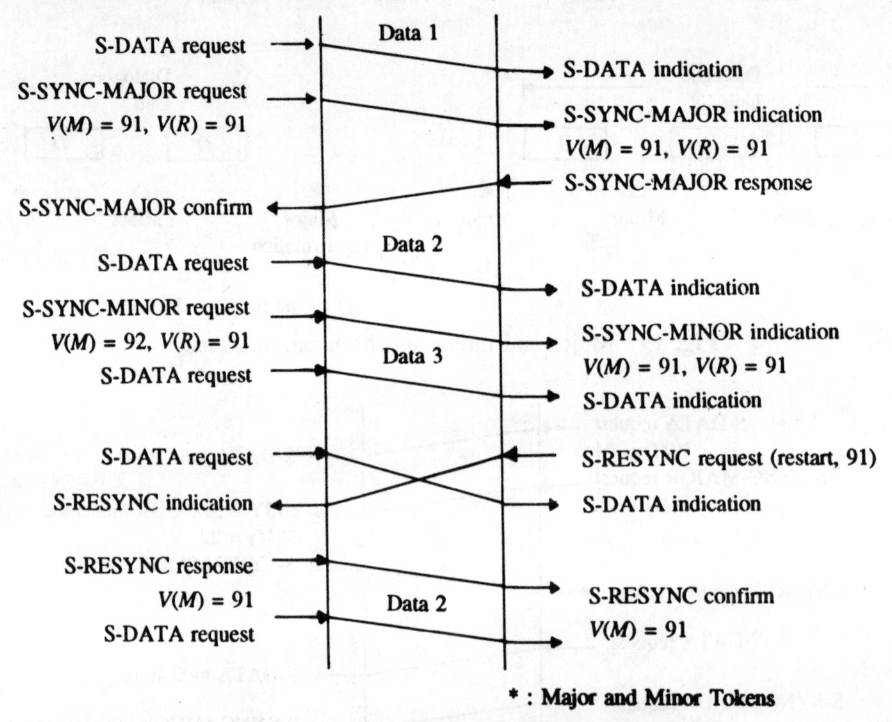

Fig. 21 Resynchronization mechanism.

If a user wishes to abandon the current dialogue unit, it can specify the abandon option in the S-RESYNC request primitive. The state after resynchronization with the abandon option is the same as that after the connection is established. The next synchronization point serial number to be used in future is indicated by the Session layer to the users. It is always greater than the last used synchronization point.

When the set option is used, the current dialogue unit is abandoned as in the abandon option but the users indicate a new synchronization point which can have any valid value. The actual use of this option is a matter of semantics to be determined by users by mutual agreement.

Activity. The Session layer allows the users to distinguish different logical pieces of work called "activities" which are performed during a session. An activity may consist of one or more dialogue units (Fig. 22). An activity has to be specifically started and ended by the users.

The start of an activity is marked by an activity start major synchronization point and the termination of an activity is marked by an activity end major synchronization point. Structuring of the interaction into activities enables one to

- interrupt and resume an activity during lifetime of a Session connection (Fig. 23a);

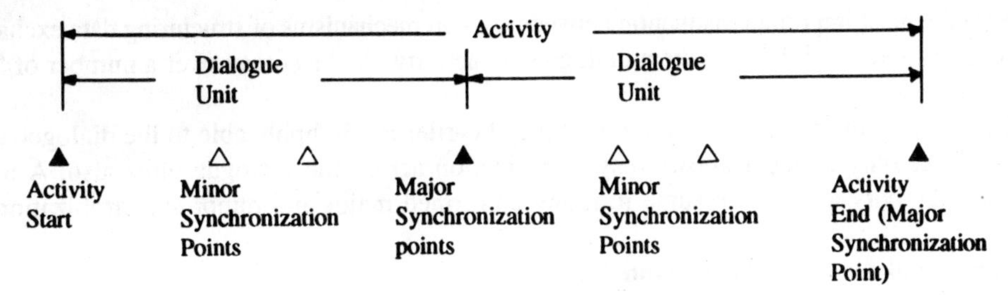

Fig. 22 Activity delimitation.

- interrupt an activity in one Session connection and resume the activity in a subsequent Session connection (Fig. 23b);
- initiate several activities though only one activity is carried out at a time (Fig. 23c).

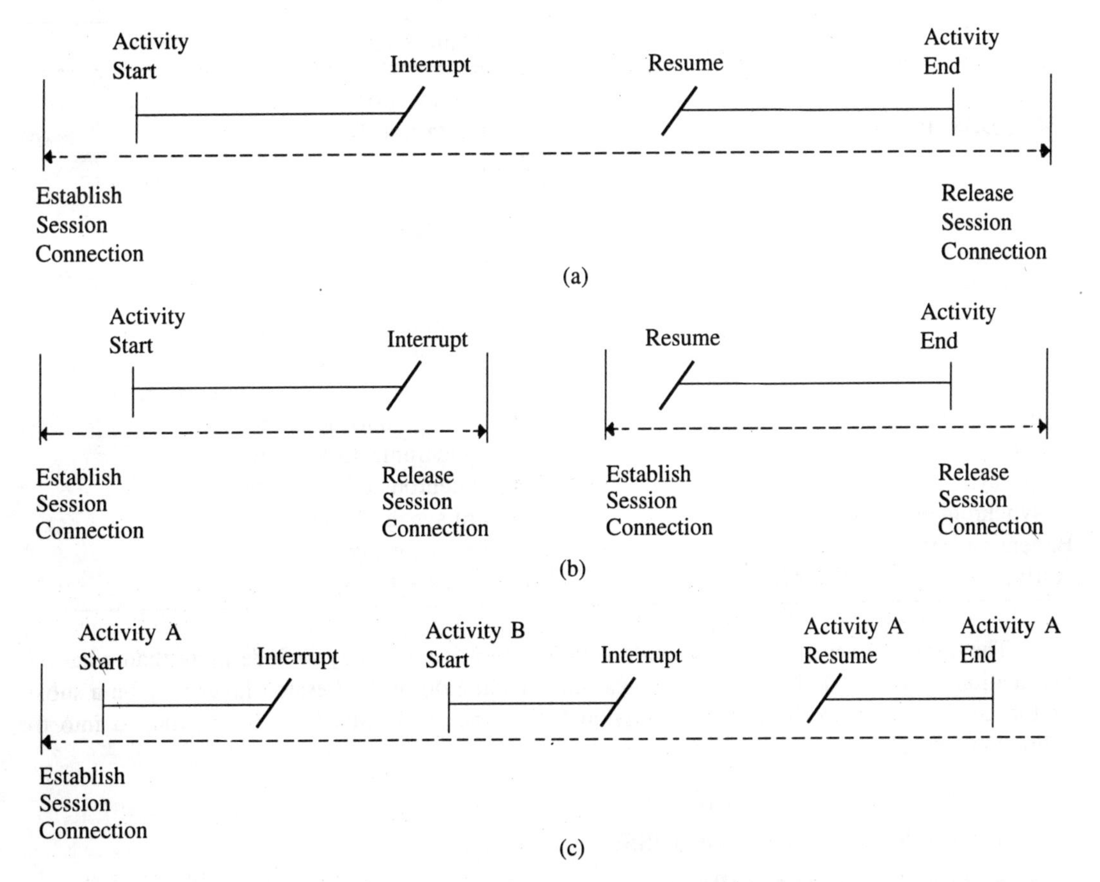

Fig. 23 (a) Resumption of an activity in the same Session connection; (b) Resumption of an activity in the subsequent Session connection; (c) Multiple activities in the same Session connection.

It is necessary to identifiy connections and activities to achieve the above objectives. Session connection identifier and activity identifier are provided for this purpose.

Note the most important distinction between the two mechanisms of structuring data exchanges—dialogue units and activities. Unlike a dialogue, an activity can be spread over a number of Session connections.

The concept of resynchronization introduced earlier is still applicable to the dialogue units of an activity. Activity concept allows resynchronization across the dialogue units also. A user can interrupt an activity and then resume it at any confirmed major and minor synchronization point.

8.2 Functional Units of the Session Layer

The Session layer consists of several functional units. Table 2 lists all the functional units and their main functions. The functional units are selected by the users during the Session connection establishment phase.

Table 2 Functional Units of the Session Layer

Functional unit	Main functions
Kernel (BCS, BSS, BAS)	Session connection Normal data transfer Orderly release User abort Provider abort
Negotiated release (BSS)	Token-controlled connection release
Half duplex (BCS, BSS, BAS)	Two-way alternate data transfer
Full duplex (BCS, BSS)	Two-way simultaneous data transfer
Expedited data	Expedited data transfer
Typed data (BSS, BAS)	Data transfer not subject to data token
Capability data exchange (BAS)	Data exchange outside an activity
Minor synchronize (BSS, BAS)	Minor synchronization points
Major synchronize (BSS)	Major synchronization points
Resynchronize (BSS)	Backward synchronization
Exceptions (BAS)	Exception reporting
Activity management (BAS)	Activity management

The complete set of functional units is quite extensive and may include more than what one application would need. Therefore, a typical implementation of the Session layer may be a subset of the complete set. In the past, the Session layer functional units have been grouped into the following subsets:

- Basic Combined Subset (BCS),
- Basic Synchronized Subset (BSS),
- Basic Activity Subset (BAS).

All the subsets require the kernel which provides the basic functionality of the Session layer. The other functional units which constitute these subsets are indicated in Table 2. It may be noted that the subsets are not negotiated as a group. The functional units must be individually requested in the S-CONNECT request primitive.

Kernel. The kernel allows use of the basic set of services that must be provided by any implementation of the Session layer. It provides Session connection establishment, orderly release, user- and provider-initiated connection abort and normal data transfer.

Negotiated Release. It provides token control over the release of a Session connection. The user who currently owns the release token can initiate the release of the connection. The other user can either accept or reject the request for release. It also provides transfer of release token from one user to the other.

Half Duplex. Two-way alternate (TWA) mode of interaction is provided by this functional unit. The communication is controlled by a data token and only the user who currently owns the token can send data. It also provides for transfer of the data token from one user to the other.

Full Duplex. This functional unit allows two-way simultaneous (TWS) mode of interaction over the Session connection. There is no data token in this mode. TWS and TWA modes are mutually exclusive and a given implementation may offer either.

Expedited Data. This functional unit allows transfer of an expedited data unit over the Session connection. This data unit results in corresponding expedited data units at the lower layers and is not flow controlled.

Typed Data. Typed data can be sent in the TWA mode without need of the data token. Single data unit can be sent even if the user does not own the data token at the time of sending the typed data unit.

Capability Data Exchange. This functional unit can be chosen only when the activity management functional unit has also been chosen. It allows transfer of limited amount of data when an activity is not in progress. This data is intended for control purpose and may enable users to ascertain whether they are capable of undertaking an activity.

Minor Synchronization. This functional unit enables insertion of minor synchronization points. It is token controlled.

Major Synchronization. This functional unit allows insertion of major synchronization points which mark the beginning and end of a dialogue unit. It is also token controlled.

Exceptions. This functional unit enables reporting of a special event by one user to the other or by the providers to the users. The special event could be a protocol/procedural error noticed by a user or provider. The users are informed of occurrence of such exceptional events so that they may initiate a recovery procedure such as resynchronization. The exception reporting is applicable in TWA mode only and to the user who does not own the token.

Activity Management. This function allows user interactions to be delimited into logically related activities. The boundaries of the activities are decided by the users. An activity can be interrupted and resumed as explained earlier. This functional unit includes GIVE-CONTROL primitives which transfer all the tokens.

8.3 Session Layer Protocol Data Units (SPDU)

The format and size of SPDU are not fixed. There are several fields which are often optional and variable in length. Therefore, each field is made capable of self-identification by having a three subfield structure of type-length-value. Figure 24a shows the general format of a SPDU. It consists of

- the SI (SPDU Identifier) field of one octet. The decimal value of the binary number identifies the type of SPDU.
- the LI (Length Indicator) field of one or three octets. It indicates the number of octets in rest of the header. "11111111" in the first octet of the LI field indicates a three octet LI field.
- the parameters field can have several forms. It may contain one or more fields having the type-length-value format (Fig. 24b). Alternatively, it may contain a PGI (Parameter Group Identifier) and a LI field indicating length of the parameter group (Fig. 24c). The LI field is followed by one or more fields having the type-length-value format. PGI and PI fields are of one octet.
- User data.

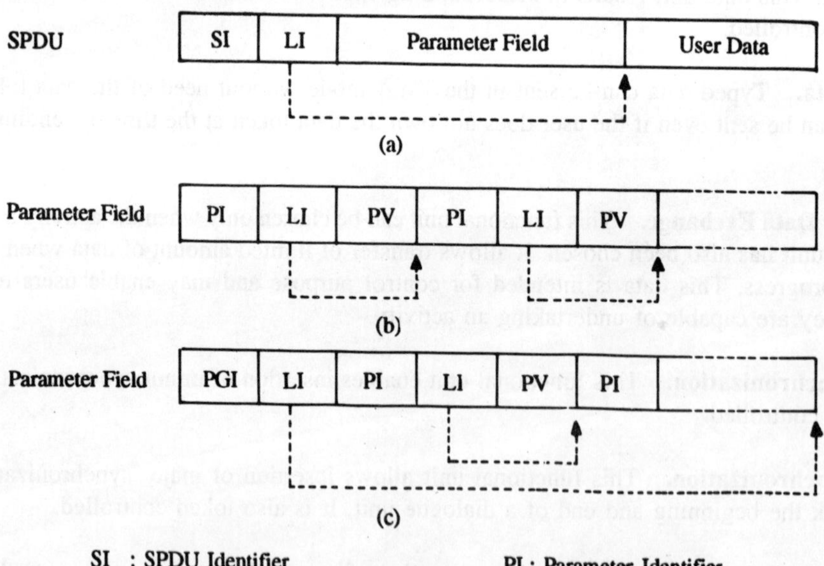

Fig. 24 The general format of SPDU.

We will not take up the description of various types of SPDUs, their options and coding rules which may be found in the ISO/CCITT documents mentioned earlier.

9 THE PRESENTATION LAYER

All along upto the Session layer, we have assumed that the user data is a string of bits (or octets) and we have addressed issues relating to the reliable transfer of bits from one end system to another. The user data represents information and our ultimate aim must be to preserve the information content of the data during transmission and to ensure that the information is well understood by the users. Computer systems may differ in their representation of data elements in the following respects:

- Character codes can be ASCII, EBCDIC or others.
- Integers can vary in length from 8 to 64 bits and use the 1's or 2's complement representation.
- Real numbers can be encoded as fixed or floating point numbers and the latter can differ in lengths of mentissa and exponent as well as using different bases for the exponent.
- Bytes may be addressed left to right or right to left.
- Data structures such as arrays, records or lists may be different.

It may, therefore, be necessary to carry out transformation of data representation when transferring information between two application programmes implemented in differing environments. The information content or semantics must be preserved during the transformation.

One possibility is to transform the data representation of one end system to that of the other end system. This approach is not very satisfactory because if there are N different end systems, $N(N - 1)$ translators would be required to match data representations of every possible pair of computers. Every new end system when interconnected would require a translator for every other computer.

A better approach is to perform translation at each end system to a common intermediate form. This would require $2N$ translations (one to and one from the internal representation of each system). This approach would also avoid the need to know the internal representation of data structures within an end system.

The sixth layer of the OSI Reference Model, the Presentation layer is primarily concerned with issues relating to external data representation. Irrespective of the internal data representation of an end system, the Presentation layer of an end system encodes the information into an external representation which is well understood by the Presentation layer of the other end systems. During the encoding process, the information content is preserved and it is not necessary for an end system to have knowledge of the internal data representation used by the other end system.

9.1 Data Type

Data typing is used in most high-level programming languages. The type essentially characterizes the data values (objects) and operations which can be applied on the values. For example, objects of boolean type can take values "true" or "false" and on these values the boolean operations such as OR, AND, NOT etc. can be applied. The data types can be divided into two categories:

1. Primitive type
2. Constructed type.

A primitive type is an elementary data unit which is not decomposable into smaller elements. Examples of primitive data types are:

- Boolean — "True" or "False"
- Integer — Signed natural numbers
- Character — Set of printable characters

- Binary coded decimal (BCD) 4-bit representation of the digits 0–9
- Floating point Real numbers with floating decimal point.

Constructed data type consists of a collection of several elements of the primitive type and/ or constructed type. As elements themselves can be of the constructed type, nested data structures are possible. A constructed object may be manipulated as a complete unit or its individual elements may be manipulated. The elements of constructed data type may all be of the same type (homogeneous) or mixed type (heterogeneous). The collection of elements may be ordered (sequence type) or unordered (set type).

9.2 Data Syntaxes

There are three data syntax types used for representing information at the Application and the Presentation layers (Fig. 25):

1. Abstract syntax
2. Local concrete syntax
3. Transfer syntax.

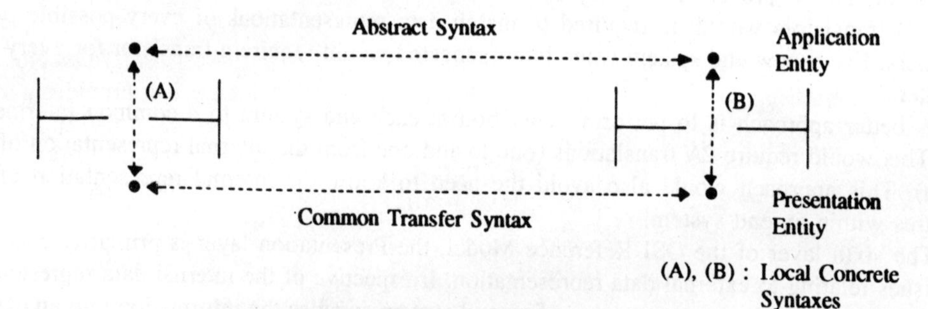

Fig. 25 Abstract, local and transfer syntaxes.

Abstract Syntax. The Application entities exchange the Application layer PDUs (APDUs). The APDUs are described by using a new approach similar to that of defining a primitive and constructed data types. In this new description, an APDU definition is called abstract syntax. It is called a syntax because the APDU has a structure. It is called abstract because it is not concerned with the representation of the information. For example, abstract syntax may define a data type called CheckingBalance, the values of which consist of integers. It does not specify whether a value of CheckingBalance data type is represented by decimal digits, binary numbers, or any other form.

Local Concrete Syntax. The local concrete syntax defines how a particular data value is represented in an end system. The local concrete syntaxes of two communicating end systems may be different or same. For example, one system may represent a value of CheckingBalance data type as a 32-bit binary number using 2s complement notation and the other system may represent the same data type as a string of decimal digits, each decimal digit represented as a 4-bit binary code.

Common Transfer Syntax. The transfer syntax defines how a particular abstract syntax is to be encoded when it is exchanged between two Presentation entities. The sending Presentation entity encodes the data values of its local concrete syntax into the transfer syntax and the receiving Presentation entity decodes the data values of the transfer syntax into its local concrete syntax. In short, an Application entity always sends or receives an APDU whose structure is defined by the abstract syntax and whose representation is in local concrete syntax.

The two Presentation entities should agree on the common transfer syntaxes to be used for transferring each type of the abstract syntax. During establishment of the Presentation connection, the originating Application entity requests a set of abstract syntaxes (AS1, AS2, ...). The Presentation entity determines the set of alternative transfer syntaxes (TSa, TSb, etc.) associated with each abstract syntax and indicates these in the connection request PPDU. The responding Presentation entity indicates the chosen transfer syntaxes for each type of the abstract syntax in its response. For example, the originating Presentation entity may indicate the possible transfer syntaxes TSa, TSb and TSc for the APDU type AS*i* (Abstract Syntax *i*). The responding entity may accept the transfer syntax TSc for the abstract syntax AS*i*.

Presentation Layer Context. Each pair of abstract syntax AS*i* and its negotiated transfer syntax TS*i* are termed as the Presentation layer context. The combined set of such contexts is called a defined context set. There is a default context which is used when the defined context set is empty.

9.3 Abstract Syntax Notation

Abstract syntaxes can be described in OSI environment with the help of Abstract Syntax Notation. One (ASN.1) which is an ISO standard, *ISO 8824, Abstract Syntax Notation 1*. It provides a means for defining data structures without specifying the data representation. It defines the primitive types from which complex constructed type data structures can be developed. The primitive type is identified by a tag which consists of a class and a primitive type identifier. There are four classes:

(i) Universal (00): This class consists of data types which are universal (application independent). These data types have been defined in the ASN.1 specification.

(ii) Application (01): This refers to the data types defined in the other international standards and are specific to an application (e.g., banking).

(iii) Context-specific (10): This indicates those data elements whose type is defined by the context e.g., a set.

(iv) Private (11): This denotes data types which are assigned by specific users, and their scope is confined to the particular abstract syntax.

For universal class, ASN.1 specifies the primitive types and constructors, a brief list of which is given below:

Identifier			
	1	:	BOOLEAN
	2	:	INTEGER
	3	:	BITSTRING
	4	:	OCTETSTRING
	6	:	OBJECT IDENTIFIER
	9	:	Real
	10	:	ENUMERATED
	16	:	SEQUENCE and SEQUENCE-OF

17 : SET and SET-OF
18 : NumericString
19 : PrintableString
22 : IA5String
23 : UTCTime
25 : GraphicString
28 : CharacterString

SEQUENCE type allows one to define an ordered list of data elements of arbitrary types. SEQUENCE-OF is similar except that all the elements must be of the same type. In SET and SET-OF types, the list is not ordered.

An example of a simple employee record expressed in ASN.1 data type notation is given below.

```
Employee      :: =  SEQUENCE                    (16)
    {
        name            PrintableString         (19)
        empNumber       INTEGER                 (2)
        salary          INTEGER                 (2)
    }
```

The above record defines the *Employee* data type as an ordered list made up of data elements called *name, empNumber* and *salary*. These elements have data types as indicated against each. The ASN.1 tags associated with these data types have been shown in parentheses. When a value of a particular type is encoded for transmission, the associated tag class and identifier are also encoded along with it. Thus, a Presentation layer entity receiving an encoded value can determine the type by examining the tag.

If the above employee record is expressed as a set, the elements may be in any order and, therefore, it is essential to distinguish the elements having the same type identifier. A Context-specific tag is assigned to the two integer data types. When the record is encoded, the Context-specific tags accompany each data element.

```
Employee      ::= SET
    {
        name                    PrintableString,
        empNumber               [0] INTEGER,
        salary                  [1] INTEGER
    }
```

With context-specific tags, the type identifier tag (2: INTEGER) is not required. It is implicit in the context in which the record is used. Therefore keyword IMPLICIT is included in the notation.

```
Employee      ::= SET
    {
        name                PrintableString,
        empNumber           [0] IMPLICIT INTEGER,
        salary              [1] IMPLICIT INTEGER
    }
```

9.4 Transfer Syntax Encoding

The abstract syntax defined above needs to be encoded by the Presentation layer into common transfer syntax. ISO 8825, *Specification of Basic Encoding Rules for ASN.1*, defines a set of rules for encoding and decoding the values of data types expressed in ASN.1. The basic encoding rules use a type-length-value form of encoding. For example, the *empNumber* value of 123456 is encoded using 5 octets as shown in Fig. 26. The first octet defines the type, the second indicates the length of the value field and the last three octets contain the data value encoded using 2s complement notation.

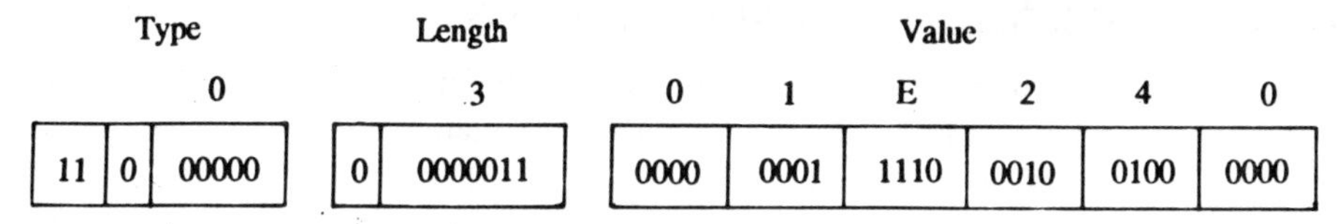

Fig. 26 Encoding of data type values.

The first two bits "11" of the first octet define the class which is private. The next "0" indicates that the value is a primitive data type and is not a constructed type. The final five bits "00000" indicate the implicit tag decimal 0 which identifies the value as *empNumber*. The first bit of the second octet indicates the extension of the length field. "0" in this position indicates the final octet of the length field. The three octet length of the value field is expressed as "0000011".

9.5 Presentation Layer Service

The Presentation layer service is documented in ISO 8822, *Presentation Service Definition*. The corresponding CCITT document is X.216. Like the Session layer, the Presentation service is provided by functional units within the Presentation layer. Some of the functional units are only for the pass-through Session service. These functional units map the Session service as the Presentation service to the user Application entity without any value addition. There are three other functional units which provide the functions exclusive to the Presentation layer.

Kernel. This functional unit is always required. It provides the Presentation connection establishment and release service. The defined context set and the default context are negotiated during connection establishment and these contexts are used during the life of the connection. The defined context cannot be modified unless the context management functional unit is chosen during establishment of the connection.

Context Management. If the context management function is chosen, the presentation entities can modify the defined context set. The user Application entities inform the new abstract syntax to be added (or deleted) and the Presentation entities negotiate the common transfer syntax while the connection is in operation. Need for context modification arises, for example, when a new file having a different syntax is required to be accessed.

Context Restoration. If the context restoration functional unit is chosen, it is possible to restore the defined context after a resynchronization occurs at the Session layer.

9.6 Presentation Layer Protocol

The Presentation layer protocol is documented in ISO 8823, *Presentation Protocol Specification*. The corresponding CCITT recommendation is X.226. The protocol defines the Presentation PDUs (PPDUs) and mechanisms for

- connection establishment and release,
- negotiation of transfer syntaxes,
- manipulation of the defined context, and
- restoration of defined context set after resynchronization.

Since many of the Presentation services are derived directly from the Session layer services, the protocol specifies direct mapping between related Presentation and Session service primitives. For these services, no PPDU is defined.

10 THE APPLICATION LAYER

The architecture of the Application layer differs considerably from that of the six layers below it. It can be viewed as consisting of several units each having its own functional units and providing a distinct set of services. As a result, there are a large number of Application protocols. In the following sections, we will confine the discussion on the Application layer to its architecture, terminology and functionality of some of its building blocks.

10.1 Application Process

An application process represents a set of resources that can be used to perform information processing. It is identified by an application-process title which is unambiguous throughout the OSI environment. An invocation of an application process refers to performing an information processing activity.

Application process can be thought of as an application program and its invocation as execution of the program. The purpose of OSI architecture is to enable the exchange of meaningful information between two application processes.

10.2 Application Entity

Application process represents resources associated with information processing and with communication. Therefore, it can be viewed as partly residing in the OSI Application layer (Fig. 27). The OSI Application entity represents that part of the application process which provides resources for OSI communication. Like the application process, the Application entity has Application-entity-title which is unambiguous in the OSI environment.

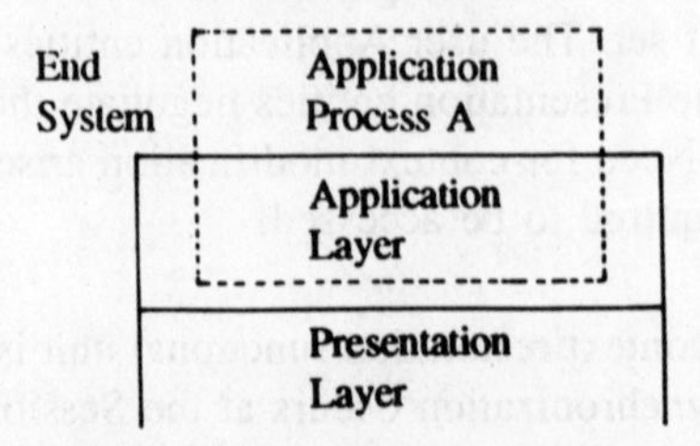

Fig. 27 Application process.

An application process may include one or more Application entities, each representing a different set of resources used for OSI communication (Fig. 28). This feature distinguishes the Application layer from the other layers which implement a single entity.

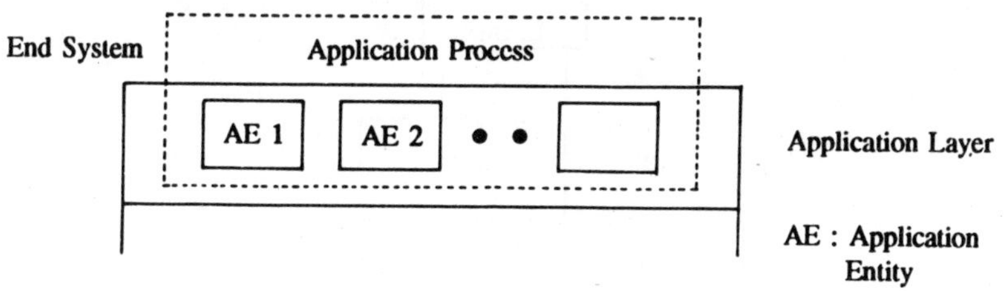

Fig. 28 Application entities.

10.3 Application Service Elements

An Application entity can be further broken down into a collection of Application Service Elements (ASE) and a user element. Each ASE provides a set of OSI communication functions for a particular purpose (Fig. 29). There are several international standard ASEs. The user element represents the part which uses the services of the ASEs to accomplish the communication objectives of the application process.

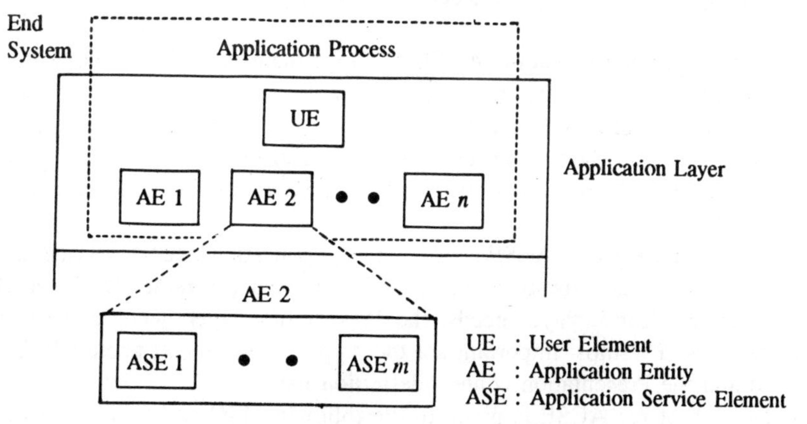

Fig. 29 Application service elements.

Each ASE is described by a service definition and a protocol specification. An ASE can use the Presentation layer service or can also call services of other ASEs in the Application entity (Fig. 30).

10.4 General Purpose Application Service Elements

There are some general purpose ASEs which provide commonly used services needed by a number of application processes. The existing general purpose service elements are:

- Association Control Service Element (ASCE)
- Reliable Transfer Service Element (RTSE)

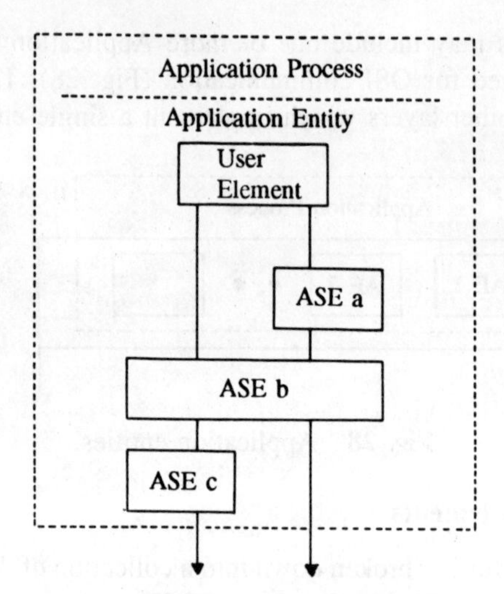

Fig. 30 Relationship between the user element, ASEs and the Presentation service.

- Remote Operations Service Element (ROSE)
- Commitment, Concurrency, and Recovery (CCR).

Association Control Service Element (ACSE). Since there is no layer above the Application layer, there is no concept of connection which we came across at each layer interface. At the Application layer, we talk of association between the communicating Application entities for the logical relationship between them. The association is controlled by an ASE called Association Control Service Element (ACSE). In other words, ACSE establishes and releases associations between Application entities.

Every Application entity uses ACSE because at the Application layer all services are connection-mode. During association establishment phase, ACSE indicates a rather large set of parameters which include the Application layer needs, the Presentation layer needs and the pass-through Session layer services. The most important are the Application context name which indicates the ASE to be used and the Presentation context definition list.

Full description of the ACSE is given in the following ISO and CCITT documents:

Service definition	ISO 8649	CCITT X.217
Protocol specification	ISO 8650	CCITT X.227

Reliable Transfer Service Element (RTSE). The word reliable here does not refer to recovery from the lost or damaged messages. Rather, it implies the capability to recover from the lost associations. When the RTSE is used in an Application entity, it has sole responsibility for utilizing the ACSE. The recovery from lost association is built on the Session layer synchronization service which is mapped by the Presentation layer. In its association request, it calls for the required functional units of the Session layer for this purpose. It indicates start activity, minor and major synchronization points and controls the tokens. The RTSE specifications are given in the following ISO and CCITT documents:

Service definition	ISO 9066-1	CCITT X.218
Protocol specification	ISO 9066-2	CCITT X.228

Remote Operations Service Element (ROSE). In many applications, communication between the application processes is interactive and involves execution of some operations by an Application entity on request from another (remote) Application entity. The Remote Operations Service Element (ROSE) provides such a capability to the user elements.

The basic ROSE model involves an invoker which sends a request for remote operation and a performer which returns a response. The response may be the result of successful operation, error indication or a reject. In some applications, a response may not be expected.

The ROSE operations may be synchronous or asynchronous. In synchronous operation, the response for the last request must be obtained before the next operation is taken up. In asynchronous operation, this wait is not required.

The invoker can be an association-initiator entity or an association-responding entity, or both. A typical example of remote operation is the electronic mail system. The mail-server process invokes the mail-user process to deliver a mail. The mail-user process invokes the mail-server process to submit the mail.

Detailed description of ROSE is given in the following ISO/CCITT documents:

Service definition	ISO 9072-1	CCITT X.219
Protocol specification	ISO 9072-2	CCITT X.229

Commitment, Concurrency, and Recovery (CCR). This ASE provides the capability to perform distributed processing transactions such as updating multiple copies of a replicated data base. During updation it maintains the data integrity and consistency in spite of outages in communication or system outages. With CCR, one system (master) is always in control and is always attempting to perform an operation (e.g., data base updation) and the other systems (slave) operates under its control.

Figure 31 shows the basic two-phase mechanism for data base updation. To update a record in the slave system, the master issues a BEGIN request which includes a unique identifier, a time out value and the data as parameters. The slave responds with the READY indication after it has stored to record safely and locked it so that no one else can read or change it. The data base is not yet updated. The master then issues COMMIT request and the slave updates the data base giving COMMIT confirmation.

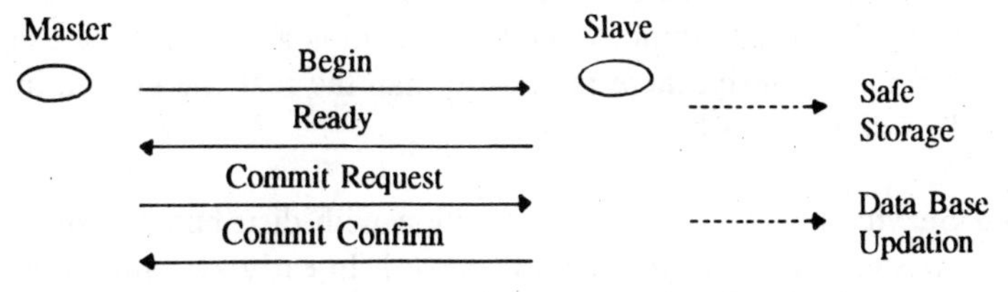

Fig. 31 Two-phase data base updation mechanism.

Two-phase operation ensures data consistency if there are multiple copies of the data base in several systems. During the first phase all the systems indicate their readiness and in the second phase all the copies of the data base are updated concurrently. If any system fails to report readiness, the master issues a ROLLBACK request to all the systems to instruct them to discard the planned update (Fig. 32).

CCR is defined in the following ISO and CCITT documents:

Service definition	ISO 9804	CCITT X.237
Protocol specification	ISO 9805	CCITT X.247

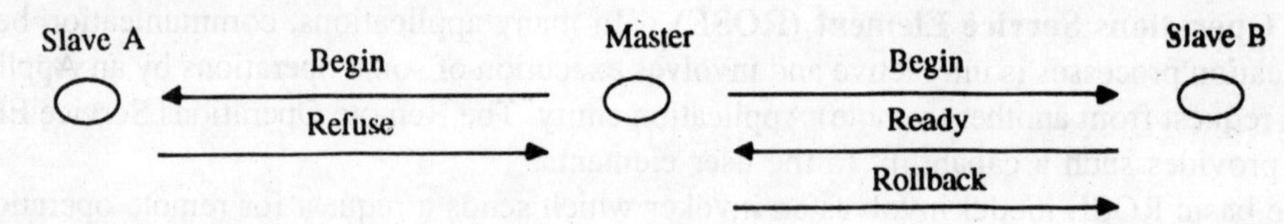

Fig. 32 Discarding a planned data base update.

10.5 Application Specific Service Elements

The general purpose ASEs described above provide service to other ASEs and to the application processes. There are also many international standards for more application-specific ASEs designed to provide specific types of services, sometimes directly to even human users. Some of the important application-specific ASEs are:

- File Transfer, Access and Management (FTAM)
- Virtual Terminal Service
- Directory Services.

These ASEs need services of the general purpose ASEs, ACSE in particular.

File Transfer, Access and Management (FTAM). File Transfer, Access and Management (FTAM) is an international standard, documented in ISO 8571. It defines an ASE which incorporates functions to support a remote file system in the OSI environment.

Users need to transfer files between the systems e.g., to submit a job for execution on a remote computer or to print a file. There is also need to access files remotely to update the shared information. The problem is that the file systems differ greatly with respect to file structure, information representation, access methods, file management information, and naming conventions. FTAM is a set of file handling protocols which overcomes this problem by defining a virtual file system which enables the users to transfer files, access individual records within a file and to manage the file attributes such as names and access permissions.

The virtual file system provides a common, standardized external appearance for all network file systems. As the name "virtual" implies, the file system is as seen by the users but it may not really exist in that form. Obviously, there is need to map the real file system into the virtual file system and this task is carried out by this ASE.

Virtual Terminal Service (VTS). Human interaction with distributed systems is usually via a terminal which consists of a display unit and a keyboard. In early systems, when networking was not prevalent, the terminal used to be dumb and was connected to the host through a dedicated link, but today a terminal can communicate with any computer connected to the network. It is, however, necessary that the computer be capable of handling the terminals which try to access it. Most communication problems are due to differences in characteristics of the many types of terminals available in the market.

The above problem is overcome by loading a virtual terminal service software which makes the host see a familiar set of terminal characteristics always. Similarly, the terminal sees a familiar terminal handler software. There is, of course, a need to define the standard virtual terminal characteristics and to map the actual terminal characteristics to the virtual terminal characteristics.

A virtual terminal is modelled as consisting of abstract objects—virtual keyboard, virtual display, a virtual cursor etc. On these abstract objects, abstract operations like character delete, clear the screen etc. can be performed. The ISO standards for VTS are:

- Virtual Terminal Service ISO 9040
- Virtual Terminal Protocol ISO 9041

Directory Service. The basic purpose of the directory service is to provide a networkwide capability to express a request in terms of a user-friendly, globally unique name and to have this name mapped into a network-specific address that can be used for any delivery to the name, e.g., to deliver electronic mail, directory service is needed to map the user-provided names to the network addresses for mail delivery. There are many additional services provided by the directory service. CCITT recommendations on the directory service are listed in X.500 series of documents. Corresponding ISO standards are in ISO 9594 series of documents.

11 SUMMARY

The Transport layer provides the required quality of end-to-end data transport service in an optimum manner. Delivery delay, residual error rate and throughput are some important quality of service parameters. The Network service is augmented by the Transport layer to meet these requirements.

The upper three layers of the OSI model, namely, the Application layer, the Presentation layer, and the Session layer, provide services for meaningful exchange of information. The Session layer incorporates functional units for implementing dialogue discipline, synchronization and recovery from failed connections. The service of these units is available as the Presentation service to the Application entity. The Presentation layer is concerned with data representation and a common transfer syntax for exchange of information. The Application layer at the top of the OSI model has to meet the diverse requirements of the application processes and, therefore, contains several Application entities. Remote LOGIN, file transfer, file access, reliable transfer of information, directory service etc. are some of the many functions and services provided by the Application layer. Its structure is very complex and application-dependent.

PROBLEM

1. In the exchange of TPDUs shown below, fill in the blanks within the parenthesis.

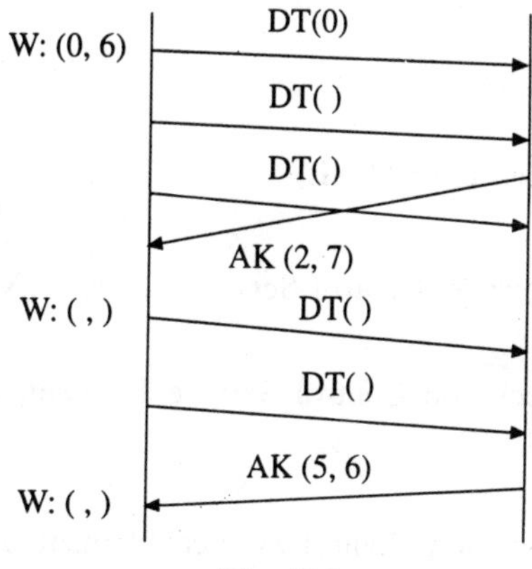

Fig. P.1

ISO Standards for the Open System Interconnection

The ISO standards for Open System Interconnection (OSI) are given below. Those standards which are still at the draft stage have not been included. Many other standards which are not covered in the text have also not been included. Relevant CCITT recommendations, wherever available, are indicated within parentheses.

THE OSI REFERENCE MODEL

ISO 7498–1 (1984)
OSI—Basic Reference Model (X.200)

ISO 7498 ADD 1 (1987)
OSI—Basic Reference Model, Addendum 1: Connectionless data transmission

ISO 7498–2 (1989)
OSI—Basic Reference Model, Part 2: Security architecture

ISO 7498–3 (1989)
OSI—Basic Reference Model, Part 3: Naming and addressing

ISO 7498–4 (1989)
OSI—Basic Reference Model, Part 4: Management framework (X.700)

THE APPLICATION LAYER

ISO 9545 (1989)
Application Layer Structure (X.207)

Association Control Service Element (ACSE)

ISO 8649 (1988)
Service Definition for the Association Control Service Element (X.217)

ISO 8649 AMD 1 (1990)
Service Definition for the Association Control Service Element, Amendment 1: Authentication during association establishment

ISO 8649 AMD 2 (1991)
Service Definition for the Association Control Service Element, Amendment 2: Connectionless-mode ACSE service

ISO 8650 (1988)
Protocol Specification for the Association Control Service Element (X.227)

ISO 8650 AMD 1 (1990)
Protocol Specification for the Association Control Service Element, Amendment 1: Authentication during association establishment

ISO 10035 (1991)
Connectionless ACSE Protocol to Provide the Connectionless-mode ACSE Service

ISO 10169–1 (1991)
Conformance Test Suite for the ACSE Protocol, Part 1: Test suite structure and test purposes

Job Transfer and Manipulation

ISO 8831 (1992)
Job Transfer and Manipulation Concepts and Services

ISO 8832 (1992)
Specification of the Full Class Protocol for Job Transfer and Manipulation

Commitment, Concurrency, and Recovery

ISO 9804 (1990)
Service Definition for the Commitment, Concurrency, and Recovery Service Element (X.237)

ISO 9805 (1990)
Protocol Specification for the Commitment, Concurrency, and Recovery Service Element (X.247)

Reliable Transfer

ISO 9066–1 (1989)
Text Communication—Reliable Transfer, Part 1: Model and service definition (X.218)

ISO 9066–2 (1989)
Text Communication—Reliable Transfer, Part 2: Protocol specification (X.228)

Remote Operations

ISO 9072–1 (1989)
Text Communication—Remote Operations, Part 1: Model, notation, and service definition (X.219)

ISO 9072–2 (1989)
Text Communication—Remote Operations, Part 2: Protocol specification (X.229)

File Transfer, Access, and Management (FTAM)

ISO 8571–1 (1988)
File Transfer, Access, and Management (FTAM), Part 1: General introduction

ISO 8571–1 AMD 1 (1992)
File Transfer, Access, and Management (FTAM), Part 1: General introduction, Amendment 1: File-store management

ISO 8571–2 (1988)
File Transfer, Access, and Management (FTAM), Part 2: Virtual filestore

ISO 8571–2 AMD 1 (1992)
File Transfer, Access, and Management (FTAM), Part 2: Virtual filestore, Amendment 1: Filestore management

ISO 8571–3 (1988)
File Transfer, Access, and Management (FTAM), Part 3: The file service definition

ISO 8571–3 AMD 1 (1992)
File Transfer, Access, and Management (FTAM), Part 3: The file service definition, Amendment 1: Filestore management

ISO 8571–4 (1988)
File Transfer, Access, and Management (FTAM), Part 4: The file protocol specification

ISO 8571–4 AMD 1 (1992)
File Transfer, Access, and Management (FTAM), Part 4: The file protocol specification, Amendment 1: Filestore management

ISO 8571–5 (1990)
File Transfer, Access, and Management (FTAM), Part 5: PICS proforma

Virtual Terminal

ISO 9040 (1990)
Virtual Terminal Service—Basic Class

ISO 9040 AMD 2 (1992)
Virtual Terminal Service—Basic Class, Amendment 2: Additional functional units

ISO 9041–1 (1990)
Virtual Terminal Protocol—Basic Class

ISO 9041 AMD 2 (1992)
Virtual Terminal Protocol—Basic Class, Amendment 2: Additional functional units

The Directory

ISO 9594–1 (1990)
The Directory, Part 1: Overview of concepts, models and services (X.500)

ISO 9594–2 (1990)
The Directory, Part 2: Models (X.501)

ISO 9594–3 (1990)
The Directory, Part 3: Abstract service definition (X.511)

ISO 9594–4 (1990)
The Directory, Part 4: Procedures for distributed operations (X.518)

ISO 9594–5 (1990)
The Directory, Part 5: Protocol specifications (X.519)

ISO 9594–6 (1990)
The Directory, Part 6: Selected attribute types (X.520)

ISO 9594–7 (1990)
The Directory, Part 7: Selected object classes (X.521)

ISO 9594–8 (1990)
The Directory, Part 8: Authentication framework (X.509)

ISO 9594–9 (1990)
The Directory, Part 9: Replication (X.525)

THE PRESENTATION LAYER

ISO 8822 (1988)
Connection-oriented Presentation Service Definition (X.216)

ISO 8822 AMD 1 (1991)
Presentation Service Definition—Amendment 1: Connectionless-mode Presentation service

ISO 8822 AMD 2 (1992)
Connection-oriented Presentation Service Definition, Amendment 2: Additional session synchronization functionality to the Presentation service user

ISO 8823 (1988)
Connection-oriented Presentation Protocol Specification (X.226)

ISO 8823 AMD 5 (1992)
Connection-oriented Presentation Protocol Specification, Amendment 5: Additional session synchronization functionality to the Presentation service user

ISO 8823–2 (1992)
Connection-oriented Presentation Protocol Specification, Part 2: PICS proforma

ISO 8824 (1990)
Specification of Abstract Syntax Notation One (ASN.1) (X.208)

ISO 8825 (1990)
Specification of Basic Encoding Rules for ASN.1 (X.209)

ISO 9576 (1991)
Connectionless Presentation Protocol to Provide the Connectionless-mode Presentation Service

THE SESSION LAYER

ISO 8326 (1987)
Basic Connection-oriented Session Service Definition (X.215)

ISO 8326 AMD 4 (1992)
Basic Connection-oriented Session Service Definition, Amendment 4: Additional synchronization functionality

ISO 8327 (1987)
Basic Connection-oriented Session Protocol Specification (X.225)

ISO 8327–2 (1992)
Basic Connection-oriented Session Protocol Specification, Part 2: PICS proforma

ISO 8327 AMD 3 (1992)
Basic Connection-oriented Session Protocol Specification, Amendment 3: Additional synchronization functionality

THE TRANSPORT LAYER

ISO 8072 (1986)
Transport Service Definition (X.214)

ISO 8072 ADD 1 (1986)
Transport Service Definition, Addendum 1: Connectionless-mode transmission

ISO 8073 (1992)
Connection-oriented Transport Protocol Specification (X.224)

ISO 8602 (1988)
Protocol for Providing the Connectionless-mode Transport Service

ISO 11570 (1992)
Transport Protocol Identification Mechanisms

THE NETWORK LAYER

ISO 8208 (1990)
X.25 Packet Level Protocol for Data Terminal Equipment

ISO 8208 ADD 1 (1992)
X.25 Packet Level Protocol for Data Terminal Equipment, Addendum 1: Alternative logical channel identifier assignment

ISO 8208 ADD 3 (1991)
X.25 Packet Level Protocol for Data Terminal Equipment, Addendum 3: Conformance requirements

ISO 8348 (1987)
Network Service Definition (X.213)

ISO 8348 ADD 1 (1987)
Network Service Definition, Addendum 1: Connectionless-mode transmission

ISO 8348 ADD 2 (1988)
Network Service Definition, Addendum 2: Network layer addressing

ISO 8348 ADD 3 (1988)
Network Service Definition, Addendum 3: Additional features of the Network service

ISO 8473 (1988)
Protocol for Providing the Connectionless-mode Network Service

ISO 8473 ADD 3 (1989)
Protocol for Providing the Connectionless-mode Network Service, Addendum 3: Provision of the underlying service assumed by ISO 8473 over subnetworks which provide the OSI Data Link service

ISO 8648 (1988)
Internal Organization of the Network Layer

ISO 8878 (1992)
Use of X.25 to Provide the Connection-mode Network Service (X.223)

ISO 8880–1 (1990)
Specification of Protocols to Provide and Support the OSI Network Service, Part 1: General principles

ISO 8880–2 (1992)
Specification of Protocols to Provide and Support the OSI Network Service, Part 2: Provision and support of the connection-mode Network service

ISO 8880–3 (1990)
Specification of Protocols to Provide and Support the OSI Network Service, Part 3: Provision and support of the connectionless-mode Network service

ISO 8881 (1989)
Use of the X.25 Packet Level Protocol in Local Area Networks

ISO 8882–2 (1992)
X.25–DTE Conformance Testing, Part 2: Data Link layer test suite

ISO 9542 (1988)
End System to Intermediate System Routing Exchange Protocol for Use in Conjunction with ISO 8473

ISO 9574
Provision of the OSI Connection-mode Network Service by Packet Mode Terminal Equipment Connected to an ISDN

TR 10029 (1989)
Operation of an X.25 Interworking Unit

ISO 10030 (1990)
End System Routing Information Exchange Protocol for Use in Conjunction with ISO 8878

ISO 10030 AMD 2 (1992)
End System Routing Information Exchange Protocol for Use in Conjunction with ISO 8878, Amendment 2: PICS proforma

ISO 10589 (1992)
Intermediate System—Intermediate System Intra-domain Routing Information Exchange Protocol for Use in Conjunction with ISO 8473

THE DATA LINK LAYER

ISO 1155 (1978)
Use of Longitudinal Parity to Detect Errors in Information Messages

ISO 1177 (1985)
Character Structure for Start/Stop and Synchronous Character Oriented Transmission

ISO 3309 (1991)
High Level Data Link Control Procedures—Frame Structure

ISO 3309 AMD 2 (1992)
High Level Data Link Control Procedures—Frame Structure, Addendum 2: Extended transparency options for start/stop transmission

ISO 4335 (1991)
High Level Data Link Control Procedures—Elements of Procedures

ISO 4335 AMD 4 (1991)
High Level Data Link Control Procedures—Elements of Procedures, Addendum 4: Multi-selective reject option

ISO 7478 (1987)
Multilink Procedures

ISO 7776 (1986)
High Level Data Link Control Procedures—X.25 LAP-B-Compatible DTE Data Link Layer Procedures

ISO 7776 AMD 1 (1986)
High Level Data Link Control Procedures—X.25 LAP-B-Compatible DTE Data Link Layer Procedures, Amendment 1: PICS proforma

ISO 7809 (1991)
High Level Data Link Control Procedures—Classes of Procedures

ISO 7809 AMD 5 (1991)
High Level Data Link Control Procedures—Classes of Procedures, Addendum 5: Connectionless class of procedures

ISO 7809 AMD 6 (1992)
High Level Data Link Control Procedures—Classes of Procedures, Addendum 6: Extended transparency options for start/stop transmission

ISO 7809 AMD 7 (1991)
High Level Data Link Control Procedures—Classes of Procedures, Addendum 7: Multi-selective reject option

ISO 8471 (1987)
HDLC—Balanced Classes of Procedures—Data Link Layer Address Resolution/Negotiation in Switched Environments

ISO 8886 (1992)
Data Link Service Definition (X.212)

THE PHYSICAL LAYER

ISO 2110 (1989)
25-Pole DTE/DCE Interface and Contact Number Assignments

ISO 2110 AMD 1 (1991)
25-Pole DTE/DCE Interface and Contact Number Assignments, Addendum 1: Data signalling rates above 20 kbps

ISO 2593 (1993)
34-Pole DTE/DCE Interface and Contact Number Assignments

ISO 4902 (1989)
37-Pole DTE/DCE Interface and Contact Number Assignments

ISO 4903 (1989)
15-Pole DTE/DCE Interface and Contact Number Assignments

ISO 7480 (1992)
Start-Stop Transmission Signal Quality at DTE/DCE Interface

ISO 9067 (1987)
Automatic Fault Isolation Procedures Using Test Loops

ISO 9543 (1989)
Synchronous Transmission Signal Quality at DTE/DCE Interface

ISO 10022 (1990)
Physical Service Definition (X.211)

LOCAL AREA NETWORKS

ISO 2382-25 (1992)
Vocabulary—Part 25: Local area networks

ISO 8802–2 (1989)
LANs, Part 2: Logical link control

ISO 8802–3 (1992)
LANs, Part 3: CSMA/CD

ISO 8802–4 (1990)
LANs, Part 4: Token passing bus access method and Physical layer specifications

ISO 8802–5 (1992)
LANs, Part 5: Token ring access method and Physical layer specifications

ISO 8802–7 (1991)
LANs, Part 7: Slotted ring access method and Physical layer specifications

ISO 9314–1 (1989)
Fibre Distributed Data Interface (FDDI)—Token Ring Physical Layer Protocol (PHY)

ISO 9314–2 (1989)
Fibre Distributed Data Interface (FDDI)—Token Ring Media Access Control (MAC)

ISO 9314–3 (1989)
Fibre Distributed Data Interface (FDDI)—Physical Layer Medium Dependent (PMD) Requirements

ISO 10039 (1991)
MAC Service Definition

TECHNICAL REPORTS

TR 8509 (1987)
Service Conventions (X.210)

TR 10172 (1991)
Network/Transport Protocol Interworking Specification

TR 9575 (1990)
OSI Routing Framework

TR 9577 (1990)
Protocol Identification in the Network Layer

TR 10029 (1989)
Operation of an X.25 Interworking Unit

TR 9578 (1990)
Interface Connectors Used in LANs

TR 10178 (1992)
Structure and Coding of LLC Addresses in LANs

TR 7477 (1985)
Arrangement for DTE to DTE Physical Connection Using V.24 Interchange Circuits

Glossary

ABM (Asynchronous Balanced Mode): Asynchronous mode of data transfer between combined stations using HDLC protocol.

Abstract syntax: A data structure, typically an Application layer PDU.

ACK (Acknowledgement): An indication that a data unit was received without any error.

ACSE (Association Control Service Element): An ASE which establishes and releases associations.

Activity: An identified operation at the Session layer; it may span one or more consecutive dialogue units.

Adaptive equalizer: An equalizer with capability of automatic equalizer adjustment for minimum intersymbol interference.

ADCCP (Advanced Data Communications Control Procedure): A bit-oriented synchronous Data Link control protocol of ANSI.

AFI (Authority and Format Indicator): The field of NSAP address which indicates how the address is formatted.

ALOHA: A contention media access control technique for local area networks.

ANSI: American National Standards Institute.

Application layer: Seventh layer of the OSI Reference Model.

Architecture: A specification that defines how the system is organized, defining functional modularity as well as protocols and interfaces for communication and cooperation among the modules.

ARM (Asynchronous Response Mode): Asynchronous mode of data transfer between primary and secondary stations using HDLC protocol.

ASE (Application Service Element): A functional unit of an Application entity, described by a service definition and a protocol specification.

ASCII (American Standard Code for Information Interchange): A code set defined by ANSI.

ASK (Amplitude Shift Keying): A modulation scheme for digital signals in which amplitude of the carrier is modulated.

Association: Roughly, the equivalent of a connection at the Application layer.

Asynchronous: Not synchronous, i.e., occurrence of an event such as transmission of a data unit is not conditionally associated with occurrence of another event such as reception of an acknowledgement.

Asynchronous transmission: Serial transmission of bytes which have start and stop bits.

Balanced transmission: A transmission mode in which the pair of conductors of transmission medium, carry signal currents which are opposite in phase with respect to a common ground.

Baseband transmission: Transmission of signals without modulation.

BCC (Block Check Character): Error detecting byte in BISYNC protocol.

BER (Bit Error Rate): Number of errors in a fixed number of transmitted bits.

BISYNC (Binary Synchronous Communications): A byte oriented Data Link protocol developed by IBM.

Bit oriented: One of the two categories of Data Link protocols, in which all the fields including control information coded at bit level.

Bit stuffing: Insertion of extra bits in a bit stream to avoid appearance of unintended control sequences.

BPSK (Binary PSK): PSK with two phase states.

Bridge: A LAN interconnection device that operates at the media access control sublayer.

Broadband: Transmission of signals after modulation and multiplexing of the modulated carriers.

BSC: Same as BISYNC.

Bus: A LAN topology in which stations are attached to a cable terminated at both ends and all the transmissions propagate along the cable to reach all the stations.

Byte: A group of bits, eight bits usually, considered as a single unit during processing.

Byte oriented: A category of Data Link protocols, in which fields are one or more bytes long and symbols from one of the code sets are used to send control information.

Carrierband: Same as single channel broadband.

CCITT (Consultative Committee for International Telephone and Telegraph): An international standards-making group.

CCR (Commitment, Concurrency, and Recovery): An ASE that provides capability of distributed processing, maintaining data integrity and consistency.

Characteristic impedance: Input impedance of an infinitely long transmission line; a transmission line is usually terminated by its characteristic impedance.

Checksum: An error detecting code based on a summation operation performed on the bits to be checked.

Circuit switching: A technique of establishing a dedicated communication path of fixed bandwidth between two end systems through a subnetwork.

CLNS (Connectionless-mode Network Service): Network service for transfer of data units by the Network entities without prior establishment of any connection.

Collision: A condition in which two frames are being transmitted at the same time over the shared medium of a LAN and are rendered unintelligible; in X.25, it implies a condition when the logical channel identifiers of the CALL REQUEST packet and INCOMING CALL packet happen to be same.

Combined station: A station in HDLC protocol that can act both as a primary and a secondary station.

Common transfer syntax: It defines how the abstract syntax of the Application layer is to be encoded for its exchange between two Presentation entities.

Concatenation: In OSI terminology, concatenation refers to mapping two or more (*N*)-PDUs into one (*N*–1)-SDU.

Connection: A relationship between two entities, which is established to provide an agreed upon service.

CONS (Connection-mode Network Service): Network service for transfer of data units by the Network entities after establishing an end-to-end Network connection.

Contention access: Media access methods in which the stations compete with one another to transmit on the medium.

CRC (Cyclic Redundance Check): An error detection scheme in which the check bits are generated as a remainder of division carried out on the data bits by a predetermined binary sequence.

Crosstalk: Unwanted appearance of signals of other channels in a communication channel.

CSMA/CD (Carrier Sense Multiple Access/Collision Detection): A media access control technique in local area networks.

CTS: Clear to send.

CUG (Closed User Group): A subset of users for whom the subnetwork services are restricted to communication among the members of the group.

Datagram: A data packet which contains source and destination addresses and is routed through the subnetwork independently.

Data Link layer: Second layer of the OSI Reference Model.

DCE (Data Circuit Terminating Equipment): This mainly refers to the subnetwork side equipment that interfaces with the Data Terminal Equipment (DTE).

DCS (Defined Context Set): A set of defined Presentation layer contexts.

Decibel: A measure of relative strengths of two signals.

Delay distortion: Distortion of a signal due to nonlinear phase characteristic of the transmission medium.

Differential encoding: An encoding scheme that results in encoded signal representing change in binary values of the digital signal rather than the absolute values.

DLE: Data Link Escape.

DNIC (Data Network Identification Code): It consists of three digit country code and single digit network code.

DSP (Domain-Specific Part): The local option portion of an N-SAP address.

DTE (Data Terminal Equipment): The user equipment that interfaces with DCE and acts as data source and data sink.

EBCDIC (Extended Binary Coded Decimal Interchange Code): A code set defined by IBM.

EIA: Electronics Industries Association.

End system: An OSI system that contains the application process. It needs to communicate with the application process in another end system.

Equalizer: A passive or active circuit that compensates for the linear distortions; attenuation equalizer is for attenuation distortion; group delay equalizer is for phase distortion.

Error control: Detection and correction of content and flow integrity errors in the transfer of data units.

Error rate: The ratio of number of data units in error to the total number of data units.

Ethernet: A 10 Mbps baseband local area network developed jointly by Xerox, Intel and DEC.

Expedited data: Data units that are to be sent as quickly as possible without regard for flow control.

Fast select: An X.25 facility for data transfer that allows inclusion of data in the CALL REQUEST and CALL CLEARING packets.

FCS (Frame Check Sequence): The field of HDLC frame containing CRC code for detecting content errors.

FDDI (Fibre Distributed Data Interface): A standard for optical fibre based local area networks.

FDM (Frequency Division Multiplexing): A technique of multiplexing signals by allotting them different frequency bands.

FEXT: Far end cross talk.

Flag: A unique bit sequence employed to delimit the opening (and closing) of a frame.

Flow control: To control transfer of data units for preventing flooding the receiver with more data units than it can handle.

Four-wire circuit: A circuit having separate pairs of wires for transmit and receive signals.

Frame: Commonly used name for the Data Link layer PDU.

FSK (Frequency Shift Keying): A modulation scheme for digital signals in which frequency of the carrier is modulated.

FTAM (File Transfer, Access, and Management): An ASE that incorporates functions to support remote file operations.

Full duplex: A transmission mode that allows simultaneous transmission of data in both the directions.

Gateway: A term generally used for intermediary devices that interconnect two networks at layer 7 of the OSI Reference Model.

GOSIP: Government OSI Profile.

Group delay: Slope of phase-frequency characteristic; measure of phase distortion.

Half duplex: A transmission mode that allows data transmission in one direction at a time.

HDLC (High-level Data Link Control): A bit-oriented Data Link protocol developed by ISO.

Headend: The end termination of a bus in local area networks.

Header: First segment of a data unit containing all information for identifying the data unit and directing it to the destination.

Host: An information processor which is generally self sufficient and does not need any supervision from other processors.

Hybrid: A balanced transformer for converting a two-wire circuit into a four-wire circuit and vice versa.

IA5 (International Alphabet 5): ISO version of ASCII.

ICI (Interface Control Information): Control information which is exchanged between adjacent layers of the OSI Reference Model.

IEEE: Institute of Electrical and Electronics Engineers.

I-Frame: Information frame of HDLC protocol.

Impulse noise: A high-amplitude short duration noise pulse.

Interface: It refers to the interface specifications of adjacent functional modules of a horizontally and/or vertically partitioned system.

Intermodulation noise: Noise caused by the nonlinear characteristics of the signal processing components.

Internetwork: A network formed by interconnecting several subnetworks using bridges, routers and gateways.

IP: Internet Protocol.

IS (Intermediate System): An OSI system that interconnects two subnetworks.

ISO: International Standards Organization.

Kernel: An essential functional unit of the Session layer to support the basic set of services.

LAN: Local Area Network.

LAP (Link Access Procedure): Asynchronous unbalanced Data Link protocol specified by CCITT.

LAP-B (Link Access Procedure-Balanced): Asynchronous balanced Data Link protocol specified by CCITT.

Layering: That aspect of network architecture which partitions the communication process into distinct hierarchy of functional layers, each having been assigned a specific function.

Level: CCITT equivalent of the layer of ISO terminology. Level is associated with a numerical identifier (e.g. level 1) and layer is associated with a descriptive identifier (e.g. Physical layer).

LLC (Logical Link Control): A sublayer of Data Link layer in local area networks for error control, flow control, sequencing and user addressing functions.

Local concrete syntax: The actual representation of data values in an end system.

Logical channel: At the packet level of X.25, multiple logical channels are derived by assigning a unique logical channel identifier on every packet of each logical channel.

LRC (Logical Redundancy Check): An error detection check which is applied on a block of bytes.

MAC sublayer: A sublayer of Data Link layer in local area networks for media access control.

Major/Minor synchronization point: An OSI Session layer service that allows the users to insert a synchronization point in a dialogue, confirmation of which indicates that data units have been safely stored or processed by the receiver.

MAN: Metropolitan Area Network.

Manchester encoding: A digital signal encoding scheme in which "1" is encoded as high to low transition and "0" is encoded as low to high transition.

MAP: Manufacturing Automation Protocol.

Media access control: The discipline followed by stations of local area networks for transmission of their signals over a common medium shared by them.

Message switching: A data switching technique in which the full message is switched through the nodes of the subnetwork using store and forward system.

MLP (Multi Link Procedure): A sublayer of level 2 of CCITT X.25 that combines several SLP sublayers to make all of them look like a single data link and provides increased reliability of operation.

Modem: A contraction of the terms modulator and demodulator; it is used for data transmission over analog telephone network.

Modulation: A process by which some characteristics of a sinusoidal signal such as amplitude, phase or frequency are varied in accordance with another signal.

Multiplexing: In OSI terminology, multiplexing at (*N*)-layer refers to mapping of more than one (*N*)-connections to one (*N*–1)-connection.

Network layer: Third layer of the OSI Reference Model.

NEXT: Near end cross talk.

Node: A subnetwork equipment that interconneets other nodes and end systems; carries out switching and routing functions.

NRM (Normal Response Mode): A synchronous mode of data transfer between primary and secondary stations using HDLC protocol.

NRZ (Non Return to Zero): One of the two ways of representing a bit as an electrical signal; a bit is represented by a high level or as a low level of the electrical signal.

NRZI: Non Return to Zero Inverted.

Octet: An eight-bit data unit.

Open system: A system that can be interconnected to other systems according to established standards.

Open system interconnection architecture: A network architecture standard developed by ISO.

OSI: Open System Interconnection.

OSI Reference Model: A seven-layered communication model defined by ISO around which the OSI architecture is built and standards are developed.

P/F: Poll/Final bit which indicates whether it is a poll command from a primary station or it is the final response of the secondary station.

Packet: A data unit; usually refers to the Network layer.

Packet switching: A data switching technique in which packets of the message are switched through the nodes of the subnetwork using store and forward system.

PAD: Packet Assembler/Disassembler.

Parity: Even or odd characteristic of the number of 1s in a bit sequence.

Parity bit: The extra bit added to a sequence of bits to fix the parity of the sequence (including the parity bit) either always even or always odd.

PCI (Protocol Control Information): Control information which is exchanged between the peer layers of the OSI Reference Model.

PCM: Pulse Code Modulation.

PDU (Protocol Data Unit): A data unit exchanged between peer layers; it consists of PCI and user data.

Peer: An entity at the same protocol layer.

Physical layer: First layer of the OSI Reference Model.

Piggybacking: Inclusion of an acknowledgement in an outgoing data unit.

PLP: Packet Layer Protocol.

Point-to-multipoint communication: Communication between a master/primary station and several slave/secondary stations.

Point-to-point communication: Communication between two stations.

Polling: The process by which a master/primary station invites a slave/secondary station to transmit.

Preamble: Bit sequence which precedes a frame for bit synchronization at the receiver.

Presentation context: An abstract syntax and a selected transfer syntax.

Presentation layer: Sixth layer of the OSI Reference Model.

Primary station: A station that controls communication of a data link by issuing commands and interpreting the responses received from the secondary stations.

Profile: A selected set of parameters or protocols.

Protocol: A set of rules and conventions for communications between peer entities.

PSK (Phase Shift Keying): A modulation scheme for digital signals in which the carrier phase is modulated.

Psophometric curve: A curve indicating sensitivity of the ear to noise at different frequencies.

PSTN: Public Switched Telephone Network.

PVC (Permanent Virtual Circuit): An X.25 connection which is always in data transfer phase.

QAM: Quadrature Amplitude Modulation.

QOS (Quality of Service): Expressing the desired data transfer characteristics in terms of communication parameters such as error rate, throughput, delay, etc.

Recommendation: A term CCITT uses for its standards.

Register insertion: A media access control technique for ring topology of local area networks.

REJ-*N* (Reject): Acknowledgement for the I-frames up to $N-1$ and request for retransmission of all the I-frames starting from N.

Relaying: Receiving and retransmitting a data unit.

Remodulator: A headend device of a broadband bus LAN; it recovers the digital signal from the inbound carrier and retransmits the signal on the outbound carrier.

Repeater: A device that receives a digital signal, regenerates it and then retransmits it.

Reset: Re-establishes a predefined state at both the ends of a connection.

Resynchronization: At the Session layer, resynchronization implies restoring the state of dialogue between the two communicating Session entities to a previously defined state.

Ring: A LAN topology in which the stations are attached to ring interface units connected in the form of a closed loop.

RNR-*N* (Receiver Not Ready): Acknowledgement for the I-frames up to $N-1$ and request for stopping further transfer of I-frames temporarily.

ROSE (Remote Operations Service Element): An ASE that provides capability of remote execution of operations by an Application entity.

Router: An internetworking device that interconnects the subnetworks at layer 3 of the OSI model.

RR-*N* (Receiver Ready): Acknowledgement for the I-frames up to $N-1$.

RTS: Request to Send.

RTSE (Reliable Transfer Service Element): An ASE which has the sole responsibility of recovering from lost associations.

RZ (Return to Zero): One of the two ways of representing a bit as an electrical signal; a bit is represented as a transition (low to high or high to low).

SAP (Service Access Point): Access point between the adjacent layers of the OSI Reference Model through which the services are requested and provided.

SDLC (Synchronous Data Link Control): A bit-oriented Data Link protocol developed by IBM.

SDU (Service Data Unit): The data unit received from the next higher layer which is sent to the peer entity in another end system.

Secondary station: A station that operates under the control of a primary station by sending responses to the received commands.

Segmenting: In OSI terminology, segmenting refers to dividing a SDU into several PDUs to accommodate the incompatible sizes of the data units.

Selecting: The process by which a master/primary station alerts one of the slave/secondary stations to receive data.

Service primitive: A generic call, between entities of adjacent layers to provide the desired service.

Session layer: Sixth layer of the OSI Reference Model.

Simplex transmission: A mode of transmission that allows transmission in one direction only.

Sliding window flow control: A flow control technique that allows numbered frames within a window to be transmitted without waiting for individual acknowledgements of the frames from the receiver.

Slotted ALOHA: An extension of ALOHA media access control technique in which the frames are transmitted in predefined time slots.

SLP (Single Link Procedure): Same as LAP-B.

SNA (System Network Architecture): IBM's proprietary architecture for networking.

SNAC (Subnetwork access control functions): A sublayer of the Network layer.

SNDC (Subnetwork-dependent convergence functions): A sublayer of the Network layer.

SNIC (Subnetwork-independent convergence functions): A sublayer of the Network layer.

Source routing: A routing mechanism of internetwork in which the sender includes the routing information of a frame in the frame structure.

Spanning tree: A graph structure of a network which interconnects all the nodes but in which there is never more than one active path connecting any two nodes.

Splitting: It refers to mapping one (N)-connection to more than one $(N-1)$-connections.

SREJ (Selective Reject): A request for selective retransmission of an HDLC I-frame.

Star topology: A LAN topology in which the stations are connected to a central switch.

Stat Mux: Statistical time division multiplexer.

STDM (Statistical Time Division Multiplexing): A variation of TDM in which time slots are not preassigned but are allotted on demand.

STE: Signalling terminal equipment.

Stop and wait flow control: A flow control mechanism which allows transmission of next frame only after acknowledgement of the previous frame is received.

Store and forward: The process of message handling by the subnetwork nodes which is based on temporarily storing the messages in a queue and then forwarding them to the next node when the turn matures.

Sublayer: A sub-division of layer functions.

Subnetwork: A subset of computer network which excludes the end systems and is concerned with routing, switching and transmission of data units between end systems.

SVC (Switched Virtual Circuit): An X.25 connection that needs to be established and released.

Synchronous: Occurrence of an event such as transmission of a data unit is conditionally associated with occurrence of another event such as reception of an acknowledgement.

Synchronous transmission: Serial transmission bits whose occurrences are marked by clock pulses.

TAX: Trunk Automatic Exchange.

TCP: Transmission Control Protocol.

TDM (Time Division Multiplexing): A technique of multiplexing signals by allotting them different time slots.

Token: An object, ownership of which allows one to perform an operation.

Token bus: A bus topology LAN in which right to transmit is controlled by a token which is circulated among the stations.

Token ring: A ring topology LAN in which right to transmit is controlled by a token.

Topology: The structure of paths and nodes of a subnetwork.

Trailer: Last segment of PDU that usually contains bits for error detection.

Transparency: Ability to send arbitrary patterns of data though some bit patterns may have special control significance.

Transport layer: Fifth layer of the OSI Reference Model.

TWA (Two-Way Alternate): A communication mode that allows only one entity to communicate at a time.

Two-wire circuit: A circuit having a common pair of wires for transmit and receive directions.

TWS (Two-Way Simultaneous): A communication mode that allows the two entities to communicate simultaneously.

User: A layer entity which utilizes the services of the next lower layer entity.

VRC (Vertical Redundancy Check): An error detection check which is applied on a block of bytes.

VTS (Virtual Terminal Service): An ASE which gives capability to support remote login from variety of terminal types.

WAN: Wide Area Network.

Wiring concentrator: A site through which all the stations are connected to form a ring.

XID: Exchange identification.

Suggested Further Reading

Bartee, T.C., *Data Communications, Networks and Systems*, Howard W. Sams & Co., Indianapolis, 1985.

Bates, P., *Practical Digital and Data Communications with LSI Applications*, Prentice-Hall, Englewood Cliffs, New Jersey, 1987.

Bertsekas, D. and R. Gallager, *Data Networks*, 2nd ed., Prentice-Hall of India, New Delhi, 1992.

Black, U., *Computer Networks: Protocols, standards, and interfaces,* 2nd ed., Prentice-Hall, Englewood Cliffs, New Jersey, 1993.

_______, *Data Communications and Distributed Networks*, 3rd ed., Prentice-Hall of India, New Delhi, 1995.

_______, *Data Link Protocols*, Prentice-Hall, Englewood Cliffs, New Jersey, 1993.

_______, *Data Networks: Concepts, theory, and practice,* Prentice-Hall, Englewood Cliffs, New Jersey, 1990.

_______, *OSI: A model for computer communications standards*, Prentice-Hall, Englewood Cliffs, New Jersey, 1991.

_______, *Physical Level Interfaces and Protocols*, IEEE Computer Society Press, Washington, D.C., 1988.

_______, *X.25 and Related Protocols*, IEEE Computer Society Press, Washington, D.C., 1991.

Bleazard, G.B., *Handbook of Data Communication*, National Computing Centre Ltd., Manchester, 1982.

Brooner, E.G. and P. Wells, *Computer Communication Techniques*, Howard W. Sams & Co., Indianapolis, 1983.

Campbell, J., *The RS-232 Solution,* BPB Publications, New Delhi, 1990.

Cole, G.D., *Computer Networking for System Programmers,* John Wiley & Sons, New York, 1990.

_______, *Implementing OSI Networks*, John Wiley & Sons, New York, 1990.

Dale, G., *Handbook of Data Communications,* National Computing Center Ltd., Manchester, 1975.

Davidson, R.P. and N.J. Muller, *Internetworking LANs Operation: Design and management*, Artech House, Dedham (Mass.), 1992.

Davies, D.W., D.L.A. Barber, W.L. Price, and C.M. Solomonides, *Computer Networks and Their Protocols,* John Wiley & Sons, New York, 1979.

Doll, D.R., *Data Communications Facilities, Networks and Systems Design,* John Wiley & Sons, New York, 1978.

Fitzgerald, J. and T.S. Eason, *Fundamentals of Data Communications,* John Wiley & Sons, New York, 1978.

Flint, D.C. *Data Ring Main,* John Wiley & Sons, New York, 1983.

Fortier, P.J., *Handbook of LAN Technology,* McGraw-Hill, New York, 1989.

Freer, J.R., *Computer Communications & Networks,* Affiliated East-West Press, New Delhi, 1988.

Friend, G.E., *Understanding Data Communications,* Howard W. Sams & Co., Indianapolis, 1988.

Gandy, M., *Microcomputer Communications,* National Computing Centre Ltd., Manchester, 1985.

Gee, K.E., *Local Area Networks,* National Computing Centre Ltd., Manchester, 1982.

Halshall, J. and S. Shaw, *OSI Explained: End-to-end computer communications standards,* John Wiley & Sons, New York, 1988.

Held, G., *Data Communications Networking Devices,* John Wiley & Sons, New York, 1992.

_______, *Token Ring Networks,* John Wiley & Sons, New York, 1994.

Jain, B.N. and A.K. Agrawala, *Open System Interconnection: Its architecture and protocols*, Elsevier, New York, 1990.

Jones, V., *MAP/TOP Networking*, McGraw-Hill, New York, 1988.

Karp, H.R., *Basics of Data Communications*, McGraw-Hill, New York, 1976.

Knightson, K., T. Knowles, and J. Larmouth, *Standards for Open System Interconnection*, McGraw-Hill, New York, 1988.

L'Anson, C. and A. Pell, *Understanding OSI Applications,* Prentice-Hall, Englewood Cliffs. New Jersey, 1993.

Lane, J.E., *Communicating with Microcomputers,* National Computing Centre Ltd., Manchester, 1981.

Loomis, M.E.S., *Data Communications,* Prentice-Hall, Englewood Cliffs, New Jersey, 1983.

Madron, T., *LANs: Applications of IEEE/ANSI 802 Standards,* John Wiley & Sons, New York, 1989.

_______, *Local Area Networks: The next generation*, John Wiley & Sons, New York, 1990.

Markley, R.W., *Data Communications and Interoperability*, Prentice-Hall, Englewood Cliffs, New Jersey, 1980.

Martin, J., *Computer Networks and Distributed Processing*, Prentice-Hall, Englewood Cliffs, New Jersey, 1981.

Martin, J. and J. Leben, *Principles of Data Communications*, Prentice-Hall International, London, 1988.

_______, *DECnet Phase V—An OSI implementation*, Digital Press, Maynard, Mass., 1992.

Martin, J. and K. Chapman, *Local Area Networks: Architectures and implementation,* Prentice-Hall, Englewood Cliffs, New Jersey, 1989.

McGovern, T., *Data Communications Concepts and Applications*, Prentice-Hall, Englewood Cliffs, New Jersey, 1988.

Miller, M.A., *Introduction to Digital and Data Communications,* Jaico Publishing House, New Delhi, 1993.

Mirchandani, S., *FDDI Technology and Applications,* John Wiley & Sons, New York, 1993.

Morgan, L., *Managing On-Line Data Communications Systems,* National Computing Centre Ltd., Manchester, 1975.

Nichols, E., S. Jocelyn, and L. Morgan, *Selection of Data Communications Equipments,* National Computing Centre Ltd., Manchester, 1979.

Nilausen, J., *Token Ring Networks,* Prentice-Hall, Englewood Cliffs, New Jersey, 1993.

Poulton, S., *Packet Switching and X.25 Networks,* Pitman Publishing Ltd., London, 1989.

Pye, C., *Networking with Microcomputers,* National Computing Centre Ltd., Manchester, 1985.

———, *What is OSI?* National Computing Centre Ltd., Manchester, 1988.

Roden, M.S., *Digital and Data Communication Systems,* Prentice-Hall, Englewood Cliffs, New Jersey, 1982.

Rose, M., *The Open Book: Practical perspective on OSI,* Prentice-Hall, Englewood Cliffs, New Jersey, 1990.

Russell, D., *The Principles of Computer Networking,* Cambridge University Press, London, 1989.

Santamaria, A. and F.J. Lopez Harnandez, *Wireless LAN Systems,* Artech House, Dedham, (Mass.), 1994.

Sarikaya, B., *Principles of Protocol Engineering and Conformance Testing*, Ellis Horwood, Chichester (UK), 1993.

Schatt, S., *Understanding Local Area Networks*, 4th ed., Prentice-Hall of India, New Delhi, 1995.

Scott, P.R.D., *Introducing Data Communications Standards,* National Computing Centre Ltd., Manchester, 1979.

———, *Modems in Data Communications,* National Computing Centre Ltd., Manchester, 1980.

Sinnema, W., *Digital, Analog and Data Communications,* Reston, (Virginia), 1982.

Sloman, M. and J. Kramer, *Distributed Systems and Computer Networks,* Prentice-Hall, Englewood Cliffs, New Jersey, 1987.

Stalling, W., *Data and Computer Communications,* Macmillan Publishing Company, New York, 1988.

———, *Handbook of Computer Communications Standards—Vol. 1: The Open System Interconnection (OSI) Model and OSI-Related Standards; Vol. 2: Local Area Networks,* Howard W. Sams & Co., Indianapolis, 1990.

———, *Local Networks,* Macmillan Publishing Company, New York, 1990.

Strom, J., *OSI in Microcomputer LANs,* National Computing Centre Ltd., Manchester, 1989.

Tanenbaum, A.S., *Computer Networks,* 2nd ed., Prentice-Hall of India, New Delhi, 1994.

Tangney, B. and D. O'Mahony, *Local Area Networks and their Applications,* Prentice-Hall, Englewood Cliffs, New Jersey, 1988.

Tomasi, W. and V.F. Alisouskas, *Telecommunications: Voice/Data with fiber optic applications,* Prentice-Hall, Englewood Cliffs, New Jersey, 1988.

Index